Ina Hoeschele

INTRODUCTION TO STATISTICAL INFERENCE

E.S. Keeping

DOVER PUBLICATIONS, INC.
New York

Bibliographical Note

This Dover edition, first published in 1995, is an unabridged and unaltered republication of the work first published by the D. Van Nostrand Company, Princeton, New Jersey, in 1962.

Library of Congress Cataloging-in-Publication Data

Keeping, E.S.
Introduction to statistical inference / E.S. Keeping.
p. cm.
Originally published: Princeton, N.J. ; Van Nostrand, 1962.
Includes bibliographical references and index.
ISBN 0-486-68502-0 (pbk.)
1. Mathematical statistics. I. Title.
QA276.K25 1995
519.5—dc20 95-81
CIP

Manufactured in the United States of America
Dover Publications, Inc., 31 East 2nd Street, Mineola, N.Y. 11501

PREFACE

This book is an outgrowth of the two parts of *Mathematics of Statistics* by John F. Kenney and myself (Van Nostrand, Part I, 3rd ed., 1954; Part II, 2nd ed., 1951). It seemed advisable to prepare a new textbook which would emphasize the inferential and decision-making aspects of statistics and which would assume a mathematical background on the part of the reader roughly intermediate between those of Parts I and II.

Part I has for several years been used as a textbook for first-year students lacking any previous knowledge of calculus but with a fairly good background of high school or junior college algebra. Part II, on the other hand, presupposes at least two years of calculus with a corresponding mathematical maturity, and is primarily intended for senior or even graduate students desiring a deeper understanding of statistical principles. The present work requires a knowledge of elementary calculus and is adapted to the usual third-year university level in mathematics. The first chapter contains some of the elements of set theory, but this material is being more and more widely taught nowadays, even in quite junior courses, and is almost essential for the understanding of the idea of probability.

Throughout the book proofs are given where possible, but in many instances the mathematical details are relegated to the Appendix. There also will be found brief treatments of topics which the student is unlikely to have encountered in his regular courses—the gamma and beta functions, Stirling's approximation, Jacobians, Bernoulli numbers, etc. Although matrix algebra is nowadays much more prominent than formerly, and is invading even freshman courses, we have included in the Appendix enough of the elements of this subject to permit the occasional use of matrix notation in the body of the book. The brevity and convenience of this notation make it well worth while for the student of statistics, at least in the more advanced parts of the subject, to spend a little time on mastering the necessary algebra.

The present book was planned as another joint effort by Professor Kenney and myself. However, as the work proceeded, the bulk of the writing was left to me, and Professor Kenney eventually decided that his name ought not in fairness to be attached to it. I am much indebted to him for his generous action and for his continuing advice and criticism throughout the period of preparation. In common with many others of my generation, I learned the elements of statistics from Kenney's *Mathematics of Statistics, Parts I and II,* and I value very highly the privilege of having collaborated with him in the later revised editions of these books.

Since the concept of probability is fundamental to statistical inference, the first chapter is concerned mainly with the elements of the calculus of probability. The treatment is heuristic rather than rigorous, but does attempt to give a reasonable foundation of the idea of probability as a measure, and to interpret probability objectively in terms of relative frequencies and subjectively in terms of betting odds.

The second chapter contains the essential statistical techniques of summarizing the data in a sample prior to making inferences about the population. The routine computations of mean, variance, median, etc., are described for the benefit of those students without any previous statistical training. The general properties of distributions, with their cumulants and cumulant generating functions, are then discussed, and illustrated by reference to a number of special probability distributions (binomial, Poisson, normal, gamma and beta, chi-square, log-normal). This leads up to the relation between a sample and the population from which it is drawn and the concepts of confidence intervals and fiducial inference.

In Chapter 6 the principles of testing hypotheses and making decisions with assigned risks of error are introduced. The method of maximum likelihood and the concept of the power of a test are dealt with. All this is crucial in any discussion of statistical inference.

After a treatment of different sampling procedures, including sequential methods, the usual exact statistical tests on samples from a normal population are discussed. This is followed by the analysis of variance (which also assumes normality as generally applied) and by a discussion of certain nonparametric methods which can be used when it is unsafe to postulate a normal population.

Bivariate (linear regression, correlation and contingency) problems are next dealt with, followed in Chapter 12 by non-linear regression and curve-fitting. Finally there is a short chapter on multivariate problems and stochastic processes, giving only the barest introduction to these extensive fields.

Sets of problems are included at the end of each chapter. These are arranged in groups according to the sections in the chapter to which they relate. Within each group the problems are roughly in order of difficulty. An attempt has been made to maintain a balance between numerical examples and questions on pure theory. For some of the numerical problems it is very desirable to have the use of a desk computer; everyone who has much to do with statistics is almost bound to acquire facility in the use of such computing machines. Much can be done with even so inexpensive and compact a device as the little pocket "Curta" calculator. Hints are provided for the solution of the more difficult problems. I am again grateful to Professor Kenney for permission to include a number of problems taken from Parts I and II of our joint work already mentioned.

The tables given in Appendix B should suffice for most statistical tests, but the student should if possible have access to more complete sets of tables, such as Pearson and Hartley's *Biometrika Tables for Statisticians,* Vol. I (Cambridge University Press, 1954) or Fisher and Yates' *Statistical Tables for Biological, Agricultural and Medical Research* (Oliver and Boyd, 5th ed., 1957).

For permission to reprint tables or portions of tables, I am indebted to the following: Sir Ronald A. Fisher, F.R.S., and Messrs. Oliver and Boyd, Ltd., Edinburgh (Tables B.3 and B.4); Professor M. S. Bartlett and the Department of Statistics, University College, London (Table B.1); Dr. A. J. Jonckheere, Professor E. S. Pearson, and the publishers of "Biometrika" (Table B.10, 8.3 and 8.5); Dr. D. Auble and the Institute for Educational Research, Indiana University (Table B.9); Dr. G. W. Snedecor and the Iowa State University Press (Table B.5); Dr. F. J. Massey and the American Statistical Association (Table B.6); Dr. J. E. Walsh and the American Statistical Association (Table B.8); Dr. F. Wilcoxon and the American Cyanamid Company (Table 10.8).

A list of references is given at the end of each chapter, but this list is not intended as even a partial bibliography. It serves merely to indicate to the student a few books or papers in which he may find a fuller treatment, or more detailed proofs, of some of the statements in the text, and also in a few cases to give the source of the numerical data used in problems.

In so vast a subject as modern mathematical statistics a textbook writer has to be selective. It is highly probable that some statisticians, looking at this book, will feel that the emphasis is misplaced here and there or that a better choice of topics could have been made. I can only plead that to me the choice has seemed reasonable for the type of student I had in mind.

It is hardly possible for me to express adequately my indebtedness to all the teachers and writers from whose lectures or papers I have derived help and inspiration. I am grateful also to the publishers' readers who examined this book in manuscript and offered valuable suggestions for improvement. In conclusion, I should like to express my appreciation of the help of Mrs. I. Maj, who coped most efficiently with the job of typing a manuscript plentifully sprinkled with mathematical symbols.

E. S. K.

FOREWORD

This book contains sufficient material for a two-semester or full-year course, with three lecture periods per week. Some less important sections, which might be omitted on a first reading, are starred.

For a one-semester course, it would be advisable to read most of Chapters 1 to 6 (omitting the starred sections) and also the first parts of Chapters 8 and 11. The instructor will naturally have the responsibility of deciding on the material which he considers most relevant to the needs of his particular students.

References to numbered equations within the same section are to the last part of the number only. Thus Eq. (1.10.6) if referred to within §1.10, would be quoted as Eq. (6). If referred to in any later section, the complete number is quoted.

Numbers enclosed in square brackets, such as [2], refer to the literature references given at the end of the chapter.

CONTENTS

Chapter 1

PROBABILITY

1.1 **Uncertainty of Statistical Inference** In everyday life we are again and again faced with the necessity of making decisions. Many of these are trivial, some may be serious, but almost always there is an element of uncertainty about the wisdom of the decision we make. Scientists in their regular work have a similar problem. They have to draw conclusions from their enquiries or experimental results, but their observations are liable to error and may from their very nature be subject to considerable irregular fluctuations. Any conclusions that the scientists draw will therefore not be rigid and unalterable, but will merely be more or less probable. The theory of probability is the groundwork of scientific inference, and as such is the subject of this chapter.

Statistics is concerned with variables that fluctuate in a more or less unpredictable way, such as the monthly total of highway fatalities in the state of New York or the yearly average yield of wheat in bushels per acre on a Saskatchewan farm. There may be assignable causes, with predictable effects on the total of fatal accidents in a particular month, but no one would expect to be able, month after month, to predict the total exactly. The essence of a statistical variable is that, to some extent at least, it is unpredictable. We call this characteristic *randomness* and will later give it a more precise definition.

Since in almost all experimental work, and particularly in the biological and social sciences, the results are influenced by a variety of conditions largely beyond the experimenter's control, there is always this element of randomness about the results of experiment. Variations in rainfall and soil composition affect plant yields; individual peculiarities affect the behavior of rats or guinea pigs. Even in the "exact" sciences, such as physics and astronomy, with their relatively high precision of measurement, there is still a residuum of unavoidable experimental error. It is the task of statistical inference to draw valid conclusions about the world around us from such limited and imperfect observations as we can make. Since these conclusions are not certain, we would like to attach to them an estimate of the probability of their truth. How this can be done in certain types of problems will be told in later chapters.

Probability has in modern atomic physics an even deeper significance. The "principle of indeterminacy," formulated originally by W. Heisenberg, lays it down as a cardinal truth of physics that certain pairs of variables, such as position and momentum, or energy and time, cannot both be measured, even in principle, with unlimited precision, but that the more accurately one of the pair is known, the less accurately can the other be determined. This has nothing to do with the limitations and errors of actual physical apparatus—it is a theoretical limitation that would hold with perfect apparatus. A consequence of this principle is that the variable representing a particle in modern quantum theory is something that, as far as it can be represented physically at all, is a probability —the probability, roughly speaking, that the particle is at a certain position in space at a certain time.

1.2 **Intuitive Idea of Probability** Everyone has a general idea of what is meant by the words "probable" and "probability." We hear the radio announcer at breakfast-time say that it will probably rain before the afternoon, and we decide to wear a raincoat to the office. It is, however, by no means a simple matter to give a definition of probability that will adequately cover all cases and serve as a satisfactory foundation for statistical inference.

Some people feel that probability refers only to a state of mind, the strength of one's belief in a proposition. In order to make probability more than merely subjective, these writers have to speak of the degree of "rational" belief in a proposition, something that should perhaps be called *credibility* rather than probability. By making some rather arbitrary assumptions it is possible to arrive at a numerical calculus of probabilities, but we shall not for the present pursue this line of thought any further. (See references [1] and [2] at the end of the chapter, and also § 1.8).

The mathematical treatment of probability arose historically out of discussions of games of chance in the 17th century (see [3], [4], [5]). If a die seems to be honest and well made, it is reasonable to suppose that it is equally likely to fall, if rolled in the customary way, with any of its six faces uppermost. This judgment merely involves a recognition of the fundamental symmetry of the die. It is not perfectly symmetrical, of course, since the faces are marked differently, but we judge that this minimal lack of symmetry will not appreciably affect the chances. Similarly if five cards are dealt from a well-shuffled deck, we feel that (unless the dealer is crooked) any specified set of five cards is about as likely as any other. In these cases it is a comparatively simple matter to calculate the probabilities of events that may be of interest—the probability for instance that a 6 turns up on a die, or that the five cards dealt from a deck of 52 are all of one suit. Trouble arises when we cannot assume the fundamental symmetry—how can we assess the probability of 6 with a loaded die?

There are many writers who feel that the only kind of probability definition that makes much sense, particularly in statistics, is based on the idea of the relative frequency with which events of interest happen in a long series of similar trials. To assess the probability of 6 with a die (loaded or not) we roll it a large number of times and count the number of times 6 turns up. The ratio of this number to the total number of rolls is the relative frequency of 6 and is an approximation to the true probability of 6. Unfortunately we cannot simply say that the probability is the limiting value of this ratio as the number, n, of trials increases, since the ratio does not tend to a limit in the strict mathematical sense.

What we can say is that, if some conditions are satisfied, then in the long run it becomes almost certain that the difference between the observed relative frequency and a fixed limit will be less than any number we like to name. The smaller this number, of course, the greater the number of trials we shall have to make to reach a state of "almost certainty." The conditions to be satisfied are, first, that the successive trials are independent (which means that the result of any trial is not influenced by what has happened on previous trials) and, second, that the context of the trials has remained essentially unaltered. This means that all the circumstances surrounding the trials are either unchanged during the whole set of observations or, if they are changed, have no appreciable effect on the trials. The decision as to what circumstances are or are not relevant is one which is often needed in experimental work, and must be made on the basis of experience.

The above statement is a special form of the "law of large numbers," which will be stated more precisely later on. As the number of trials increases, the relative frequency of the event in question converges stochastically (or, as it is sometimes expressed, "converges in probability") to a limiting value, which is defined as the probability of the event. For a detailed, semipopular discussion of this concept, see [6].

1.3 **Events** We shall take the point of view that probability relates to *events*, which are phenomena that may be observed either to happen or not to happen in a particular context. Thus if two dice are rolled repeatedly, one event in which we may be interested is a total of seven spots turning up. This event has a definite probability within the context of rolls of these particular dice. The context includes not only the rolls that have actually occurred, but all those that might conceivably occur if we had unlimited time and patience to go on rolling the dice. Another type of event is a measurement, for example, of an intelligence quotient for a 10-year-old Negro boy. The observed value, say 114, to the nearest whole number, is one that has a probability significance within the context of all such measurements on boys in the United States or only those on 10-year-old boys, or perhaps only those on all 10-year-old Negro boys. The particular context depends on what probability we are interested in.

Events may be classified as *simple* or *compound*. A compound event is one that can be decomposed into a set of simple events, whereas a simple event cannot be decomposed any further. The occurrence of 6 in a throw of a die is a simple event. The occurrence of 7 with two dice is a compound event, because it can be split up into six simple events, each of which corresponds to the same compound event, namely, 6 and 1, 5 and 2, 4 and 3, 3 and 4, 2 and 5, 1 and 6. Here the first number in each pair represents the number of spots shown by the first die and the second number that shown by the second die.

A simple event may be represented by a point in a suitable "space." The space corresponding to the number of spots on the upper face of a die consists of six isolated points, numbered 1, 2, 3, 4, 5, 6, on the axis of real numbers. Any observed simple event, such as a 6, is represented by one of these points. The compound event, "a throw of at least 4," is represented by the three points labelled 4, 5, and 6. The space representing all the possibilities in a particular situation is often denoted by $\mathscr{U}$, meaning the universal set of all possibilities.

The space, $\mathscr{U}$, corresponding to the possible outcomes with two dice, is a square lattice of 36 points, as in Figure 1. The x-coordinate represents the first die and the y-coordinate the second die. The compound event, "the total of spots shown is 7," is represented by the set of six points which are ringed in the figure.

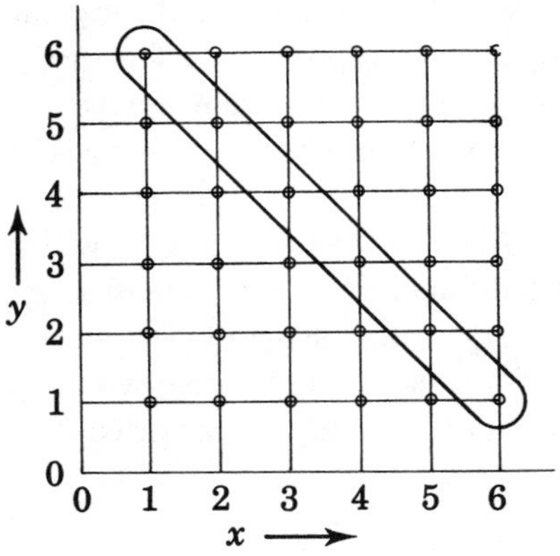

FIG. 1 SIMPLE AND COMPOUND EVENTS, WITH TWO DICE

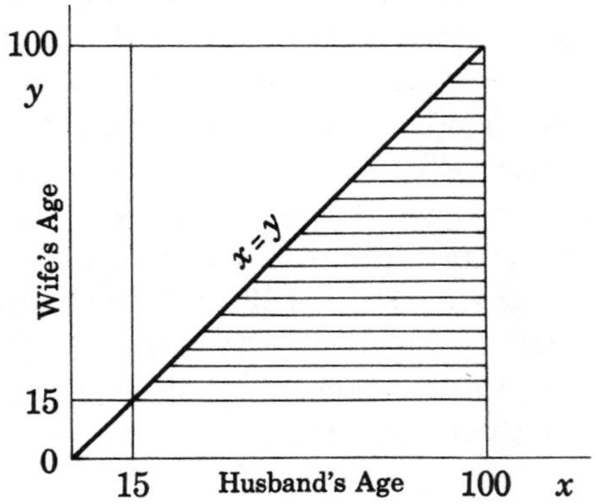

FIG. 2 COMPOUND EVENT IN CONTINUOUS SPACE

An insurance company may be interested in the age distribution of married couples. If x years represents the husband's age and y years the wife's age, the space $\mathscr{U}$ for the event, "the husband is x years old and the wife is y years old," is a region of the x-y plane between limits (say 15 and 100) for both variables. The compound event, "the husband is older than the wife," will be represented by the shaded area in Figure 2, below the line $x = y$.

1.4 **Elements of Set Theory Applied to Events** We shall denote events by letters, A, B, $C \ldots$, or sometimes by A_1, A_2, $A_3 \ldots$. These events will be represented in diagrams by regions (or points) of the appropriate space. For convenience the diagrams will be drawn as though the space were continuous, although it may often consist in reality of a set of discrete points (as in the example of the two dice, Figure 1).

The basic idea of using diagrams like those in Figures 3 and 4 seems to be due to the 18th-century Swiss mathematician Euler. Refinements were made by the British logician John Venn (1834–1883), and such diagrams are now usually called Venn diagrams.

The contrary event to A (i.e., the event "A does not happen") will be represented by $\tilde{A}$, which may be read as "A-tilde," or "not-A." The events A and $\tilde{A}$ together make up the whole of the appropriate space $\mathscr{U}$. Hence $\tilde{A}$ is often called the *complement* of A (see Figure 3a).

If A and B are events within the same space $\mathscr{U}$, A is said to *imply* B (or, symbolically $A \subset B$) if whenever A occurs B necessarily occurs. The region corresponding to A is nowhere outside the region corresponding to B (see Figure 3b). For example, the event, "the number of spots shown by a die is 4," implies the event, "the number of spots shown is even." If $A \subset B$ and $B \subset A$, the events A and B are *equivalent* (symbolically, $A = B$).

The event, "both A and B" (i.e., the simultaneous occurrence of both events), is called the *intersection* of A and B and is represented by $A \cap B$. In a Venn diagram it is represented by the intersection of the areas corresponding to A and B (Figure 3c).

The event, "A and/or B" (i.e., the occurrence of at least one of the events A and B), is called the *union* of A and B, and is denoted by $A \cup B$. It is represented by the whole area that is included in either the A area or the B area (Figure 3d). If the events A and B are such that they cannot both happen at the same time, they are said to be *disjoint*, or *mutually exclusive*. In this case $A \cup B$ is represented by the sum of the areas of A and B (Figure 3e), and the union is then often denoted by $A + B$. The symbol "+" here denotes a logical sum and means "either . . . or". It is not the plus sign of arithmetic.

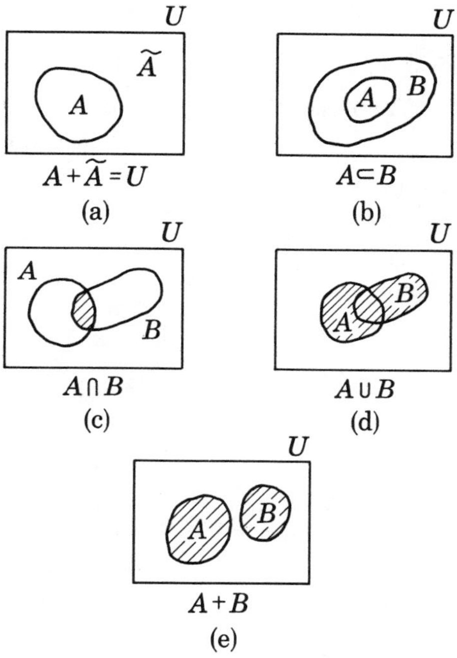

FIG. 3 VENN DIAGRAMS

The above notation is readily extended to a finite number of events. The intersection of the events $A_1, A_2 \ldots A_n$ is denoted by

$$A_1 \cap A_2 \cap A_3 \ldots \cap A_n, \quad \text{or} \quad \bigcap_{k=1}^{n} A_k,$$

while their union is denoted by $A_1 \cup A_2 \ldots \cup A_n$, or $\bigcup_{k=1}^{n} A_k$. If the events are disjoint, the union is often denoted by $\sum_{k=1}^{n} A_k$.

The event $A + \tilde{A}$, or $\mathscr{U}$, may be interpreted as a sure event (one that is bound to happen). It is represented by the whole of the appropriate space. The event $A \cap \tilde{A}$, or $\mathscr{E}$ may be interpreted as an impossible event. The set of points representing $\mathscr{E}$ is said to be *null*, or *empty*.

1.5 **Some Theorems on Union and Intersection of Events** The following theorems may readily be proved from the definitions of the preceding section.

They are most easily appreciated by reference to the corresponding Venn diagrams (Figures 3 and 4).

THEOREM 1.1 *If* $A \subset B$, *then* $\tilde{B} \subset \tilde{A}$.

THEOREM 1.2 $A \cap B = B \cap A$ *and* $A \cup B = B \cup A$. In algebraic language, both operations are commutative.

THEOREM 1.3 $(A \cap B) \cap C = A \cap (B \cap C) = A \cap B \cap C$,
and $(A \cup B) \cup C = A \cup (B \cup C) = A \cup B \cup C$.

Both operations are associative.

THEOREM 1.4 $A \cap (B \cup C) = (A \cap B) \cup (A \cap C)$.
$A \cup (B \cap C) = (A \cup B) \cap (A \cup C)$.

See Figure 4, a and b. The two operations are distributive with respect to one another.

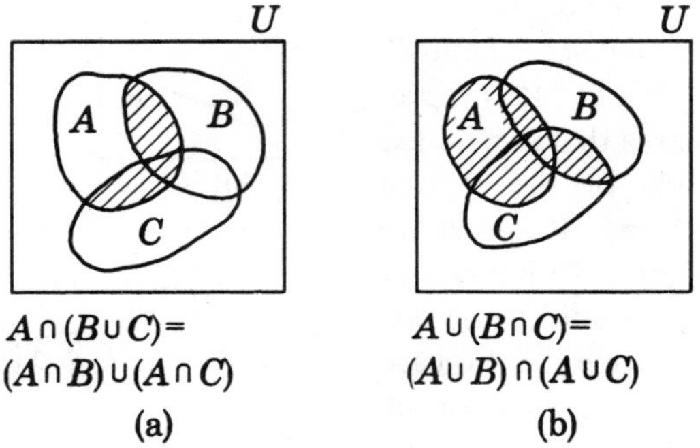

FIG. 4 VENN DIAGRAMS FOR THREE EVENTS

THEOREM 1.5 $(\widetilde{A \cap B}) = \tilde{A} \cup \tilde{B}$.
$(\widetilde{A \cup B}) = \tilde{A} \cap \tilde{B}$.

It will be observed that in Theorems 1.2 to 1.5 there is a perfect duality between union and intersection. If in any theorem the symbols of union and intersection are interchanged, another true theorem results.

For further elementary discussion of sets, see [7].

1.6 **Probability as a Measure** An event, as we have seen, corresponds to a subset of the universal set $\mathscr{U}$ of all possibilities in the particular situation. Suppose we can in some reasonable way assign a *weight* (a non-negative number) to each point or element of area of $\mathscr{U}$, so that the total of these weights is 1. Then the weight assigned to an event A will be the sum of the weights of all the points (or elements of area) which make up A. This is called the *measure* of A. We then define the probability of A as equal to its measure.

It may be noted that the concept of measure has a much wider meaning than this in modern mathematics. Probability measure is only one among many types of measure.

The assignment of weights in an actual problem will depend upon the information available and on an analysis of all the possibilities. Thus if we have no reason to doubt the accuracy of a given die and the honesty with which it is rolled, it seems reasonable to assign the same weight to each of the six logically possible outcomes. In other words we take the weights as each equal to $\frac{1}{6}$, since they must add up to 1. The measure (and therefore the probability) of the event, "the number of spots is even," is $\frac{3}{6}$, since this event includes the three simple events, 2, 4 and 6.

Suppose that there are three horses—A, B, C—in a race, and that, on form, I judge that A is twice as likely to win as either B or C, but that the chances of B and C are about equal. I would then assign weights $\frac{1}{2}$, $\frac{1}{4}$, $\frac{1}{4}$ to A, B, C, respectively. The probability of the event, "either A or B wins," would be $\frac{1}{2} + \frac{1}{4} = \frac{3}{4}$. The probability of the event, "either B or C wins," would be $\frac{1}{4} + \frac{1}{4} = \frac{1}{2}$.

The techniques of calculating probabilities, once the weights have been assigned, will occupy the major part of this chapter.

1.7 **Properties of a Probability Measure** The basic properties of the probability measure $P(A)$ of an event A are

(1.7.1) $$P(A) = 1 \quad \text{if} \quad A = \mathscr{U}.$$

This means that *some* event in $\mathscr{U}$ is bound to happen.

(1.7.2) $$0 \leq P(A) \leq 1 \quad \text{for every } A \text{ in } \mathscr{U}.$$

(1.7.3) $$P(A \cup B) = P(A) + P(B) - P(A \cap B).$$

The first two follow immediately from the definition. The third may be appreciated by reference to Figure 3, c and d. In reckoning $P(A) + P(B)$, the weight of every element of the intersection is counted twice. The sum is therefore greater than $P(A \cup B)$ by the measure of this intersection.

If A and B are disjoint,

(1.7.4) $$P(A + B) = P(A) + P(B).$$

This is called the *addition law* for probabilities.

Since A and $\tilde{A}$ are disjoint, and $A + \tilde{A} = \mathscr{U}$, it follows from equations (1) and (4) that

(1.7.5) $$P(A) + P(\tilde{A}) = 1.$$

This rule is often useful when it happens to be easier to calculate $P(\tilde{A})$ than $P(A)$. We can find $P(A)$ by calculating $1—P(\tilde{A})$.

Since the impossible event $\mathscr{E}$ is the complement of the sure event, we see that

(1.7.6) $$P(\mathscr{E}) = 1 - P(\mathscr{U}) = 0.$$

The probability of an impossible event is zero.

It should be understood that because $P(A) = 0$ it does not necessarily follow that A is impossible. If the set $\mathscr{U}$ is represented by the points inside a square of unit side, it would be reasonable in some contexts to make the measure of any sub-region of the square equal to the area of that region. The measure of any set of isolated points or finite line segments would then be zero, although we could not say that these points or lines correspond to impossible events. If a property holds within a given region, except for a set of points of measure zero, it is said to hold *almost everywhere*. The statement, therefore, that $P(A) = 0$ does not imply that A is impossible. It does imply that the measure of the points corresponding to A is zero. In the same way, a probability of 1 does not necessarily mean that the event in question is absolutely sure, but only that the exceptional points in the appropriate space have a measure zero.

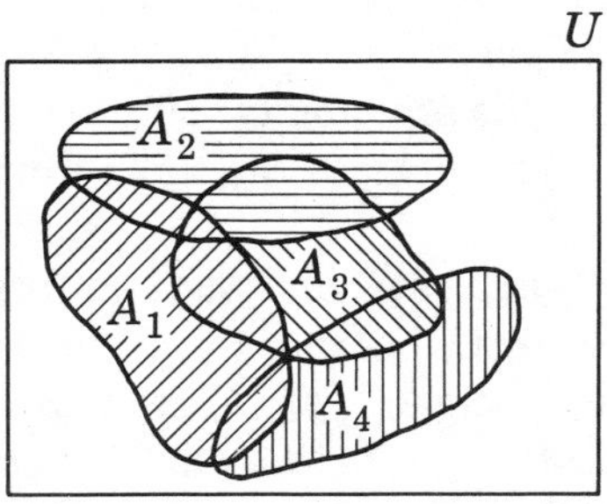

FIG. 5 UNION OF SEVERAL EVENTS

THEOREM 1.6 *If $A \subset B$, then $P(A) \leq P(B)$.*

To prove this, let C be the part of B not included in A; then $B = C + A$. By equation (4) above,

$$P(B) = P(C) + P(A) \geq P(A),$$

since $P(C) \geq 0$ by (2).

THEOREM 1.7 *For n events $A_k(k = 1, 2 \ldots n)$,*

$$P\left(\bigcup_{k=1}^{n} A_k\right) \leq \sum_{k=1}^{n} P(A_k)$$

It is readily seen from a Venn diagram (see Figure 5) that

$$\bigcup_{k=1}^{n} A_k = A_1 + \tilde{A}_1 \cap A_2 + \tilde{A}_1 \cap \tilde{A}_2 \cap A_3 + \ldots + \tilde{A}_1 \cap \tilde{A}_2 \ldots \cap \tilde{A}_{n-1} \cap A_n.$$

These disjoint sets (for the case $n = 4$) are shaded differently in the figure. But $\tilde{A}_1 \cap A_2 \subset A_2$, $\tilde{A}_1 \cap \tilde{A}_2 \cap A_3 \subset A_3$, etc.; therefore, by Theorem 1.6, $P(\tilde{A}_1 \cap A_2) \leq P(A_2)$, etc. It follows that

$$P\left(\bigcup_{k=1}^{n} A_k\right) \leq P(A_1) + P(A_2) + \ldots + P(A_n) = \sum_{k=1}^{n} P(A_k).$$

This theorem may be extended to include infinite sets of events. The equality sign holds for disjoint events.

THEOREM 1.8 *The extension of Eq. (3) to three events gives*

$$(1.7.7)\quad P(A \cup B \cup C) = P(A) + P(B) + P(C) - P(A \cap B) - P(A \cap C) - P(B \cap C) + P(A \cap B \cap C)$$

This is easily proved by writing D for the event $B \cup C$ and applying Eq. (3). Note that, by Theorem 1.4, $A \cap D = (A \cap B) \cup (A \cap C)$.

The result may be further extended to n events as follows:

$$(1.7.8)\qquad P\left(\bigcup_{j=1}^{n} A_j\right) = \sum_j P(A_k) - \sum_{jk}' P(A_j \cap A_k) + \sum_{jkl}'' P(A_j \cap A_k \cap A_l) - \dots + (-1)^{n-1} P\left(\bigcap_1^n A_j\right),$$

where $\sum'_{jk}$ means that the sum is over all j and k with $j \neq k$, $\sum''_{jkl}$ means that the sum is over all j, k, l with no two of these equal, and so on.

1.8 **Interpretation of Probability in Terms of Betting Odds** A recent book by L. J. Savage [2] emphasizes the personal aspect of probabilities. Savage argues that the rational man acts as if there exists for him, corresponding to each situation in which he has to make a decision, a set of probabilities and a set of *utilities*. The probabilities relate to the various states or aspects of the world that may be supposed to exist (relevant to the particular situation). The utilities measure in some way the values that will accrue to him from each particular decision for each particular unknown state of the world. He acts in such a way as to make the utility he expects to get as great as possible.

The probabilities, being a personal matter, can be assessed by presenting the man with a suitable bet. If there are just two relevant states of the world, s_1 and s_2, with probabilities p_1 and p_2, and if the man is offered betting odds of C to 1 against s_1, he will take the bet, provided $p_2/p_1 < C$. By varying C until the bet is accepted, the values of p_1 and p_2 can be determined (remembering that $p_1 + p_2 = 1$). When the English physicist, Sir John Cockcroft, said recently, speaking about the British atomic reactor, Zeta, "I am 90% certain that the neutrons were produced by a thermonuclear reaction," he meant presumably that he would be willing to bet as high as 9 to 1 that the reaction was in fact thermonuclear. When a man bets on a horse which is quoted at odds of 6 to 1 against, it may be concluded that he regards the probability of its winning as at least $\frac{1}{7}$. $\left(\text{If } p_1 = \frac{1}{7}, p_2 = \frac{6}{7}, \frac{p_2}{p_1} = 6.\right)$ At least the "rational man" would so act.

1.9 **Interpretation of Probability in Terms of Relative Frequency** If an event A is assigned the probability p, in a particular context, one interpretation of this probability is that in a long series of n similar trials in this context, the event would happen in approximately np trials and fail to happen in the remainder. As already mentioned, this is a special case of the law of large numbers. Where there is no other natural or reasonable method of assigning probabilities, the experimental method of counting the number of times (r) that the event happens, and using the approximation r/n for p, is usually applicable. Insurance companies estimate the chance that an insured person will survive to a given age by a study of the records of such persons in the past. Actually, the context has not, in this particular example, remained quite constant. Improvements in public health, and new drugs, have in recent years greatly increased the chances of survival of individuals, so that the probabilities assessed from the records of, say, the past 50 years tend to be too low. However, there is no way, except from the records, to assess these probabilities. The companies should, of course, use the most recent records that are available.

	B	$\tilde{B}$	
A	a	b	r_1
$\tilde{A}$	c	d	r_2
	c_1	c_2	n

FIG. 6 TWO-BY-TWO FREQUENCY TABLE

If, as the outcome of every trial, we can say that a particular event A has or has not happened, and also at the same time that another event B has or has not happened, there are clearly four possibilities altogether as regards the two events jointly, namely, the events denoted by $A \cap B$, $A \cap \tilde{B}$, $\tilde{A} \cap B$, and $\tilde{A} \cap \tilde{B}$. These are mutually exclusive, or disjoint.

If the corresponding frequencies for these compound events are a, b, c, and d, (the sum of all these being n), the relative frequencies are a/n, b/n, etc. The relative frequency of A is $(a + b)/n = r_1/n$, where r_1 is the total frequency in the first row of the two-by-two table (Figure 6), since both events $A \cap B$ and $A \cap \tilde{B}$ imply that A happens. If we denote the relative frequency of A by $f(A)$, then

$$f(A) = f(A \cap B) + f(A \cap \tilde{B}). \tag{1.9.1}$$

The event $A \cup B$ includes all cases in which either A or B (or both) happen, that is, it includes $A \cap B$, $A \cap \tilde{B}$ and $\tilde{A} \cap B$, but not $\tilde{A} \cap \tilde{B}$. The total frequency for $A \cup B$ is therefore $a + b + c = r_1 + c_1 - a$, where r_1 is the total frequency in the first row and c_1 is the total frequency in the first column. Since $f(B) = c_1/n$, we have the rule

$$f(A \cup B) = f(A) + f(B) - f(A \cap B) \tag{1.9.2}$$

If we regard the relative frequencies (for large n) as approximations to the corresponding probabilities, we arrive at the basic law for probabilities given in (1.7.3).

1.10 **Conditional Probability** In the table of Figure 6, the event A happens in r_1 cases altogether, and in a of these cases the event B also happens. We can therefore state that the relative frequency of B, when it is known that A happens, is given by

(1.10.1) $$f(B|A) = a/r_1 = a/n \div r_1/n = f(B \cap A)/f(A),$$

where it is assumed, of course, that r_1 is not zero.

This rule also can be assumed to apply to probabilities, and we can in fact define the *conditional probability* of B, given that A happens, by

(1.10.2) $$P(B|A) = \frac{P(B \cap A)}{P(A)}, \qquad P(A) \neq 0$$

In the same way,

(1.10.3) $$P(A|B) = \frac{P(A \cap B)}{P(B)}, \qquad P(B) \neq 0$$

The two events A and B are said to be *independent* if

(1.10.4) $$P(A \cap B) = P(A) \cdot P(B)$$

If this is so, then from (2) and (3) $P(B|A) = P(B)$ and $P(A|B) = P(A)$. This means that the probability of A or of B does not depend at all upon whether the other happens, in agreement with the intuitive idea of independence. Equation (4) is often called the *multiplication law* for probabilities.

With more than two events the situation becomes rather complicated. Three events A, B, C are independent if each pair (AB, AC and BC) are independent and if also

(1.10.5) $$P(A \cap B \cap C) = P(A) \cdot P(B) \cdot P(C).$$

This implies that *four* probability conditions have to hold. These may be $P(A|B) = P(A)$, $P(A|C) = P(A)$, $P(B|C) = P(B)$ and $P(C|A \cap B) = P(C)$. There are also five other conditions (given by interchanging the letters A, B, C) which hold when the first four hold.

The general relationship which replaces (5) when A, B, C are not independent may be written

(1.10.6) $$P(A \cap B \cap C) = P(A) \cdot P(B|A) \cdot P(C|A \cap B), \; P(A) \neq 0, \; P(A \cap B) \neq 0$$

The fact that three events may be pairwise independent without being completely independent may be illustrated by the following example [8]. Imagine four similar discs in a bowl, numbered respectively 112, 121, 211, 222, and suppose one disc is picked at random. Let the events A, B, C be "the first digit on the disc picked is 1," "the second digit on this disc is 1," and "the third digit on this disc is 1." Then it is easy to see that $P(A) = P(B) = P(C) = \frac{1}{2}$, $P(A \cap B) = P(A \cap C) = P(B \cap C) = \frac{1}{4}$, but $P(A \cap B \cap C) = 0$. The three pairs AB, AC, and BC are therefore independent but the condition (5) is not satisfied, so that A, B, C are not all three independent.

1.11 **Elements of Combinatorial Analysis** Most ordinary calculations in elementary probability are based on the assumption that every simple element of the finite set $\mathscr{U}$ of all possible outcomes has the same weight. The probability measure of an event A is therefore proportional to the *number* of elements of $\mathscr{U}$ included in A. Thus to find the probability of the event, "a total of 11 spots with two dice," we assume that all of the 36 ordered pairs of numbers which can represent the fall of the two dice have an equal weight. Two of these, namely (6, 5) and (5, 6), correspond to a sum of 11, and the probability of this event, on the basis of our assumption, is therefore 2/36. In more complicated situations the calculation of the number of elements included in a particular subset of $\mathscr{U}$ will often involve the mathematics of permutations and combinations. We therefore recall briefly a few definitions and theorems, without giving proofs of the latter.

THEOREM 1.9 *The number of ordered arrangements (permutations) of n distinguishable objects is* $1 \cdot 2 \cdot 3 \ldots \cdot n$. *This number is denoted by $n!$* (read "factorial n").

THEOREM 1.10 *The number of ways of selecting and arranging in order r out of n distinguishable objects is $n!/(n-r)!$.* This is often denoted by $(n)_r$, a notation due to Feller [9]. When $r = n$, the result should reduce to $n!$, so that we must agree to define $0!$ as 1.

THEOREM 1.11 *The number of ways of arranging in order n_1 objects all alike of one kind, n_2 all alike of a second kind, and so on, up to k kinds of objects, is $n!/(n_1!\,n_2! \ldots n_k!)$, where $\Sigma n_i = n$.*

THEOREM 1.12 *The number of ways of picking r out of n distinguishable objects, regardless of the order in which they are arranged, is called the number of combinations. It is given by* $\binom{n}{r} = \dfrac{n!}{r!(n-r)!} = (n)_r/r!$. The symbol $\binom{n}{r}$ may be read "n above r." Other notations such as $C(n, r)$ or ${}^nC_r$ are also met with, but the one used here seems to be increasingly common. From the definition it follows immediately that $\binom{n}{0} = \binom{n}{n} = 1$. The symbol $\binom{n}{r}$ for $r > n$ is defined as 0.

Since $\binom{n}{r} = n(n-1)(n-2)\ldots(n-r+1)/r!$, the notation can be extended by writing

$$\begin{aligned}\binom{-n}{r} &= (-n)(-n-1)(-n-2)\ldots(-n-r+1)/r! \\ &= (-1)^r n(n+1)(n+2)\ldots(n+r-1)/r! \\ &= (-1)^r\binom{n+r-1}{r}.\end{aligned}$$

THEOREM 1.13 *The binomial theorem for a positive integral index may be written*

$$(1+x)^n = 1 + nx + \frac{n(n-1)}{2!}x^2 + \ldots + x^n$$

$$= \sum_{r=0}^{n} \binom{n}{r} x^r$$

$$= \sum_{r=0}^{\infty} \binom{n}{r} x^r$$

since all the coefficients $\binom{n}{r}$ *vanish when* $r > n$.

THEOREM 1.14 *The binomial theorem for a negative integral index may be written*

$$(1+x)^{-n} = 1 - nx + \frac{n(n+1)}{2!}x^2 - \ldots$$

$$= \sum_{r=0}^{\infty} (-1)^r \binom{n+r-1}{r} x^r$$

$$= \sum_{r=0}^{\infty} \binom{-n}{r} x^r$$

The two theorems 1.13 and 1.14 are therefore formally identical with the substitution of $-n$ for n. The first expansion however corresponds to a finite number of terms and the second to an infinite series.

1.12 **Sampling from a Finite Population** The process of picking a set of r objects out of a given set of n objects is often called *sampling*. The n objects constitute a "population," or "universe," and the r objects constitute the sample. If every object in the population has an equal chance to be picked for a particular sample, this sample is said to be *random*. Other kinds of sampling are of course possible. For instance, we could arrange the population in some order and pick every tenth object. This would be a *systematic sample*. In most problems of statistical inference, where it is required to infer properties of a population from those of a sample, it is understood that the sampling is random. Sometimes, for convenience or even for increased accuracy, some special scheme of sampling may be adopted, but if we are to make valid inferences from the sample to the population there must be an element of randomness about the choice of the sample. Sampling procedures are discussed more fully in Chapter 7.

There are two ways in which we can choose our random sample. We may pick an object, make a note of it, and put it back before picking the next object. This is called "sampling with replacements," and of course implies that the same object can appear more than once in the same sample. In fact, since there are n possible choices for each item in the sample, the number of ways of picking the sample (taking the order into account) is n^r. On the other hand, when an object is picked for the sample it may be put on one side and removed from the population. It is not then available to be picked again in the same sample. This is the situation envisaged in Theorems 1.10 and 1.12, and is called "sampling without replacements." The number of ways of picking the sample (taking order into account) is now $(n)_r = n(n-1)\ldots(n-r+1)$.

EXAMPLE 1 The number of possible five-digit numbers (including those beginning with one or more zeros) is 10^5. The number with all five digits different is $(10)_5 = 30{,}240$. The probability that a five-digit number selected at random will have all its digits different is therefore 0.3024.

EXAMPLE 2 The probability that in a class of 25 students no two will have the same birthday is, by a similar argument, $(365)_{25}/(365)^{25}$. On the assumption that all days in the year are equally likely as birthdays (and ignoring leap years), the number of possible arrangements of birthdays among the 25 people is $(365)^{25}$ and the number of arrangements with all birthdays different is $(365)_{25}$. The probability may be written as

$$p = \left(1 - \frac{1}{365}\right)\left(1 - \frac{2}{365}\right)\cdots\left(1 - \frac{24}{365}\right)$$

As a rough approximation, writing $\log_e(1 - x) \approx -x$, we have

$$\log_e p \approx -\frac{1 + 2 + \ldots + 24}{365} = -0.823$$

giving $p \approx 0.44$. The exact result, which may be evaluated by means of a table of logarithms of factorials (e.g., Glover's Tables [10] or Biometrika Tables [11]), is 0.4315. It is rather surprising to most people that the chance of at least two coincident birthdays in a group of this size should be as high as it is, namely 0.5685 ($= 1 - p$).

EXAMPLE 3 The probability of holding precisely three aces in a hand at bridge is the ratio of the number of possible hands containing three aces to the number of hands altogether. The basic assumption is that every completely specified hand of 13 cards that can be dealt from a deck of 52 cards is just as likely as every other, which is probably reasonable if the deck is more thoroughly shuffled than is customary in actual play.

The total number of possible hands is $\binom{52}{13}$, which is a very large number, about 635 billions. The number with three aces is given by multiplying the number of ways of picking the three aces, namely $\binom{4}{3}$, by the number of ways of picking the 10 other cards in the hand out of the 48 cards in the deck which are not aces, namely $\binom{48}{10}$. The required probability is therefore $\binom{4}{3}\binom{48}{10}\Big/\binom{52}{13}$, which reduces to 0.0412.

EXAMPLE 4 Why does it pay, in the long run, to bet even money on seeing 6 at least once in four rolls of a die, but not to bet even money on seeing double-6 at least once in 24 rolls with two dice?

This problem was solved early in the history of probability theory. It was posed by the Chevalier de Méré, a courtier and amateur mathematician at the French court around 1650, to the celebrated mathematician Blaise Pascal. Since the chance of 6 in a single throw with one die is six times as great as the chance of double-6 with two dice, it seemed only natural to him that the number of throws necessary for an even chance of seeing the second event should be just six times the number necessary for the same chance of seeing the first event. However, calculations did not seem to bear this out, and Pascal was able to show him that his supposition was in fact not true.

It is easier to find the probability of the complementary event, no 6 at all, than that of at least one 6. The probability of not seeing 6 in a single roll is 5/6, and if the rolls are independent the probability of not-6 on four successive rolls is $\left(\frac{5}{6}\right)^4$. The probability of at least one 6 is therefore $1 - \left(\frac{5}{6}\right)^4 = 0.516$, which is a better than even chance.

Similarly, the chance of not seeing double-6 in all 24 successive rolls with two dice is $(35/36)^{24}$ and the chance of at least one double-6 is therefore $1 - (35/36)^{24} = 0.491$. This is a less than even chance. The two chances are so nearly equal, however, that it seems unlikely that the difference could be detected empirically in ordinary play.

EXAMPLE 5 A card is drawn from an ordinary deck, looked at, and replaced, and the deck is shuffled. How many times should this be done in order to have a 90% chance of seeing the ace of spades at least once?

The same argument as in Example 4 leads to the conclusion that the chance of not seeing the ace of spades in n successive draws is $(51/52)^n$. If this is put equal to 0.1, we have an equation for n.

Inverting both sides and taking common logs, we obtain

$$n(\log 52 - \log 51) = \log 10 = 1$$

so that $n = \dfrac{1}{0.0084} = 119$.

In problems such as this, n must necessarily be an integer, so that the probability cannot always be adjusted exactly to a pre-assigned value. By taking the next highest integral value of n, we ensure that the probability will be at least equal to the value given.

1.13 **The Indicator Function** The idea of a function plays an important role in probability theory. A *function* is a rule which takes us from one set (called the *domain* of the function) to another set (called the *range*). To each element of the domain the function assigns one element of the range. Thus the function x^2

may have as domain the whole set of real numbers, positive and negative and zero, and if so its range is the set of positive real numbers and zero. To every value of x there is just one value of x^2. The function is said to be defined on the domain.

In probability theory, a function may be defined on the set of points such as a_i which belong to the universal set $\mathcal{U}$ of all possibilities (within the given context). If its range consists of the set $R = (r_1, r_2 \ldots r_k)$ and if the function assigns to each a_i the value r_j, we may write the functional relation as $f(a_i) = r_j$.

It often happens that the function assigns to several elements of $\mathcal{U}$ the same element r_j. If so, we can define a probability measure on the space R by giving to each element r_j the sum of the weights of all the points a_i which are such that $f(a_i) = r_j$. If A denotes the set of all these points a_i, then $P(A)$ is the weight attached to r_j, where P is the original probability measure on $\mathcal{U}$. The measure defined in this way on R is called the *measure induced by the function f*.

A particularly simple and useful example of a function is the *indicator function* I_A, defined on the whole space $\mathcal{U}$. This has just two values in its range, 1 when the point a_i belongs to A (a subset of $\mathcal{U}$) and 0 when it belongs to $\tilde{A}$. The indicator function corresponding to an event A may be thought of as 1 whenever A happens and 0 when it does not happen.

The indicator function of the event $A \cap B$ is given by

(1.13.1) $$I_{A \cap B} = I_A \cdot I_B$$

since I_A and I_B are both not zero only for points lying in the intersection of A and B. Similarly for the union of A and B,

(1.13.2) $$I_{A \cup B} = I_A + I_B - I_{A \cap B}$$

as is easily verified by checking the values of the right hand side for the different regions making up $A \cup B$ in a Venn diagram.

If the whole space $\mathcal{U}$ is partitioned into a set of disjoint events $A_1, A_2, \ldots A_n$, and if a function X is defined on all points of $\mathcal{U}$ by the relation $X(A_j) = x_j$, where $x_1, x_2 \ldots x_n$ are real numbers, then X is called a *simple random variable*, or a *variate*. From the definition of the indicator function, it follows that

(1.13.3) $$X = \sum_{j=1}^{n} x_j I_{A_j}$$

Thus for two dice the space of possibilities consists of 36 points (Figure 1). If the variate X is the sum of spots shown by the two dice, x_j may take any one of the eleven values 2, 3, 4 . . . 12. The set A_1 consists of the single point (1, 1). The set A_2 consists of two points (1, 2) and (2, 1), and so on. A variate may thus be thought of as a *mapping* from the space of possibilities $\mathcal{U}$ (the domain) into the axis of real numbers (the range). All the elements of A_1 are mapped into the point x_1, all the elements of A_2 into the point x_2, and so on. See Figure 7. We shall for the most part adhere to the convention of representing a variate by a capital letter such as X and the numerical values in its range by small letters such as x. We can then, for instance, speak of the probability that X takes the value x, or the set of values between x_1 and x_2.

This distinction between a variate X and a numerical value x (which the variate may take) is one which the student should try to get clear in his mind. The variate is a function *on* the space of events *to* the real axis. It associates with each possible event A_j a real number x_j, which may be the obvious number connected with the event (as in the example of the sum of spots shown by two

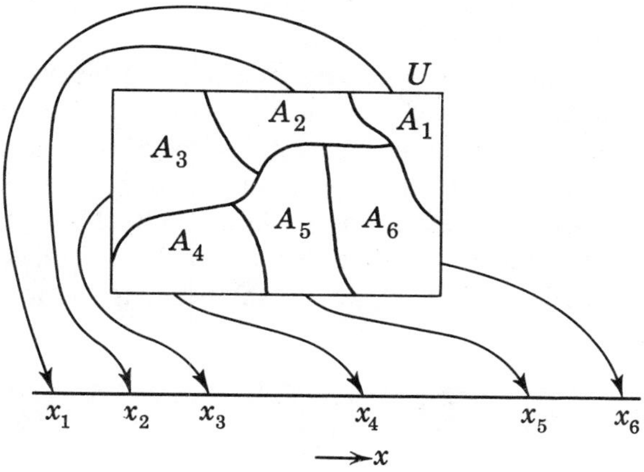

FIG. 7 VARIATE X AS A MAPPING INTO THE REAL AXIS

dice) or a more or less arbitrary number (as when we denote a male birth by 1 and a female birth by 0). The domain of X is the set of all possible events A_j in the particular context considered, and the range of X is the corresponding set of real numbers. The range, of course, may include only a discrete set of numbers or may include all real numbers in a finite interval or even the whole of the real axis. The domain is often called the "sample space" or the "possibility space." A point in this space (or a set of points) is a possible event of the type considered.

1.14 **Expectation** If $P(A_j)$ is the probability of the event A_j, the expectation of the variate X is

$$E(X) = \sum_{j=1}^{n} x_j P(A_j) \tag{1.14.1}$$

EXAMPLE 6 If 10,000 tickets are sold in a lottery and if there is one prize of \$1000 and ten prizes of \$50, what is the expectation of the worth of a single ticket?

If X is the worth of a ticket in dollars it is a variate which takes three values, namely, 1000, 50 and 0. The probabilities corresponding to these, on the assumption that the winning tickets are picked by a purely random process, are 1/10,000, 1/1000 and 9989/10,000 respectively. The expectation of X is therefore \$(1000/10,000 + 50/1000), or 15 cents. If the price of the ticket were 15 cents the lottery would be "fair," in the sense that the price would be equal to the expectation. Actual lotteries and gambling games are not fair in this sense, since a substantial percentage of the money raised goes to the organizers (or the "bank") and is not available for prizes.

EXAMPLE 7 Johnny is collecting a set of 12 kinds of coupon, one coupon being found in each packet of a particular breakfast cereal. If the family buys a new packet on Monday of each week, how long should he expect to have to wait before the set is complete? It is assumed that the different kinds of coupon are distributed at random in packets of the cereal.

Suppose that on a particular Monday, after opening the new packet, Johnny has collected x different kinds of coupon $(1 \le x \le 11)$. The chance that the packet to be opened in one week's time contains one of the kinds he already has is $x/12$, and the chance that it has a new kind is $1 - x/12$. The chance that he has to wait two weeks before getting a new kind is $x/12(1 - x/12)$. The chance that he has to wait r weeks and then gets a new kind is $(x/12)^{r-1}(1 - x/12)$. Denoting this probability by $p(r, x)$, we obtain as the expectation of r for a given x the expression

$$\sum_{r=1}^{\infty} rp(r, x) = \left(1 - \frac{x}{12}\right) \sum_{1}^{\infty} r\left(\frac{x}{12}\right)^{r-1}$$

$$= \left(1 - \frac{x}{12}\right)\left[1 + 2\frac{x}{12} + 3\left(\frac{x}{12}\right)^2 + \dots\right] = \frac{12}{12 - x}$$

The total expected time before the set is complete is therefore

$$\sum_{1}^{11} \frac{12}{12 - x} = 12\left(\frac{1}{11} + \frac{1}{10} + \dots + 1\right) = 36.2 \text{ weeks}$$

THEOREM 1.15 *If I_A is the indicator function for the set A, $E(I_A) = P(A)$.*

By the definition of I_A it takes only the two values, 1 for A and 0 for $\tilde{A}$. Its expectation is therefore $1 \cdot P(A) + 0 \cdot P(\tilde{A})$.

THEOREM 1.16 *If X and Y are variates defined over the same space $\mathscr{U}$ of possibilities, $E(X + Y) = E(X) + E(Y)$.*

If the space $\mathscr{U}$ is subdivided into disjoint sets $A_j (j = 1, 2 \dots n)$ and also into disjoint sets $B_k (k = 1, 2 \dots m)$, then, by Eq. (1),

$$E(X) = \sum_j x_j P(A_j), \qquad E(Y) = \sum_k y_k P(B_k)$$

Now the event A_j may be separated into disjoint sets $A_j \cap B_1, A_j \cap B_2 \dots A_j \cap B_m$ (see Figure 8), so that $A_j = \sum_k A_j \cap B_k$. Similarly, $B_k = \sum_j A_j \cap B_k$. Therefore,

$$E(X) + E(Y) = \sum_j x_j P\left(\sum_k A_j \cap B_k\right) + \sum_k y_k P\left(\sum_j A_j \cap B_k\right)$$

By the addition law for disjoint events,

$$P\left(\sum_k A_j \cap B_k\right) = \sum_k P(A_j \cap B_k), \quad \text{and} \quad P\left(\sum_j A_j \cap B_k\right) = \sum_j P(A_j \cap B_k)$$

Hence,

$$E(X) + E(Y) = \sum_j \sum_k (x_j + y_k)P(A_j \cap B_k).$$

By definition, $E(X + Y) = \sum_j\sum_k (x_j + y_k)P(A_j \cap B_k)$ since for all points in the intersection of A_j and B_k, the variable $X + Y$ has the value $x_j + y_k$.

The result can be extended to any finite number of variates.

1.15 **Independent Variates** We have defined the concept of independence for events, in § 1.10. Classes of events $\mathscr{A}$, $\mathscr{B}$, $\mathscr{C}$ . . . are independent if A_j, B_k, C_l . . . are independent, where A_j is any member of $\mathscr{A}$, B_k any member of $\mathscr{B}$, and so on. Random variables X, Y, Z . . . are independent if the partitions of $\mathscr{U}$ on which they are defined are independent, that is, if $X = \sum x_j I_{A_j}$, $Y = \sum y_k I_{B_k}$, etc., where each A_j is independent of each B_k, etc.

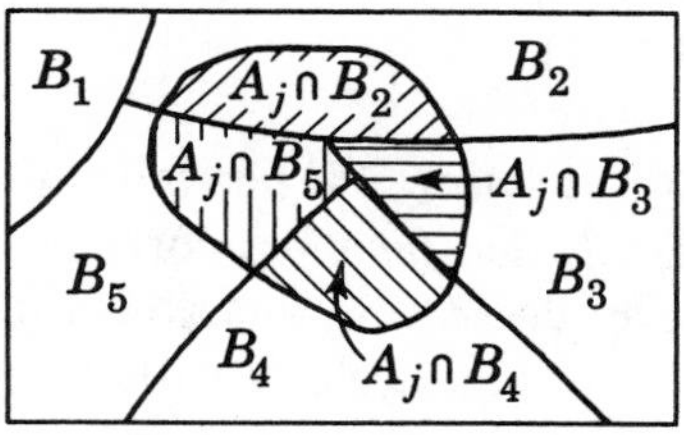

FIG. 8 DISJOINT SUBSETS OF A SET A

THEOREM 1.17 *If X and Y are independent,*

$$E(XY) = E(X)\cdot E(Y)$$

By definition, $E(XY) = \sum_j \sum_k x_j y_k P(A_j \cap B_k)$. Since A_j and B_k are independent, $P(A_j \cap B_k) = P(A_j)P(B_k)$. Therefore,

$$\begin{aligned} E(XY) &= \sum_j \sum_k x_j y_k P(A_j)P(B_k) \\ &= \left[\sum_j x_j P(A_j)\right]\left[\sum_k y_k P(B_k)\right] \\ &= E(X)\cdot E(Y). \end{aligned}$$

This result also can be extended to any finite number of variates.

1.16 **Continuous Probability** In some problems we cannot divide the whole space of possibilities $\mathscr{U}$ into a finite (or even a denumerable) set of regions A_j in each of which X takes a definite value x_j. What happens is that X varies continuously over some interval and there is a probability $f(x)\,dx$ that it takes a value between x and $x + dx$. In such a case $f(x)$ is called a *probability density*, or simply a *density*. It is a single-valued, non-negative function, integrable over its whole domain of definition (say from $x = a$ to $x = b$), and such that $\int_a^b f(x)\,dx = 1$.

It is customary to take the domain of x as the whole real axis from $-\infty$ to $+\infty$, and to put $f(x) = 0$ over any intervals which correspond to impossible values of X. It may happen, for instance, that X is by nature non-negative, in which case $f(x) = 0$ for all negative values of x.

If the event A corresponds to a value of x between α and β, the probability of A is defined by

$$P(A) = \int_{\alpha}^{\beta} f(x)\, dx$$

Note that here the function $f(x)$, defined on the axis of real numbers, induces a measure on $\mathscr{U}$, the measure of the interval dx being $f(x)\, dx$. The probability that $X \leq x$ is given by

$$F(x) = \int_{-\infty}^{x} f(u)\, du \tag{1.16.1}$$

This function, $F(x)$, is called the *cumulative distribution function* (often simply the *distribution function*) corresponding to the density $f(x)$. $F(x)$ is a non-negative, never-decreasing function, with values lying between 0 and 1, inclusive.

For a discrete distribution, in which X takes the distinct values x_j ($j = 1, 2, 3 \ldots$) with probabilities $f(x_j)$, the distribution function is defined as

$$F(x) = \sum_{x_j \leq x} f(x_j) \tag{1.16.2}$$

This is a step-function (see Figure 9a). It increases by a finite amount $f(x_j)$ at each point x_j, but remains constant in between. At each point x_j, $F(x)$ has the value at the top of the riser, and so is continuous on the right but not on the left. Figure 9b shows a typical distribution function for a continuous variate.

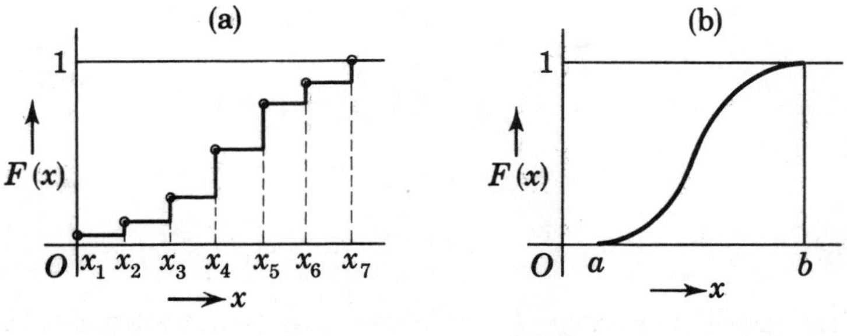

FIG. 9 (a) DISCRETE DISTRIBUTION FUNCTION

FIG. 9 (b) CONTINUOUS DISTRIBUTION FUNCTION

EXAMPLE 8 (Bertrand's problem) What is the probability that if a chord is drawn at random in a circle of radius a, the length of the chord will be greater than a?

This illustrates a difficulty that often arises in such problems. There is no unique answer unless the words "at random" are more carefully defined.

Suppose we pick a point A anywhere on the circle and draw the diameter AOA' through A. Any chord AB will make an angle θ with AA' which lies between $-\pi/2$ and $\pi/2$. If this angle is between $-\pi/3$ and $\pi/3$ (so that AB is between AC', and AC), the condition $l > a$ is satisfied. The probability required is therefore $\int_{-\pi/3}^{\pi/3} f(\theta)\, d\theta / \int_{-\pi/2}^{\pi/2} f(\theta)\, d\theta$, where $f(\theta)$ is the probability density for θ. On the assumption that all possible values of θ are equally likely, $f(\theta) = 1/\pi$, and the probability reduces to 2/3.

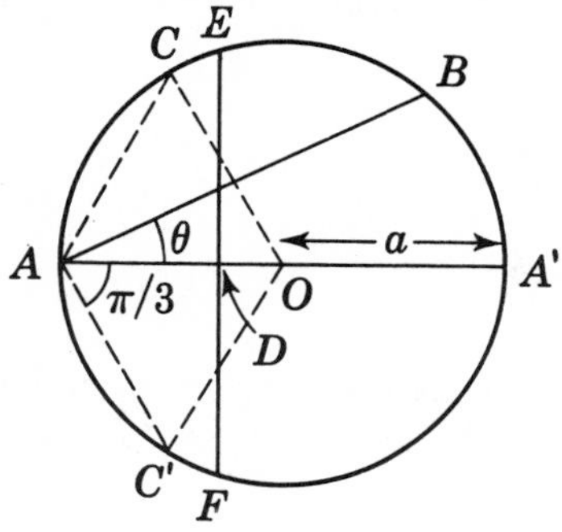

FIG. 10 BERTRAND'S PROBLEM

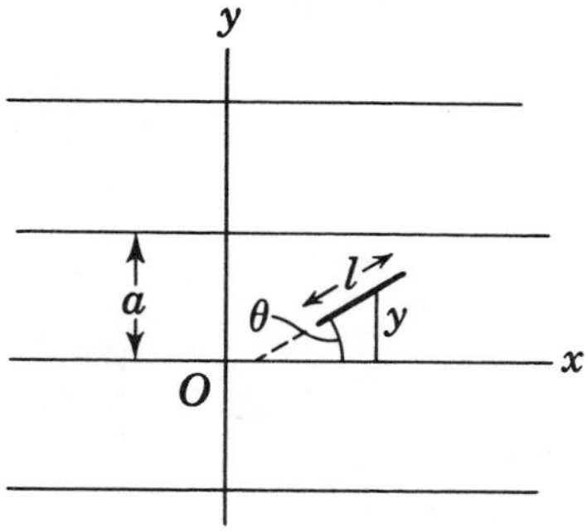

FIG. 11 BUFFON'S PROBLEM

Another possible procedure for drawing the chord "at random" would be to select A as before and pick a point D on AO. The perpendicular EF to AO passing through D defines a chord of the circle. If $OD = x$, EF will be greater than a whenever $x < \sqrt{3}a/2$. If the probability density for x is $f(x)$, the required probability is $\int_0^{\sqrt{3}a/2} f(x)\, dx / \int_0^a f(x)\, dx$. On the assumption that all possible values of x are equally likely, $f(x) = 1/a$, and the probability reduces to $\sqrt{3}/2 = 0.866$.

The two answers correspond to different induced measures, depending on the way we conceive the chord drawn at random. It would not be possible to settle the question by resort to experiment, because in devising any experimental set-up for drawing a random chord (such as tossing a straight piece of wire, longer than the diameter of the circle, on to the table-top on which the circle is drawn) we would need to choose one particular interpretation of the random process.

EXAMPLE 9 (Buffon's problem) Parallel straight lines, a distance a apart, are ruled on a horizontal table. A straight piece of wire, or needle, of length $l < a$, is tossed at random on to the table. What is the probability that it crosses a line?

If we take the x-axis along one of the lines and the y-axis perpendicular, it is easy to see that the x-coordinate of the centre of the needle is immaterial. It is the y-coordinate of the centre and the angle θ made by the needle with the x-axis which matter. (Fig. 11).

The needle will cross the nearest line if the distance y of its centre from this line is less than $\frac{1}{2}l \sin \theta$. The domain of y is from 0 to $a/2$ and that of θ from 0 to π. If the joint probability density for y and θ is $f(y, \theta)$, the required probability is

$$p = \int_0^\pi \int_0^{\frac{1}{2} l \sin \theta} f(\theta, y)\, dy\, d\theta \Big/ \int_0^\pi \int_0^{a/2} f(\theta, y)\, dy\, d\theta$$

The assumption of random tossing means that we regard all possible values of θ and y as equally likely. If so, $f(\theta, y)$ is constant and the probability becomes

$$p = \int_0^\pi \int_0^{\frac{1}{2} l \sin \theta} dy\, d\theta \Big/ \int_0^\pi \int_0^{a/2} dy\, d\theta$$

$$= \frac{l}{\pi a} \int_0^\pi \sin \theta\, d\theta = 2l/(\pi a)$$

This result suggests an empirical method of approximating to the value of π. Various trials of the method have been made from time to time (see, for example, reference [12]).

1.17 **Random Numbers** The importance of ensuring that a sample is effectively random if a valid inference is to be made from the sample to the population, has already been mentioned. Various tests of randomness have been devised and some will be mentioned later (see §§ 10–13 and 10–14). The choosing of a random sample, even from an artificial population of cards, balls, discs, or the like, is not easy, since the mechanical shuffling or mixing may be far from adequate, cards may tend to stick together, balls may not be equally smooth, and so on. When it is necessary to pick random samples from a crop in the ground or from a group of experimental animals, the task is much harder. There is a natural tendency to pick what seem to be *typical* rather than truly random samples.

Experience has shown that the best method is to use a set of *random numbers*. These numbers have usually been obtained by some mechanical process, such as a very carefully made roulette wheel, and have been thoroughly tested for randomness. They are generally arranged in sets of two or four digits and grouped in thousands. A short extract from one such table [13] is given in the Appendix, Table B1. The largest table up to the present time contains a million random digits [14] and was prepared because of the increasing need for very large blocks of random numbers in some modern statistical techniques.

Random numbers may be used to simulate the results of a chance experiment, such as tossing a coin. Instead of actually tossing a real coin (assumed to have a probability for heads equal to 0.5), one may open a table of random numbers, jab a forefinger anywhere on the page and start reading random numbers (in pairs) from the place so indicated. Whenever the number is between 00 and 49 inclusive it is read as "head", and when it is between 50 and 99 inclusive as "tail". Thus the succession of numbers 55, 58, 79, 50, 56, 01, 51, 65, 92, 32, 21, 66, 35, 18, 65, 08 . . . will be recorded as the following sequence of heads and tails: T T T T T H T T T H H T H H T H . . . . This arrangement, which the author obtained on his first trial, is a random sequence and could easily happen with an actual coin, but it is not what we would write down naturally if we tried to forecast the result of such an experiment. Although random, it is hardly typical.

In order to use random numbers to draw samples from a given population it is necessary to allot numbers, or blocks of numbers, to elements in the population. Thus if we have a population of 600 from which we want to draw a sample of 20, we number the members of the population (the data for each individual may be entered on a numbered card, for example) and then we read off consecutive random numbers in groups of three digits, ignoring all numbers over 600 and disregarding repetitions. The numbers so obtained, such as 284, 444, 323, 424, 358, 521, 406, 565, 457, 078 . . . , represent the individuals selected for the sample. If the population consists of several classes, the members of any one class being practically alike, and we want a sample from this population, we simply have to know how many members (or what proportion of members) there are in each class in the population, and allot a block of random numbers to each class. The size of the block should be proportional to the number of members in the class. Every random number that belongs to a particular block indicates a member of that class drawn for the sample. Thus, if there are five classes in the population, say A, B, C, D, E, numbering respectively 80, 200, 450, 240 and 30 individuals, we can allot blocks of four-digit random numbers as follows: A(0000 to 0799), B(0800 to 2799), C(2800 to 7299), D(7300 to 9699), E(9700 to 9999). The size of each block of numbers (800 for A, 2000 for B, etc.) is proportional to the size of the corresponding class in the population. The following set of random numbers: 6469, 7152, 0256, 6137, 0458, 0968, 9610, 5778, 8500, 8981, would indicate a sample C, C, A, C, A, B, D, C, D, D, consisting of two A's, one B, four C's and three D's.

Random numbers may also be used in many other ways. One common requirement in experimental work is to randomize the order of a group of objects, such as plots of land in a block, where the plots are to have different treatments. To randomize nine objects, or in other words, to form a random permutation of the integers 1 to 9, we can read off any set of consecutive random digits, ignoring zeros and repetitions. Thus, the set 3 4 7 6 6 4 5 5 6 6 4 9 0 1 5 6 6 3 6 8 8 0 2 gives the order 3 4 7 6 5 9 1 8 2. Modifications of this method can be used with groups of larger size. The important thing in randomization is to use an impersonal, objective method and not to trust to intuition.

Although Table B.1 is known to satisfy several tests of randomness, any limited collection of random numbers is bound to show some peculiarities. Accordingly, the table should not be used over and over again. If very large blocks of random numbers are needed, recourse should be had to larger tables such as [13] and [14].

PROBLEMS

A (§§ 1.7–1.10)

1. Suppose that of a group of people surveyed, 30% own both a house and a car, 40% own a car but not a house, 10% a house but not a car, and 20% own neither. Illustrate by a Venn diagram. What is the percentage that own either a house or a car or both? What is the fraction of car owners that are also house owners?

2. Let A stand for the event that a man chosen at random from a certain population is overweight and B for the event that he is over 50. Write down the symbols for the probabilities that (a) he is not overweight, (b) if he is overweight he is also over 50, (c) if over 50 he is not also overweight.

3. If $P(A) = \frac{1}{3}$, $P(B) = \frac{3}{4}$ and $P(A \cup B) = 11/12$, what is $P(A \cap B)$? Find also $P(A|B)$ and $P(B|A)$.

4. If A and B are independent events, and if $P(A) = \frac{1}{3}$ and $P(B) = \frac{3}{4}$, what is $P(A \cup B)$? *Hint:* Usé Eq. (1.10.4).

5. Prove from the definition of conditional probability that $P(A \cap B \cap C) = P(A) \cdot P(B|A) \cdot P(C|A \cap B)$.

6. Write out the detailed proof of Eq. (1.7.11).

7. Two good dice are rolled simultaneously. Let A denote the event "the sum shown is 8" and B the event "the two show the same number." Find $P(A)$, $P(B)$, $P(A \cap B)$, $P(A \cup B)$, $P(A|B)$ and $P(B|A)$.

B (§ 1.11)

1. How many five-digit numbers are there with every digit odd? How many with no digit lower than 6?

2. How many arrangements can be made of the letters of the word "caught" if the vowels are always together and in the same order?

3. Four strangers board a bus in which there are six empty seats. In how many different ways can they be seated?

4. Six papers are set in an examination, two of them in mathematics. In how many different orders can the papers be given if the two mathematics papers are not to be successive?

5. Show that the number of ways in which p positive and n negative signs ($p \geq 0$, $0 \leq n \leq p + 1$) may be placed in a row so that no two negative signs shall be together is $\binom{p+1}{n}$. *Hint:* With the positive signs placed in a row, there are $p + 1$ possible places for the first negative sign, p for the second, and so on. The negative signs are not distinguishable.

6. Prove that $\binom{n}{r} = \binom{n}{n-r}$.

7. Prove that $r\binom{n}{r} = n\binom{n-1}{r-1}$. Hence, show that $\sum_{r=1}^{n} r\binom{n}{r} = n2^{n-1}$.

Hint: $\sum_{r=1}^{n}\binom{n-1}{r-1} = \sum_{r=0}^{n-1}\binom{n-1}{r}$. Put $x = 1$ in Theorem 1.13.

8. If $\binom{n}{12} = \binom{n}{8}$, what is n?

9. At a long dinner table the host and hostess sit opposite each other at the ends. In how many ways can $2n$ guests be arranged (n on a side) so that two particular guests do not sit together? *Hint:* Place these two guests first.

10. Prove that $\sum_{r=0}^{k}\binom{m}{r}\binom{n}{k-r} = \binom{m+n}{k}$, $k \leq m$. *Hint:* Use Theorem 1.13 and the identity $(1 + x)^{m+n} = (1 + x)^m(1 + x)^n$.

11. Use Problem 10 to prove that

(a) $\sum_{r=0}^{k} \binom{k}{r}^2 = \binom{2k}{k}$

(b) $\binom{m+1}{k} = \binom{m}{k} + \binom{m}{k-1}$

C (§ 1.12)

1. If four cards are drawn at random from a deck of 52 cards, what is the probability that there will be one card of each suit?

2. A bag contains nine white and three black balls. If five balls are drawn, without replacement, what is the probability that at least two are black?

3. What is the chance that a bridge hand of 13 cards contains both the ace and king of spades?

4. A batch of 1,000 lamps is known to have 5% defectives. If five lamps are chosen at random and tested, what are the probabilities that (a) none is defective (b) there are exactly two defectives?

5. A factory produces screws, put up in boxes of 100. Boxes are inspected by taking 20 screws at random and rejecting the box if any defects are found in the sample. What is the probability of passing a box that contains two defective screws?

6. Calculate the probability of throwing a 6 with an ordinary die at least once in six trials.

7. A room has three lamp sockets. From a collection of 10 light bulbs, of which only six are good, I select three at random and put them in the sockets. 'What is the probability that I shall have light? *Hint:* Find the probability of *not* getting light, i.e., of selecting three bad bulbs.

8. *A* and *B* take turns in throwing two dice, the winner being the first to throw 9. Show that if *A* has the first throw, their respective chances of winning are in the ratio 9/8. *Hint: A* may win on 1st, 3rd, 5th . . . throws, and these possibilities are mutually exclusive.

9. Cards are dealt from a well-shuffled deck until an ace appears. Show that the probability that exactly n cards will be dealt before the first ace is $4(48)!\,(51-n)!/[52!(48-n)!]$.

10. (*The matching problem*). A man writes four letters and addresses four envelopes. His secretary puts the letters in the envelopes at random. Show that the probability that at least one letter gets into its right envelope is $1 - 1/(2!) + 1/(3!) - 1/(4!)$. Generalize for n letters. *Hint:* Use Eq. (1.7.12). Let A_j denote the event that the jth letter gets into the right envelope. The probability required is $P(A_1 \cup A_2 \cup A_3 \cup A_4)$. For n letters the result is close to $1 - e^{-1} = 0.632$ (see Appendix A.1). The approximation is correct to the third figure for $n \geq 6$.

11. In a gambling game, a player may deal 10 cards from a well-shuffled bridge deck and wins if, at any stage of the dealing, the number on a card is the same as the number of cards dealt. (Face cards are assigned the number zero). Find the probability that the dealer will win. *Hint:* This is a slightly more general matching problem (see Hint to Problem 10). Show that $P(A_j) = 4\cdot 51!/52!$, $P(A_j \cap A_k) = 4^2\cdot 50!/52!$, etc.

12. Ten absent-minded professors, each with a hat, attend a meeting and each man leaves with one of these hats chosen at random. What is the approximate probability that no one gets his own hat? What is the probability that exactly nine men get their own hats?

13. A bridge player and his partner have nine spades between them. What are the respective probabilities that the other four spades are split between the opponents 4–0, 3–1, 2–2?

14. Twelve cards have been dealt, six down and the other six showing a jack, two kings, a 7, a 5 and a 4. What is the probability that the next card will be a 4 or less, ace counting low? *Hint:* The six cards down do not affect the answer.

15. From an urn containing 10 balls, numbered from 1 to 10, balls are drawn one at a time and placed in a straight row of holes also numbered 1 to 10. If each ball is placed in its proper hole, what is the probability that there will not be an empty hole between two filled ones at any time of the drawing? *Hint:* When k holes are filled, there are two favorable positions for the next ball—unless the k holes include an end hole, when there is only one favorable position. Multiply the probabilities of success for $k = 1$ to 9.

16. Prove that the probability that some one of the four hands of cards in a particular bridge deal contains all 13 cards of a suit is about 1 in 40 thousand millions. (More hands of this character have been reported than would be expected. This fact may be due to imperfect shuffling in actual play.)

D (§§ 1.13–1.15)

1. A bag contains five nickels and a quarter, all wrapped separately so as to be indistinguishable. A boy is allowed to draw one coin at a time and keep it until he draws the quarter, when he must stop. What is his expectation?

2. A tosses six pennies and agrees to pay B \$6 if either six heads or six tails appear and \$5 if either five heads or five tails appear. In every other case he takes B's stake. How much should this stake be to make the game fair? *Hint:* If the stake is \$$x$, calculate B's expectation and put it equal to zero.

3. A coin is tossed until a head appears and the number of tails obtained is recorded. Find the probability of getting x tails before the first head, and the expected value of x. *Hint:* $1 + 2r + 3r^2 + \ldots = (1 - r)^{-2}, r < 1$.

4. In an infinite series of independent trials of an event with constant probability p of success in a single trial, what is the expectation of the number of failures preceding the first success?

5. From a deck of 13 spades a person draws cards one at a time, replacing each time, until he draws the ace. What is the expectation of the number of cards drawn?

6. What is the mathematical expectation of the sum of points on n dice, tossed at the same time?

7. A tosses 3 pennies and B two, and the winner is the one with the greatest number of heads. The winner takes the combined stakes. If there is a tie they continue tossing until a decision is reached. How much money should A put up on a game, to each dollar that B puts up, to make the game fair? (A game is a set of tosses leading to a decision. Theoretically, a game might go on indefinitely, but the probability of this is zero.)

8. (*The Petersburg Paradox*). A tosses a coin repeatedly, having agreed to give B \$$2^n$ if n tails appear before the first head. (Thus B receives \$1 if the first toss is a head, \$2 if one tail precedes a head, and so on). If, however, 10 tails appear in succession before a head, the game stops there and B receives \$$2^{10}$. What sum should B pay A for this privilege? *Note:* The paradox arises from the fact that if the game is allowed to go on indefinitely, B's expectation is infinite. This seems contrary to common sense and has been the subject of much discussion. Even in the limited game, B would be foolish to pay a sum equal to his expectation, unless he intends to play the game a few thousand times: See, e.g., [3] and [9].

9. Five cards are drawn at random from a deck without replacement, looked at, and then replaced. This is done 1,000 times. How often would you expect to get: (a) 5 of one suit, (b) 4 of one suit, (c) 3 of one suit, 2 different, (d) 3 of one suit, 2 of another, (e) 2 of each of two suits, 1 different, (f) 2 of one suit, 3 different? *Hint:* The expected number in each case is 1,000 times the probability of that combination.

E (§ 1.16)

1. Let A_1, A_2, A_3, A_4, A_5 be sub-sets of the two-dimensional x-y plane, defined as follows:

A_1 = set of (x, y) such that $x \leq 2, y \leq 4$, denoted by $\{x \leq 2, y \leq 4\}$
$A_2 = \{x \leq 2, y \leq 1\}$, $A_3 = \{x \leq 0, y \leq 4\}$,
$A_4 = \{x \leq 0, y \leq 1\}$, $A_5 = \{0 < x \leq 2, 1 < y \leq 4\}$.

Given that $P(A_1) = 7/8$, $P(A_2) = 1/2$, $P(A_3) = 3/8$, and $P(A_4) = 1/4$, find $P(A_5)$. Illustrate by a diagram.

2. A circle of diameter 8 in. is drawn in the interior of a square of side 12 in. A penny (diameter $\frac{3}{4}$ in.) is dropped at random on the square, which is lying on a horizontal table. If only those cases are counted when the penny lies wholly inside the square, what is the probability that it is also wholly inside the circle? *Hint:* The *center* of the penny is equally likely to fall anywhere within the area open to it. Calculate the possible area and the favorable area.

3. The floor of a large room is made of hardwood, laid in strips 1 in. wide, with cracks between of negligible width. A coin of diameter $1\frac{1}{2}$ in. is dropped on the floor. What is the probability that it touches three strips?

4. A third method of drawing a chord "at random" in a circle (see Example 8 above, § 1.16) is to select the *center* of the chord at random and then draw the chord. When the center is determined, so is the whole chord. If the center is equally likely to be anywhere within the given circle, show that the answer to Bertrand's problem is $\frac{3}{4}$.

5. A thin stick of length a is broken into three pieces. What is the probability that these pieces can be arranged to form a triangle? *Hint:* No piece may be longer than $a/2$. If x, y, z are the three lengths, they satisfy the condition $x + y + z = a$, which represents the part of a plane contained in the positive octant. Find the area of the part of this plane corresponding to the given condition.

6. A diamond of value V is broken into two pieces. If the value of a diamond varies as the square of its weight, what is the expected value of the broken diamond? *Hint:* If w is the original weight, the probability that one piece has a weight x to $x + dx$ is dx/w, $0 < x < w$.

REFERENCES

[1] Jeffreys, H., *Theory of Probability*, Oxford Univ. Press, 1948.
[2] Savage, L. J., *The Foundations of Statistics*, Wiley, 1954.
[3] Todhunter, I., *History of the Mathematical Theory of Probability from the Time of Pascal to that of Laplace*, MacMillan, 1865 (reprinted by Chelsea, 1958).
[4] David, F. N., "Studies in the History of Probability and Statistics," *Biometrika*, **42**, 1955, 1–15.
[5] Kendall, M. G., "Studies in the History of Probability and Statistics," *Biometrika*, **43**, 1956, 1–14.
[6] von Mises, R., *Probability, Statistics and Truth*, 2nd ed., George Allen and Unwin, 1957.
[7] Davis, R. L. (ed.), *Elementary Mathematics of Sets, with Applications*, Mathematical Assn. of Amer., 1958.
Christian, R. R., *Introduction to Logic and Sets*, Ginn, 1958.
[8] Uspensky, J. V., *Introduction to Mathematical Probability*, McGraw-Hill, 1937.
[9] Feller, W., *An Introduction to Probability Theory and its Applications, Vol. I* (2nd ed.), Wiley, 1957.
[10] Glover, J. W., *Tables of Applied Mathematics in Finance, Insurance, Statistics*, Wahr, 1923.
[11] Pearson, E. S., and Hartley, H. O., *Biometrika Tables for Statisticians, Vol. I*, Cambridge Univ. Press, 1954.
[12] Coolidge, J. L., *An Introduction to Mathematical Probability*, Oxford Univ. Press, 1925.
[13] Kendall, M. G., and Smith, Babington B., *Tables of Random Sampling Numbers*, Cambridge Univ. Press, 1946.
[14] The Rand Corporation, *A Million Random Digits*, Free Press, 1955.

Chapter 2

FREQUENCY DISTRIBUTIONS, FRACTILES AND MOMENTS

2.1 **Frequency Distribution in a Sample** To facilitate applications of statistical inference, the data supplied by observation or experiment are usually organized and summarized to expose their essential characteristics. Some of the methods and techniques for extracting the essential information supplied by data, which are to be regarded as constituting a sample, will be presented in this chapter. Incidentally, we might point out that statistical inference is not concerned solely with data that have already been obtained. An important branch of modern statistical theory deals with the design of experiments and shows the experimenter how to arrange his work so as to be able subsequently to extract the maximum of information from a limited amount of data.

Data obtained in an experiment, or as the result of an enquiry or questionnaire, are often presented in a table. A common form is the *frequency table*, in which one column gives observed values x of a random variable X and the other gives the frequency with which each of these values was obtained. Recall that X is a function on the space of events to the real axis. Its domain, if discrete, is the set of events A_j and its range is the set of distinct real numbers x_j. If X is continuous, its range usually includes all real numbers in some interval. In this case the range is divided up into convenient sub-intervals and the frequencies corresponding to the various sub-intervals (or *classes*) are entered in the table.

Table 2.1 records some data obtained by D. A. S. Fraser [1] in tossing a crudely made plastic die. The variable X is here the number of spots observed, and is of course discrete. The frequencies f of the six values observed in the first 400 tosses are given. The third column will be referred to later.

TABLE 2.1

x	f	F
1	73	73
2	83	156
3	80	236
4	57	293
5	41	334
6	66	400
	400	

Table 2.2 similarly presents a distribution of weights, measured to the nearest pound, for a sample of 1000 eight-year-old girls in Glasgow. The weight may be regarded as a continuous variable over a range of about 40 lb, divided into 10 sub-intervals, each of width 4 lb. Since a measured weight of 39 lb would be recorded for any child between 38.5 and 39.5 lb, the real limits for these sub-intervals (often called the *class boundaries*) are 27.5 lb, 31.5 lb, 35.5 lb, and so on. The central values of the sub-intervals are called *class-marks* and will be denoted by x_c. The upper class boundaries (ends of sub-intervals) will be denoted by x_e.

TABLE 2.2

Measured Class Limits	Class-Mark (x_c)	Frequency (f)	Upper Class Boundary (x_e)	Cumulative Frequency F
28 – 31 lb	29.5 lb	1	31.5 lb	1
32 – 35	33.5	14	35.5	15
36 – 39	37.5	56	39.5	71
40 – 43	41.5	172	43.5	243
44 – 47	45.5	245	47.5	488
48 – 51	49.5	263	51.5	751
52 – 55	53.5	156	55.5	907
56 – 59	57.5	67	59.5	974
60 – 63	61.5	23	63.5	997
64 – 67	65.5	3	67.5	1000
		1000		

2.2 **Cumulative Frequency Distributions** It is often convenient to present the data of a frequency table in a slightly different form, recording for suitable values of x the total number of items in the sample which have an observed X equal to or less than x. For the discrete distribution of Table 2.1, the third column gives these cumulative frequencies (accumulated by adding the ordinary frequencies one by one from the top of the column downwards). They are denoted by F.

For a grouped distribution like that of Table 2.2, it is usually convenient to choose the upper class boundaries x_e as the selected values of x corresponding to F. There will be no measured values actually coinciding with x_e (since the measured values are all recorded to the nearest unit and the x_e to half a unit) and therefore the cumulative frequency gives the number of items with X *less than* x_e.

2.3 **Graphical Representation** A frequency table for a discrete distribution may be represented graphically by drawing ordinates equal to the frequency on a convenient scale at the various values of x. Thus Figure 12 corresponds to Table 2.1. The tops of the ordinates may be joined by straight lines, but these are merely to assist the eye and have no significance at intermediate values of x.

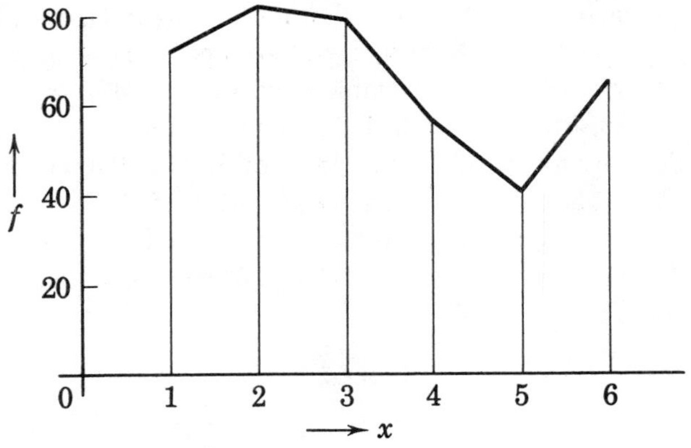

FIG. 12 FREQUENCY GRAPH FOR DISCRETE VARIATE

A continuous distribution, grouped in classes, is usually represented graphically by a histogram, as in Figure 13 (for the data of Table 2.2). The rectangles are drawn with bases corresponding to the true class intervals and with heights proportional to the frequencies. With all the class intervals equal, as in this example, the *areas* of the rectangles also represent the corresponding frequencies. In some tables the class intervals are not all equal, and then the heights must be suitably adjusted to make the areas proportional to the frequencies.

If the mid-points of the tops of the rectangles are joined by straight lines, the result is a *frequency polygon*, which may also be regarded as representing the data. The frequency polygon for Table 2.2. is shown dotted in Figure 13.

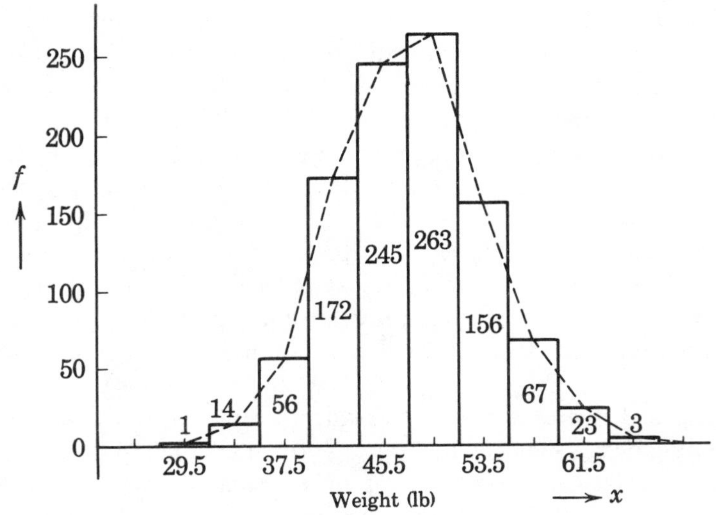

FIG. 13 HISTOGRAM AND FREQUENCY POLYGON FOR CONTINUOUS VARIATE

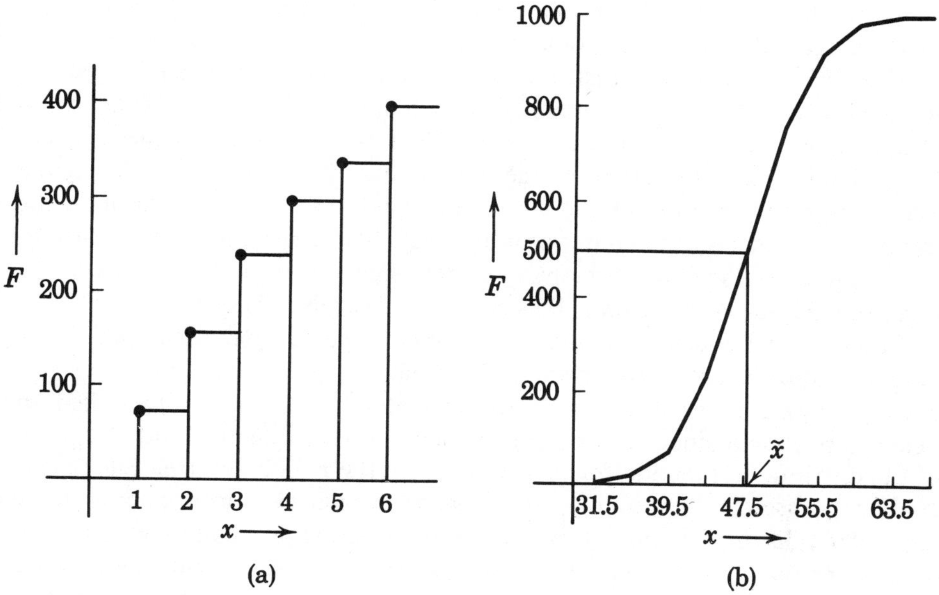

(a) DISCRETE VARIATE (b) CONTINUOUS VARIATE

FIG. 14 CUMULATIVE FREQUENCY POLYGONS

Graphs of the cumulative frequencies in Tables 2.1 and 2.2 are shown in Figure 14. The first is a step-diagram, similar to Figure 9a, in which at each value of x the frequency of values equal to or less than x is plotted. In Figure 14b there are no measured values equal to the values at the ends of the intervals, and the plotted points represent the frequency of values less than x.

2.4 **Frequency Curves and Ogives** The observed data usually refer to a sample of finite size which may be considered as representative of a very large, or practically infinite, population. The data of Table 2.1 refer to a sample of 400 tosses out of the indefinitely large number of tosses which could conceivably be made with this particular die, given unlimited time and patience, before the die finally wears out. The population of eight-year-old Glasgow schoolgirls, from which the sample of Table 2.2 was taken, is not infinite but is certainly large. With a very large population and a continuous variate, we can imagine the class intervals as being very short while still containing many observed values in each class (we must suppose that the measurements are correspondingly accurate). The frequency polygon will then approximate to a smooth *frequency curve* which represents the distribution of the variable X in the population. The area under this curve, between two fixed values a and b, represents the total number of individuals in the population with values of X between a and b. If instead of the total frequency we consider the curve as giving the *relative* frequency, (the *proportion* of values in this interval) the curve then represents the probability density in the population. The area between $X = a$ and $X = b$ is the *probability* that a random item from the population has a value between a and b.

One task of practical statistics is to determine whether a given sample can reasonably be regarded as coming from a population of a particular kind. Usually some plausible assumption is made about the form of the frequency curve (or probability density curve) for the population, and parameters defining the exact shape and position of the curve are calculated so as to make it fit the frequency polygon for the sample as well as possible. Some test is then applied to find out whether the fit can reasonably be regarded as satisfactory. This process is called curve-fitting, and some illustrations will be given later.

Just as the frequency polygon for a very large sample, with small class intervals, approximates to a smooth frequency curve, so the cumulative frequency polygon for such a sample approximates to a smooth curve called an *ogive*. If relative frequencies are used, the ogive becomes identical with the graph of the distribution function for the variate X in the population (see Fig. 9b).

For a discrete distribution, the relative frequency corresponding to a particular value of X is an approximation to the actual probability of this value of X in the population sampled. If some prior hypothesis is made about these probabilities it may be tested by noting the agreement of the observed relative frequencies with the predicted values. We might, for example, use the data of Table 2.1 to test the hypothesis that all faces of this particular die are equally likely to turn up (see § 10.2).

2.5 **The Median and Other Fractiles** The *median* of a sample of size N is the value of x for which the cumulative frequency is equal to $N/2$. In other words, it is that value of x which is exceeded by half the members of the sample. For this reason it is often used as an *average*, that is, a single (more or less central) value which may be regarded as in some sense typical of the whole sample. For a small sample the median is the middle one when the items are arranged in order (if the number of items is even, the median is usually taken half-way between the two middle ones).

The median of a population with distribution function $F(x)$ is that value $\tilde{x}$ for which $F(\tilde{x}) \doteq 0.5$. The median is therefore easily marked on an ogive or on a cumulative frequency polygon. In Fig. 14b, $\tilde{x}$ is the abscissa of the point on the polygon with ordinate $N/2 = 500$. If this point lies (as it will usually) on one of the straight sides of the polygon, the abscissa may be calculated by linear interpolation between the values at the beginning and end of this side. Thus in Table 2.2, $N/2 = 500$. The value 488 of F corresponds to an x of 47.5 lb and the value 751 to an x of 51.5 lb. The value $\tilde{x}$ corresponding to $F = 500$ is therefore $47.5 + \dfrac{500 - 488}{751 - 488} \times 4 = 47.68$ lb. The assumption underlying this computation is that the items of the sample in any class may be regarded as having values of x which are distributed approximately uniformly over the whole class interval.

For a discrete distribution, such as that of Table 2.1, the median is not a very precise quantity. It is obvious that $\tilde{x} = 3$, but there are 80 individual values all equal to 3 and only 156 out of 400 are definitely less than 3.

The values of x which correspond to cumulative frequencies of $N/4$ and $3N/4$ are called the *first quartile* and the *third quartile* respectively (often denoted by Q_1 and Q_3). One quarter of the sample is below Q_1 and three quarters below Q_3. Similarly, *deciles* (corresponding to tenths) and *percentiles* (corresponding to hundredths) may be defined. The third decile, D_3, for instance, is the value of x corresponding to a cumulative frequency of 3/10, and might equally well be called the thirtieth percentile, P_{30}. Since these points correspond to certain specified *fractions* of the distribution they are collectively called *fractiles* (or sometimes *quantiles*).

In general, the kth percentile (P_k) corresponds to a cumulative frequency equal to $k\%$ of N. The number k is called the *percentile rank* of P_k. It may be calculated by interpolation from a table such as Table 2.2.

Fractiles are often computed in order to obtain a measure of the spread or dispersion of a distribution. The more a distribution is concentrated around a central value, the less as a rule will be the distance between Q_1 and Q_3, or between say P_7 and P_{93}, so that the differences $Q_3 - Q_1$ or $P_{93} - P_7$ may be taken as measuring the dispersion. Deciles are often used by psychologists in assessing the performance of a student on some aptitude or achievement test. If, for example, a certain student's score is known to lie between D_8 and D_9 as determined from a large group taking the test, we can say that this student is better than eight-tenths of the group but is not in the top tenth, on this particular test.

2.6 **Fractiles as Statistics** The median and the other fractiles belong to the class of "statistics," which in this sense (as a plural word) means quantities calculated from experimental or observational data for a sample and used to make estimates about the population from which the sample is drawn. The median of a sample is one statistic which gives information about the population, and the interquartile range, $Q_3 - Q_1$, is another. However there are some statistics which are better than others for the purpose of giving reliable information. As an average, the median has the disadvantage that it does not use all the data available in the sample. It depends only on the order of the observations and not directly on their actual size. The median of the numbers 1, 7, 11, 12, 19, 26, and 34 is 12, since this is the middle number, but any other set of seven numbers arranged in ascending order with 12 in the middle would have the same median.

Statistics differ also in the extent of their sampling fluctuations, that is, in the extent to which their numerical values vary from one sample to another, of the same size and drawn from the same population. Other things being equal, a statistic with the least possible sampling fluctuation will be preferred. It turns out that in most situations the median has a greater sampling fluctuation than the arithmetic mean, which belongs to the class of statistics known as *moments*, and in general the arithmetic mean will be the statistician's preferred average. The mean has the further advantage that it lends itself more readily than the median to mathematical manipulation. A fuller discussion of the relative merits of several different types of average may be found, for example, in [2].

2.7 **Moments** Suppose that a discrete variate X can take k distinct values x_i $(i = 1, 2 \ldots k)$, and that f_i individuals in a sample have the value x_i. The total size of the sample is $\sum f_i = N$. The rth *moment of X about zero* is defined by

$$m'_r = \mathrm{N}^{-1} \sum_{i=1}^{k} f_i x_i^r \qquad (2.7.1)$$

Obviously $m'_0 = 1$. The most important case is when $r = 1$; the statistic m'_1 is called the *arithmetic mean*, and it will be convenient to denote it simply by m. The notation $\bar{X}$ is also commonly used, and serves to indicate that the arithmetic mean is a quantity of the same physical nature as X. If X is a length measured in centimeters, then $\bar{X}$ will also be a length in centimeters.

To calculate m for a distribution such as that of Table 2.1, it is merely necessary to form a column of values of fx and total it (see column 3 of Table 2.3). Here $m = \dfrac{1308}{400} = 3.27$.

TABLE 2.3

x	f	fx	fx^2	fx^2
1	73	73	73	73
2	83	166	332	664
3	80	240	720	2160
4	57	228	912	3648
5	41	205	1025	5125
6	66	396	2376	14256
	400	1308	5438	25926

The fourth and fifth columns of the table give the second and third moments respectively. Here $m'_2 = \dfrac{5438}{400} = 13.595$, and $m'_3 = \dfrac{25926}{400} = 64.815$.

In dealing with a grouped distribution such as that of Table 2.2, the moments are calculated on the assumption that all the individuals in a class can be regarded as having the central value (or class-mark) of that class. Although this is not actually true, the errors caused by grouping are not usually serious unless the grouping is very coarse. (For a method of correction, see § 5.10.)

The numerical labor of the calculation may generally be substantially reduced by suitable coding. If one of the class-marks (x_0) is chosen near the center of the table, and a new variable, u, is defined by

$$u = \frac{(x - x_0)}{c} \qquad (2.7.2)$$

where c is the class-interval, the values of u will as a rule be much simpler to work with than the original x values. As an illustration, Table 2.4 shows the procedure for calculating the first three moments about zero from the data of Table 2.2, where x_0 has been taken as 45.5 and c as 4. The values of m'_r $(r = 0,1, 2, 3)$ are given in the last row of the table. The last column is a check column (suggested by Charlier). The check depends on the identity:

$$\sum f_i(u_i + 1)^3 = \sum f_i u_i^3 + 3 \sum f_i u_i^2 + 3 \sum f_i u_i + \sum f_i. \tag{2.7.3}$$

TABLE 2.4

x	u	f	fu	fu^2	fu^3	$f(u+1)^3$
29.5	−4	1	−4	16	−64	−27
33.5	−3	14	−42	126	−378	−112
37.5	−2	56	−112	224	−448	−56
41.5	−1	172	−172	172	−172	0
45.5	0	245	0	0	0	245
49.5	1	263	263	263	263	2104
53.5	2	156	312	624	1248	4212
57.5	3	67	201	603	1809	4288
61.5	4	23	92	368	1472	2875
65.5	5	3	15	75	375	648
		1000	553	2471	4105	14177
m'_r		1	0.553	2.471	4.105	

Thus 14,177 = 4105 + 3(2471) + 3(553) + 1000.
The values of $f(u + 1)^3$ are found by multiplying each f by the cube of the u *in the next line* (since the values of u increase by 1 as we go down the column).

The arithmetic mean of the original variate X is found by multiplying the moment m'_1 (in terms of u) by c and adding x_0. Here

$$m = 4(0.553) + 45.5 = 47.712 \text{ lb.}$$

2.8 **Moments about the Mean. Variance, skewness and kurtosis** The rth moment about the mean of the variate X is defined as:

$$m_r = N^{-1} \sum f_i(x_i - m)^r \tag{2.8.1}$$

As before we see that $m_0 = 1$. Also, putting $r = 1$, we have

$$m_1 = N^{-1} \sum f_i x_i - m = 0 \tag{2.8.2}$$

for every sample, so that m_1 does not tell us anything about the sample. However, m_2, m_3, m_4 . . . are statistics which are often used for expressing the characteristics of a sample and making inferences about the population.

Historically they have been of great importance in the development of mathematical statistics, but the related k-statistics (see § 5.4) are now recognized to be more convenient.

The second moment, m_2, is often called the *variance*. It is given by

$$(2.8.3)\qquad \begin{aligned} m_2 &= N^{-1} \sum f_i(x_i - m)^2 \\ &= N^{-1}(\sum f_i x_i^2 - 2m \sum f_i x_i + m^2 \sum f_i) \\ &= m'_2 - m^2 \end{aligned}$$

since $\sum f_i x_i = Nm$, and $\sum f_i = N$.

The positive square root of the variance is called the *standard deviation*, denoted usually by s. It is the most widely-used measure of the spread of a sample distribution. Many authors, however, prefer to define the variance as the second k-statistic, k_2, which is related to m_2 by means of the equation

$$(2.8.4)\qquad k_2 = \frac{Nm_2}{N-1} = (N-1)^{-1} \sum f_i(x_i - m)^2$$

This definition has some advantages from the point of view of statistical inference, and we shall adopt it in this book. Of course, in large samples there is little difference between k_2 and m_2.

The spread of a distribution is most naturally and easily measured by the *range* of the variate, that is, by the difference $x_N - x_1$, where the measured values $x_1, x_2 \ldots x_N$ are supposed to be arranged in increasing order of size. The range, however, is not very convenient mathematically and is apt to be sensitive, in large samples, to sampling fluctuations. The standard deviation makes use of all the information in the sample, and is generally the most reliable measure of spread, even though it is a little more troublesome to calculate. The meaning of the standard deviation is perhaps most easily grasped by noting that in a good many common types of distribution, which are more or less symmetrical about a central value and tail off in both directions, roughly two-thirds of all the variates (in a rather large sample) will lie within an interval of x, extending for one standard deviation on either side of the mean. If the sample consists of several hundred individuals, the standard deviation will usually be (roughly) one-sixth of the range. This is worth remembering, as a guard against gross errors (such as misplacing a decimal point) in the calculation of the standard deviation. Reasons for these statements will appear when the normal distribution is discussed (see §§ 3.13 and 8.21).

The third moment, m_3, and the fourth moment, m_4, depend on the shape of the frequency polygon representing the distribution. Because of the cancelling of positive and negative third powers, a symmetrical distribution will have $m_3 = 0$. A distribution with a long tail extending out to the right will usually have a positive m_3, while one with a long tail out to the left will usually have a negative m_3. This is because the positive values of $(x - m)^3$ tend to outweigh the negative values, or vice versa. The statistic m_3 may therefore be used as a measure of the *skewness* of the distribution. It is not true however that if $m_3 = 0$ then the distribution is necessarily symmetrical [3]. To have a measure independent of the units of x, it is customary to divide m_3 by $(m_2)^{3/2}$. The skewness is then a pure number.

The fourth moment, m_4, divided by $m_2{}^2$ is a measure of *kurtosis*. This word, which means "peakedness," was adopted because in some common types of distribution a high value of m_4 is associated with a high central peak in the frequency polygon. However, the value of m_4 is very much dependent on the shape of the tails [4] and may have little to do with any central peak.

As we shall see later, the k statistics k_3 and k_4 are more convenient than m_3 and m_4 for measuring skewness and kurtosis respectively. We are usually interested in estimating these characteristics for a population, and the k-statistics give better estimates than the moments.

2.9 **Moments for a Probability Distribution** In many problems of statistical inference, samples are drawn from a population which is assumed to be described by a known type of mathematical distribution function. There may be one or more parameters occurring in this function; the values of these parameters are not known but can be estimated from the samples. One of the more common methods of estimation makes use of the relation between the moments of a sample and the moments of the population. For the present, the population will be thought of as infinitely large and characterized by its distribution function $F(x)$. The random variable X may be discrete or continuous, and in the latter case a density function $f(x)$ will exist (see § 1.16). Since $F(x)$ is the probability that $X \leq x$, the distribution of X in the population is often called a probability distribution.

If X is discrete, and if the probability is p_j that it takes the value x_j, the expectation of X^r is defined as

(2.9.1)
$$\mu'_r = E(X^r) = \sum_j x_j{}^r p_j$$

This is called the rth *moment of X about zero*. If there are infinitely many possible values of j, it is assumed that the sum converges.

The first moment, μ'_1, is the expectation of X and is often called the *population mean*. It will be denoted by μ, without prime or subscript, and corresponds to the sample statistic $m'_1(= m)$ previously defined in § 2.7. We shall adopt as far as possible the very useful convention of distinguishing between sample and population by the use of Latin letters such as m for the former and Greek letters such as μ for the latter.

If X is continuous, the definition corresponding to equation (1) is

(2.9.2)
$$\mu'_r = E(X^r) = \int_{-\infty}^{\infty} x^r f(x)\,dx$$

provided that the integral exists.

The expectation of $(X - \mu)^r$ is called the rth *moment about the mean*, and is given by

(2.9.3) $$\mu_r = E(X - \mu)^r = \sum_j (x_j - \mu)^r p_j$$

when X is discrete, or by

(2.9.4) $$\mu_r = E(X - \mu)^r = \int_{-\infty}^{\infty} (x - \mu)^r f(x)\, dx$$

when X is continuous. Since $\sum_j p_j = 1$ and $\int_{-\infty}^{\infty} f(x)\, dx = 1$, it is obvious that $\mu_0 = 1$ in all cases. Also,

(2.9.5) $$\mu_1 = E(X - \mu) = E(X) - \mu = 0$$

for any kind of distribution that possesses a mean. The lowest useful value of r is 2, and μ_2 is a very important descriptive parameter for the population, known as the *population variance*. The square root of μ_2 is usually denoted by σ and is called the *population standard deviation*. It is a measure of the spread or dispersion of the population about its mean. The sample standard deviation s defined in § 2.8 provides an estimate of σ.

Using the binomial theorem, we may write

(2.9.6) $$(x - \mu)^r = \sum_{q=0}^{r} (-1)^q \binom{r}{q} x^{r-q} \mu^q$$

This allows us to express the moments μ_r in terms of the moments μ'_r (which are usually easier to calculate). We have, assuming that moments up to the rth exist,

(2.9.7) $$\mu_r = E(X - \mu)^r = \sum_{q=0}^{r} (-1)^q \binom{r}{q} \mu^q E(X^{r-q})$$
$$= \sum_{q=0}^{r} (-1)^q \binom{r}{q} \mu^q \mu'_{r-q}$$

Thus, for example,

(2.9.8) $$\mu_2 = 1 \cdot \mu'_2 - 2\mu\mu'_1 + 1 \cdot \mu^2 \mu'_0$$
$$= \mu'_2 - \mu^2$$

(note that $\mu'_1 = \mu$ and $\mu'_0 = 1$)

(2.9.9) $$\mu_3 = \mu'_3 - 3\mu'_2\mu + 3\mu'_1\mu^2 - \mu'_0\mu^3$$
$$= \mu'_3 - 3\mu'_2\mu + 2\mu^3$$

(2.9.10) $$\mu_4 = \mu'_4 - 4\mu'_3\mu + 6\mu'_2\mu^2 - 4\mu'_1\mu^3 + \mu'_0\mu^4$$
$$= \mu'_4 - 4\mu'_3\mu + 6\mu'_2\mu^2 - 3\mu^4$$

The quantities μ_3/σ^3 and μ_4/σ^4 are measures of skewness and kurtosis, respectively, for the population.

EXAMPLE 1 Assuming that a well-made die has a probability $\frac{1}{6}$ of falling with any specified face uppermost, the expectation of the number of spots on the upper face is

(2.9.11)
$$\mu = \sum_{j=1}^{6} j \cdot \tfrac{1}{6} = \tfrac{1}{6} \cdot 21 = 3.5$$

The variance of this number is

(2.9.12)
$$\begin{aligned} \mu_2 &= \mu'_2 - \mu^2 \\ &= \sum_{j=1}^{6} j^2 \cdot \tfrac{1}{6} - \mu^2 \\ &= \tfrac{1}{6} \cdot 91 - (3.5)^2 \\ &= 2.917 \end{aligned}$$

so that $\sigma = 1.71$. Note that the sum of the integers from 1 to n is given by $\frac{1}{2}n(n + 1)$ and the sum of the squares of the integers from 1 to n is

$$\tfrac{1}{6}n(n + 1)(2n + 1).$$

EXAMPLE 2 Suppose that a continuous variate X may take values between 0 and 2, with a density function

$$f(x) = x, \qquad 0 \le x \le 1$$
$$f(x) = 2 - x, \qquad 1 \le x \le 2$$

The graph is triangular, with a vertex of height 1 at $x = 1$.

The expectation of X is obviously 1 (from considerations of symmetry) but may be calculated from the relation

$$\begin{aligned} \mu &= \int_0^2 x f(x)\, dx \\ &= \int_0^1 x^2\, dx + \int_1^2 x(2 - x)\, dx \\ &= \tfrac{1}{3} + \tfrac{2}{3} = 1 \end{aligned}$$

The variance of X is $\mu_2 = \mu'_2 - \mu^2$, where $\mu'_2 = \int_0^2 x^2 f(x)\, dx = \frac{7}{6}$. Therefore, $\mu_2 = \frac{1}{6}$ and $\sigma = 0.408$.

2.10 **Generating Functions** It is often convenient mathematically to consider functions which serve to "generate" moments or other characteristics of a population. When such a function of a real variable (say h) is expanded in powers of h, the coefficients of h, $h^2/2!$, $h^3/3!$ . . . form the set of moments or other quantities. It is in this sense that these quantities may be thought of as generated by the function.

For a discrete variate the *moment generating function* (m.g.f.) is defined by

(2.10.1)
$$M(h) = E(e^{hX}) = \sum_j e^{hx_j} p(x_j)$$
$$= \sum_j \left(1 + hx_j + \frac{h^2}{2!} x_j^2 + \ldots\right) p(x_j)$$
$$= 1 + h\mu'_1 + \frac{h^2}{2!}\mu'_2 + \ldots = \sum_{r=0}^{\infty} \frac{\mu'_r h^r}{r!}$$

where it is assumed that the indicated sum converges.

For a continuous variate the m.g.f. is defined by

(2.10.2)
$$M(h) = \int_{-\infty}^{\infty} e^{hx} f(x)\, dx$$

provided that the integral converges, and, if so, this reduces to the same series as in equation (1) above.

EXAMPLE 3 For a symmetrical die, the values of x_j are 1, 2, 3, 4, 5, 6, and $p(x_j)$ is $\frac{1}{6}$ for each of them. Therefore,

$$M(h) = \tfrac{1}{6}(e^h + e^{2h} + \ldots + e^{6h})$$
$$= \tfrac{1}{6}e^h(e^{6h} - 1)(e^h - 1)^{-1}$$

Written out in powers of h, this function becomes

$$\frac{1}{6}\left(6 + S_1 h + S_2 \frac{h^2}{2!} + S_3 \frac{h^3}{3!} + \ldots\right)$$

where $S_1 = 1 + 2 + \ldots + 6 = 21, \quad S_2 = 1^2 + 2^2 + \ldots + 6^2 = 91,$
$S_3 = 1^3 + 2^3 + \ldots + 6^3 = 441$, etc.

Therefore, $\mu'_1 = S_1/6 = 7/2, \quad \mu'_2 = S_2/6 = 91/6$, etc.

EXAMPLE 4 For the continuous rectangular distribution specified by

$$f(x) = \frac{1}{2a}, \quad -a < x < a$$
$$f(x) = 0, \quad x < -a \quad \text{or} \quad x > a$$

we obtain

(2.10.3)
$$M(h) = \frac{1}{2a}\int_{-a}^{a} e^{hx}\, dx = \frac{1}{2ah}(e^{ah} - e^{-ah})$$
$$= \frac{1}{ah}\sinh ah.$$

Expressed as a power series,

$$M(h) = 1 + \frac{a^2h^2}{3!} + \frac{a^4h^4}{5!} + \ldots$$

so that the odd moments are all zero and the even moments are given by

$$\mu'_2 = \frac{a^2}{3}, \qquad \mu'_4 = \frac{a^4}{5}, \qquad \text{etc.}$$

These are the coefficients of $h^2/2!$, $h^4/4!$, etc. in the above series. Since the mean of the distribution is zero, the moments μ_r are the same as the moments μ'_r.

A fuller discussion of moment generating functions may be found in [5].

*2.11 **Factorial Moments** For a variate X which is discrete and takes values spaced at unit intervals it is sometimes convenient to use *factorial moments,* defined by

(2.11.1)
$$\mu'_{(r)} = \sum_j (x_j)_r p(x_j)$$

where $(x_j)_r = x_j(x_j - 1)(x_j - 2) \ldots (x_j - r + 1)$.

For the die of Example 1 in § 2.9 we have $\mu'_{(1)} = 21/6$, $\mu'_{(2)} = 70/6$, $\mu'_{(3)} = 210/6$, etc. The highest non-zero moment is $\mu'_{(6)} = 5!$.

The *factorial moment generating function* (f.m.g.f.) is given by

(2.11.2)
$$G(h) = \sum_j (1 + h)^{x_j} p(x_j)$$

For the die just mentioned this becomes

$$G(h) = \frac{1}{6} \sum_{j=1}^{6} (1 + h)^j$$

$$= 1 + \frac{7h}{2} + \frac{35}{6} h^2 + \frac{35}{6} h^3 + \frac{7}{2} h^4 + \frac{7}{6} h^5 + \frac{1}{6} h^6$$

2.12 **Cumulants** If the *logarithm* (to base *e*) of the m.g.f. can be expanded as a series of powers of h (which converges in some interval including $h = 0$) in the form

(2.12.1)
$$K(h) = \log_e M(h) = \kappa_1 h + \kappa_2 \frac{h^2}{2!} + \kappa_3 \frac{h^3}{3!} + \ldots = \sum_{r=1}^{\infty} \kappa_r \frac{h^r}{r!}$$

then the coefficients κ_r of $h^r/r!$ are called the *cumulants* of the distribution (κ is the Greek letter kappa) and $K(h)$ is called the *cumulant generating function* (c.g.f.). The cumulants play an important role in sampling theory, as was first pointed out by Sir R. A. Fisher, who emphasized their advantages over moments. As will be seen shortly, the first cumulant is the same as μ'_1 (the population mean) and the second and third are the same as μ_2 and μ_3 respectively. The higher cumulants differ, however, from the corresponding moments.

If the origin is taken at the population mean (so that $\mu'_1 = 0$ and μ'_r is therefore the same as μ_r for all r), we have

(2.12.2) $$M(h) = 1 + \sum_{r=2}^{\infty} \mu_r \frac{h^r}{r!}$$

Also, by the definition of $K(h)$,

(2.12.3) $$\frac{dK(h)}{dh} = \frac{1}{M(h)} \frac{dM(h)}{dh} = \kappa_1 + \kappa_2 h + \kappa_3 \frac{h^2}{2!} + \kappa_4 \frac{h^3}{3!} + \dots$$

From the definition of $M(h)$ in Eq. (2),

(2.12.4) $$\frac{dM(h)}{dh} = \mu_2 h + \mu_3 \frac{h^2}{2!} + \mu_4 \frac{h^3}{3!} + \dots$$

so that, from Eqs. (2), (3) and (4),

$$\left(1 + \mu_2 \frac{h^2}{2!} + \mu_3 \frac{h^3}{3!} + \dots\right)\left(\kappa_1 + \kappa_2 h + \kappa_3 \frac{h^2}{2!} + \kappa_4 \frac{h^3}{3!} + \dots\right) = \mu_2 h + \mu_3 \frac{h^2}{2!} + \mu_4 \frac{h^3}{3!} + \dots$$

By equating coefficients of corresponding powers of h on the two sides of this equation, we find

(2.12.5) $$\kappa_1 = 0, \kappa_2 = \mu_2, \kappa_3 = \mu_3, \kappa_4 = \mu_4 - 3\mu_2^2, \dots$$

The most common measure of kurtosis for a population is $\kappa_4/\kappa_2^2 = (\mu_4/\mu_2^2) - 3$.

If the origin from which x is measured is changed from $x = 0$ to $x = a$, any given value x is changed to $x - a$. This will make no difference to any of the moments about the mean (since the mean will also be changed to $\mu - a$), and therefore will not change any of the cumulants from κ_2 on. The first cumulant, however, becomes $\kappa_1 - a$. In the above derivation a was taken as the population mean μ, so that before the shift we should have

(2.12.6) $$\kappa_1 = \mu.$$

If the scale of measurement is altered so that a value previously recorded as x now becomes bx, the effect on $M(h)$ is to replace h by bh. The rth moment (whether about the origin or the mean) and the rth cumulant are multiplied by b^r. For the f.m.g.f., a change of origin has the effect of multiplying $G(h)$ by $(1 + h)^{-a}$ and a change of scale replaces $1 + h$ by $(1 + h)^b$.

The most important property of moment generating functions is that if $X_1, X_2 \dots X_N$ are independent variates, and if L is a linear function of the X's given by $L = c_1X_1 + c_2X_2 + \dots + c_NX_N$ (the c's being arbitrary constants, not all zero), then the m.g.f. of L is

$$M(h) = \prod_{j=1}^{N} M_j(c_j h) \tag{2.12.7}$$

i.e., it is the product of the m.g.f.'s for the separate variates, with $c_j h$ substituted for h. From the definition of $K(h)$ it follows that

$$K(h) = \sum_{j=1}^{N} K_j(c_j h) \tag{2.12.8}$$

so that the separate c.g.f.'s are *added.* To find the c.g.f. for a sum of independent variates all we have to do is to add the c.g.f.'s for the variates taken separately. This is the principal reason for introducing cumulants.

In § 2.8 the k-statistics were briefly mentioned. These are simply related to the cumulants of the population. In fact, the expectation of the rth k-statistic k_r is the corresponding cumulant κ_r. The k_r are therefore often used as estimates of the cumulants κ_r (at any rate for $r \leq 4$). See § 5.8.

*2.13 **Characteristic Functions** For some distributions the moment generating function does not exist. If, however, we replace h by ih ($i = \sqrt{-1}$) in the definition (2.10.1) and define the *characteristic function* by

$$C(h) = \sum_j e^{ihx_j} p(x_j) \tag{2.13.1}$$

or

$$C(h) = \int_{-\infty}^{\infty} e^{ihx} f(x)\, dx \tag{2.13.2}$$

then $C(h)$ always exists, as a complex number, for any distribution for which the $p(x_j)$, or $f(x)$, are defined. It may be written as a series:

$$C(h) = 1 + ih\mu'_1 - \frac{h^2}{2!}\mu'_2 - \frac{ih^3}{3!}\mu'_3 + \frac{h^4}{4!}\mu'_4 + \ldots \tag{2.13.3}$$

and so may be regarded as generating moments in the same sort of way as $M(h)$. It may be noted that corresponding to (2) there is a reciprocal relation

$$2\pi f(x) = \int_{-\infty}^{\infty} e^{-ihx} C(h)\, dh \tag{2.13.4}$$

The density function $f(x)$ is said to be the *Fourier transform* of $C(h)$. Tables of the Fourier transform (such as those in [6]) may be useful in finding $C(h)$, given $f(x)$, or vice versa.

2.14 **Bienaymé's Theorem** If $X_1, X_2 \ldots X_N$ are pairwise independent variates (see § 1.10), and if L is a linear combination given by $L = \sum_j c_j X_j$, then the variance of L is

$$V(L) = \sum_j c_j^2 V(X_j) \tag{2.14.1}$$

If, for example, all the c_j are equal to 1, the theorem states that the variance of a sum of pairwise independent variates is equal to the sum of their variances.

To prove this, let $E(X_j)$ be $\mu_{(j)}$, where the j is placed in parentheses to show that $\mu_{(j)}$ is not a jth moment but the first moment of the jth variate. By Theorem 1.16,

$$E(L) = \sum_j c_j \mu_{(j)}$$

From the definition of variance.

$$\begin{aligned}(2.14.2)\qquad V(L) &= E[L - E(L)]^2 \\ &= E\left[\sum_j c_j(X_j - \mu_{(j)})\right]^2 \\ &= E\sum_j c_j^2(X_j - \mu_{(j)})^2 + E\sum_{(j \neq k)}\sum c_j c_k (X_j - \mu_{(j)})(X_k - \mu_{(k)})\end{aligned}$$

(Note that in forming the square of a sum of N quantities there are N terms in which each quantity is squared, and $N(N - 1)$ terms in which each quantity is multiplied by a different one. These latter are the double sum above for which $j \neq k$).

Now $V(X_j) = E(X_j - \mu_{(j)})^2$. Also we may define the *covariance* of two distinct variates, X_j and X_k, by

$$(2.14.3)\qquad C(X_j, X_k) = E[(X_j - \mu_{(j)})(X_k - \mu_{(k)})]$$

Equation (2) then states that

$$(2.14.4)\qquad V(L) = \sum_j c_j^2 V(X_j) + \sum_{j \neq k} c_j c_k C(X_j, X_k)$$

If X_j and X_k are independent, it follows from Theorem 1.17 that

$$\begin{aligned}E[(X_j - \mu_{(j)})(X_k - \mu_{(k)})] &= E(X_j - \mu_{(j)}) \cdot E(X_k - \mu_{(k)}) \\ &= 0\end{aligned}$$

since $E(X_j) = \mu_{(j)}$ by definition. Equation (4) therefore reduces to Eq. (1).

Variates which are such that their covariance is zero are said to be *uncorrelated.* It is sufficient for this theorem that the variates should be pairwise uncorrelated, and they need not be independent in the full sense. (See § 1.10.)

The *Pearson coefficient of correlation* between two variates X_j and X_k is defined by

$$(2.14.5)\qquad \rho_{jk} = \frac{C(X_j, X_k)}{[V(X_j) \cdot V(X_k)]^{1/2}}$$

It is a pure number with range from -1 to $+1$ inclusive, and is zero when X_j and X_k are uncorrelated. If we write $V(X_j) = \sigma_j^2$, and $C(X_j, X_k) = \rho_{jk}\sigma_j\sigma_k$, equation (4) above becomes

$$(2.14.6)\qquad V(L) = \sum_j c_j^2 \sigma_j^2 + \sum_{j \neq k} c_j c_k \rho_{jk} \sigma_j \sigma_k$$

From the definition it is obvious that $\rho_{jk} = \rho_{kj}$.

The coefficient of correlation is important in problems involving two or more variates, and its properties will be discussed more fully in chapter 11. The method of calculating a sample statistic r for estimating ρ is given in § 11.14.

2.15 **Markov's Inequality** This states that if X is a non-negative variate with expectation μ, and if λ is any real positive number,

$$P(X \geq \lambda) \leq \mu/\lambda. \tag{2.15.1}$$

To prove this, we note that the set of all possible values of X can be divided into two sub-sets X_1 and X_2, where X_1 contains all values $\geq \lambda$ and X_2 all values $< \lambda$. By definition,

$$\begin{aligned} \mu = E(x) &= \int_0^\infty xp(x)\,dx \\ &= \int_0^\lambda xp(x)\,dx + \int_\lambda^\infty xp(x)\,dx \\ &\geq \int_\lambda^\infty xp(x)\,dx \end{aligned} \tag{2.15.2}$$

since $p(x)$ is never negative.

But since in this last integral x is not less than λ,

$$\int_\lambda^\infty xp(x)\,dx \geq \lambda \int_\lambda^\infty p(x)\,dx = \lambda P(X \geq \lambda). \tag{2.15.3}$$

From (2) and (3),

$$\mu \geq \lambda P(X \geq \lambda)$$

which is equivalent to (1).

2.16 **Chebyshev's Inequality** (attributed also to Bienaymé) This is a deduction from Markov's inequality and states that for a variate X, possessing first and second moments,

$$P(|X - \mu| \geq \lambda) \leq \frac{\sigma^2}{\lambda^2} \tag{2.16.1}$$

where $E(X) = \mu$, $V(X) = \sigma^2$.

If in (2.15.1) we substitute for X the non-negative quantity $(X - \mu)^2$, for which the expectation is σ^2, we obtain

$$P[(X - \mu)^2 \geq \lambda^2] \leq \frac{\sigma^2}{\lambda^2}$$

But to say that $(X - \mu)^2 \geq \lambda^2$ is the same as to say that $|X - \mu| \geq \lambda$, whence the theorem follows.

EXAMPLE 5 If X is the sum of spots showing up on two good dice, it is easily calculated that $E(X) = 7$ and $V(X) = 35/6$. (The two dice are supposed to fall independently, and Bienaymé's theorem, with Theorem 1.16, gives the result.)

The probability that $|X - 7| \geq 4$ is therefore $\leq 35/96$. The actual probability is 1/6, so that the inequality is not very sharp. However, Chebyshev's inequality is of great theoretical importance because of its wide applicability to a variety of distributions.

Several similar inequalities, usually requiring further restrictions on the variate, are known. See [7].

2.17 **The Joint Distribution of Two Variates** We have seen that the expectation and the variance can be readily obtained for a sum of random variables, if the separate expectations and variances, as well as the covariances, are known. If the variates are independent, the moment generating function and cumulant generating function for the sum are easily found from those for the individual variates (§ 2.12). If, however, the distribution function itself is required, the calculation is usually more difficult, even for independent variates.

Let us first suppose that X and Y are continuous variates. If $f(x, y)\,dx\,dy$ is the probability that at the same time X takes the value x (to $x + dx$) and Y the value y (to $y + dy$), then $f(x, y)$ is called the *joint probability density* for X and Y. The density for X alone, regardless of Y, is

$$g(x) = \int_{-\infty}^{\infty} f(x, y)\,dy \tag{2.17.1}$$

and the density for Y alone, regardless of X, is

$$h(y) = \int_{-\infty}^{\infty} f(x, y)\,dx \tag{2.17.2}$$

The variates X and Y are independent if, and only if,

$$f(x, y) = g(x)h(y) \tag{2.17.3}$$

The distribution functions for X and Y, respectively, are

$$G(x) = \int_{-\infty}^{x} g(u)\,du, \quad H(y) = \int_{-\infty}^{y} h(v)\,dv \tag{2.17.4}$$

while the joint distribution function is

$$F(x, y) = \int_{-\infty}^{x} \int_{-\infty}^{y} f(u, v)\,du\,dv \tag{2.17.5}$$

If the variates X and Y are discrete, there is no density function but the distribution functions exist. The joint distribution function is

$$F(x, y) = \sum_{x_i \leq x} \sum_{y_i \leq y} f(x_i, y_i). \tag{2.17.6}$$

Whether X and Y are continuous or discrete, the necessary and sufficient condition for independence is

$$F(x, y) = G(x)H(y). \tag{2.17.7}$$

*** 2.18 The Distribution Function for a Sum of Two Independent Variates** Let $Z = X + Y$ and let $P(z)$ be the distribution function for Z. By definition, $P(z)$ is the probability that $X + Y \leq z$. If we plot possible values of X and Y, using rectangular coordinates in the plane, the region corresponding to $X + Y \leq z$ is all that part of the plane lying below and to the left of the line $X + Y = z$ (see Figure 15). For any given y, the probability that $X \leq z - y$ is $G(z - y)$,

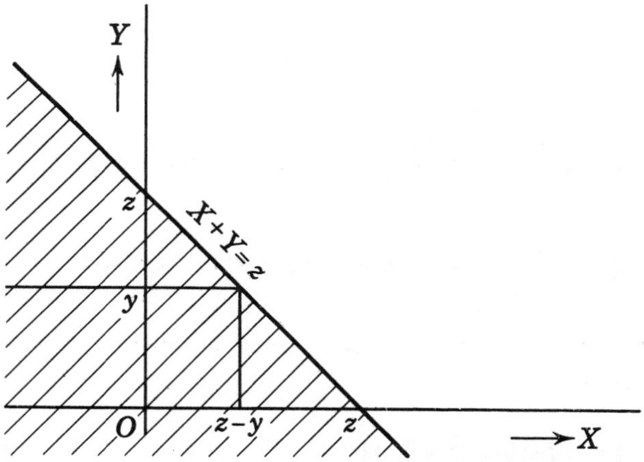

FIG. 15 SPACE OF THE VARIATE Z = X + Y

so that the required probability is obtained by multiplying $G(z - y)$ by the probability for y and integrating over all y.

$$P(z) = \int_{-\infty}^{\infty} G(z - y)h(y)\, dy \tag{2.18.1}$$

By differentiating with respect to z we obtain

$$p(z) = \int_{-\infty}^{\infty} g(z - y)h(y)\, dy \tag{2.18.2}$$

This is called the *convolution* of the density functions g and h. It is the density function for Z.

EXAMPLE 6 Let X have a uniform distribution on (0, 1) and Y a symmetrical triangular distribution on (0, 2). Then Z has a distribution, which we wish to find, on (0, 3), since z cannot take any values outside this interval.

The given density functions are

$$(2.18.3) \qquad \begin{cases} g(x) = 1, 0 < x < 1 \\ h(y) = y, 0 \leq y \leq 1;\ h(y) = 2 - y, 1 \leq y \leq 2 \end{cases}$$

Since $g(z - y)$ is 1 for values of y between $z - 1$ and z, and is 0 for all other values,

$$(2.18.4) \qquad p(z) = \int_{z-1}^{z} h(y)\, dy$$

Now, from the definition of $h(y)$, $p(z)$ has different expressions for the three intervals of z, namely, (0, 1), (1, 2) and (2, 3). In fact,

$$p(z) = \int_0^z y\, dy = \tfrac{1}{2}z^2, 0 < z < 1$$

$$(2.18.5) \qquad p(z) = \int_{z-1}^{1} y\, dy + \int_1^z (2 - y)\, dy = 3(z - \tfrac{1}{2}) - z^2, 1 < z < 2$$

$$p(z) = \int_{z-1}^{2} (2 - y)\, dy = \tfrac{1}{2}(3 - z)^2, 2 < z < 3$$

The graph of $p(z)$ is formed of parts of three parabolas, joined together. It is symmetrical about $z = 1\frac{1}{2}$.

2.19 **Joint Distribution of k Variates** The notation and definitions of § 2.17 may be extended to three or more variates. The variates $X_1, X_2 \ldots X_k$ are independent if and only if the joint distribution function is equal to the product of the separate distribution functions,

$$(2.19.1) \qquad F(x_1, x_2 \ldots x_k) = F_1(x_1)F_2(x_2) \ldots F_k(x_k)$$

If the variates are continuous, possessing density functions $f_1(x_1), f_2(x_2) \ldots f_k(x_k)$, the joint density function is

$$(2.19.2) \qquad f(x_1, x_2 \ldots x_k) = f_1(x_1)f_2(x_2) \ldots f_k(x_k)$$

As we shall see later, this relation is very useful in the theory of sampling. A sample of k items is selected at random from a large population characterized by a given or assumed distribution function. Each individual item may be regarded as an independent choice from this distribution, and the probability density of the sample actually selected is the product of the probability densities for the separate items. Since the various items all have the same density function $f(x)$, the joint probability density for the sample with values $x_1, x_2, \ldots x_k$ is

$$(2.19.3) \qquad f(x_1, x_2 \ldots x_k) = f(x_1)f(x_2) \ldots f(x_k)$$

PROBLEMS

A (§§ 2.1–2.8)

1. Criticise the following statements (occasionally seen in examination answers): "Median $= N/2$," "$Q_1 = N/4$."

2. Construct a histogram showing the age distribution of deaths of infants under one month from the following table, taken from an official publication of the United States Government:

Age at Death	*Frequency*
Under 1 day	26,665
1 day	8,364
2 days	6,344
3 to 6 days	12,375
1 week	10,911
2 weeks	7,717
3 weeks but under 1 month	6,212
	78,588

Hint: "1 day" means anything over one day but under two days; "3 to 6" means over three but under seven, and so on. Take the month as 30 days long. The *areas* of the rectangles, not the heights, represent the corresponding frequencies.

3. Construct a cumulative frequency table and a cumulative frequency polygon for the data in Problem 2. What was the approximate probability (at the time these data were collected) that an infant who lived for less than a month died in the first week?

4. The following table gives the results of 280 tests made on a certain kind of coal for ash content:

Percentage Ash	*Frequency*
3.0– 3.9	1
4.0– 4.9	7
5.0– 5.9	28
6.0– 6.9	78
7.0– 7.9	84
8.0– 8.9	45
9.0– 9.9	28
10.0–10.9	7
11.0–11.9	2
	280

Calculate the median and the first and third quartiles. Find the percentile rank of an ash content of 8.5%. *Hint:* Form a percentage cumulative frequency table, corresponding to the values of x_e. The percentile rank of 8.5 is the corresponding percentage cumulative frequency.

5. Calculate D_9 and P_{70} for the data of Problem 2; also the percentile rank of 10 days. State in words what each of these statistics means.

6. For a set of 15 ungrouped sample measurements we find $\sum x = 480$, $\sum x^2 = 15{,}735$. Find the mean and standard deviation of X.

7. For a sample of size 2, show that $m_2 = (x_1 - x_2)^2/2$, and $s\,(= k_2^{1/2}) = |x_1 - x_2|$.

8. Calculate the first three moments about zero for the distribution of Question 4. Then obtain the mean, variance (k_2), the standard deviation ($k_2^{1/2}$) and the moment measure of skewness ($m_3/m_2^{3/2}$) for this distribution. *Hint:* Use the coded variable $u = x - 7.45$.

9. Prove that if we have a set of N_1 values $x_{1i}(i = 1, 2 \ldots N_1)$ and another set of N_2 values $x_{2j}(j = 1, 2 \ldots N_2)$ of the variate X, then the mean of the combined set $\bar{x}$ is given in terms of the two separate means $\bar{x}_1$ and $\bar{x}_2$ by the relation $(N_1 + N_2)\,\bar{x} = N_1\bar{x}_1 + N_2\bar{x}_2$.

B (§§ 2.9–2.13)

1. A variate X has the density function $f(x) = c\,(12x + x^2 - x^3)$, $0 \leq x \leq 4$, $f(x) = 0$ for all other values of x. Find c and calculate the mean, standard deviation and skewness for this distribution. Sketch the curve for $f(x)$.

2. A variate X has the density function $f(x) = x/2\ (0 \leq x \leq 1)$, $f(x) = 1/2\ (1 \leq x \leq 2)$ and $f(x) = (3 - x)/2\ (2 \leq x \leq 3)$. Find the variance and the moment measure of kurtosis of X. *Hint:* This distribution is symmetrical about $x = 3/2$. Calculate moments for $u = x - 1\frac{1}{2}$.

3. A continuous variate has the density function $f(x) = Cx^{1/2}(1 - x)^{3/2}$, $0 \leq x \leq 1$. Show that this function vanishes with infinite slope at $x = 0$, vanishes with zero slope at $x = 1$ and has a maximum at $x = 1/4$. Sketch the curve. Calculate C and find the mean and variance. *Hint:* Put $x = \sin^2\theta$ and use a reduction formula for integration.

4. The density function for X is $f(x) = cx^2e^{-x}(x \geq 0)$. Calculate c and also the mean and variance of X. Find the cumulant generating function for X.

5. Find the moment generating function for the triangular distribution of Example 2, § 2.9.

6. A distribution has the m.g.f. $M(h) = (q + pe^h)^N$, where p and q are constants and $p + q = 1$. Find the c.g.f. and the first four cumulants.

7. From a point on the circumference of a circle of radius a, a chord is drawn in a random direction. Show that the expected value of the length of the chord is $4a/\pi$ and that the variance of the length is $2a^2(1 - 8/\pi^2)$. *Hint:* See Example 8, § 1.16. Take the density $f(\theta)$ as $1/\pi$.

8. If a variate can take any value from 0 to 1 with equal probability, show that its standard deviation is $\sqrt{3}/6 = 0.289$. A set of two-digit random numbers such as that in Appendix B.1 may be regarded as giving approximate random choices from the interval (0, 1), number 43 for instance being read as 0.43. Use the result of Problem A-7 to obtain an estimate of s from 50 samples, each of size two, taken from Table B.1 and compare with the theoretical value.

9. The *Cauchy distribution* is defined by the density function $f(x) = \dfrac{a}{\pi}\cdot\dfrac{1}{(x - b)^2 + a^2}$, $-\infty < x < \infty$, $a > 0$. Show that the mean and variance of this distribution do not exist, but that the mean is b if the improper integral defining it is interpreted as the Cauchy principal value (see Appendix A.3).

10. The *Pareto distribution* is defined by $f(x) = \dfrac{\alpha}{\beta}\left(\dfrac{\beta}{x}\right)^{\alpha+1}$ $(x > \beta)$, and $f(x) = 0$ otherwise. Show that the rth moment exists only if $\alpha > r$. Find the expectation and variance of x if $\alpha > 2$.

11. Find the median and the mode of the distribution with density $f(x) = abx^{a-1}(1 + bx^a)^{-2}$, $b > 0$, $a > 1$, $0 \leq x < \infty$. *Hint:* The median is that value $\tilde{x}$ for which $\int_0^{\tilde{x}} f(x)\,dx = \int_{\tilde{x}}^{\infty} f(x)\,dx$. The mode is that value $\hat{x}$ for which $f(x)$ is a maximum.

12. Prove that the characteristic function of the *Laplace distribution*, with density $f(x) = \frac{1}{2}e^{-|x|}(-\infty < x < \infty)$ is $C(h) = (1 + h^2)^{-1}$. Calculate the variance of this distribution.

13. A discrete variate X has a distribution in the population defined by $f(x) = \theta(1 - \theta)^x$, for $x = 0, 1, 2 \ldots$, where θ is a parameter with a value between 0 and 1. Calculate the probability of a sample of N, in which N_0 have $x = 0$, N_1 have $x = 1$, etc. and $\sum N_i = N$.

Find for what value of θ this probability is a maximum. (The value so defined is called a *maximum likelihood estimator* of θ. See further in § 6.1. *Hint:* The sample mean $\bar{x} = (N_1 + 2N_2 + 3N_3 + \ldots)/N$.

14. Prove the statement (2.12.7). *Hint:* Since the variates X_j are independent, the joint density function is the product of the separate density functions. Therefore,

$$M_L(h) = \int \ldots \int e^{hL} f(x_1, x_2 \ldots x_N)\, dx_1 \ldots dx_N$$
$$= \int \exp[hc_1x_1] f_1(x_1)\, dx_1 \int \exp[hc_2x_2] f_2(x_2)\, dx_2 \ldots \int \exp[hc_N x_N] f_N(x_N)\, dx_N$$

where $f_j(x_j)$ is the density function for X_j.

C (§§ 2.14–2.19)

1. If X is the number of spots showing up in a single throw with a good die, show that the Markov inequality gives $P(X \geq 5) \leq 0.7$ and the Chebyshev inequality gives $P(|X - 3.5| \geq 2) \leq 35/48$. What are the true values of these probabilities?

2. A discrete variate X can take only the values $x = 1, 2, 3 \ldots$ with probability 2^{-x} (this is a *geometric distribution*). Prove that Chebyshev's inequality gives $P(|X - 2| \geq 2) \leq \frac{1}{2}$. What is the true probability? *Hint:* $1 + 4r + 9r^2 + 16r^3 + \ldots = (1 + r)/(1 - r)^3$, for $r < 1$.

3. For the two distributions with density functions

$$\text{(a) } f(x) = 1, \ (0 \leq x \leq 1)$$
$$\text{(b) } f(x) = e^{-x} (x \geq 0)$$

calculate $P(|x - \mu| \geq 2\sigma)$ and compare with the value given by Chebyshev's inequality.

4. If X and Y are independent random variables with density functions $f(x) = C_1x^m e^{-x/2}$, $g(y) = C_2y^n e^{-y/2}$ respectively, show that the density function of $W = X + Y$ is $h(w) = C_3 w^{m+n+1} e^{-w/2}$. *Hint:* See Appendix A.6.

5. If X and Y are variates with joint density function $f(x, y)$ and if $U = Y/X$, use the method of § 2.18 to find the density function for U. Show that $h(u) = \int_{-\infty}^{\infty} |x| f(x, ux)\, dx$. *Hint:* Draw the line $Y = uX$ in the X-Y plane, and show the areas corresponding to $U \leq u$. Note that $U \leq u$ implies $Y \leq uX$ if $X > 0$ but $Y \geq uX$ if $X < 0$. Find the distribution function for U and differentiate to get $h(u)$.

6. In Problem C-5 above, suppose that X and Y are independently and uniformly distributed on the interval (0, 1), so that $f(x, y) = 1$ everywhere inside a unit square and $f(x, y) = 0$ outside. Prove that

$$h(u) = 0, \quad u < 0$$
$$h(u) = 1/2, \quad 0 < u < 1$$
$$h(u) = 1/(2u^2), \quad u > 1$$

REFERENCES

[1] Fraser, D. A. S., *Statistics: An Introduction*, Wiley, 1958.

[2] Kenney, J. F., and Keeping, E. S., *Mathematics of Statistics, Part I*, Van Nostrand, 1954.

[3] Elkin, J. M., "Asymmetrical Distributions with Zero Third Moments," *Amer. Math. Monthly*, **62**, 1955, pp. 37–38.

[4] Kaplansky, I., "A Common Error Concerning Kurtosis," *J. Amer. Stat. Ass.*, **40**, 1945, p. 259.

[5] Curtiss, J. H., "Generating Functions in the Theory of Statistics," *Amer. Math. Monthly*, **48**, 1941, pp. 374–386.

[6] Campbell, G., and Foster, R., *Fourier Integrals for Practical Applications*, Van Nostrand, 1948. See also *Tables of Integral Transforms, Vol. I* (Bateman Manuscript Project), McGraw-Hill, 1954.

[7] Savage, I. R., *Probability Inequalities of the Tshebycheff Type*, National Bureau of Standards Report 1744, 1952.

Chapter 3

THE BINOMIAL, POISSON AND NORMAL DISTRIBUTIONS

3.1 **The Binomial (or Bernoulli) Distribution** Suppose that all the individuals in a population are divided in imagination into two sets according as they have, or do not have, a certain attribute A. Such a division is called a "dichotomy" (a cutting in two)—every individual belongs either to the one set or to the other. The attribute A is often conventionally called a "success;" it may, for example, be "head" in a population of coin-tosses or "male" in a population of children. We assume that there is a definite probability θ that an individual chosen at random from the population has the attribute A. This probability may be estimated by taking a sample of N individuals and noting the number X which are A's. The ratio X/N is called the *relative frequency of success*, and will be denoted by p. It is, of course, a random variable and, as an estimate of θ, is more reliable the larger the sample size.*

The binomial distribution is concerned with the variation of X or p among samples of size N from a population characterized by the parameter θ. It is assumed that the probability of success is unchanged by the process of selecting an individual for the sample, so that we must assume either that the population is infinite or, if it is finite, that the sampling is done "with replacements" (see § 1.12). Furthermore, each item for the sample is supposed to be chosen independently of all the rest.

Under these conditions the probability that the first x individuals selected will all be A's is θ^x and the probability that the next $N - x$ will all be not-A's is $(1 - \theta)^{N-x}$. The probability of a set of x A's followed by a set of $N - x$ not-A's is $\theta^x(1 - \theta)^{N-x}$, and this is also the probability for any other pre-selected arrangement of x A's and $N - x$ not-A's. However, we are not interested in the precise arrangement of A's and not-A's, merely in the *total number* of A's in the sample. Hence we can combine together the probabilities for all the $\binom{N}{x}$ permutations of x successes and $N - x$ failures, and state that the probability of x successes, no matter in what order successes and failures occur, is given by

$$b(x, N, \theta) = \binom{N}{x} \theta^x(1 - \theta)^{N-x} \tag{3.1.1}$$

*To stick to our Greek and Latin convention, the symbol for the probability should be π instead of θ, to correspond with the sample statistic p. But the risk that π may be misinterpreted as 3.14159 . . . is serious.

where $x = 0, 1, 2 \ldots N$. This is the binomial distribution, discussed by James Bernoulli (1654–1705) in a book published in 1713, after his death. It gives the probability that X has a value exactly x. It is called binomial because, if we write $1 - \theta$ as ϕ, $b(x, N, \theta)$ is the term containing θ^x in the expansion of the binomial $(\phi + \theta)^N$. The probability that $X = x$ is of course the same as the probability that $p = x/N$.

TABLE 3.1

x	$512\, b(x, 9, \frac{1}{2})$	$512\, B(x, 9, \frac{1}{2})$
0	1	512
1	9	511
2	36	502
3	84	466
4	126	382
5	126	256
6	84	130
7	36	46
8	9	10
9	1	1
	512	

EXAMPLE 1 For a good coin we may take $\theta = \frac{1}{2}$, so that the probability of x heads in nine tosses will be

$$b(x, 9, \tfrac{1}{2}) = \frac{9!}{x!(9-x)!}(\tfrac{1}{2})^9, \quad x = 0, 1, 2 \ldots 9$$

By giving x all possible values we obtain Table 3.1, in which the probabilities have been multiplied by $2^9 = 512$ so as to avoid fractions.

The *cumulative binomial probability* is usually defined as

$$(3.1.2) \qquad B(x, N, \theta) = \sum_{u=x}^{N} b(u, N, \theta)$$

It is the probability of *at least* x successes. Values for $N = 9$ and $\theta = \frac{1}{2}$ are given in Table 3.1. The probability of at least six heads in nine tosses, for example, is $130/512 = 0.254$. Note that the distribution function, as defined in § 1.16, is $1 - B(x + 1, N, \theta)$.

The binomial distribution, even though X is discrete, may be represented by a histogram in which rectangles of unit base, centered at $x = 0, 1 \ldots N$, are drawn with heights equal to $b(x, N, \theta)$. The histogram for

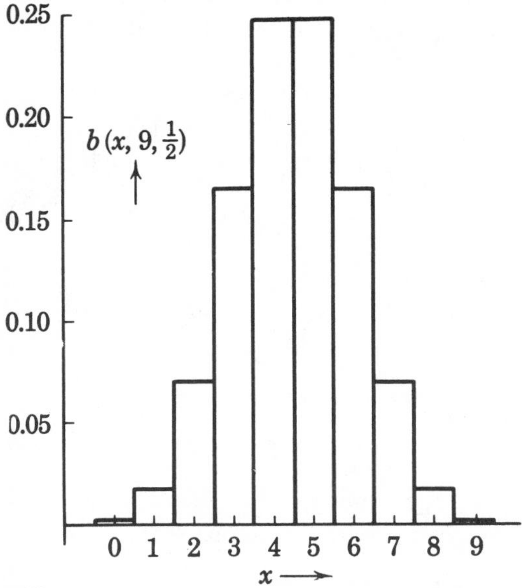

FIG. 16 BINOMIAL DISTRIBUTION, $\theta = 0.5$

Table 3.1 is shown in Figure 16. Since all the observations are actually at the centers of the intervals, there is no grouping error (see § 2.7).

Numerical values of $b(x, N, \theta)$ for given x, N and θ may be calculated by means of logarithms of factorials. Seven-figure tables of $\log n!$ for $n = 1$ to 1000 are given in Glover's Tables [1] and *Biometrika* Tables [2]. Extensive tables of the cumulative binomial distribution are now available (see [3] and [4]). Separate terms of $b(x, N, \theta)$ may be obtained from these tables, if desired, by differencing successive entries, since $b(x, N, \theta) = B(x, N, \theta) - B(x + 1, N, \theta)$, but for most practical purposes the cumulative probabilities are more useful. It is not necessary to tabulate values for θ beyond 0.5, since

$$B(x, N, 1 - \theta) = 1 - B(N - x + 1, N, \theta) \tag{3.1.3}$$

3.2 **Recursion Formula for Binomial Probabilities** From Eq. (3.1.1) we see, by cancelling common factors, that

$$\frac{b(x, N, \theta)}{b(x - 1, N, \theta)} = \frac{N - x + 1}{x} \cdot \frac{\theta}{1 - \theta} = 1 - \frac{x - (N + 1)\theta}{x(1 - \theta)} \tag{3.2.1}$$

The ratio of $b(x, N, \theta)$ to $b(x - 1, N, \theta)$ is therefore greater than, or less than, 1, according as $x < (N + 1)\theta$ or $x > (N + 1)\theta$. The values of b increase with x as long as x is below $(N + 1)\theta$ and decrease with x when x is above $(N + 1)\theta$. If $x = (N + 1)\theta$, the probabilities that $X = x$ and that $X = x - 1$ are equal. This is the case for Example 1 when $x = 5$. In general the probability is a maximum when X is equal to the integer next below $(N + 1)\theta$.

3.3 **Moments and Cumulants for the Binomial Distribution** In calculating moments, etc., for the binomial distribution, it is convenient to use the concept of indicator function, defined in § 1.13. If the event A_j is that of selecting the jth item for the sample, and if $I_{A_j} = 1$ when A_j is a success and $I_{A_j} = 0$ when A_j is a failure, the number of successes X is simply $\sum_{j=1}^{N} I_{A_j}$. By Theorem 1.15, $E(I_{A_j}) = P(A_j)$, which is θ for each item. Therefore,

$$\mu = E(X) = \sum_j E(I_{A_j}) = \sum_j P(A_j) = N\theta \tag{3.3.1}$$

which gives the mean of the binomial distribution. The variance σ^2 is similarly obtainable from Bienaymé's Theorem (§ 2.14), according to which

$$\sigma^2 = V(X) = \sum_j V(I_{A_j}) \tag{3.3.2}$$

where

$$V(I_{A_j}) = E(I_{A_j} - \theta)^2 = E(I_{A_j}^{\ 2}) - 2\theta E(I_{A_j}) + \theta^2$$

But $I_{A_j}^{\ 2}$ takes exactly the same values as I_{A_j}, namely, 0 and 1, so that $E(I_{A_j}^{\ 2}) = E(I_{A_j}) = \theta$. Therefore, $V(I_{A_j}) = \theta - \theta^2$, and from equation (2),

$$\sigma^2 = \sum_j (\theta - \theta^2) = N\theta(1 - \theta) \tag{3.3.3}$$

which is the required variance. Since p differs from X only by the constant factor $1/N$, it follows that

$$E(p) = \frac{1}{N} E(X) = \theta \tag{3.3.4}$$

$$V(p) = \frac{1}{N^2} V(X) = \frac{\theta(1-\theta)}{N} \tag{3.3.5}$$

The moment generating function for I_{A_j} is

$$M_j(h) = E(\exp[hI_{A_j}]) = \sum \exp[hI_{A_j}]P(I_{A_j})$$
$$= e^0 \cdot (1-\theta) + e^h \cdot \theta$$

since I_{A_j} is either 0 or 1, with probabilities $1 - \theta$ and θ respectively. Therefore,

$$M_j(h) = 1 - \theta + \theta e^h$$

The m.g.f. for X is therefore given—see Eq. (2.12.7)—by

$$M(h) = [M_j(h)]^N = (1 - \theta + \theta e^h)^N \tag{3.3.6}$$

The cumulant generating function is

$$K(h) = \log M(h) = N \log (1 - \theta + \theta e^h) \tag{3.3.7}$$
$$= N \log \left(1 + \theta h + \frac{\theta h^2}{2!} + \frac{\theta h^3}{3!} + \ldots\right)$$

If the logarithm is expanded in a series of powers of h, the first four successive cumulants are found to be

$$\begin{aligned} \kappa_1 &= N\theta = \mu \\ \kappa_2 &= N\theta(1-\theta) = \sigma^2 \\ \kappa_3 &= N\theta(1-\theta)(1-2\theta) = \sigma^2(1-2\theta) \\ \kappa_4 &= N\theta(1-\theta)(1-6\theta+6\theta^2) = \sigma^2(1-6\theta+6\theta^2) \\ &= \sigma^2\left(1 - \frac{6\sigma^2}{N}\right) \end{aligned} \tag{3.3.8}$$

The *skewness* is $(\kappa_3/\kappa_2)^{3/2} = (1 - 2\theta)/\sigma$ and therefore is zero only when $\theta = \frac{1}{2}$. For small values of θ the distribution has a positive skewness, which diminishes, however, as N increases, since σ is proportional to $N^{1/2}$.

The *kurtosis* is

$$\frac{\kappa_4}{\kappa_2{}^2} = \frac{1}{\sigma^2} - \frac{6}{N} \tag{3.3.9}$$

If $\theta = 1/2$, $\sigma^2 = N/4$, and $\kappa_4/\kappa_2{}^2 = -2/N$.

By differentiating with respect to θ the expression for the rth moment μ_r, namely,

$$\mu_r = \sum_{x=0}^{N} \binom{N}{x} \theta^x (1-\theta)^{N-x}(x - N\theta)^r \tag{3.3.10}$$

we may obtain a *recursion formula* for the moments, namely,

$$\mu_{r+1} = \theta(1-\theta)\left[Nr\mu_{r-1} + \frac{d\mu_r}{d\theta}\right] \tag{3.3.11}$$

From this formula, starting with $\mu_0 = 1$ and $\mu_1 = 0$, all subsequent binomial moments may be calculated in turn by giving r the values 1, 2, 3 . . . .

A still simpler recursion formula is that for the cumulants,

$$\kappa_{r+1} = \theta(1-\theta)\frac{d\kappa_r}{d\theta}, \qquad r \geq 1 \tag{3.3.12}$$

from which, starting with $\kappa_1 = N\theta$, the higher cumulants may readily be obtained. (See hint to Problem A-7.)

3.4 **The Bernoulli Law of Large Numbers** Let p_N be the relative frequency of success in N independent trials, the probability of success θ being the same in all trials. Since $E(p_N) = \theta$, we have by Chebyshev's inequality, § 2.16:

$$P(|p_N - \theta| \geq \lambda) \leq \frac{\sigma^2}{\lambda^2} \tag{3.4.1}$$

Here $\sigma^2 = \theta(1-\theta)/N$, and since, for all θ between 0 and 1, $\theta(1-\theta) \leq 1/4$, Eq. (1) becomes

$$P(|p_N - \theta| \geq \lambda) \leq 1/(4N\lambda^2) \tag{3.4.2}$$

For any fixed $\lambda > 0$ and any given $\varepsilon > 0$, we can always take N so large that $1/(4N\lambda^2) < \varepsilon$. This means that for large enough N the probability that p_N differs from θ by any fixed amount, however small, can be made as near to zero as we like. This is sometimes expressed as "p_N converges in probability to the value θ." Note that this is not the same thing as ordinary mathematical convergence (see § 1.2). The law expressed by equation (2) is a form of the *weak law of large numbers*.

The number N given by this equation may be quite large when λ and ε are small. Thus if $\lambda = 0.01$ and $\varepsilon = 0.001$, we find that $N > 2{,}500{,}000$. As we shall see in § 3.12, the approximation of the binomial distribution by the normal distribution permits us to find a much smaller N satisfying the requirements. The value above is certainly sufficient, but not necessary.

* 3.5 **Non-Bernoulli Sampling** The two chief variants from the true Bernoulli (binomial) sampling scheme described above are (1) the *Poisson scheme* in which the probability of success θ_j at the jth trial varies from one trial to

another, and (2) the *hypergeometric scheme* (sampling without replacements from a finite population) in which the probability at any stage depends on the results of the previous trials.

For the Poisson scheme, we have instead of (3.3.1),

$$\mu = E(X) = \sum_j P(A_j) = \sum \theta_j \tag{3.5.1}$$

Also,

$$\sigma^2 = \sum_j V(I_{A_j}) = \sum_j (\theta_j - \theta_j^2) \tag{3.5.2}$$

If

$$\theta = \frac{1}{N} \sum \theta_j$$

and if

$$\sigma_\theta^2 = \frac{1}{N} \sum_j (\theta_j - \theta)^2 = \frac{1}{N} \sum \theta_j^2 - \theta^2,$$

we have

$$\begin{aligned} \sigma^2 &= N\theta - N(\sigma_\theta^2 + \theta^2) \\ &= N\theta(1 - \theta) - N\sigma_\theta^2 \end{aligned} \tag{3.5.3}$$

Here θ is the mean of the θ_j, and, σ_θ^2 is their mean square difference from the mean. This shows that in the Poisson scheme the variance of X is *less* than it would be if the probability of success were constant over all trials.

In the hypergeometric scheme, suppose we have a finite population of size M, in which the number of "successes" is S and the number of "failures" is F, with $S + F = M$. The probability of success at the *first* trial is $\theta = S/M$. The total number of possible different samples of size N is $\binom{M}{N}$. The number containing x successes and $N - x$ failures is $\binom{S}{x}\binom{F}{N - x}$, so that the probability of exactly x successes in a sample of size N is

$$h(x, N, M, S) = \frac{\binom{S}{x}\binom{F}{N - x}}{\binom{M}{N}} \tag{3.5.4}$$

This may be written

$$h(x, N, M, S) = \frac{S!\,F!\,N!\,(M - N)!}{x!\,(S - x)!\,(N - x)!\,(F - N + x)!\,M!} \tag{3.5.5}$$

When M is very large and θ not too near 0 or 1 this approximates to $b(x, N, \theta)$.

The expression on the right of (5), when multiplied by a constant, is equal to the coefficient of u^x in the series expansion of the hypergeometric function $F(\alpha, \beta, \gamma, u)$ with $\alpha = -N$, $\beta = -S$, $\gamma = F - N + 1$. Hence the name of the distribution [5]. In calculating the higher moments of the hypergeometric distribution it is simplest to obtain first the factorial moments (see § 2.11). The first two factorial moments are

$$\text{(3.5.6)} \qquad \begin{cases} \mu'_{(1)} = \sum_{x=1}^{N} x h(x, N, M, \theta) \\ \mu'_{(2)} = \sum_{x=2}^{N} x(x-1) h(x, N, M, \theta) \end{cases}$$

and from these we easily obtain

$$\text{(3.5.7)} \qquad \begin{cases} \mu = \mu'_{(1)} \\ \sigma^2 = \mu'_{(2)} - \mu^2 + \mu \end{cases}$$

From Eqs. (5) and (6),

$$\text{(3.5.8)} \quad \mu'_{(1)} = \frac{S!F!N!(M-N)!}{M!} \sum_{x=1}^{N} \frac{1}{(x-1)!(S-x)!(N-x)!(F-N+x)!}$$

$$= \frac{S!F!N!(M-N)!}{M!} \sum_{x=0}^{N-1} \frac{1}{x!(S-1-x)!(N-1-x)!(F-N+1+x)!}$$

$$= \frac{SN}{M} \sum_{x=0}^{N-1} \frac{(S-1)!F!(N-1)!(M-1-N+1)!}{x!(S-1-x)!(N-1-x)!(F-N+1+x)!(M-1)!}$$

$$= \frac{SN}{M} \sum_{x=0}^{N-1} h(x, N-1, M-1, S-1)$$

$$= \frac{SN}{M} = N\theta$$

Similarly, we may calculate

$$\text{(3.5.9)} \quad \mu'_{(2)} = \frac{S!F!N!(M-N)!}{M!} \sum_{x=2}^{N} \frac{1}{(x-2)!(S-x)!(N-x)!(F-N+x)!}$$

$$= \frac{S(S-1)N(N-1)}{M(M-1)}$$

$$\sum_{x=0}^{N-2} \frac{(S-2)!}{x!(S-2-x)!} \frac{F!(N-2)!(M-2-N+2)!}{(N-2-x)!(F-N+2+x)!(M-2)!}$$

$$= \frac{S(S-1)N(N-1)}{M(M-1)} \sum_{x=0}^{N-2} h(x, N-2, M-2, S-2)$$

$$= \frac{N\theta(S-1)(N-1)}{M-1}$$

From Eq. (7), we have

$$E(X) = \mu = N\theta = \frac{NS}{M} \tag{3.5.10}$$

and

$$\begin{aligned} V(X) = \sigma^2 &= \mu'_{(2)} + N\theta(1 - N\theta) \\ &= N\theta\left[\frac{(S-1)(N-1)}{M-1} + \frac{M - NS}{M}\right] \\ &= \frac{N\theta}{M(M-1)}(M-S)(M-N) \\ &= N\theta(1-\theta)\frac{M-N}{M-1} \end{aligned} \tag{3.5.11}$$

The mean is therefore exactly the same as in pure binomial sampling, but the variance is less because of the factor $(M - N)/(M - 1)$. For M large compared with N this correcting factor is nearly 1.

3.6 **The Poisson Distribution for Rare Events** If in a binomial distribution the probability of success is very small (so that the event "success" may be said to be rare) but the size of the sample N is so large that the expected number of successes in the sample is moderate (say between 0.1 and 10), the probability of exactly x successes is given *approximately* by

$$p(x, \mu) = \mu^x \frac{e^{-\mu}}{x!} \tag{3.6.1}$$

where $\mu = N\theta$. The theoretical distribution given *exactly* by Eq. (1), for all integral values of x from 0 onwards, is called the *Poisson distribution.*

The true (binomial) probability of x successes is

$$\begin{aligned} b(x, N, \theta) &= \frac{N!}{x!(N-x)!}\theta^x(1-\theta)^{N-x} \\ &= \frac{N(N-1)\dots(N-x+1)}{x!}\left(\frac{\mu}{N}\right)^x\left(1 - \frac{\mu}{N}\right)^{N-x} \\ &= \frac{\mu^x}{x!}\left(1 - \frac{1}{N}\right)\left(1 - \frac{2}{N}\right)\dots\left(1 - \frac{x-1}{N}\right)\left(1 - \frac{\mu}{N}\right)^{N-x} \end{aligned} \tag{3.6.2}$$

where each of the x factors $N, N - 1 \dots (N - x + 1)$ has been divided by one of the factors of N^x.

In this equation we suppose that x is a fixed number but that N tends to infinity and θ to zero in such a way that $N\theta$ tends to the fixed value μ. All the factors $1 - \frac{1}{N}, 1 - \frac{2}{N} \dots 1 - \frac{x-1}{N}, \left(1 - \frac{\mu}{N}\right)^{-x}$ tend to the value 1, but the limit of $\left(1 - \frac{\mu}{N}\right)^N$ is the number $e^{-\mu}$ (see mathematical Appendix A.1). Hence $\lim_{\substack{N\to\infty\\ \theta\to 0}} b(x, N, \theta) = p(x, \mu)$.

Instances of the Poisson distribution might be the number of blind births per year in a large city, the number of occurrences of hands containing four aces in an evening of bridge at a club, or the number of typographical errors per page in professionally typed material. The numbers of births, hands of bridge or typed symbols will be large, and the probability of the rare event described (blind birth, hand with four aces or error) is small, but there may well be a few such events in each instance. Considering the births, for example, we assume that these are independent, that the probability of a blind birth remains constant, and that the total number of births per year in the region considered is approximately constant. If so, the annual number of blind births in the region, as recorded over a period of several years, should fluctuate approximately in accordance with a Poisson distribution.

3.7 **Moments and Cumulants of the Poisson Distribution** For the theoretical distribution given by Eq. (3.6.1), x may take all integral values 0, 1, 2 . . . . It may be noted that

$$\sum_{x=0}^{\infty} p(x, \mu) = e^{-\mu} \sum_{0}^{\infty} \frac{\mu^x}{x!} = 1 \tag{3.7.1}$$

as it should be, since

$$\sum_{0}^{\infty} \frac{\mu^x}{x!} = e^{\mu}.$$

The expectation of X is

$$\begin{aligned} E(X) &= \sum_{0}^{\infty} x p(x, \mu) \\ &= \mu \sum_{1}^{\infty} \frac{\mu^{x-1} e^{-\mu}}{(x-1)!} = \mu e^{-\mu} e^{\mu} = \mu \end{aligned} \tag{3.7.2}$$

The moment and cumulant generating functions may be found from those of the binomial (§ 3.3) by writing $\theta = \mu/N$ and letting N tend to infinity. Thus

$$\begin{aligned} M(h) &= \lim_{N\to\infty} \left[1 + \mu \frac{(e^h - 1)}{N}\right]^N \\ &= \exp[\mu(e^h - 1)], \text{ by Appendix A.1} \end{aligned} \tag{3.7.3}$$

$$\begin{aligned} K(h) &= \log M(h) = \mu(e^h - 1) \\ &= \mu\left(h + \frac{h^2}{2!} + \frac{h^3}{3!} + \dots\right) \end{aligned} \tag{3.7.4}$$

All the cumulants are therefore equal to μ. In particular, the variance is μ, the skewness is $\kappa_3/\kappa_2^{3/2} = \mu \cdot \mu^{-3/2} = \mu^{-1/2}$, and the kurtosis is $\kappa_4/\kappa_2^2 = \mu^{-1}$.

3.8 **Tables of the Poisson Function** Tables of $p(x, \mu)$ for values of μ between 0.1 and 15 may be found in Biometrika Tables [2]. More extensive tables have been calculated by Molina and by Kitagawa [6]. The former has also tabulated the cumulative probabilities:

(3.8.1) $$P(c, \mu) = \sum_{x=c}^{\infty} p(x, \mu)$$

As an approximation to the cumulative binomial for moderately small θ, the cumulative Poisson function $P(x, \mu)$ is improved by subtracting a term proportional to $p(x - 1, \mu)$. As shown by Gram and Charlier,

(3.8.2) $$B(x, N, \theta) \approx P(x, \mu) - \tfrac{1}{2}\theta(x - 1 - \mu)p(x - 1, \mu)$$

EXAMPLE 2 For $N = 10$, $\theta = 0.1$, and $x = 3$, we have $\mu = 1$, $B(3, 10, 0.1) = 0.07019$, $P(3, 1) = 0.08030$. The correcting term is $-0.05p(2, 1) = -0.00920$, which makes the approximation 0.07110, and so improves it considerably.

* 3.9 **The Poisson Distribution of Random Events** The clicks heard in a Geiger counter at a chosen location may be regarded as produced by independent events—the passage of cosmic rays or particles from a radioactive source through the counter. Also the result is practically independent of the precise time of observation t (assuming that the radioactive source is relatively long-lived). Consider, again, the arrival of incoming calls at a telephone switchboard. Except in special circumstances of national or local excitement, the calls may be regarded as practically independent of one another. The hypothesis that they are also independent of t is more dubious since there are slack times during the day, holidays, etc.) but one five-minute period will probably be very like another during the regular office hours, Monday to Friday.

These are examples of sequences of independent physical events, each of which has a well-defined probability of occurring in an interval of time δt (from t to $t + \delta t$), where this probability, although it depends of course on δt, may be considered independent of t. On these assumptions it is a simple matter to show that the distribution of the number of events X occurring in a fixed interval T is a Poisson distribution. Even though the assumptions may not be fully justified, the distribution seems in practice to be very nearly Poisson. Certainly telephone engineers have found that calculations based on this distribution are very useful in designing switchboards to accommodate expected telephone traffic.

The proof that the distribution is Poisson goes as follows. Let $p(t)$ be the probability that exactly one event (such as a click) occurs in a time interval of length t. Also let $q(t)$ be the probability that no such events occur and $r(t)$ the probability that more than one event occurs, in the interval t. Since these three possibilities are mutually exclusive and exhaustive,

(3.9.1) $$p(t) + q(t) + r(t) = 1$$

It is reasonable to suppose that $q(0) = 1$, since no events will occur in an interval of zero length. We assume also that $q'(t)$ tends to the value $-a$ $(a > 0)$ as $t \to 0$, where $q'(t) = d\,q(t)/dt$. This means that $q(t)$ decreases as t increases from zero. Furthermore, we will suppose that $r(t)/t \to 0$ as $t \to 0$, which means that the probability of more than one event in the interval of length t tends to zero even more rapidly than t itself. (If the probability of one event in a very short interval is small, the probability of two or more events in the same short interval will be of a higher order of smallness). With these assumptions we can show that the probability of exactly x events in the interval t is given by

$$p(x, at) = (at)^x e^{-at}/x! \tag{3.9.2}$$

which is the Poisson distribution with parameter at.

Let X denote the variate "number of events (such as clicks) occurring in an interval of length t" and let n be any fixed positive integer. Subdivide the interval into n non-overlapping sub-intervals each of length t/n. Let E be the event "in exactly x of these sub-intervals just one click occurs" and F the event "two or more clicks occur in at least one of the sub-intervals." Then if E occurs and not F, the value of X will be x; and if the value of X is x, either E or F must occur. That is,

$$E \cap \tilde{F} \subset (X = x) \subset E \cup F \tag{3.9.3}$$

By Theorems 1.6 and 1.7,

$$P(E \cap \tilde{F}) \le P(X = x) \le P(E \cup F) \le P(E) + P(F) \tag{3.9.4}$$

But

$$\begin{aligned} P(E) &= P(E \cap F) + P(E \cap \tilde{F}) \\ &\le P(F) + P(E \cap \tilde{F}) \end{aligned}$$

so that

$$P(E) - P(F) \le P(E \cap \tilde{F}) \tag{3.9.5}$$

From (4) and (5), we obtain

$$P(E) - P(F) \le P(X = x) \le P(E) + P(F) \tag{3.9.6}$$

We will now show that $P(F) \to 0$ as $n \to \infty$, from which it follows that $P(X = x) = P(E)$. Let F_i be the event "two or more clicks occur in the ith sub-interval." Then $F = \bigcup_i F_i$ and

$$P(F) = P\left(\bigcup_i F_i\right) \le \sum_i P(F_i) \tag{3.9.7}$$

In the notation of Eq. (1), $P(F_i) = r(t/n)$, and since this is the same for each subinterval,

$$\sum_i P(F_i) = nr\left(\frac{t}{n}\right) = t\,\frac{r(t/n)}{t/n} \tag{3.9.8}$$

which, by the assumptions made above, tends to the value 0 as $n \to \infty$. Thus from Eqs. (7) and (8) we see that $P(F) \to 0$, as stated.

Now $P(E)$ is the probability of exactly x successes in n independent trials of an event (namely the occurrence of just one click in a sub-interval of length t/n) of which the probability of success in a single trial is $p(t/n)$. By the binomial theorem,

$$P(E) = \binom{n}{x} p^x(1 - p)^{n-x} \tag{3.9.9}$$

where p stands for $p(t/n)$. Also by the assumption regarding $q(t)$ and its derivative,

$$q'(0) = \lim_{t\to 0} \frac{q(t) - q(0)}{t} = \lim_{t\to 0} \frac{q(t) - 1}{t} = -a \tag{3.9.10}$$

so that

$$q(t) = 1 - at + t\varepsilon(t) \tag{3.9.11}$$

where $\varepsilon(t) \to 0$ as $t \to 0$. Applying this result to the interval t/n, we have

$$q\left(\frac{t}{n}\right) = 1 - \frac{at}{n} + \frac{t}{n}\varepsilon\left(\frac{t}{n}\right) \tag{3.9.12}$$

By (1),

$$\begin{aligned} p\left(\frac{t}{n}\right) &= 1 - q\left(\frac{t}{n}\right) - r\left(\frac{t}{n}\right) \\ &= \frac{at}{n} - \frac{t}{n}\varepsilon\left(\frac{t}{n}\right) - r\left(\frac{t}{n}\right) \\ &= \frac{t}{n}\left[a - \varepsilon\left(\frac{t}{n}\right) - \frac{r(t/n)}{t/n}\right] \\ &= \frac{b_n t}{n} \end{aligned} \tag{3.9.13}$$

where $b_n = a - \varepsilon(t/n) - r(t/n)/(t/n)$, which tends to the value a as $n \to \infty$. Therefore $np(t/n) \to at$ (a fixed, positive number) as $n \to \infty$. But this is just the condition for the Poisson approximation to hold; therefore,

$$P(E) \to e^{-at}\frac{(at)^x}{x!} = p(x, at) \tag{3.9.14}$$

Moreover, the probability that $X = x$ lies between $P(E) - P(F)$ and $P(E) + P(F)$. As $n \to \infty$ both these extremes tend to the value $p(x, at)$, and therefore so does $P(X = x)$. This is the Poisson distribution for random events.

The quantity a is the expected number of events (clicks) in unit time; it may be estimated from the ratio N/T where N is the total number of clicks occurring in a fairly long interval T.

3.10 **The Normal Distribution as an Approximation to the Binomial** If the binomial probabilities $b(x, N, \theta)$ are plotted against x for different values of N, it will be found that, as N increases, the histograms so drawn approximate more and more closely to a symmetrical bell-shaped curve known variously as the *normal*, or *Gaussian, curve*, or the *curve of error*. The normal distribution, represented by this curve, plays a central part in statistical theory. Since the range of the binomial variable X, and its expectation, both increase with N, the histograms get wider and flatter and move further to the right as N increases (see Figure 17, which shows outlines of the histograms for $\theta = \frac{1}{2}$, $N = 9$, 16 and

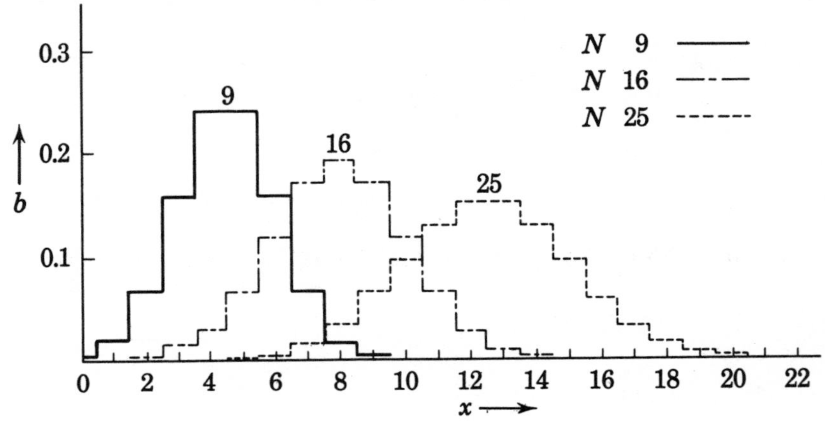

FIG. 17 THREE BINOMIAL DISTRIBUTIONS WITH INCREASING N

25). To avoid this it is convenient to use instead of X the *standardized variate:*

$$Z = (X - \mu)/\sigma \tag{3.10.1}$$

which is the difference of X from its expectation expressed in units of the standard deviation, and at the same time to multiply the ordinates $b(x, N, \theta)$ by σ. Since σ is proportional to the square root of N, the effect of the change of scale is to compress the histogram horizontally and extend it vertically, and the change of origin keeps the center at $z = 0$ for all values of N. An outline of the histogram for $N = 50$ is shown in Figure 18, along with the limiting normal curve. Almost the whole of the distribution lies within about three standard deviations on either side of the mean, between $z = \pm 3$. The approximation to the normal curve is much better, for moderate values of N, when θ is near 0.5 than when it is near 0 or 1.

The probability that $X = x$ is given by the binomial expression

$$P(X = x) = b(x, N, \theta) = \frac{N!}{x!(N - x)!}\theta^x\phi^{N-x}$$

where $\phi = 1 - \theta$. Taking logs (to base e), we have

$$\log P = \log N! - \log x! - \log(N - x)! + x \log \theta + (N - x)\log \phi \tag{3.10.2}$$

With the use of Stirling's approximation for the logarithm of $n!$ (see Appendix A.2), namely,

(3.10.3) $$\log n! \approx (n + \tfrac{1}{2})\log n - n + \tfrac{1}{2}\log(2\pi),$$

Eq. (2) becomes

(3.10.4) $$\log P \approx (N + \tfrac{1}{2})\log N - (x + \tfrac{1}{2})\log x - (N - x + \tfrac{1}{2})\log(N - x) - \tfrac{1}{2}\log(2\pi) + x\log\theta + (N - x)\log\phi$$

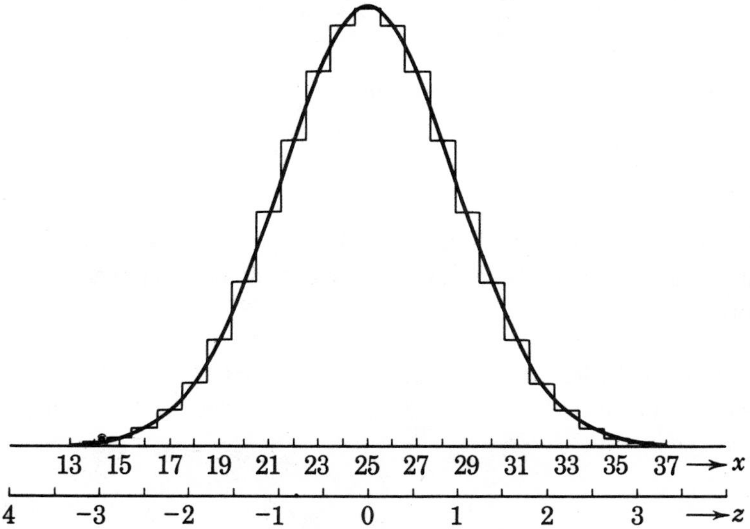

FIG. 18 STANDARDIZED BINOMIAL DISTRIBUTION AND NORMAL CURVE ($\theta = 0.5$)

If we now change to the standardized variate, putting $z = (x - \mu)/\sigma = (x - N\theta)(N\theta\phi)^{-1/2}$, so that

(3.10.5) $$x = N\theta + (N\theta\phi)^{1/2}z$$

and

(3.10.6) $$N - x = N\phi - (N\theta\phi)^{1/2}z$$

we find, from Eq. (4),

(3.10.7) $$\begin{aligned}\log P \approx &-\tfrac{1}{2}[\log N + \log\theta + \log\phi + \log(2\pi)]\\ &-[N\theta + \tfrac{1}{2} + (N\theta\phi)^{1/2}z]\log[1 + z(\phi/N\theta)^{1/2}]\\ &-[N\phi + \tfrac{1}{2} - (N\theta\phi)^{1/2}z]\log[1 - z(\theta/N\phi)^{1/2}]\end{aligned}$$

Expanding the logarithms in series and arranging the terms on the right of Eq. (7) in powers of $N^{-1/2}$, we finally obtain after some manipulation

(3.10.8) $$\log P \approx -\tfrac{1}{2}\log(2\pi N\theta\phi) - \frac{z^2}{2} + \frac{z}{2N^{1/2}}[(\theta/\phi)^{1/2} - (\phi/\theta)^{1/2}] + \frac{z^3}{6N^{1/2}}[(\phi^3/\theta)^{1/2} - (\theta^3/\phi)^{1/2}] + \dots$$

where the unwritten terms are of order N^{-1}. For large N the terms of order $N^{-1/2}$ may be neglected (and they vanish identically when $\theta = \phi = \frac{1}{2}$). If so,

$$\log P \approx -\tfrac{1}{2}\log(2\pi N\theta\phi) - \frac{z^2}{2}$$

so that

$$P \approx \frac{1}{\sqrt{2\pi N\theta\phi}}\, e^{-z^2/2} = \frac{1}{\sigma\sqrt{2\pi}}\, e^{-(x-\mu)^2/(2\sigma^2)} \tag{3.10.9}$$

The limiting form for σP is therefore given by

$$\lim_{N\to\infty} \sigma P = \frac{1}{\sqrt{2\pi}}\, e^{-z^2/2} \tag{3.10.10}$$

The function on the right is the standardized form of the normal distribution and will usually be denoted by $\phi(z)$. It is tabulated in Appendix B (Table 2). For more extensive tables see references [7] and [8].

3.11 **Approximation of the Cumulative Binomial by the Cumulative Normal Distribution** The distribution function for the standardized normal distribution is

$$\Phi(z) = \int_{-\infty}^{z} \phi(u)\, du \tag{3.11.1}$$

where the integral is improper but converges (see Appendix A.3). This function represents the area under the standardized curve from $-\infty$ up to the given value z. It is tabulated in Appendix B and in references [7] and [8], although some of these tables give instead of $\Phi(z)$ the integral $\int_0^z \phi(u)\, du$, which merely differs from $\Phi(z)$ by 0.5. (Because of the symmetry of $\phi(u)$ about $u = 0$, $\int_{-\infty}^{0} \phi(u)\, du$ is one-half $\int_{-\infty}^{\infty} \phi(u)\, du$ and this latter integral is equal to 1 (see Appendix A.7).

The binomial probability that $X \geq x$ is given exactly by $B(x, N, \theta)$. This is approximately equal to the probability that $Z \geq z$, where Z has a normal distribution and $z = (x - \mu)/\sigma$. However, because the binomial distribution is discrete while the normal distribution is continuous, a better approximation is given by putting $z = (x - (1/2) - \mu)/\sigma$. In Figure 19, the probability $B(x, N, \theta)$ is given by the sum of the area of the shaded rectangles, and the bases of these rectangles extend from $x - 1/2$ on. It can be shown [9] that the error involved is less than $0.140/\sigma$.

Various attempts have been made to give a better approximation to the value of z which is such that

$$B(x, N, \theta) = 1 - \Phi(z) \tag{3.11.2}$$

A simple approximation is that of Freeman and Tukey [10], namely,

$$z \approx 2\left[\sqrt{x(1-\theta)} - \sqrt{(N-x+1)\theta}\right] \tag{3.11.3}$$

and a more elaborate one is due to Camp and Paulson,

$$z \approx \frac{a}{3b^{1/2}} \tag{3.11.4}$$

where
$$a = 9 - x^{-1} - [9 - (N-x+1)^{-1}]\left[\frac{(N-x+1)\theta}{x(1-\theta)}\right]^{1/3}$$

$$b = x^{-1} + (N-x+1)^{-1}\left[\frac{(N-x+1)\theta}{x(1-\theta)}\right]^{2/3}$$

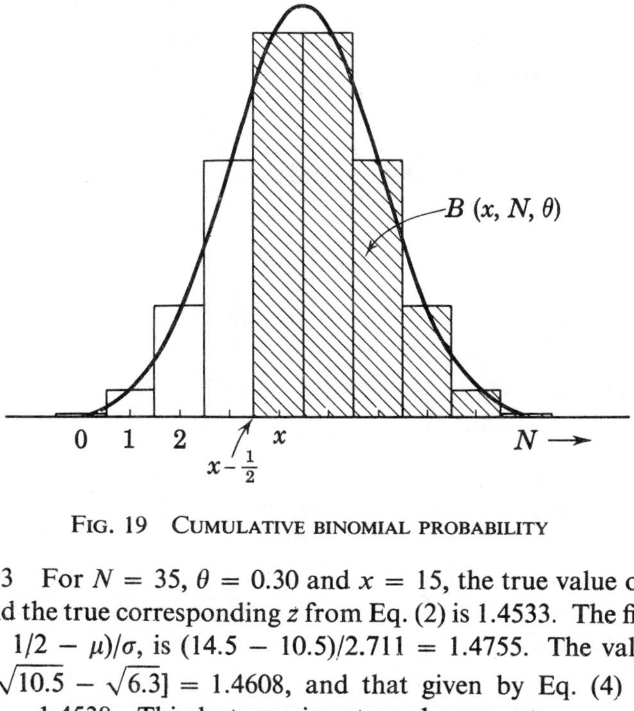

FIG. 19 CUMULATIVE BINOMIAL PROBABILITY

EXAMPLE 3 For $N = 35$, $\theta = 0.30$ and $x = 15$, the true value of $B(x, N, \theta)$ is 0.07307, and the true corresponding z from Eq. (2) is 1.4533. The first approximation, $(x - 1/2 - \mu)/\sigma$, is $(14.5 - 10.5)/2.711 = 1.4755$. The value given by Eq. (3) is $2[\sqrt{10.5} - \sqrt{6.3}] = 1.4608$, and that given by Eq. (4) is $(1.3829)/3(0.10054)^{1/2} = 1.4538$. This last one is extremely accurate.

To approximate the binomial probability that $x_1 \le X \le x_2$, namely, $B(x_1, N, \theta) - B(x_2 + 1, N, \theta)$, we can in the same way use $z_1 = (x_1 - 1/2 - \mu)/\sigma$ and $z_2 = (x_2 + 1/2 - \mu)/\sigma$ and take as the first approximation

$$P(x_1 \le X \le x_2) \approx \int_{z_1}^{z_2} \phi(u)\, du \tag{3.11.5}$$

Closer approximations can be obtained by using equation (2) with (3) or (4) for $B(x_1, N, \theta)$ and $B(x_2 + 1, N, \theta)$ separately.

3.12 Bernoulli's Law of Large Numbers (Using the Normal Distribution) We saw in § 3.4 that, from Chebyshev's inequality,

$$P\left(\left|\frac{X}{N} - \theta\right| \geq \lambda\right) \leq (4N\lambda^2)^{-1} \tag{3.12.1}$$

which means that by making N large enough we can be practically certain that X/N will differ from θ by as small an amount as we please. However, if we use the normal approximation to the cumulative binomial we can achieve the same result with a much smaller value of N.

If $x_1 = N(\theta - \lambda)$ and $x_2 = N(\theta + \lambda)$, x_1 and x_2 need not be integers when λ is arbitrary. The probability that $|X/N - \theta| < \lambda$, which is the probability that X lies *between* x_1 and x_2, is, for large N, close to the probability that Z lies between z_1 and z_2, where $z_1 = -N\lambda/\sigma$ and $z_2 = N\lambda/\sigma$. This probability is $\int_{z_1}^{z_2} \phi(u)\,du = 2\Phi(z_2) - 1$, since $z_1 = -z_2$. The required probability $P(|X/N - \theta| \geq \lambda)$ is therefore $1 - (2\Phi(z_2) - 1) = 2(1 - \Phi(z_2))$. In order to make this less than some fixed ε for a given λ, we have to choose z_2 so that $\Phi(z_2) > 1 - \varepsilon/2$. This provides a lower limit for N.

Thus if $\lambda = 0.01$ and $\varepsilon = 0.001$, we find that $z_2 > 3.29$. Since $\sigma^2 = N\theta(1 - \theta)$, we have then

$$N^{1/2}\lambda > 3.29[\theta(1 - \theta)]^{1/2} \tag{3.12.2}$$

For all values of θ, $\theta(1 - \theta) \leq \frac{1}{4}$, so that this inequality will certainly be satisfied if $N^{1/2}\lambda > 1.645$, and therefore if $N > 27{,}060$. This is considerably better than the bound on N (2,500,000) obtained in § 3.4 by the use of Chebyshev's inequality.

EXAMPLE 4 If $\theta = \frac{3}{5}$ and $N = 600$, what is the probability that the relative frequency of success will differ from $\frac{3}{5}$ by less than 0.01?

Here $\sigma = [N\theta(1 - \theta]^{1/2} = 12$ and $N\theta = 360$. Also $P(|X/N - 0.6| < 0.01)$ $= P(354 < X < 366) \approx P(z_1 < Z < z_2)$, where $Z = (X - 360)/12$, $z_1 = -0.458$, $z_2 = 0.458$. This last probability is $2\Phi(0.458) - 1 = 0.353$, which is the probability required.

3.13 Properties of the Normal Distribution The probability density for the standardized normal distribution is

$$\phi(z) = (2\pi)^{-1/2}e^{-z^2/2}, \quad -\infty < z < \infty \tag{3.13.1}$$

This is an even function, $\phi(z) = \phi(-z)$, with a maximum value 0.3989 at $z = 0$. The quartiles ϕ_3 and ϕ_1 are at $z = \pm 0.6745$, since

$$\int_0^{0.6745} \phi(z)\,dz = 0.25 \tag{3.13.2}$$

The probability is therefore 0.5 that z has a value between -0.6745 and $+0.6745$. There is an even chance that a normal variate lies between the mean plus 0.6745 times the standard deviation and the mean minus 0.6745 times the standard deviation.

The *moment generating function* for the standard normal distribution is

(3.13.3)
$$M(h) = \int_{-\infty}^{\infty} e^{hz}\phi(z)\,dz$$
$$= (2\pi)^{-1/2}\int_{-\infty}^{\infty} e^{hz-(1/2)z^2}\,dz$$

If we change the variable of integration from z to u, where $u = z - h$, this becomes

(3.13.4)
$$M(h) = (2\pi)^{-1/2}e^{(1/2)h^2}\int_{-\infty}^{\infty} e^{-(1/2)u^2}\,du$$
$$= e^{(1/2)h^2}$$
$$= 1 + \tfrac{1}{2}h^2 + \frac{1}{2!}(\tfrac{1}{2}h^2)^2 + \ldots$$

The moments about the mean are therefore $\mu_2 = 1$, $\mu_4 = 3$, $\mu_6 = 15$, etc. The odd-order moments are all zero, as is obvious from the symmetry of the distribution about $z = 0$.

The *cumulant generating function* is

(3.13.5)
$$K(h) = \log M(h) = \tfrac{1}{2}h^2$$

The only non-zero cumulant is therefore

(3.13.6)
$$\kappa_2 = 1$$

which expresses the fact that the variance is unity, as of course it must be for a standardized distribution. For the non-standardized normal distribution, with mean μ and variance σ^2, we have, by § 2.11,

(3.13.7)
$$\kappa_1 = \mu, \quad \kappa_2 = \sigma^2$$

The great simplicity of the system of cumulants for the normal distribution is one reason for the importance of this distribution in statistical theory.

The normal curve is asymptotic to the z-axis as $z \to \pm\infty$. Practically, it almost touches the axis beyond $z = \pm 4$. Table 3.2 gives the proportion of area beyond $z = \pm z_0$ for a few selected values of z_0, and therefore represents the probability that a standard normal variate will have a value outside the given interval. This table will be useful later on in problems of estimation where the variate concerned may be regarded as approximately normal.

3.14 **Probability Graph Paper** The graph of the distribution function $\Phi(z)$ is a roughly S-shaped curve, looking something like the cumulative frequency polygon of Figure 14b, but of course smooth throughout. If the scale on the graph paper is properly adjusted, this curve may be straightened. In Figure 20, the data of Table 2.2 are plotted on special probability graph paper [11]. The scale along the axis of x is uniform, but on the axis representing percentage cumulative frequency the scale is compressed in the middle and extended near the top and bottom so that the polygon of Figure 14b becomes almost a straight line. The points marked in the diagram are the values of $100F/N$, or, in this case, $F/10$, plotted against corresponding values of x_e. The fact that these points apparently lie close to a straight line is good presumptive evidence that the distribution of the variate X (in the population from which the sample is taken) is approximately normal. A method of testing this presumption will be discussed later.

TABLE 3.2

z_0	$P(\lvert z\rvert > z_0)$	z_0	$P(\lvert z_0\rvert > z_0)$
1	0.3173	0.6745	0.5
1.5	0.1336	1.2816	0.2
2	0.0455	1.6449	0.1
2.5	0.0124	1.9600	0.05
3	0.0027	2.3263	0.02
3.5	0.00046	2.5758	0.01
4	0.00006	3.2905	0.001

If a straight line is drawn by eye as evenly as possible between the plotted points, one can make a quick rough estimate of the median, quartiles, etc. for the distribution (by noting the values of x corresponding to percentage cumulative frequencies of 50, 25, 75, etc.). One can also estimate readily the probabilities that chosen values of x will be exceeded in the population.

* 3.15 **The Angular Transformation for Binomial Variates** If a variate is thought to be binomial, transformations such as those given in equations (3.11.3) and (3.11.4) will aid in the approximation to normality. That is, the quantity on the right hand side of each of these equations is approximately a standard normal variate. However, a different transformation is often used with a different purpose in mind, namely to make the variance more nearly constant (independent of θ). The so-called *angular transformation* which is appropriate here, and which was suggested by Fisher, is

$$A = \sin^{-1}(p^{1/2}) \tag{3.15.1}$$

where $p = X/N$, the observed proportion of successes, and A is the angle (in degrees) whose sine is $p^{1/2}$. A table of values of A for given p may be found in [12].

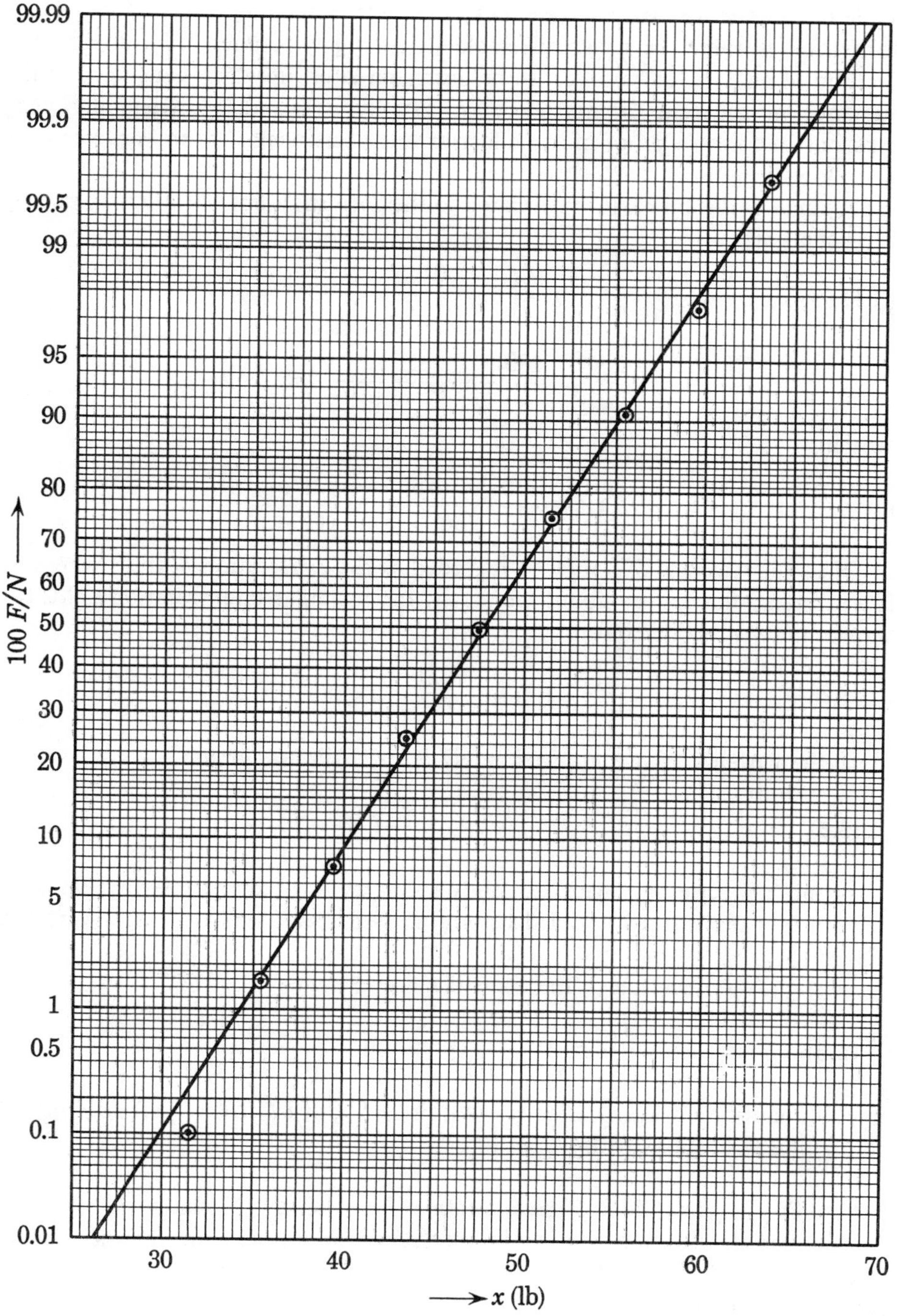

Fig. 20 Data of Table 2.2 plotted on probability graph paper

As we have seen, the variance of p is $\theta(1-\theta)/N$, but it turns out that the variance of A is approximately $821/N$, whatever θ may be. The angular transformation is therefore often advisable in testing experimental results by the method of analysis of variance, to be discussed later, in which constancy of the variance in different circumstances is usually assumed. A decidedly non-rigorous but simple proof of the effect of this transformation on variance goes as follows:

A small variation δA in A is related to the corresponding variation δp in p by the equation

(3.15.2)
$$\delta A \approx \frac{dA}{dp}\,\delta p = \frac{180}{\pi}\,\frac{1}{2p^{1/2}(1-p)^{1/2}}\,\delta p$$

so that

(3.15.3)
$$(\delta A)^2 \approx \left(\frac{90}{\pi}\right)^2 \frac{(\delta p)^2}{p(1-p)} = \frac{821(\delta p)^2}{p(1-p)}$$

Now the variance of p may be regarded as the expectation of $(\delta p)^2$ where δp is a sampling fluctuation about the mean, and a similar interpretation holds for the variance of A. Therefore

(3.15.4)
$$V(A) \approx \frac{821}{p(1-p)}\,V(p)$$

Since the variance of p is approximately $p(1-p)/N$, we obtain

(3.15.5)
$$V(A) \approx \frac{821}{N}$$

A more precise argument [13] shows that as $N \to \infty$ the distribution of A does in fact tend to a normal distribution with mean $\sin^{-1}\theta^{1/2}$ and variance $821/N$.

A slight modification of the transformation, namely,

(3.15.6)
$$A = \frac{1}{2}\left[\sin^{-1}\left(\frac{X}{N+1}\right)^{1/2} + \sin^{-1}\left(\frac{X+1}{N+1}\right)^{1/2}\right]$$

gives a quantity A whose variance is within $\pm 6\%$ of $821/(N+\frac{1}{2})$ for almost all binomial distributions with $N\theta \geq 1$.

*** 3.16 The Square Root Transformation for Poisson Variates** If X is a Poisson variate with mean μ, we know that $V(X) = \mu$ and the skewness of X, given by $\kappa_3/\kappa_2^{3/2}$, is $\mu^{-1/2}$. A transformation which serves to stabilize the variance approximately is

(3.16.1)
$$Y = X^{1/2}$$

By a similar argument to that in § 3.15, $\delta Y \approx \frac{1}{2}X^{-1/2}\,\delta X$ and $(\delta Y)^2 \approx \frac{1}{4X}(\delta X)^2$,

so that $V(Y) \approx \frac{1}{4X}V(X) \approx \frac{1}{4}$, since $X \approx \mu$ (which is the expectation of X).

More precisely [13], if

$$\begin{cases} Y = \sqrt{X+\alpha}, & X \geq -\alpha \\ Y = 0, & X < -\alpha \end{cases} \tag{3.16.2}$$

the distribution of $Y - \sqrt{\mu + \alpha}$ tends as $\mu \to \infty$ to a normal distribution with mean 0 and variance $\frac{1}{4}$. If $\alpha = 0$, this means that $E(Y) \to \mu^{1/2}$. Actually, for large μ and $\alpha = 0$,

$$E(Y) \approx (\mu - \tfrac{1}{4})^{1/2}, \quad V(Y) \approx \tfrac{1}{4}$$

and the skewness of Y is $-1/(2\mu^{1/2})$ approximately, which indicates that the normality is not greatly improved by the transformation (the skewness is halved numerically).

Bartlett [14] found that if $\alpha = \frac{1}{2}$, the variance is usually considerably nearer to $\frac{1}{4}$ for moderate values of μ than if $\alpha = 0$, and he recommended the use of the transformation $Y = \sqrt{(x + \frac{1}{2})}$. Johnson and Anscombe [15] recommend $\alpha = \frac{3}{8}$.

Anscombe [16] has pointed out that a better transformation, if we are interested in normalizing, is

$$Y = X^{2/3} \tag{3.16.3}$$

The variance of Y is about $4\frac{\mu^{1/3}}{9}$ and the skewness is of order μ^{-1}. The expectation of Y is $(\mu - \frac{1}{6})^{2/3}$, so that if we want a normal approximation to the Poisson cumulative function $P(c, \mu)$, we may write

$$P(c, \mu) = 1 - \Phi(z) \tag{3.16.4}$$

where

$$z \approx \frac{(c - \frac{1}{2})^{2/3} - (\mu - \frac{1}{6})^{2/3}}{\frac{2}{3}\mu^{1/6}} \tag{3.16.5}$$

The term $c - \frac{1}{2}$ is used instead of c as a correction for continuity, like that discussed in § 3.11.

Example 5 Let $\mu = 4$, $c = 6$. The true probability that $X \geq 6$ is $P(6, 4) = 0.2149$. The value of z given by Eq. (5) is $z = [(5.5)^{2/3} - (23/6)^{2/3}]/[(2/3)\cdot 4^{1/6}] = 0.7935$ and the corresponding probability is 0.2138, which is a fairly good approximation to the truth.

PROBLEMS

A. (§§ 3.1–3.3)

1. In a binomial distribution with $\theta = \frac{2}{5}$, calculate the probability that the number of successes in six trials is either 1, 2 or 3.

2. If ten "good" coins are tossed, what is the probability of (a) at least three heads, (b) not more than three heads?

3. Show that the greatest value of $\binom{2n}{x}$ for positive integral values of x occurs when $x = n$. What does this tell us about the binomial distribution with $\theta = \frac{1}{2}$?

4. An antiaircraft battery in England during World War II had on the average three out of five successes in shooting down "flying bombs" that came within range. What was the chance that if eight bombs came within range, not more than two got through the barrage without being shot down?

5. A and B play a game in which A's chance of winning is $\frac{2}{3}$. In a series of eight such games, supposedly independent, what is the chance that A will win at least six?

6. If the probability of success in a single trial is 0.01, how many independent trials are necessary in order to have the probability of at least one success greater than $\frac{1}{2}$? *Hint:* Find n so that $1 - (0.99)^n > \frac{1}{2}$.

7. Prove the relation of Eq. (3.3.12) for successive binomial cumulants.

Hint: $\kappa_r = \left[d^r \dfrac{K(h)}{dh^r}\right]_{h=0}$, so that $\dfrac{d\kappa_r}{d\theta} = N\left[\dfrac{d^r}{dh^r}\left(\dfrac{e^h - 1}{1 - \theta + \theta e^h}\right)\right]_{h=0}$.

Also, $\kappa_{r+1} = \left[d^{r+1} \dfrac{K(h)}{dh^{r+1}}\right]_{h=0} = N\left[\dfrac{d^r}{dh^r}\left(\dfrac{\theta e^h}{1 - \theta + \theta e^h}\right)\right]_{h=0}$

8. In a series of n trials of a binomial distribution the numbers of successes and failures are n_1 and n_2. Calculate the covariance of n_1 and n_2 and the variance of the difference $n_1/n - n_2/n$. *Hint:* $C(n_1, n_2) = E(n_1n_2) - E(n_1)E(n_2)$. For second part, see § 2.14.

B. (§§ 3.4–3.5)

1. If 1,000 trials are made of an event with probability of success $\frac{1}{3}$ in each trial, find the Bernoulli upper limit for the probability that the proportion of successes will differ from $\frac{1}{3}$ by as much as 0.05.

2. How many trials must be made of an event with binomial probability of success $\frac{1}{2}$ in each trial, in order to be assured (by the Bernoulli law) with probability at least 0.9 that the relative frequency of success will be between 0.48 and 0.52?

3. Suppose that in a Poisson sampling scheme the probability of success on the jth trial is always either 0 or 1, and that in N trials there are N_1 cases of $\theta_j = 0$ and N_2 of $\theta_j = 1$. Show that the formula of Eq. (3.5.3) for the variance of the number of successes reduces, as it should, to zero.

4. If p_N is the proportion of successes in N independent trials, the probability of success at the jth trial being θ_j, prove that p_N converges in probability to θ, where $\theta = \dfrac{1}{N}\sum \theta_j$. *Hint:* Show that $P(|p_N - \theta| \geq \lambda)$ can be made arbitrarily small for any fixed $\lambda > 0$.

5. Two persons are picked at random from a group of five persons, consisting of three men and two women. Let X represent the number of men in the sample picked. Write down the probabilities for the possible values of X. Calculate the expectation and variance of X and so verify the formulas of Eqs. (3.5.10) and (3.5.11).

C. (§§ 3.6–3.9)

1. A Poisson distribution is such that the probability is the same for $X = 1$ and for $X = 2$. What is this probability?

2. A liquid culture medium contains on the average μ bacteria per milliliter. Many samples are taken, of 1 ml each, and the total number of bacteria in each sample is counted. Assuming that the distribution is Poisson and that 10% of the samples are free from bacteria, estimate μ.

3. If on the average the proportion of defective fuses in a large consignment is 0.015, calculate the approximate probability that in a box of 200 fuses there will not be more than 2 defective.

4. A seed distributor finds that on the average 5% of his seeds will not germinate. He puts them up in packages of 100 and guarantees 90% germination. Find an approximate expression for the probability that a given package will violate the guarantee.

5. Suppose that the number of telephone calls received by an operator in a particular 5-min interval, say from 9:30 a.m. to 9:35 a.m., is a Poisson variate with mean 4. Find the probability that on a future working day the operator will receive in this interval of time (a) not more than one call, (b) six or more calls.

6. A retailer with limited storage space finds that, on the average, he sells two boxes of parrot food per week. He replenishes his stock every Monday morning so as to start the week with four boxes on hand. What are the probabilities that (a) he sells his entire stock in a week, (b) he is unable to fill at least one order? With how many boxes should he start the week so as to have a probability at least 0.99 of being able to fill all orders? *Hint:* Assume a Poisson distribution of sales with mean 2, and find the probability of x or more sales.

7. Show that if X is a Poisson variate with mean μ, then $E(X^2) = \mu E(X + 1)$.

8. The mean absolute deviation (m.a.d.) about the mean for a variate X is defined as $E(|X - \mu|)$. Show that for a Poisson variate with $\mu = 1$, the m.a.d. is $2/e$ times the standard deviation.

9. Prove that the sum of two independent Poisson variates with means μ_1 and μ_2 is Poisson with mean $\mu_1 + \mu_2$. *Hint:* Use Eqs. (2.12.8) and (3.7.4).

D. (§§ 3.10–3.14)

1. From the tables in Appendix B.2, write down the values of $\phi(1.75)$, $\phi(-0.64)$, $\Phi(2.07)$ and $\Phi(-1.63)$.

2. (a) Determine z so that $\int_{-z}^{z} \phi(u)\, du = \frac{3}{4}$; (b) If $\Phi(z) = 0.43$, calculate z. *Hint:* Use linear interpolation on the tabular values.

3. A variate X is distributed normally with mean 12 and standard deviation 2. Find the probability that X lies between 9.5 and 13.0.

4. A sample of size 1500 is normally distributed with $\mu = 75$ and $\sigma = 10$. Find (a) the value of X such that the corresponding cumulative frequency (F) is 800, (b) the number of items in the sample with $X < 80$.

5. The median of a normal distribution is 89.0 and the first quartile is 75.5. What is the standard deviation?

6. An electric railway company operating a subway uses thousands of light bulbs in its underground stations. On the morning of January 1, 1960, the company put into service 5,000 new bulbs. Assuming that the distribution of length of life for these bulbs is normal, with a mean of 50 days and a standard deviation of 19 days, how many of them would need to be replaced by midnight on January 31, 1960?
How many by March 9, 1960? (Count January 1 as a full day.)

7. A collection of human skulls is divided into three classes according as a certain length-breadth index X is (a) under 75, (b) from 75 to 80, (c) over 80. These are called, respectively, dolichocephalic (long-headed), mesocephalic (medium) and brachycephalic (short-headed). Assuming that the distribution of X is normal, and that out of 50 skulls examined the numbers in the three classes are 29, 19, and 2, find the mean and standard deviation of X for this collection. *Hint:* Find the values of z for $X = 75$ and 80 from the corresponding values of $\Phi(z)$.

8. The mean height of soldiers in a regiment containing 1,000 men is 68.22 in., with a standard deviation of 3.29 in. If the distribution is normal, how many men over 6 ft tall would you expect to find in the regiment?

9. Verify that the points of inflexion of the standard normal curve are at $z = \pm 1$, and that the tangents to the curve at these points meet the z-axis at $z = \pm 2$. *Hint:* At the points of inflection the second derivative of $\phi(z)$ vanishes.

10. Use Eq. (3.11.2) to find an approximation to the probability of at least 7 successes in 20 independent trials when the probability of success in each trial is $\frac{1}{4}$. Also calculate the approximations given by Eqs. (3.11.3) and (3.11.4) and compare with the true value, 0.2142.

11. Calculate an approximation to the probability that in 1,000 binomial trials, with probability of success $\frac{1}{2}$ in each trial, the number of successes will be *outside* the limits 481 to 519 inclusive. What is the upper bound on this probability given by Chebyshev's inequality?

12. A normal distribution with mean μ and variance σ^2 is truncated at $X = a$ and all values less than a are discarded. Show that the mean of the truncated distribution is at $\mu + \sigma\phi(\alpha)/[1 - \Phi(\alpha)]$, where $\alpha = (a - \mu)/\sigma$. *Hint:* $\int_\alpha^\infty v\phi(v)\,dv = -\int d\phi(v) = \phi(\alpha)$.

13. Give an alternative "proof" of the theorem that the limiting form of the standardized binomial variate, as $N \to \infty$, is the standardized normal variate, by showing that the c.g.f. of the binomial tends to that of the normal as $N \to \infty$, and assuming that a distribution is uniquely determined by its c.g.f. *Hint:* The c.g.f. for the binomial is $N \log(1 - \theta + \theta e^h)$. For the standardized binomial we must subtract $\mu h/\sigma$ and replace e^h by $e^{h/\sigma}$, where $\mu = N\theta$ and $\sigma^2 = N\theta(1 - \theta)$ (see § 2.12). Expand the logarithm in powers of h and show that $K(h) \to h^2/2$ as $N \to \infty$.

14. Use Stirling's approximation (Appendix A.2) to prove that if the probability of success in a single trial is $\frac{1}{2}$, then in a series of n binomial trials the probability of exactly x successes is $(2/\pi n)^{1/2} \exp[-2(x - n/2)^2/n]$, neglecting terms of order $1/n$.

E. (§§ 3.15–3.16)

1. Use the method of § 3.15 to show that if the variance of X is approximately proportional to X^2 then the transformation $Y = \log X$ produces a variate with approximately constant variance.

2. Show that if the variance of X is approximately proportional to $(1 - X^2)^2$ then a suitable transformation for producing a variate with approximately constant variance is $Y = \frac{1}{2}\log[(1 + X)/(1 - X)]$.

3. The following table gives the percentage damage by boll weevils on cotton plants treated in various ways, there being five replications for each treatment. Use the simple angular transformation, Eq. (3.15.1), to obtain corresponding values of A. Calculate the estimated variance (k_2) for each treatment, both for the original variate X and for the angular variate A. Does the variance appear to be more nearly constant (as between treatments) after the transformation than before?

	Treatments				
Replications	1	2	3	4	5
1	18	17	27	34	42
2	18	14	12	27	42
3	14	14	17	23	25
4	10	8	12	26	24
5	11	9	11	15	22

REFERENCES

[1] Glover, J. W., and Wahr, G. (ed.), *Tables of Applied Mathematics in Finance, Insurance, Statistics*, Ann Arbor, 1923.

[2] Pearson, E. S., and Hartley, H. O. (ed.), *Biometrika Tables for Statisticians*, Cambridge Univ. Press, 1954.

[3] *Tables of the Cumulative Binomial Distribution*, U.S. Army Ordnance Corps, 1952. This gives to 7 decimal places the values of $B(x, N, \theta)$ for $N = 1(1)100$, $\theta = 0.01(0.01)0.5$, and all x concerned. The notation 0.01(0.01)0.5 means that values are tabulated at intervals of 0.01 from 0.01 to 0.5 inclusive.

[4] *Harvard Tables of the Cumulative Binomial Distribution*, Harvard Univ. Press, 1956. In this 5-place table $N = 1(1)50(2)100(10)200(20)500(50)1000$, and $\theta = 0.01(0.01)0.50$. Values are also given for some other fractional values of θ, namely, 16ths and 12ths.

[5] See, for example, Ford, L. R., *Differential Equations*, McGraw-Hill, 1955, p. 155.

[6] Molina, E. M., *Poisson's Exponential Binomial Limit*, Van Nostrand, 1947. Table I gives 6- or 7-place values of $p(x, \mu)$ for $\mu = 0.001(0.001)0.01(0.01)0.30(0.1)15.0(1)100$ and for appropriate x. Table II gives $P(c, \mu)$ for the same values of μ and for appropriate c.

Kitagawa, T., *Tables of Poisson Distribution*, Baifukan, Tokyo, 1952. Gives 7- or 8-decimal values of $p(x, \mu)$ for $\mu = 0.001(0.001)1.000(0.01)10.00$.

[7] Kelley, T. L., *Statistical Tables*, Harvard Univ. Press, 1948. Gives x and $\phi(x)$ to 8 places for $\Phi(x) = 0.5(0.0001)1$.

[8] Federal Government Work Projects Administration, *Tables of Probability Functions, Vol. II*, National Bureau of Standards, 1942. Gives $\phi(x)$ and $2\Phi(x) - 1$ to 15 places for $x = 0(0.0001)1(0.001)7.8$. An auxiliary table continues to $x = 10$ (to 7 significant figures).

[9] Raff, M. S., "On Approximating the Point Binomial," *J. Amer. Stat. Ass.*, **51**, 1956, pp. 293–303.

[10] Freeman, M. F., and Tukey, J. W., "Transformations Related to the Angular and the Square Root," *Ann. Math. Stat.*, **21**, 1950, pp. 607–611.

[11] Probability graph paper (both ordinary and logarithmic) is obtainable from the Codex Book Co., Inc., Norwood, Mass.

[12] Fisher, R. A., and Yates, F., *Statistical Tables for Biological, Agricultural and Medical Research*, Oliver and Boyd, 4th ed., 1953.

[13] Curtiss, J. H., "On Transformations Used in Analysis of Variance," *Ann. Math. Stat.*, **14**, 1943, pp. 107–122.

[14] Bartlett, M. S., "The Use of Transformations," *Biometrics*, **3**, 1947, pp. 39–52.

[15] Anscombe, F. J., "The Transformation of Poisson, Binomial and Negative-Binomial Data," *Biometrika*, **35**, 1948, pp. 246–254.

[16] Anscombe, F. J., Discussion following a paper by Hotelling, *J. Roy. Stat. Soc.*, **B15**, 1953, p. 229.

Chapter 4

OTHER PROBABILITY DISTRIBUTIONS

4.1 **Reasons for Studying Probability Distributions** The main purpose of studying distributions, such as those in this chapter and in Chapter 3, is to be able to draw inferences about populations which we can sample. An empirical sampling distribution will be more or less irregular, but its form may suggest that the population distribution is closely normal or Poisson or of some other well-known mathematical type. In the next chapter we shall discuss the procedures for finding the parameters of a theoretical curve to make it fit the observed distribution as closely as possible. Once having done this, we can proceed to make mathematical deductions about the population and perhaps test these by further observations.

The most important reason, however, for studying probability distributions is their use in testing statistical hypotheses and assessing the significance of experimental results. The practical statistician calculates certain statistics from his observational data and then uses prepared tables in order to decide whether or not to accept a particular hypothesis or to judge an observed result as significant. The preparation of these tables requires an exact knowledge of the distribution of the statistic concerned, in random samples from a population of known type. When these distributions have been found, the necessary tables can be calculated, nowadays usually with the help of electronic digital computers. For reasons of mathematical convenience, the population is generally assumed to be normal. We shall come across later in this book several examples of the distribution of statistics derived from samples taken from normal populations, and we shall find that these distributions are closely related to one or other of the distributions studied in the present chapter.

4.2 **The Rectangular (Uniform) Distribution** This has already been mentioned in Example 4, § 2.10. The continuous variate X has the probability density $f(x) = (\beta - \alpha)^{-1}$, $\alpha < x < \beta$, and $f(x) = 0$ for $x > \beta$ and $x < \alpha$. The density function is discontinuous at $x = \alpha$ and $x = \beta$ (see Figure 21). Moments of all orders may easily be calculated.

An interesting property of continuous variates is that if $F(x)$ is the distribution function of X, and if we make the transformation

$$Y = F(X) \tag{4.2.1}$$

then Y has a uniform distribution, with $f(y) = 1$, on the interval (0, 1).

Since any distribution function has a range from 0 to 1, the values of Y must obviously be confined to the interval 0 to 1. Let $G(y)$ be the distribution function of Y, and let a and b be values of Y corresponding to values a' and b' of X. Then

$$a = F(a'), \qquad b = F(b')$$

The probability that Y lies between a and b is the same as the probability that X lies between a' and b', which is $F(b') - F(a')$. Therefore,

(4.2.2)
$$G(b) - G(a) = F(b') - F(a') = b - a$$

where $0 < a < b < 1$.

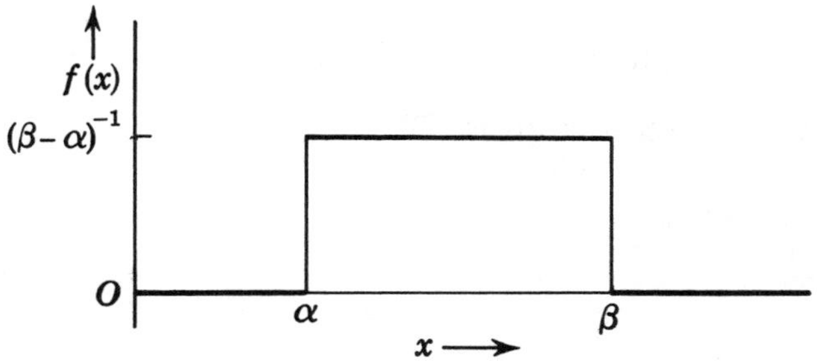

FIG. 21 RECTANGULAR DISTRIBUTION

If we replace a by y and b by $y + \Delta y$, this becomes $\dfrac{G(y + \Delta y) - G(y)}{\Delta y} = 1$, or, in the limit as $\Delta y \to 0$,

(4.2.3)
$$\frac{dG(y)}{dy} = 1, \qquad 0 < y < 1$$

The derivative of $G(y)$ is the density function of Y, namely, $g(y)$, so that

(4.2.4)
$$g(y) = 1, \qquad 0 < y < 1$$

which shows that the distribution of Y is rectangular. The transformation expressed by Eq. (1) is called the *probability transformation.* It is sometimes useful, in proving general theorems about continuous distributions, to be able to transform them to so simple a distribution, mathematically speaking, as the rectangular one.

4.3 **Distribution Function of a Transformed Variate** If $u(x)$ is a given function of x, and if x is a value assumed by a random variable X, then u may be regarded as a value of a new random variable U. The distribution function of U is

(4.3.1)
$$G(u) = P(U \le u) = \int_{u(x) \le u} f(x)\, dx$$

where $f(x)$ is the probability density for X and the integral is taken over all values of x such that $u(x) \leq u$. The density function $g(u)$ is found by differentiating $G(u)$ with respect to u. If the variate X is discrete, the integral in Eq. (1) must be replaced by a sum.

EXAMPLE 1 If $f(x) = 1, 0 < x < 1$, and $u(x) = -2 \log x$, then $u(x) < u$ if and only if $x > e^{-u/2}$. Therefore,

$$G(u) = \int_{e^{-u/2}}^{1} dx = 1 - e^{-u/2}$$

and

$$g(u) = \tfrac{1}{2}e^{-u/2}, \qquad 0 < u < \infty \tag{4.3.2}$$

This distribution is illustrated in Figure 22.

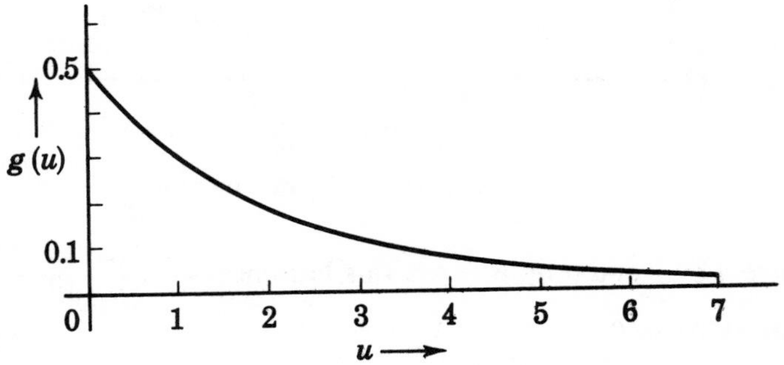

FIG. 22 EXPONENTIAL DISTRIBUTION

EXAMPLE 2 Suppose $f(x) = (2/9)(x + 1)$, $-1 \leq x < 2$, and $u(x) = x^2$. The range of u is $0 \leq u < 4$, but in the interval from 0 to 1 there are two values of x corresponding to any given u (e.g., $u = \frac{1}{4}$ for $x = \frac{1}{2}$ or $x = -\frac{1}{2}$). Since $x = \pm\sqrt{u}$ and runs from -1 to 2, the interval of x corresponding to $u(x) < u$ is from $-\sqrt{u}$ to $\sqrt{u}$ as long as $u \leq 1$ but only from -1 to $\sqrt{u}$ when $u > 1$. Therefore,

$$G(u) = \frac{2}{9}\int_{-\sqrt{u}}^{\sqrt{u}} (x + 1)\, dx$$

$$= \tfrac{4}{9}\sqrt{u}, \qquad 0 \leq u \leq 1$$

and

$$G(u) = \frac{2}{9}\int_{-1}^{\sqrt{u}} (x + 1)\, dx$$

$$= \tfrac{1}{9}(u + 2\sqrt{u} + 1), \qquad 1 < u < 4$$

The corresponding density functions are

$$g(u) = \frac{2u^{-1/2}}{9}, \qquad 0 \le u \le 1$$

(4.3.3)

$$g(u) = \frac{(1 + u^{-1/2})}{9}, \qquad 1 < u < 4$$

EXAMPLE 3 If $f(x) = \phi(x) = (2\pi)^{-1/2}e^{-x^2/2}$ (so that X is a standard normal variate), and if $u(x) = \frac{1}{2}x^2$, the range of u is from 0 to ∞ and its domain is from $-\infty$ to ∞. Therefore,

$$G(u) = \int_{-\sqrt{2u}}^{\sqrt{2u}} \phi(x)\, dx = 2\Phi(\sqrt{2u}) - 1$$

and

(4.3.4)
$$g(u) = \frac{dG(u)}{du} = (\pi)^{-1/2}u^{-1/2}e^{-u}$$

(see Appendix A.9 on differentiation under the sign of integration).

This distribution is a special case of the gamma distribution discussed in the next section.

4.4 The Gamma Distribution The distribution with density function

(4.4.1)
$$f(x) = e^{-x}\frac{x^{\alpha-1}}{\Gamma(\alpha)}, \qquad 0 \le x < \infty$$

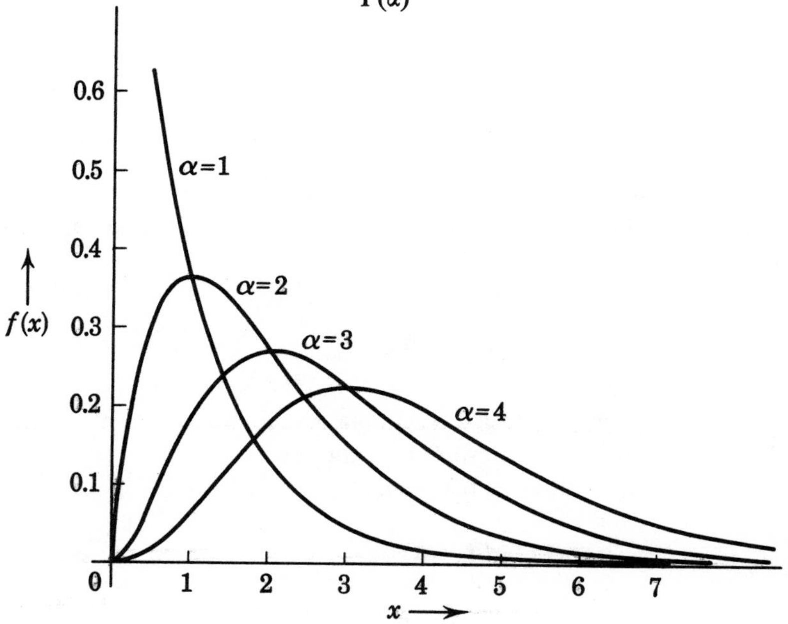

FIG. 23 GAMMA DISTRIBUTION

where $\alpha > 0$, is called the gamma distribution (see Appendix A.5). The gamma function, $\Gamma(\alpha)$, is a kind of generalized factorial with the basic property $\Gamma(\alpha + 1) = \alpha\Gamma(\alpha)$. Since $\Gamma(\frac{1}{2}) = \pi^{1/2}$ (see Appendix A.7), the gamma distribution with parameter $\alpha = \frac{1}{2}$ is the one obtained in example 3 above, given by Eq. (4.3.4). The form of the distribution, for a few values of α, is shown in Figure 23.

The m.g.f. is

$$M(h) = \int_0^\infty e^{hx} f(x)\, dx \tag{4.4.2}$$

On making the substitution $u = (1 - h)x$, and supposing that $0 < h < 1$, we obtain

$$\begin{aligned} M(h) &= \frac{1}{\Gamma(\alpha)} \int_0^\infty e^{-u} u^{\alpha-1} (1 - h)^{-\alpha}\, du \\ &= (1 - h)^{-\alpha} \end{aligned} \tag{4.4.3}$$

by the definition of $\Gamma(\alpha)$ in Eq. (A.5.1).

The c.g.f. is therefore

$$\begin{aligned} K(h) &= -\alpha \log(1 - h) \\ &= \alpha\left(h + \frac{h^2}{2} + \frac{h^3}{3} + \dots\right) \end{aligned} \tag{4.4.4}$$

so that the first few cumulants are

$$\kappa_1 = \alpha, \quad \kappa_2 = \alpha, \quad \kappa_3 = 2\alpha, \quad \kappa_4 = 6\alpha \dots \tag{4.4.5}$$

and in general $\kappa_r = \alpha\Gamma(r)$. The skewness is $\dfrac{\kappa_3}{{\kappa_2}^{3/2}} = 2\alpha^{-1/2}$ and the kurtosis is $\dfrac{\kappa_4}{{\kappa_2}^2} = 6\alpha^{-1}$.

The gamma distribution has therefore a single parameter which is at the same time the mean and the variance. As the parameter increases, the distribution becomes more nearly symmetrical.

A somewhat more general two-parameter distribution, with density function

$$f(x) = e^{-x/\beta} \frac{x^{\alpha-1}}{\beta^\alpha \Gamma(\alpha)} \tag{4.4.6}$$

where $\alpha > 0$, $\beta > 0$, is also called a gamma distribution. The rth cumulant is $\kappa_r = \alpha\beta^r\Gamma(r)$. Equation (1) corresponds to the special case $\beta = 1$.

The distribution function for the gamma variate of Eq. (6) is

$$\begin{aligned} F(x) &= \int_0^x f(u)\, du \\ &= \frac{1}{\Gamma(\alpha)} \int_0^{x/\beta} e^{-v} v^{\alpha-1}\, dv \end{aligned} \tag{4.4.7}$$

where we have made the substitution $u = \beta v$. The function of y and α, defined by

$$\Gamma_y(\alpha) = \int_0^y e^{-v} v^{\alpha-1}\, dv \tag{4.4.8}$$

is called the *incomplete gamma function*, and has been extensively tabulated [1]. Pearson's tables actually give the ratio of the incomplete to the complete gamma function, namely,

$$I(u, \alpha - 1) = \frac{\Gamma_y(\alpha)}{\Gamma(\alpha)} \tag{4.4.9}$$

with $u = y\alpha^{-1/2}$.

Thus, to find the value of $F(x)$ in Eq. (7) for any given x, we should look up in the tables the value of $I(x\beta^{-1}\alpha^{-1/2}, \alpha - 1)$.

It may be noted that the Poisson cumulative probability $P(c, \mu)$ (see § 3.8) is expressible in terms of the incomplete gamma function. In fact (see Problem B-11),

$$P(c, \mu) = \frac{\Gamma_\mu(c)}{\Gamma(c)} = I(u, c - 1) \tag{4.4.10}$$

with $u = \mu c^{-1/2}$

It was shown in § 2.12 that in order to find the cumulant generating function of a sum of independent variates we simply have to sum the individual c.g.f.'s. The sum of n independent gamma variates of the type described by Eq. (1), with parameters $\alpha_1, \alpha_2 \ldots \alpha_n$, has therefore, by Eq. (4), the c.g.f.

$$K(h) = -\sum_i \alpha_i \log(1-h)$$

and this is the c.g.f. of a gamma variate with parameter $\sum_i \alpha_i$. On the assumption that a distribution is completely determined by its c.g.f., this shows that the sum of n independent gamma variates is a gamma variate. The assumption in question is justified for the distributions likely to occur in statistical theory.

4.5 **The Beta Distributions** The two-parameter distribution with density function

$$f(x) = x^{\alpha-1} \frac{(1 - x)^{\beta-1}}{B(\alpha, \beta)}, \qquad 0 \le x \le 1 \tag{4.5.1}$$

where $\alpha > 0$, $\beta > 0$, and $B(\alpha, \beta)$ is the beta function (see Appendix A.6), is called the *beta distribution*. The somewhat similar type of distribution with density function

$$g(x) = x^{\alpha-1} \frac{(1 + x)^{-\alpha-\beta}}{B(\alpha, \beta)}, \qquad 0 \le x < \infty \tag{4.5.2}$$

where $\alpha > 0$, $\beta > 0$, may also be called a beta distribution, and will be referred to as the *beta-prime distribution*.

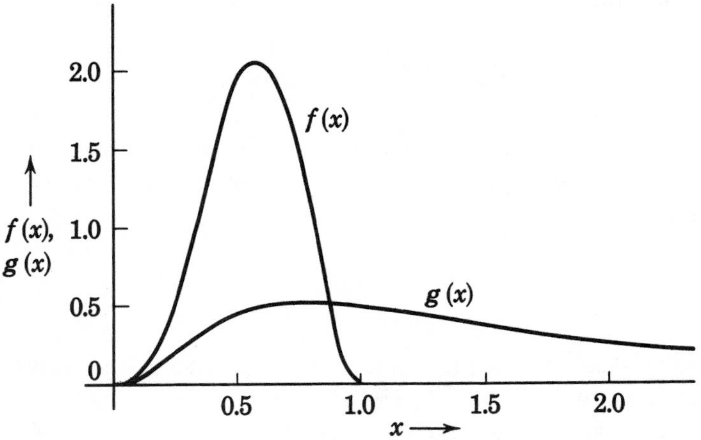

FIG. 24 BETA AND BETA-PRIME DISTRIBUTIONS

The principal property of the beta function,

$$B(\alpha, \beta) = \frac{\Gamma(\alpha)\Gamma(\beta)}{\Gamma(\alpha + \beta)} \tag{4.5.3}$$

is proved in the Appendix. The general shape of the distributions given by Eqs. (1) and (2) is illustrated in Figure 24, for $\alpha = 4$, $\beta = 3$. The curve of $f(x)$ is tangential to the x-axis at $x = 0$ and $x = 1$ if α and β are both greater than 2. The curve of $g(x)$ is tangential at $x = 0$ if $\alpha > 2$.

The rth moment about zero of the beta distribution is

$$\begin{aligned} \mu'_r &= \int_0^1 x^r f(x)\, dx \\ &= \frac{B(\alpha + r, \beta)}{B(\alpha, \beta)} \\ &= \frac{\Gamma(\alpha + r)}{\Gamma(\alpha)} \cdot \frac{\Gamma(\alpha + \beta)}{\Gamma(\alpha + \beta + r)} \\ &= \frac{(\alpha + r - 1)(\alpha + r - 2) \ldots \alpha}{(\alpha + \beta + r - 1)(\alpha + \beta + r - 2) \ldots (\alpha + \beta)} \end{aligned} \tag{4.5.4}$$

and similarly for the beta-prime distribution

$$\begin{aligned} \mu'_r &= \int_0^\infty x^r g(x)\, dx \\ &= \frac{B(\alpha + r, \beta - r)}{B(\alpha, \beta)} \\ &= \frac{(\alpha + r - 1)(\alpha + r - 2) \ldots \alpha}{(\beta - 1)(\beta - 2) \ldots (\beta - r)} \end{aligned} \tag{4.5.5}$$

if $\beta > r$. The means of these distributions are therefore at $\alpha/(\alpha + \beta)$ and $\alpha/(\beta - 1)$ respectively. The cumulants, as far as they exist, may be calculated from Eq. (4) and (5) and the relations given in §§ 2.9 and 2.12.

The distribution function for the beta variate is

(4.5.6) $$F(x) = [B(\alpha, \beta)]^{-1} \int_0^x u^{\alpha-1}(1-u)^{\beta-1}\,du$$

for $0 \le x \le 1$. This integral is called the *incomplete beta function*, $B_x(\alpha, \beta)$, and has been tabulated by Karl Pearson and his associates [2]. The tables give the ratio of the incomplete to the complete beta function, namely,

(4.5.7) $$I_x(\alpha, \beta) = \frac{B_x(\alpha, \beta)}{B(\alpha, \beta)}$$

and so the value of $F(x)$ in Eq. (6) can be read directly from these tables.

For the beta-prime distribution,

(4.5.8) $$G(x) = [B(\alpha, \beta)]^{-1} \int_0^x u^{\alpha-1}(1+u)^{-\alpha-\beta}\,du$$

On putting $1 + x = y^{-1}$ and $1 + u = v^{-1}$, this becomes

(4.5.9) $$G(x) = [B(\alpha, \beta)]^{-1} \int_y^1 v^{\beta-1}(1-v)^{\alpha-1}\,dv$$

$$= [B(\alpha, \beta)]^{-1} \int_0^{1-y} (1-w)^{\beta-1}w^{\alpha-1}\,dw$$

$$= I_{1-y}(\alpha, \beta)$$

4.6 **The Chi-Square Distribution** Let $X_1, X_2 \ldots X_n$ be n independent normal variates with means $\mu_1 \ldots \mu_n$ and variances $\sigma_1^2 \ldots \sigma_n^2$. Let the standardized variate corresponding to X_i be

(4.6.1) $$Z_i = \frac{X_i - \mu_i}{\sigma_i}$$

Then, as shown in Example 3 of § 4.3, the variable $\frac{1}{2}Z_i^2$ has the gamma distribution with parameter $\alpha = \frac{1}{2}$. We have also seen in § 4.4 that $\sum_{i=1}^n (\frac{1}{2}Z_i^2)$ is a gamma variate with parameter $\sum \alpha_i$, in this case $n/2$. If then we denote $\sum Z_i^2$ by χ^2, the variate $\frac{\chi^2}{2}$ has the density function

(4.6.2) $$f\left(\frac{\chi^2}{2}\right) = \left(\frac{\chi^2}{2}\right)^{(n/2)-1} \frac{e^{-\chi^2/2}}{\Gamma(n/2)}$$

The density function for χ^2 itself is $g(\chi^2)$, where

(4.6.3) $$g(\chi^2)\,d(\chi^2) = f\left(\frac{\chi^2}{2}\right) d\left(\frac{\chi^2}{2}\right)$$

so that

$$g(\chi^2) = \left(\frac{\chi^2}{2}\right)^{(n/2)-1} \frac{e^{-\chi^2/2}}{2\Gamma(n/2)} \tag{4.6.4}$$

A distribution with this density function is called the chi-square (χ^2) distribution. The number n is called the *number of degrees of freedom.* It is the number of independent normal variates whose squares are added to produce χ^2.

The chi-square distribution is an important one in statistical theory, being much used for testing the goodness of fit of a theoretical curve to an empirical distribution and for testing certain types of statistical hypotheses. Examples of these uses will be given later. Meanwhile we list a few properties of this distribution.

The shape of the curve of $g(\chi^2)$, plotted against χ^2, depends on the value of n. The curves for different n look like the gamma distributions of Figure 23. Since the rth cumulant for $\chi^2/2$ is $\kappa_r = (n/2)\Gamma(r)$, the rth cumulant for χ^2 is $2^r\kappa_r = 2^{r-1}(r-1)!\,n$.

The expectation, variance and skewness are therefore given by

$$\begin{cases} \kappa_1 = E(\chi^2) = n \\ \kappa_2 = V(\chi^2) = 2n \\ \gamma_1 = \dfrac{\kappa_3}{\kappa_2^{\,3/2}} = 2\left(\dfrac{2}{n}\right)^{1/2} \end{cases} \tag{4.6.5}$$

The distribution function for χ^2 is

$$G(u) = \int_0^u g(\chi^2)\, d\chi^2 \tag{4.6.6}$$

In the statistical applications, we are usually interested in the area of the chi-square curve to the *right* of a particular value u (the shaded area in Figure 25).

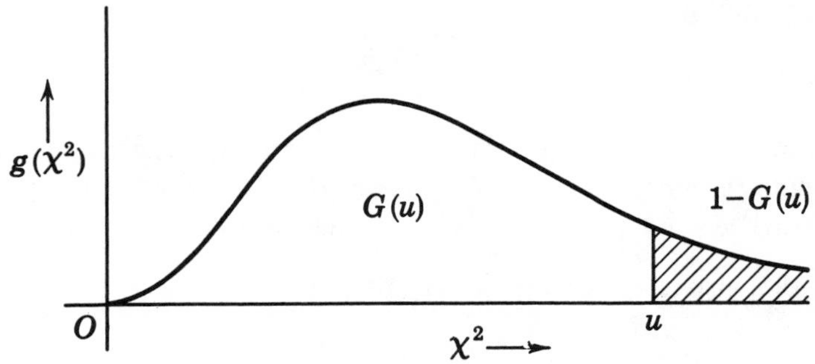

FIG. 25 CHI-SQUARE DISTRIBUTION

This is equal to $1 - G(u)$. The table in Appendix B gives values of u corresponding to selected values of $1 - G(u)$ for all n from 1 to 30. More extensive tables may be found in [3].

The function $G(u)$ is converted by the transformation $\chi^2 = 2v$ into an incomplete gamma function:

$$(4.6.7)\qquad G(u) = \int_0^{u/2} v^{(n/2)-1} e^{-v} \frac{dv}{\Gamma(n/2)} = \frac{\Gamma_{u/2}(n/2)}{\Gamma(n/2)} = I\{u(2n)^{-1/2}, (n/2) - 1\}$$

by Eq. (4.4.9), so that the tables [1] may also be used to find $G(u)$.

For large n, it was shown by Fisher that $(2\chi^2)^{1/2} - (2n - 1)^{1/2}$ is approximately a standard normal variate, so that if we need values beyond the scope of the table in the Appendix we can put

$$(4.6.8)\qquad (2\chi^2)^{1/2} \approx z + (2n - 1)^{1/2}$$

where z is given by $\Phi(z) = G(\chi^2)$.

Thus, suppose in Figure 25 that $n = 30$ and the shaded area is 0.05. The value of u given by the table of χ^2 is 43.773. The corresponding z, for $\Phi(z) = 0.95$, is 1.645, so that

$$(2\chi^2)^{1/2} \approx 1.645 + (59)^{1/2} = 9.326$$

which gives $\chi^2 = 43.49$.

A still better approximation is that of Wilson and Hilferty [4], namely,

$$(4.6.9)\qquad \left(\frac{\chi^2}{n}\right)^{1/3} = 1 - \frac{2}{9n} + z\left(\frac{2}{9n}\right)^{1/2}$$

For the case $n = 30$, $z = 1.645$, this gives $\left(\frac{\chi^2}{30}\right)^{1/3} = 1 - 1/135 + 1.645/(135)^{1/2}$ $= 1.1341$, from which $\chi^2 = 43.76$. This is very close to the true value.

*** 4.7 Theorems on the Chi-Square Distribution** The following theorems are sometimes useful in establishing the distributions of particular statistics. The proofs are either sketched briefly or are omitted altogether.

THEOREM 4.1 *If Y_i $(i = 1, 2 \ldots n)$ is one of a set of orthogonal linear functions* (see Appendix A.10) *of the independent variates $X_j (j = 1, 2 \ldots n)$, and if the X_j are normal with mean* 0 *and variance* 1, *then the distribution of $\sum_i Y_i^2$ is chi-square with n degrees of freedom.*

We first note that the distribution of any one of the Y_i is normal. This follows from Eq. (2.11.8) and the assumption that the distribution is determined by its cumulant generating function. The further assumption that the different Y_i are orthogonal to each other implies that they are independent and that

$$(4.7.1)\qquad \sum_i Y_i^2 = \sum_j X_j^2$$

and, since $\sum_j X_j^2$ is a chi-square variate, by § 4.6, so is $\sum_i Y_i^2$.

Examples of orthogonal linear transformations are

$$\begin{cases} Y_1 = (2)^{-1/2}(X_1 + X_2) \\ Y_2 = (2)^{-1/2}(X_2 - X_1) \end{cases} \tag{4.7.2}$$

and

$$\begin{cases} Y_1 = 2^{-1}(X_1 + X_2 + X_3 + X_4) \\ Y_2 = 2^{-1/2}(X_1 - X_2) \\ Y_3 = 6^{-1/2}(X_1 + X_2 - 2X_3) \\ Y_4 = 12^{-1/2}(X_1 + X_2 + X_3 - 3X_4) \end{cases} \tag{4.7.3}$$

For each Y_i the sum of the squares of the coefficients of the X_j is 1, and for any two different Y's the sum of the products of the coefficients, pair by pair, is 0. This is the distinguishing characteristic of an orthogonal linear transformation.

The fact that the sum of the squares of the coefficients of Y_i is 1 shows that the variance of Y_i is 1 (the same as the variance of the X_j). Also the expectation of Y_i is obviously 0, so that the Y_i are standard normal variates. Each Y_i^2 is therefore a chi-square variate with one degree of freedom. Note that $Y_1^2 = (\sum X_j)^2/n = n\bar{X}^2$.

THEOREM 4.2 *The sum of two independent chi-square variates with n_1 and n_2 degrees of freedom is a chi-square variate with $n_1 + n_2$ degrees of freedom.*

This follows from the corresponding property for gamma variates.

THEOREM 4.3 (Fisher's Theorem) *If $A = \sum_{j=1}^{n} X_j^2$ and $B = \sum_{i=1}^{h} Y_i^2$, where the Y_i are orthogonal linear functions of the independent standard normal variates X_j, then $A - B$ is a chi-square variate with $n - h$ degrees of freedom, and is independent of B.*

Note that by Theorem 4.2 and the distribution of Y_i^2, the quantity B is a chi-square variate with h degrees of freedom. Since $A = \sum_{i=1}^{n} Y_i^2$, the difference $A - B$ is a sum of $n - h$ of the Y_i^2 and is therefore a chi-square variate with $n - h$ degrees of freedom. Fisher's theorem states that $A - B$ and B are distributed independently [5].

THEOREM 4.4 (Cochran's Theorem) *If $A = \sum_{j=1}^{n} X_j^2$ and if $A = q_1 + q_2 + \ldots + q_k$, where the q's are quadratic forms in the X_j with $n_1, n_2 \ldots n_k$ degrees of freedom respectively, then a necessary and sufficient condition that the q's are independent chi-square variates with $n_1, n_2, \ldots n_k$ degrees of freedom is*

$$n_1 + n_2 + \ldots + n_k = n$$

A quadratic form is an expression of the type $q = \sum_{ij} a_{ij}X_iX_j$, where the a_{ij} are real numbers. To say that the form has r degrees of freedom (or is of rank r) means that the largest non-zero determinant which can be formed from the matrix a_{ij} has r rows and columns. See Appendix A.20 and reference [6].

EXAMPLE 4 If $\bar{X} = (1/n) \sum X_i$, then as shown above, $n\bar{X}^2$ is a chi-square variate with one d.f. We may write

$$A = \sum X_i^2 = \sum (X_i - \bar{X})^2 + n\bar{X}^2 \tag{4.7.4}$$

It follows that $\sum (X_i - \bar{X})^2$, which is $n - 1$ times the sample variance, is distributed as χ^2 with $n - 1$ degrees of freedom, independently of $n\bar{X}^2$.

* 4.8 **The Log-Normal Distribution** It sometimes happens that if a variate X (which takes only positive values) is markedly skew in its distribution, log X is much more nearly normal. This may be tested readily by plotting the cumulative percentage frequency for a good-sized sample against the corresponding X on special logarithmic probability graph paper. This paper has a logarithmic scale along one axis and a probability scale (like that in Figure 20) along the other. If the resulting points lie nearly on a straight line, the distribution of X in the population may be taken as log-normal.

Some examples of distributions which have been found to be nearly log-normal are the sizes of silver particles in a photographic emulsion, the survival times of bacteria in given strengths of disinfectant, the effective lengths of life of some types of industrial equipment, the blood pressures of human beings, the magnitudes of maximum annual floods for a given river, and even the numbers of words in a sentence written by George Bernard Shaw.

Let $Y = \log_e X$, and suppose that the distribution of Y is normal with mean α and variance β. Let $f(x)$ and $g(y)$ be the density functions for X and Y respectively. Then

$$f(x)\,dx = g(y)\,dy = g(y)\frac{1}{x}\,dx \tag{4.8.1}$$

so that

$$f(x) = x^{-1}g(y) = x^{-1}(2\pi\beta)^{-1/2}e^{-(y-\alpha)^2/2\beta} \tag{4.8.2}$$

The rth moment about 0 of the variate X is

$$\mu'_r = \int_0^\infty x^r f(x)\,dx \tag{4.8.3}$$

$$= \int_{-\infty}^\infty e^{ry}g(y)\,dy$$

since $x = e^y$ if $y = \log_e x$ Carrying out the integration, we obtain

$$\mu'_r = \exp\left(r\alpha + \frac{r^2\beta}{2}\right) \tag{4.8.4}$$

The mean of X is therefore

$$\mu'_1 = \mu = \exp(\alpha + \tfrac{1}{2}\beta) \tag{4.8.5}$$

and the variance is

(4.8.6)
$$\mu'_2 - (\mu'_1)^2 = \sigma^2 = \exp(2\alpha + 2\beta) - \exp(2\alpha + \beta) = \mu^2\eta^2$$

where $\eta^2 = e^\beta - 1$. The quantity η is the ratio σ/μ, which is called the coefficient of variation. The skewness of the distribution is given by

(4.8.7)
$$\gamma_1 = \eta^3 + 3\eta$$

If $Y = \log_{10} X = c \log_e X$, where $c = 0.4343$ approximately, and if α and β now refer to $\log_{10} X$, the Eqs. (5), (6) and (7) will need to be modified by writing α/c for α and β/c^2 for β.

Various modifications of the simple log-normal distribution have been suggested. A full discussion may be found in [7], and a table of critical values of the distribution in [8]. The logarithmic transformation is often used to stabilize variance in situations where the observational data fall into groups with different means and where in each group the standard deviation is roughly proportional to the mean. The transformed variates will in this case have approximately constant variance.

4.9 **Families of Theoretical Distributions** The process of curve-fitting was at one time very popular among statisticians—much more so than it is today—and whole families of theoretical distributions were invented to fit (as it was hoped) almost any kind of empirical distribution that might turn up. One such family (including eight principal types of curve and a variety of special cases) was devised by Karl Pearson. Another idea, due to the Norwegian statisticians Gram and Charlier, was to use the normal distribution, modified by adding terms proportional to the 1st, 2nd, 3rd . . . derivatives of the normal density function. The coefficients of these terms turn out to be either zero or else simply expressible in terms of the cumulants of the distribution. A brief discussion of the Pearson family of curves and of the Gram-Charlier series may be found in [9].

4.10 **The Central Limit Theorem** We close this chapter with a short description of a famous theorem which plays a central role in the theory of statistical inference, and accounts very largely for the importance of the normal distribution in theoretical investigations.

Let $X_1, X_2 \dots X_n$ be independent random variates all having the same distribution with mean μ and variance σ^2, but not necessarily normal. Let the standardized variate corresponding to X_j be

(4.10.1)
$$Z_j = \frac{X_j - \mu}{\sigma}$$

and let Y_n be defined by

(4.10.2)
$$Y_n = \frac{\sum Z_j}{n^{1/2}} = n^{1/2}\bar{Z}$$

where $\bar{Z}$ is the arithmetic mean of the Z_j. Then the theorem (in its simplest and least general form) states that as $n \to \infty$, Y_n tends to a standard *normal* variate, or, in symbols,

$$P(Y_n \leq y) \to \Phi(y) \tag{4.10.3}$$

The point of the theorem is that, no matter what the original distribution of Z_j may be (provided of course that X_j possesses a mean and variance), the mean of a large enough sample will have a nearly normal distribution.

The cumulant generating function for Z_j will be

$$K_j(h) = \frac{h^2}{2} + O(h^3)^* \tag{4.10.4}$$

since the coefficients of h and of $h^2/2!$ are 0 and 1 respectively. Since Y_n is a linear function of the Z_j, with coefficients all equal to $n^{-1/2}$, the c.g.f. for Y_n will be

$$\begin{aligned} K_j(h) &= \sum_j K_j(hn^{-1/2}) \\ &= n\left[\frac{h^2}{2n} + O\left(\frac{h^3}{n^{3/2}}\right)\right] \\ &= \frac{h^2}{2} + \text{terms of order } n^{-1/2} \end{aligned} \tag{4.10.5}$$

As $n \to \infty$, $K(h) \to h^2/2$, which is the c.g.f. for a standard normal variate. This suggests the result, which is indeed true, that

$$\lim_{n\to\infty} P\left(n^{1/2}\frac{\bar{X} - \mu}{\sigma} \leq y\right) = \Phi(y) \tag{4.10.6}$$

It is not necessary that the X_j should all have the same distribution. If $E(X_j) = \mu_j$ and $V(X_j) = \sigma_j^2$, and if $M_n = \sum_j \mu_j$ and $S_n^2 = \sum_j \sigma_j^2$, then (as proved by Lindeberg)

$$\lim_{n\to\infty} P\left(\frac{\sum X_j - M_n}{S_n} \leq y\right) = \Phi(y) \tag{4.10.7}$$

provided that the following condition holds for every $\varepsilon > 0$:

$$\lim_{n\to\infty} \frac{1}{S_n^2} \sum_{j=1}^{n} \int (x - \mu_j)^2 f_j(x)\, dx = 0 \tag{4.10.8}$$

where $f_j(x)$ is the density function for X_j and where the integral is taken over all values of x such that $|x - \mu_j| > S_n\varepsilon$. This condition implies that $S_n \to \infty$ but $\sigma_j/S_n \to 0$, as $n \to \infty$, for every value of j. In other words, the total sum of variances tends to infinity but the proportional contribution of each individual

*The notation $O(h^3)$ means terms of the *order* of h^3. This includes all terms proportional to h^3 or to any higher power of h.

variate to this sum tends to zero. If the variate X_j is discrete instead of continuous the integral in Eq. (8) must be replaced by a sum.

Lyapunov proved that if the absolute third moment exists for each of the X_j, so that for any n a finite R_n exists, given by

$$R_n^{\ 3} = \sum_1^n \int_{-\infty}^{\infty} |x - \mu_j|^3 f_j(x)\, dx \tag{4.10.9}$$

then the condition for the central limit theorem, Eq. (7), to hold is the simple one

$$\lim_{n \to \infty} \frac{R_n}{S_n} = 0 \tag{4.10.10}$$

It is not even necessary that the X_j should be independent. It is sufficient that X_i and X_j should be independent for $|i - j| > m$, where m is some fixed number. This means that if the variates are arranged in some natural order, consecutive or nearly consecutive members may be dependent, provided that all widely separated ones are independent.

Sums of random variables whose distributions do not have a finite second moment may not show any tendency to approach normality. If the X_j have a *Cauchy* distribution, given by

$$f(x) = [\pi(1 + x^2)]^{-1} \tag{4.10.11}$$

then the distribution of $\bar{X}(= n^{-1} \sum_j X_j)$ is the same Cauchy distribution, no matter how large n may be.

For a fuller discussion of the Central Limit Theorem see [10] and [11].

PROBLEMS

A. (§§ 4.1–4.3)

1. Show that the m.g.f. for the uniform distribution on the interval (0, 1) has the form $M(h) = 1 + h/2! + h^2/3! + \ldots.$ Write down the expression for the c.g.f., expand it as far as the term in h^4, and so obtain the first four cumulants.

2. What transformation will change the variate X to one having a uniform distribution on (0, 1), if the density function for X is $f(x) = (x - 1)/2$, $1 \le x < 3$, and $f(x) = 0$ for $x < 1$ and $x > 3$?

3. If $f(x) = 2xe^{-x^2}$, $x \ge 0$, find the density function for U, where $U = X^2$.

4. If X has the density function $f(x)$, $x \ge 0$, what is the density function for $U = aX^2 + b$, where $a > 0$? *Hint:* $u \ge b$; $0 \le X \le \left(\dfrac{u - b}{a}\right)^{1/2}$, when $U < u$. To get $g(u)$, use Appendix A.9.

5. If X has the density function $f(x) = 2x$, $0 \le x < 1$, find the distribution of $U = (3X - 1)^2$. *Hint:* For $U < u$, $0 \le u \le 1$, X goes from $\dfrac{1 - u^{1/2}}{3}$ to $\dfrac{1 + u^{1/2}}{3}$; for $1 \le u < 4$, X goes from 0 to $\dfrac{1 + u^{1/2}}{3}$.

6. If X has the continuous distribution function $F(x)$, what are the distribution functions of (a) e^X (b) $\sin X$ (c) $F(X)$? *Hint:* $P(e^X \leq x) = P(X \leq \log x) = F(\log x)$.

B. (§§ 4.4–4.5)

1. Show that $\binom{n}{r} = \dfrac{\Gamma(n+1)}{\Gamma(r+1)\Gamma(n-r+1)} = \dfrac{1}{rB(n-r+1, r)}$.

2. Show that the gamma distribution, Eq. (4.4.6), has a single mode at $x = \beta(\alpha - 1)$ if $\alpha > 1$ and touches the x-axis at the origin if $\alpha > 2$.

3. Evaluate as gamma functions the following integrals:
(a) $\int_0^\infty e^{-x}x^{2/3}\,dx$ (b) $\int_{-2}^\infty e^{-3x}(1 + x/2)^7\,dx$. *Hint:* In (b) put $1 + x/2 = y/6$.

4. Find the constant K if $K\int_{-\infty}^{\infty} z^2(1 + z^2)^{-N/2}\,dz = 1$. *Hint:* Put $z = \tan\theta$ and use Eq. (A.6.3).

5. Show that

$$\int_0^{\pi/2} \cos^p\theta\,d\theta = \int_0^{\pi/2} \sin^p\theta\,d\theta = \tfrac{1}{2}B\left(\frac{p+1}{2}, \frac{1}{2}\right).$$

6. Use Eq. (A.6.4) to show that

$$B(m, n) = \int_0^1 \frac{x^{m-1} + x^{n-1}}{(1 + x)^{m+n}}\,dx$$

Hint: In Eq. (A.6.4) divide the domain of integration into two parts, 0 to 1 and 1 to ∞. In the second part put $y = 1/x$.

7. Prove that $B_x(\beta, \alpha) = B(\alpha, \beta) - B_{1-x}(\alpha, \beta)$ and that therefore $I_x(\beta, \alpha) = 1 - I_{1-x}(\alpha, \beta)$. *Hint:* In Eq. (4.5.6), put $u = 1 - v$.

8. Show that the expectation of the positive square root of a beta variate with parameters α and β is $\Gamma(\alpha + \frac{1}{2})\Gamma(\alpha + \beta)/[\Gamma(\alpha)\Gamma(\alpha + \beta + \frac{1}{2})]$ and that the expectation of the positive square root of a one-parameter gamma variate is $\Gamma(\alpha + \frac{1}{2})/\Gamma(\alpha)$.

9. The harmonic mean of a variate X may be defined as the reciprocal of the expectation of $1/X$. If the probability density for X is $f(x) = x^n e^{-x}/\Gamma(n+1)$, $0 \leq x < \infty$, $n > 0$, show that the harmonic mean of X is equal to n.

Find the harmonic mean of the distribution with density

$$f(x) = x^{m-1}(1 + x)^{-m-n}/B(m, n), \quad 0 \leq x < \infty, \quad m > 1, \quad n > 2.$$

10. If X is a Poisson variate with mean M, and M itself a gamma variate with parameters α and β, show that the probability that $X = x$ is given by

$$P(X = x) = \Gamma(x + \alpha)\beta^x/[\Gamma(\alpha)x!\,(1 + \beta)^{\alpha+x}]$$
$$= (-1)^x\binom{-\alpha}{x}\beta^x(1 + \beta)^{-\alpha-x}.$$

Since this is, for each $x(0, 1, 2\ldots)$, a term in the binomial expansion of $[(1 + \beta) - \beta]^{-\alpha}$, the distribution is called "negative binomial." See § 1.11. *Hint:* Integrate the joint distribution of X and M over all values of u.

11. Prove that the cumulative Poisson probability $P(c, \mu)$ may be expressed in terms of the incomplete gamma function by the relation $P(c, \mu) = \Gamma\mu(c)/\Gamma(c) = I(u, c - 1)$, where $u = \mu c^{-1/2}$. *Hint:* Taylor's theorem may be written

$$f(a + h) = f(a) + hf'(a) + \ldots + \frac{h^{c-1}}{(c-1)!}f^{(c-1)}(a)$$
$$+ \frac{h^c}{(c-1)!}\int_0^1 f^{(c)}(a + th)(1 - t)^{c-1}\,dt$$

Put $f(x) = e^x$, $a = 0$, $h = \mu$, and divide through by e^μ. Note that

$$\sum_0^{c-1} e^{-\mu} \frac{\mu^x}{x!} = 1 - P(c, \mu).$$

C. (§§ 4.6–4.9)

1. Show that the c.g.f. of a standardized one-parameter gamma variate tends to the value $h^2/2$ as the parameter tends to infinity. Hence show that the c.g.f. of the standardized chi-square distribution tends to this value as $n \to \infty$. (See Problem D.13 of Chapter 3.)

2. Prove that if $t = (\chi^2 - n)/(2n)^{1/2}$, the density function for t is given by $f(t) = K(t + c)^{c^2-1}e^{-ct}$, $-c \le t < \infty$, where $c^2 = n/2$ and $K = (c)^{c^2}e^{-c^2}/\Gamma(c^2)$. (This distribution is known as Pearson's Type III. See § 4.9. The skewness is $2/c = (8/n)^{1/2}$).

3. Show that the probability that $\chi^2 \ge c$ may be written $1 - I(c/(2n)^{1/2}, (n-2)/2)$, where I is Pearson's incomplete gamma function.

4. If $y\,dx$ is the probability that X lies between x and $x + dx$ and if y is given by the solution of the differential equation $dy/dx = y(a - x)/(bx + c)$, show that (for suitable values of the constants a, b, c) a certain linear function of X has the χ^2 distribution with n degrees of freedom, where $n = 2(1 + a/b + c/b^2)$. *Hint:* The arbitrary constant in the solution of the differential equation is determined by the condition $\int y\,dx = 1$, from $-c/b$ to ∞. Put $V = 2(bX + c)/b^2$ and show that V is a χ^2-variate.

5. If $Y = \log_e X$, and Y is a standard normal variate, write down the density function for X, and calculate the expectation and variance of X by integration.

6. If $\log_e X$ is normally distributed with mean 1 and variance 4 calculate the probability that X lies between 1/2 and 2.

REFERENCES

[1] Pearson, K., *Tables of the Incomplete Gamma Function*, Biometric Lab., Cambridge Univ. Press, 1922.

[2] Pearson, K., *Tables of the Incomplete Beta Function*, Biometric Lab., Cambridge Univ. Press, 1934.

[3] Pearson, E. S., and Hartley, H. O., *Biometrika Tables for Statisticians, Vol. I*, Cambridge Univ. Press, 1954. Table 7 gives $1 - G(u)$ for known u and n, and Table 8 gives u for known $1 - G(u)$ and n.

[4] Wilson, E. B., and Hilferty, M. M., "The Distribution of Chi Square," *Proc. Nat. Acad. Sci.*, **17**, 1931, pp. 684–688.

[5] Fisher, R. A., "Applications of Student's Distribution," *Metron*, **5**, 1926, pp. 90–104.

[6] Cochran, W. G., "The Distribution of Quadratic Forms in a Normal System," *Proc. Camb. Phil. Soc.*, **30**, 1933–4, pp. 178–191.

[7] Aitchison, J., and Brown, J. A. C., *The Lognormal Distribution*, Cambridge Univ. Press, 1957.

[8] Moshman, J., "Critical Values of the Log-normal Distribution," *J. Amer. Stat. Ass.*, **48**, 1953, pp. 600–605.

[9] Kenney, J. F., and Keeping, E. S., *Mathematics of Statistics Part II*, Van Nostrand, 1951.

[10] Kendall, M. G., and Stuart, A., *Advanced Theory of Statistics, Vol. I*, Chas. Griffin, 1958.

[11] Gnedenko, B. V., and Kolmogorov, A. N., *Limit Distributions for Sums of Independent Random Variables*, Addison-Wesley, 1954.

Chapter 5

SAMPLE AND POPULATION

5.1 **Inferences from Sample to Population** As we have already seen, the data in most statistical problems relate to a sample drawn from some parent population, or universe (as it is sometimes called). Various characteristics of the sample, such as the mean, median, standard deviation or skewness, may be calculated from the data, and they serve to give a concise description of the sample itself. Their more important use, however, is to enable us to make statements about the population. Such statements, of course, being of the nature of inductive inferences, cannot be made with complete certainty, but only with more or less probability of being true. Nevertheless it is worth while to be able to state, for example, that the mean of a particular population may be taken as lying between 21.7 and 25.8, with a probability of 0.90 that this statement is true. We shall see in the present chapter how some estimates of this sort are arrived at.

The population characteristics in which we are interested are usually parameters which occur in the distribution of some variate. If, for example, the population is assumed to be normal, as far as a particular variate is concerned, the density function for this variate will contain two parameters, μ and σ, which are the population mean and standard deviation respectively. These may be estimated from the characteristics of a sample, such as the median and the range, for instance, or the sample mean and the sample standard deviation.

When a sample is used to make inferences about the population, we generally assume that the sample is *random*. This usually means (when the population is finite) that every individual in the population has an equal chance of being included in the sample. More generally, if X is the random variable which is under consideration and which has a distribution function $F(x)$ in the population, and if $X_1, X_2, \ldots, X_N$ are measured values of X on sample items from the population, the sampling is *random* if all the $X_i (i = 1, 2 \ldots N)$ are independent random variables (see § 1.13), each with the same distribution function as X itself. The probability that the observed sample has values equal to or less than $x_1, x_2, \ldots, x_N$ for the respective items is then $F(x_1) \cdot F(x_2) \ldots F(x_N)$.

It is usually desirable that sampling should be as nearly random as possible, although this is often hard to achieve in practice. Even if the sampling is not purely random, it is still possible to make valid inferences, provided that the respective probabilities of being included in the sample are known for all members of the population. In a scheme described as *stratified sampling*, for instance, the whole population is divided into classes (or strata), each of which is sampled separately. The sizes of the various strata must, however, be known, and within each stratum the sampling must be random. For valid statistical inferences there must always be present somewhere this quality of randomness. A fuller discussion of some common sampling procedures will be found in Chapter 7.

5.2 **Point Estimation and Interval Estimation** Sampling theory deals with questions like the following: given a random sample of N variates from a certain population, what can we say about the parameters that define the distribution of such variates within the population? There are two distinct questions that we may ask about any one parameter, namely, what is the best value to use for it and how reliable is this best value? The first question is one of *point estimation*—we want a single value which in some sense is the "best" estimate we can make of the parameter (various criteria are possible for judging the goodness of an estimate and they do not always agree in their choice of the best). The second question has to do with the interval in which we can confidently expect the true but unknown value of the parameter to lie, and is said to be a problem of *interval estimation*. We may, for instance, be able to say on the basis of a sample that the best estimate we can make of the population mean is 159 lb and that we feel 90% confident that the true value is somewhere between 150 lb and 168 lb. This interval (150 lb to 168 lb) is called a *confidence interval*, with *confidence coefficient* 90% or 0.90. The confidence interval is a random variate, calculated from the sample and having a probability distribution, whereas of course the population mean, although unknown, is not a random variate at all in the usual sense. We should not therefore speak of the probability that the population mean lies in a given interval but rather of the probability that the given interval includes the population mean.

To say that a confidence interval for a parameter has a confidence coefficient of 0.90 means that the statement "this interval includes the true value" has a probability equal to 0.90 of being correct. In other words, if we continue to make similar statements on the basis of many other samples from the same population, using the same estimation procedure, about 90% of these statements will be true.

The concept of confidence intervals is one of the main contributions to statistical theory by J. Neyman and E. S. Pearson [1]. A somewhat different concept, leading in many cases (although not in all) to identical results, is that of *fiducial intervals*, due to Sir Ronald A. Fisher [2]. In this view it is permissible to attach a fiducial probability $f(\theta)$ to the parameter θ, although this is not to be interpreted in the ordinary (frequency) sense of probability. The idea is that $\int_{\theta_1}^{\theta_2} f(\theta)\, d\theta$ is a measure of our belief that θ lies between θ_1 and θ_2 (the Latin word "fiducia" means trust). The fiducial probability, like the confidence interval, is calculated from the known sampling distribution of the statistic used to estimate θ (see § 5.4).

5.3 **Confidence Belts** For simplicity we consider a population defined by a single parameter θ and we suppose that a statistic T, derived from a sample of size N, is used to estimate θ. The statistic T is often called an *estimator* of θ. The distribution of T, for given θ, is supposedly known. That is, for any admissible value of θ (say in the range from α to β) we can calculate the probability that T will lie between two given values, t_1 and t_2.

In the diagram, Figure 26, t is plotted as abscissa and θ as ordinate. The possible values of T for a sample drawn from a population with a given value of θ, say θ', lie along the line AB. On this line we can mark two points, at t_1 and t_2, such that the probability that $T < t_1$ is a fixed value ε, say 0.05, and the probability that $T > t_2$ is also ε.

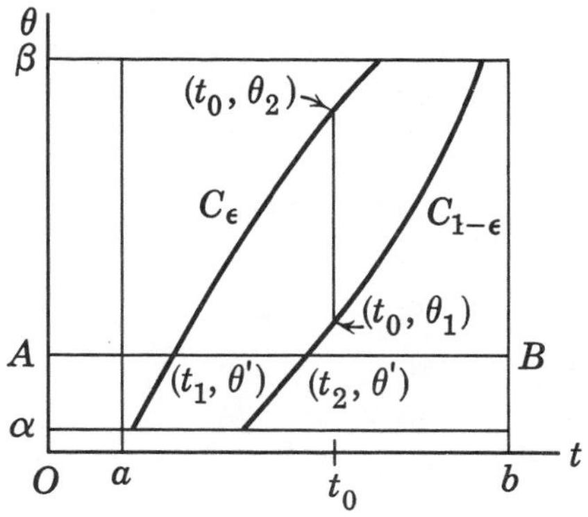

FIG. 26 CONFIDENCE BELT

If $F(t|\theta)$ is the distribution function for T, with given θ, these probability statements may be written

$$F(t_1|\theta') = \varepsilon, \qquad F(t_2|\theta') = 1 - \varepsilon \tag{5.3.1}$$

where it is assumed that $0 < \varepsilon < \frac{1}{2}$.

If we now imagine that there are a great many hypothetical populations with values of θ between α and β, and that for each one the appropriate values of t_1 and t_2 are calculated, the points so obtained will lie on curves something like those marked C_ε and $C_{1-\varepsilon}$ in the diagram. Since t is supposed to be an estimate of θ, it is reasonable to assume that both curves represent one-valued monotone-increasing functions. (If t is any sort of a reasonable estimate, it should increase as θ increases.)

The region bounded by the two curves and by the lines $\theta = \alpha$ and $\theta = \beta$ is called a *confidence belt*, with confidence coefficient $1 - 2\varepsilon$. This belt can theoretically be constructed from a knowledge of the function $F(t|\theta)$ alone.

Now suppose that for one particular random sample (of size N) we obtain a value t_0 of T. The value of θ for the population is unknown, except for the fact that it must lie between α and β. If at t_0 we draw an ordinate cutting the curves C_ε and $C_{1-\varepsilon}$ at $\theta = \theta_2$ and $\theta = \theta_1$ respectively, then all points on this ordinate between θ_1 and θ_2 lie inside the confidence belt. We see that θ_1 is the lower bound of values of θ such that $F(t_0, \theta) < 1 - \varepsilon$, and θ_2 is the upper bound of values of θ such that $F(t_0, \theta) > \varepsilon$. We can therefore assert, on the basis of our sample value t_0, that θ lies between θ_1 and θ_2, and the probability that this claim is true is $1 - 2\varepsilon$. The values θ_1 and θ_2 are the lower and upper *confidence limits* for θ, corresponding to the observed t_0, and $1 - 2\varepsilon$ is the *confidence coefficient*. The smaller the value of ε the more confidence we shall feel in the rightness of our claim, but of course the smaller we make ε the wider will our belt become, and therefore the greater will be the interval $\theta_2 - \theta_1$. We can increase our confidence in a statement only by making the statement vaguer.

In the above illustration it was assumed, for convenience, that the variable T concerned was continuous. If the variable is discrete, the curves C_ε and $C_{1-\varepsilon}$ will be stepped, as in Figure 27, which relates to the Poisson distribution of X.

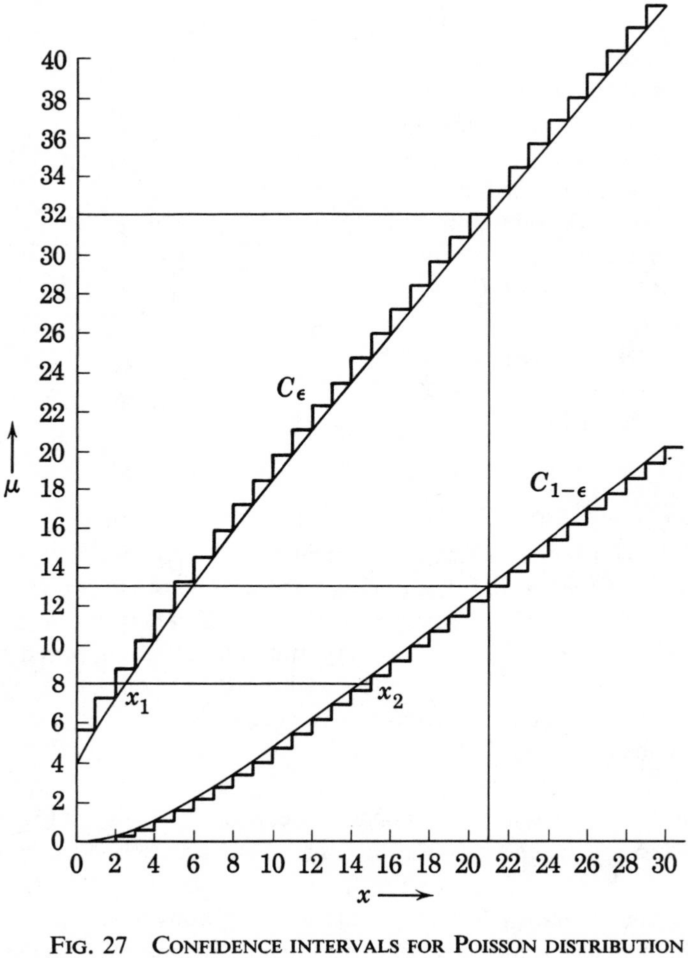

FIG. 27 CONFIDENCE INTERVALS FOR POISSON DISTRIBUTION

For any given μ, there will in general be a value of x_2 such that $P(x_2, \mu) < \varepsilon$ and $P(x_2 - 1, \mu) > \varepsilon$, where $P(x_2, \mu)$ is the cumulative Poisson function

$$P(x_2, \mu) = \sum_{x=x_2}^{\infty} e^{-\mu} \frac{\mu^x}{x!}$$

It will happen for some values of μ that there is an x_2 such that $P(x_2, \mu) = \varepsilon$ exactly. As μ increases through such a value, x_2 jumps by a unit. The horizontal portions of the stepped curve represent these values of μ. Similar considerations apply to the curve of x_1, which is such that $P(x_1, \mu) > 1 - \varepsilon$ and $P(x_1 + 1, \mu) < 1 - \varepsilon$.

The diagram (for $\varepsilon = 0.025$) shows that if a single sample value of 21 is observed for a Poisson variate X, the population mean μ may be taken, with 95% confidence, to lie between 13 and 32.

* 5.4 **Fiducial Inference** If $F(t|\theta)$ is the distribution function of T, and if t_k is a value such that

(5.4.1.) $$P(T \leq t_k) = F(t_k|\theta) = k$$

then in a fraction $1 - k$ of all samples drawn from a population with parameter θ the statistic T will exceed the critical value t_k. This value t_k is a function of θ, say $K(\theta)$, and θ is the inverse function of t_k, say $K^{-1}(t_k)$. Equation (1) may therefore be written in either form—

(5.4.2) $$P\{T \leq K(\theta)\} = k$$

or

(5.4.3) $$P\{\theta \geq K^{-1}(t_k)\} = k$$

provided $K(\theta)$ is a strictly monotone-increasing function of θ.

The form of Eq. (3) is the one preferred by Fisher and expresses what he calls a *fiducial probability* for θ. This does not depend on any assumption about the distribution of θ prior to the examination of any samples.

If we suppose that the statistic T has a continuous distribution, then, as we have seen in § 4.2, the transformed variate

(5.4.4) $$Y = F(T|\theta)$$

has a uniform (rectangular) distribution on the interval 0 to 1. This means that for any fixed number k, between 0 and 1,

(5.4.5) $$P\{Y \leq k\} = k$$

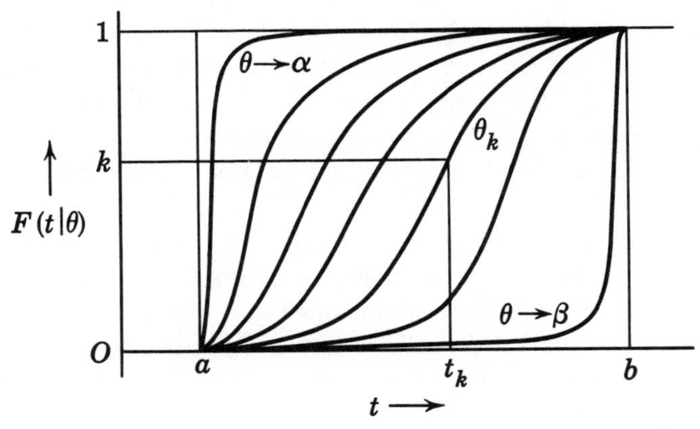

FIG. 28 FIDUCIAL INFERENCE

Let us assume that the possible values of T form an interval (a, b) and that the possible values of θ form an interval (α, β).

For any given θ we can plot $F(t|\theta)$ against t, as in Figure 28; $F(t|\theta)$ is the probability that $T \leq t$. In most cases we shall find that for a fixed t, say t_k, the values of F decrease as θ increases. The nearer θ is to α, the nearer $F(t_k|\theta)$ will be to 1, and the nearer θ is to β the nearer $F(t_k|\theta)$ to 0. If so, the equation $F(t_k|\theta) = k$ determines uniquely a value θ_k such that $\theta > \theta_k$ when $F(t_k|\theta) < k$.

Equation (5) may therefore be written

(5.4.6) $$P(\theta > \theta_k) = k = F(t_k|\theta_k)$$

or

(5.4.7) $$P(\theta \leq \theta_k) = 1 - k = 1 - F(t_k|\theta_k)$$

The quantity $1 - F(t_k|\theta_k)$ is the fiducial distribution function for θ. Actually θ_k in Eq. (6) is a random variable, determined by the relation $F(t_k|\theta_k) = k$ for a given k and for the observed value t_k of the random variable T. The probability statement really concerns this random variable θ_k. By twisting the inequality from the form in Eq. (2) to the form in Eq. (3) we can make a probability statement apparently about θ, but this does not convert θ into a random variable (see further in [3]).

5.5 **Confidence and Significance** The determination of confidence intervals is closely related to the estimation of significance. A problem that sometimes arises in statistics is that of judging whether a population parameter differs appreciably from some value which has been fixed beforehand, perhaps from some purely theoretical considerations. Suppose the theoretical value is θ_0 and the point estimate from a sample is $\hat{\theta}$. We need to assess the significance of the difference $\hat{\theta} - \theta_0$. If this difference is greater (numerically) than a certain amount we shall say that the difference is significant, if less, that it is non-significant. Obviously, it is impossible to draw a hard-and-fast line between significance and non-significance—there will be border-line cases which are difficult to classify—but statisticians in general accept the following convention: if the probability of obtaining by chance a sample with a difference numerically as great as $\hat{\theta} - \theta_0$ is less than 0.05, the observed difference is *significant*; if the probability is less than 0.01, the difference is *highly significant*; if the probability is greater than 0·05 the difference is *non-significant*. In border-line cases the statistician will usually prefer to suspend judgment and perhaps try to get a larger sample.

If, having obtained the sample estimate $\hat{\theta}$, we calculate the corresponding 95% confidence interval for θ, stretching say from θ_1 to θ_2, there will be a 5% probability that this interval will *not* include θ_0. In other words, if θ_0 lies outside the 95% confidence interval the difference $\hat{\theta} - \theta_0$ will be regarded as significant. Similarly if θ_0 lies outside the 99% confidence interval the difference will be considered highly significant.

Sometimes the statistician is faced with the question of significance for the difference of the estimates given by two separate samples. In such a case he will choose zero as the hypothetical value for this difference, and test whether the observed difference is significantly different from zero. A method of doing this is to construct the confidence interval for the observed difference and see whether this interval includes the value zero.

Some more general considerations on the testing of hypotheses will be discussed in Chapter 6.

5.6 **Desirable Properties of an Estimator** Let us suppose that we are using the statistic T, derived from a random sample of N observations $x_1, x_2 \ldots x_N$, in order to estimate the parameter θ which occurs in the distribution function of the population. The estimator T is said to be *consistent* if, as N increases indefinitely, T tends (stochastically) to the value θ. That is, for any given $\varepsilon > 0$,

(5.6.1) $$P(|T - \theta| > \varepsilon) \to 0 \text{ as } N \to \infty$$

This is an obvious common-sense requirement. We should expect a very large sample to give us practically the population value of the quantity we are trying to estimate.

A simple test for determining consistency is provided by Chebyshev's inequality, § 2.16. If T is such that $E(T) \to \theta$ and $V(T) \to 0$ as $N \to \infty$, then it follows from Eq. (2.16.1) that T is a consistent estimator of θ.

The estimator T is said to be *unbiased* if (even for finite N), $E(T) = \theta$, whatever other parameters may occur in the distribution function. If $E(T)$ merely tends to θ as $N \to \infty$, T is *asymptotically unbiased.* It is generally desirable to use an unbiased estimator where possible, but sometimes other considerations are more important.

The reliability of the estimate furnished by an estimator is measured by the reciprocal of its sampling variance. The smaller this variance the more reliable the estimate will be. The *efficiency* of the estimator T is given by comparing the variance of T with that of the estimator T_0 which, of all possible consistent statistics which might be used to estimate θ, is the one with minimum variance. That is, the efficiency of $T = V(T_0)/V(T)$. A statistic with an efficiency of 1 (usually expressed as 100%) is said to be *most efficient.*

We shall now consider some estimators which are used to estimate the moments, cumulants and other parameters of a population. It will be convenient to start with a finite population.

5.7 **Sampling from a Finite Population** Many of the results of sampling theory can be obtained by supposing that a random sample of size N is drawn from a population of size M. This enables us to use the theory of combinations. Results for an infinite population can usually be obtained by letting $M \to \infty$.

If X is the variate measured, the arithmetic mean of X for a sample of size N is

(5.7.1) $$m = N^{-1} \sum_{j=1}^{N} X_j$$

where X_j is the jth item in the sample.

The corresponding quantity for the population is

$$\mu = M^{-1} \sum_{\alpha=1}^{M} X_\alpha \tag{5.7.2}$$

where X_α is the αth item in the population. Some of the X_α will of course be the same as the X_j. We shall use X_α, however, to mean any item from the population and X_j to mean one that is also in the sample.

As an estimator of μ, m is clearly consistent, since when N becomes equal to M (it cannot get any larger than M), m becomes equal to μ. With a finite population, the expectation of a statistic such as m is defined as its average over all possible different samples of size N that could be drawn from the population. The number of these samples is $\binom{M}{N}$. We shall now show that this average for m is equal to μ, and therefore m is an unbiased estimator of μ.

TABLE 5.1

Sample No.	Sample Items	Mean (m)
1	2,5	3.5
2	2,5	3.5
3	2,7	4.5
4	2,10	6.0
5	2,21	11.5
6	5,5	5.0
7	5,7	6.0
8	5,7	6.0
9	5,10	7.5
10	5,10	7.5
11	5,21	13.0
12	5,21	13.0
13	7,10	8.5
14	7,21	14.0
15	10,21	15.5
		125.0

The number of samples in which any particular X_α occurs is equal to $\binom{M-1}{N-1}$, since the remaining $N - 1$ items in the sample can be picked from any of the other $M - 1$ items in the population. This X_α contributes X_α/N to the value of m for each sample in which it occurs, and therefore its contribution to the average m over all samples is $(X_\alpha/N)\binom{M-1}{N-1}\Big/\binom{M}{N} = X_\alpha/M$. Summing over all $_\alpha$, we obtain for the average m the amount $\sum_\alpha X_\alpha/M$, which is μ. Therefore,

$$E(m) = \mu \tag{5.7.3}$$

As an illustration involving small numbers, suppose $M = 6$ and $N = 2$. Then $\binom{M}{N} = 15$. Let the values of X_α in the population be 2, 5, 5, 7, 10 and 21. The population mean is $\mu = 50/6 = 8.33$. The 15 possible samples of size 2 and their separate values of m are given in Table 5.1. The average m over all 15 samples is $125/15 = 8.33$, which is the same as μ. (In this illustration two of the population items have the same value of X, but they count as different items in enumerating the samples.)

A precisely similar proof may be carried through for the pth moment of X about the origin, denoted by m'_p for the sample and by μ'_p for the population.

A more convenient notation for m'_p, suggested by Tukey [4], is the angle bracket $\langle p \rangle$. With this notation,

$$\langle p \rangle = N^{-1} \sum_j X_j^p \tag{5.7.4}$$

$$E(\langle p \rangle) = \mu'_p = M^{-1} \sum_\alpha X_\alpha{}^p \tag{5.7.5}$$

Let us now consider a *pair* of items X_α, X_β from the population. The subscripts distinguish them as different items, but their actual values may happen to be the same (like the two 5's in the illustration above). Each such pair appears in $\binom{M-2}{N-2}$ different samples. We define the angle bracket $\langle pq \rangle$ and the corresponding population parameter μ'_{pq} by

$$\langle pq \rangle = [N(N-1)]^{-1} \sum_{i,j}{}' X_i{}^p X_j{}^q \tag{5.7.6}$$

$$\mu'_{pq} = [M(M-1)]^{-1} \sum_{\alpha,\beta}{}' X_\alpha{}^p X_\beta{}^q \tag{5.7.7}$$

where the sum in Eq. (6) is over the $N(N-1)$ pairs X_i, X_j in a single sample of size N. The sign $\sum'$ here indicates that the sum is to be taken over all *different* values of the subscripts.

By considering the contribution of each pair of items X_α, X_β to the average of $\langle pq \rangle$, we can readily obtain the result

$$E(\langle pq \rangle) = \mu'_{pq} \tag{5.7.8}$$

The angle brackets such as $\langle pq \rangle$ are therefore unbiased estimators of the corresponding population parameters. This is true also of brackets with three, four, or more symbols.

* 5.8 **Fisher's k-Statistics** Unfortunately, the sample moments about the mean (of second or higher orders) are not unbiased estimators of the corresponding population parameters. However we can define a set of statistics, called k-statistics, each of which does have this relationship to the corresponding population parameter. When the population is infinite, these parameters become identical with the *cumulants* discussed in § 2.12.

In order to calculate the k-statistics for a sample systematically, it is convenient to start with sums of powers of the X_j. Let

$$\text{(5.8.1)} \quad \begin{cases} S_1 = \sum X_i \;\; = N\langle 1\rangle \\ S_2 = \sum X_i^2 = N\langle 2\rangle \\ S_3 = \sum X_i^3 = N\langle 3\rangle \end{cases}$$

etc.

Then, as shown in Appendix A.11,

$$\text{(5.8.2)} \quad \begin{cases} N(N-1)\langle 11\rangle = S_1^2 - S_2 \\ N(N-1)\langle 12\rangle = S_1 S_2 - S_3 \\ N(N-1)\langle 13\rangle = S_1 S_3 - S_4 \\ N(N-1)\langle 22\rangle = S_2^2 - S_4 \end{cases}$$

etc.

Also,

$$\text{(5.8.3)} \quad \begin{cases} (N-2)\langle 111\rangle \;\; = S_1\langle 11\rangle - 2\langle 12\rangle \\ (N-2)\langle 112\rangle \;\; = S_1\langle 12\rangle - \langle 22\rangle - \langle 13\rangle \\ (N-3)\langle 1111\rangle = S_1\langle 111\rangle - 3\langle 112\rangle \end{cases}$$

etc.

The k-statistics may then be defined in terms of these brackets:

$$\text{(5.8.4)} \quad \begin{cases} k_1 = \langle 1\rangle \\ k_2 = \langle 2\rangle - \langle 11\rangle \\ k_3 = \langle 3\rangle - 3\langle 12\rangle + 2\langle 111\rangle \\ k_4 = \langle 4\rangle - 4\langle 13\rangle - 3\langle 22\rangle + 12\langle 112\rangle - 6\langle 1111\rangle \end{cases}$$

Generalized k-statistics [4] may similarly be defined, such as

$$\text{(5.8.5)} \quad \begin{cases} k_{11} = \langle 11\rangle \\ k_{12} = \langle 12\rangle - \langle 111\rangle \\ k_{22} = \langle 22\rangle - 2\langle 112\rangle + \langle 1111\rangle \end{cases}$$

etc.

These serve as checks on the calculation of the k-statistics, since the following relations hold:

$$(5.8.6)\qquad \begin{cases} k_{11} = k_1^{\,2} - \dfrac{k_2}{N} \\[2ex] k_{12} = k_1 k_2 - \dfrac{k_3}{N} \\[2ex] k_{22} = \dfrac{N-1}{N+1}\left(k_2^{\,2} - \dfrac{k_4}{N}\right) \end{cases}$$

Each k is an unbiased estimator of the corresponding quantity for the population, which will be denoted by κ'. The κ''s are defined like the k's, except that $\langle p\rangle$ is replaced by μ'_p, $\langle pq\rangle$ by μ'_{pq}, etc. For an infinitely large population they become identical with the cumulants as previously defined.

The k-statistics are expressible in terms of the sample moments about the mean, discussed in § 2.8. The relations are:

$$(5.8.7)\qquad \begin{cases} k_2 = \dfrac{N}{N-1}\, m_2 \\[2ex] k_3 = \dfrac{N^2}{(N-1)(N-2)}\, m_3 \\[2ex] k_4 = \dfrac{N^2}{(N-1)(N-2)(N-3)}\left[(N+1)m_4 - 3(N-1)m_2^{\,2}\right] \end{cases}$$

It is not, however, necessary to find the moments first. The systematic procedure of Eqs. (1) to (4) will give the k-statistics directly.

The κ'_r are similarly expressible in terms of the population moments (with M substituted for N). Thus

$$(5.8.8)\qquad \begin{cases} \kappa'_2 = \dfrac{M}{M-1}\, \mu_2 \\[2ex] \kappa'_3 = \dfrac{M^2}{(M-1)(M-2)}\, \mu_3 \\[2ex] \kappa'_4 = \dfrac{M^2}{(M-1)(M-2)(M-3)}\left[(M+1)\mu_4 - 3(M-1)\mu_2^{\,2}\right] \end{cases}$$

When $M \to \infty$, $\kappa_2 = \mu_2$, $\kappa_3 = \mu_3$, $\kappa_4 = \mu_4 - 3\mu_2^{\,2}$, in agreement with the definitions given in Eq. (2.12.5).

* 5.9 **Computation of the k-Statistics** As an illustration of the arithmetic involved, we will consider the data of Table 2.2 already used to calculate some of the moments in Chapter 2. If we use an auxiliary variable $u = (x_c - 45.5)/4$,

where x_c is a class-mark, then u takes only integral values from -4 to 5 and the calculations of sums of powers are greatly shortened. The whole work can be carried through in terms of u, and at the end we can convert back to the original x values.

For this sample we first find

$$N = 1000, \qquad S_1 = \sum fu = 553$$
$$S_2 = \sum fu^2 = 2471$$
$$S_3 = \sum fu^3 = 4105$$
$$S_4 = \sum fu^4 = 18{,}407$$

Then

$$\langle 1 \rangle = 0.553$$
$$\langle 2 \rangle = 2.471$$
$$\langle 3 \rangle = 4.105$$
$$\langle 4 \rangle = 18.407$$
$$\langle 11 \rangle = \frac{(553)^2 - 2471}{999{,}000} = 0.3036$$
$$\langle 12 \rangle = \frac{(553)(2471) - 4105}{999{,}000} = 1.3637$$
$$\langle 13 \rangle = \frac{(553)(4105) - 18{,}407}{999{,}000} = 2.2539$$
$$\langle 22 \rangle = \frac{(2471)^2 - 18{,}407}{999{,}000} = 6.0935$$
$$\langle 111 \rangle = \frac{(553)(0.3036) - 2.7274}{998} = 0.1655$$
$$\langle 112 \rangle = \frac{(553)(1.3637) - 6.0935 - 2.2539}{998} = 0.7473$$
$$\langle 1111 \rangle = \frac{(553)(0.1655) - 2.2419}{997} = 0.0896$$

$$(5.9.1) \qquad \begin{cases} k_1 = & 0.553 \\ k_2 = & 2.471 - 0.3036 = 2.1674 \\ k_3 = & 4.105 - 4.0911 + 0.3310 \\ = & 0.345 \\ k_4 = & 18.407 - 9.016 - 18.280 + 8.968 - 0.538 \\ = & -0.459 \end{cases}$$

Also,

$$\begin{cases} k_{11} = 0.3036 \\ k_{12} = 1.3637 - 0.1655 = 1.1982 \\ k_{22} = 6.0935 - 1.4946 + 0.0896 = 4.6885 \end{cases}$$

and the checks of Eq. (5.8.6) hold, apart perhaps from a small rounding-off error in the last decimal place.

Finally we can convert the k's back to the original units (pounds) by writing

$$\text{(5.9.2)} \qquad \begin{cases} k_1 = 4(0.553) + 45.5 = 47.71 \text{ lb} \\ k_2 = 4^2(2.1674) \qquad = 34.68 \text{ lb}^2 \\ k_3 = 4^3(0.345) \qquad = 22.1 \text{ lb}^3 \\ k_4 = -4^4(0.459) \qquad = -118 \text{ lb}^4 \end{cases}$$

Using these as estimators of the cumulants for the population from which the sample was taken, we have the following estimated values of the population parameters:

$$\text{(5.9.3)} \qquad \begin{cases} \kappa_1 = \mu = 47.71 \text{ lb} \\ \kappa_2 = \sigma^2 = 34.68 \text{ lb}^2 \ (\sigma = 5.889 \text{ lb}) \\ \dfrac{\kappa_3}{\kappa_2^{3/2}} = \gamma_1 = 0.108 \\ \dfrac{\kappa_4}{\kappa_2^{2}} = \gamma_2 = -0.098 \end{cases}$$

* 5.10 **Sheppard's Corrections** The error due to grouping the frequencies at the mid-points of the class-intervals, in the computation of the k-statistics, may be approximately allowed for by using some corrections first suggested by Sheppard. These corrections are applied to the even-order k-statistics only, and are given by the relation

$$\text{(5.10.1)} \qquad (k_r)_c = k_r - c^r \frac{B_r}{r}, \qquad r \geq 2$$

where $(k_r)_c$ is the corrected value of k_r, c is the class-interval, and B_r is the rth Bernoulli number (see Appendix A.12). For the first two even-order k-statistics these corrections are

$$\text{(5.10.2)} \qquad \begin{cases} (k_2)_c = k_2 - \dfrac{c^2}{12} \\ (k_4)_c = k_4 + \dfrac{c^4}{120} \end{cases}$$

In practice the corrections are most easily applied to the k_r, as first obtained in the u units (for which $c = 1$). Thus from Eq. (5.9.1) the corrected values are

$$(k_2)_c = 2.1674 - 0.0833 = 2.0841$$
$$(k_4)_c = -0.459 + 0.008 = -0.451$$

Using these, our estimated κ_2 and κ_4 become

$$\kappa_2 = 33.34 \text{ lb}^2, \qquad \kappa_4 = -115 \text{ lb}^4$$

and instead of the values given in Eq. (5.9.3) we find the following estimates:

(5.10.3)
$$\begin{cases} \sigma = 5.774 \text{ lb} \\ \gamma_1 = 0.114 \\ \gamma_2 = -0.104 \end{cases}$$

Sheppard's corrections should not be used unless the frequency curve appears to have a single mode and tails off gradually at both ends. Moreover, unless the sample consists of at least several hundred items, the uncertainties due to sampling fluctuation are likely to overshadow the corrections. When the corrections are applicable, however, their use will generally (although not invariably) improve the estimates of the population parameters, and they are so easy to apply that it is usually worth while to take the slight additional trouble involved.

5.11 **Variance and Covariance of the k-Statistics** As before, when dealing with a finite population, we interpret the expectation of a statistic as its average taken over all possible different samples of the same size. The variance of k_1 will then be defined as

(5.11.1) $$V(k_1) = E(k_1{}^2) - \{E(k_1)\}^2$$

By the results obtained in § 5.8 we know that

(5.11.2) $$E(k_1) = \kappa'_1 = \mu$$

Also, from the first equation of (5.8.6),

(5.11.3) $$k_1{}^2 = k_{11} + \frac{k_2}{N}$$

so that

(5.11.4) $$\begin{aligned} E(k_1{}^2) &= E(k_{11}) + N^{-1}E(k_2) \\ &= \kappa'_{11} + N^{-1}\kappa'_2 \end{aligned}$$

It follows that

(5.11.5) $$V(k_1) = \kappa'_{11} + N^{-1}\kappa'_2 - (\kappa'_1)^2$$

The corresponding equation to (3) for the population is

(5.11.6) $$(\kappa'_1)^2 = \kappa'_{11} + \kappa'_2/M$$

so that Eq. (5) becomes

(5.11.7) $$V(k_1) = \kappa'_2(N^{-1} - M^{-1})$$

For an infinitely great population, $M^{-1} \to 0$, and we have the simple result

(5.11.8) $$V(k_1) = \kappa_2 N^{-1} = \frac{\sigma^2}{N}$$

This measures the sampling fluctuation in the value of the arithmetic mean k_1. In practice we generally do not know σ^2 except insofar as we can estimate it from the one sample which gives us k_1. If we replace σ^2 by the corresponding unbiased estimator k_2, we have as an estimator of the variance of k_1 the statistic

(5.11.9) $$\hat{V}(k_1) = \frac{k_2}{N}$$

The square root of $\hat{V}(k_1)$ is called the *standard error* of k_1. In terms of the sums of powers of X, defined in § 5.8,

(5.11.10) $$\hat{V}(k_1) = \frac{S_2}{N} - \frac{S_1{}^2 - S_2}{N(N-1)} = \frac{S_2 - S_1{}^2/N}{N-1}$$

The variance of k_2 is similarly given by

(5.11.11) $$V(k_2) = E(k_2{}^2) - \{E(k_2)\}^2$$

From the last equation of (5.8.6),

$$E(k_2{}^2) = \frac{N+1}{N-1} E(k_{22}) + \frac{1}{N} E(k_4) = \frac{N+1}{N-1} \kappa'_{22} + \frac{1}{N} \kappa'_4$$

and also,

$$E(k_2) = \kappa'_2$$

Since

$$\kappa'_{22} = \frac{M-1}{M+1}\left[(\kappa'_2)^2 - \frac{\kappa'_4}{M}\right]$$

we find, after a little rearrangement, that

(5.11.12) $$V(k_2) = \frac{M-N}{(M+1)(N-1)}\left[2(\kappa'_2)^2 + \kappa'_4\left(1 - \frac{1}{M} - \frac{1}{N} - \frac{1}{MN}\right)\right]$$

which for an infinitely great population reduces to

$$V(k_2) = \frac{2}{N-1}\kappa_2^2 + \frac{1}{N}\kappa_4 \tag{5.11.13}$$

If the parent population is *normal,* $\kappa_4 = 0$ and $\kappa_2 = \sigma^2$. In this case,

$$V(k_2) = \frac{2\sigma^4}{N-1} \tag{5.11.14}$$

In order to find the standard error of k_2, the unknown population parameters must be replaced by sample estimators. It is easily verified that, for $M \to \infty$,

$$E\left\{\frac{2}{N+1}k_2^2 + \frac{N-1}{N(N+1)}k_4\right\} = \frac{2}{N-1}\kappa_2^2 + \frac{1}{N}\kappa_4 \tag{5.11.15}$$

so that we can take the expression in braces on the left-hand side as an unbiased estimator of $V(k_2)$. Therefore,

$$\hat{V}(k_2) = \frac{2}{N+1}k_2^2 + \frac{N-1}{N(N+1)}k_4 \tag{5.11.16}$$

and the square root of this is the standard error of k_2.

The *covariance* of k_1 and k_2 may be defined as

$$C(k_1, k_2) = E(k_1k_2) - E(k_1)\cdot E(k_2) \tag{5.11.17}$$

By the second equation of (5.8.6)

$$E(k_1k_2) = E(k_{12}) + \frac{E(k_3)}{N}$$

so that

$$\begin{aligned} C(k_1, k_2) &= \kappa'_{12} + \frac{\kappa'_3}{N} - \kappa'_1\kappa'_2 \\ &= \left(\kappa'_1\kappa'_2 - \frac{\kappa'_3}{M}\right) + \frac{\kappa'_3}{N} - \kappa'_1\kappa'_2 \\ &= \kappa'_3(N^{-1} - M^{-1}) \end{aligned} \tag{5.11.18}$$

For an infinitely great population,

$$C(k_1, k_2) = \kappa_3 N^{-1} \tag{5.11.19}$$

which is zero for a *normal* population.

The first two k-statistics are therefore uncorrelated in samples from a normal population, although this is not true for skew populations. Since k_3 is an unbiased estimator of κ_3,

$$\hat{C}(k_1, k_2) = k_3 N^{-1} \tag{5.11.20}$$

5.12 The Distribution of the Sample Mean The arithmetic mean of a sample of N observations, $X_1, X_2 \ldots X_N$, is the first k-statistic k_1. As mentioned previously, the expectation of k_1 in samples from a finite population is the population mean μ (which is κ'_1). That is,

$$E(k_1) = \kappa'_1 = \mu \tag{5.12.1}$$

Also the variance of k_1, from Eqs. (5.11.7) and (5.8.8), is given by

$$V(k_1) = \kappa'_2 \frac{M - N}{MN} = \frac{M - N}{(M - 1)N} \sigma^2 \tag{5.12.2}$$

which for an infinite population becomes

$$V(k_1) = \sigma^2/N \tag{5.12.3}$$

Similar arguments to those used in § 5.11 (based on relations between angle brackets and k-statistics) can be used to obtain the higher moments of the distribution of k_1, but the calculations soon become quite complicated. It turns out that the skewness is given by

$$Sk(k_1) = \frac{M - 2N}{M - 2} \left[\frac{M - 1}{N(M - N)} \right]^{1/2} \gamma_1 \tag{5.12.4}$$

and the kurtosis by

$$\begin{aligned} Ku(k_1) = [&(M - 1)(M^2 - 6MN + M + 6N^2)\gamma_2 \\ &- 6M(MN + M - N^2 - 1)] \\ &\div [N(M - 2)(M - 3)(M - N)] \end{aligned} \tag{5.12.5}$$

where $\gamma_1 = \kappa_3/\kappa_2^{3/2}$ and $\gamma_2 = \kappa_4/\kappa_2^2$.

For an infinite population these reduce to

$$Sk(k_1) = \frac{\gamma_1}{N^{1/2}} \tag{5.12.6}$$

$$Ku(k_1) = \frac{\gamma_2}{N} \tag{5.12.7}$$

It is evident from Eqs. (6) and (7) that for large enough samples the skewness and kurtosis of the distribution of k_1 will be nearly zero, whatever the corresponding quantities for the population (as long as they are finite). This suggests that the mean of a large sample from almost any kind of population will have a distribution close to normal, and in fact, if certain conditions are satisfied, this result follows from the Central Limit Theorem (see § 4.10).

If the parent population is normal, the mean of a sample of size N is also normally distributed, with variance σ^2/N, whether N is large or small. (For a proof of this, see § 8.2.) If the parent population is not normal but has a finite variance σ^2, the variance of the sample mean is still σ^2/N and for large N the distribution is *approximately* normal. For a parent population not too wildly skew, a sample size of 30 or more will usually give a satisfactory approximation to normality.

As an illustration with even smaller sample size, a decidedly skew population was constructed by writing a number from 0 to 24 on each of 1000 circular metal-edged cardboard tags. The frequency diagram for this population is shown in Figure 29. (There were 106 tags, for example, marked 4.) The numbered discs were put into a goldfish bowl and well mixed. A sample of 10 discs

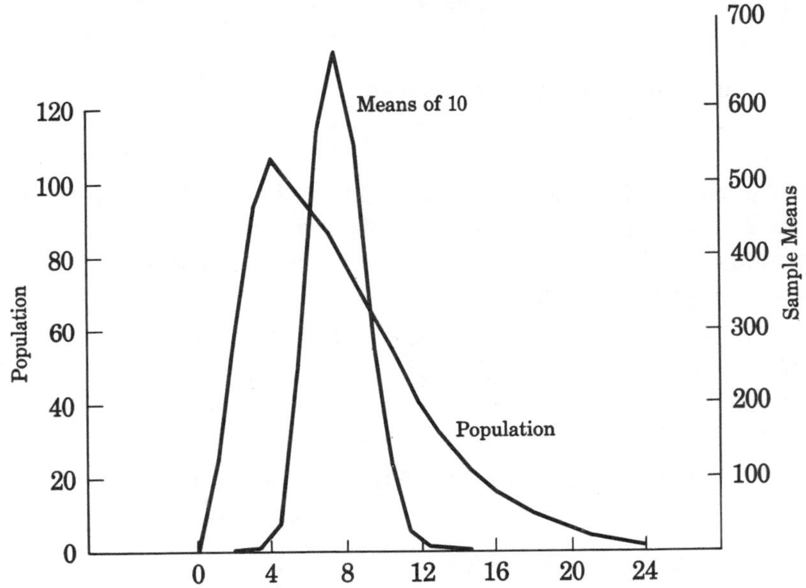

FIG. 29 FREQUENCY POLYGONS FOR A SKEW POPULATION AND FOR THE MEANS OF SAMPLES OF 10

was drawn and the numbers were noted before the discs were replaced. This was done repeatedly, and over a considerable period of time 2500 sample means were obtained. These were grouped in classes 3.0 to 3.9, 4.0 to 4.9, etc. and the first few k-statistics were calculated. The frequency polygon of the distribution of these sample means is shown in Figure 29 along with that for the parent population (the two polygons have different vertical scales, one shown on the right of the diagram and one on the left). The much more symmetrical nature of the distribution of means is obvious at a glance. Table 5.2 gives for comparison (a) the actual characteristics (population parameters) for the parent population of 1000 discs, (b) the theoretical characteristics for the distribution of mean in all possible samples of 10, (c) the estimated values for these characteristics derived from the k-statistics of 2500 actual samples, (d) the approximate standard errors for these estimates. For the skewness and kurtosis the standard errors relate to a *normal* parent population (see Chapter 8) and are not very reliable.

Characteristic	Population	Population of Means (Theoretical)	Population of Means (Estimated)
Mean	7.601	7.601	7.640 ± 0.028
Variance	19.57	1.939	2.006 ± 0.058
Skewness	0.896	0.279	0.381 ± 0.049
Kurtosis	0.508	0.042	0.095 ± 0.098

It will be seen that in all cases the estimated values agree with the theoretical values within about once or twice the standard error. The difference for the skewness is slightly more than twice its standard error.

5.13 **Confidence Interval for the Mean (in Large Samples)** If we apply the procedure of § 5.3 to the statistic m (the sample mean) as used to estimate the parameter μ (the population mean), we obtain a confidence belt which for fairly large samples is of almost uniform width. For a given value of μ, the expected value of m will be μ and its variance will be σ^2/N. If the sample size is large enough for the distribution of the mean to be regarded as normal, or if the parent population is *known* to be normal, the sample mean for given μ will, with probability 0.95, lie between $\mu - 1.96\sigma N^{-1/2}$ and $\mu + 1.96\sigma N^{-1/2}$. It follows

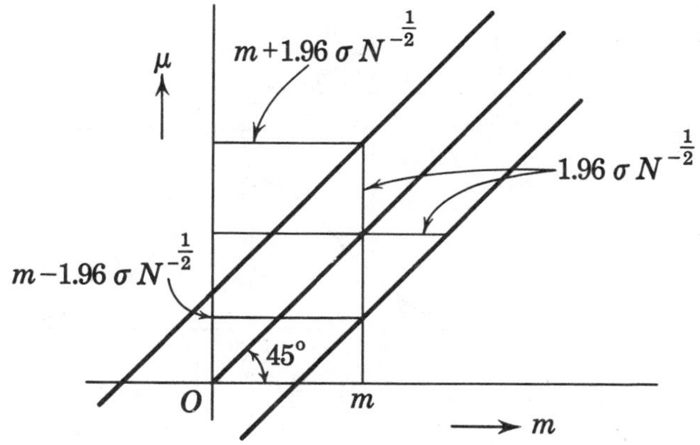

FIG. 30 CONFIDENCE BELT FOR THE SAMPLE MEAN WITH KNOWN POPULATION VARIANCE

that for a given m, the 95% confidence interval for μ lies between $m - 1.96\sigma N^{-1/2}$ and $m + 1.96\sigma N^{-1/2}$, if σ is known (Figure 30). If σ is not known, it may be replaced by an estimate such as the sample standard deviation. There is, however, a better procedure available when σ has to be estimated from a fairly small sample and when the parent population can be taken as normal. This procedure will be described in § 8.5.

EXAMPLE 1 For a sample of 345 11-year-old boys, the mean weight was found to be 74.71 lb and the standard deviation 10.65 lb. Calculate 98%

confidence limits for the mean weight in the population of 11-year-old boys from which this sample was taken.

Here $m = 74.71$ lb, and $s = 10.65$ lb. Using s to estimate σ and noting that for the standard normal law the 98% limits are at ± 2.326, we find for μ the confidence limits $74.71 \pm 2.326(10.65)/(345)^{1/2} = 74.71 \pm 1.33$ lb, or 73.38 to 76.04 lb.

EXAMPLE 2 The variable X is the lifetime (in days) of test pieces of metal sheet immersed in tap water, before failure due to corrosion. From a large number of trials the mean value (μ) of X was found to be 875, with a standard deviation of 85. For further routine testing, how large should the samples be if the average life (m) from such a sample is to differ from μ by not more than 5%, with probability 0.90?

Since 5% of μ is 43.75, the requirement is that $P(|m - \mu| \leq 43.75) = 0.90$. Assuming a normal distribution, the probability 0.90 corresponds to a standardized variate of 1.645. Therefore, $(43.75)/(\sigma N^{-1/2}) = 1.645$, with $\sigma = 85$. This gives $N = 10$.

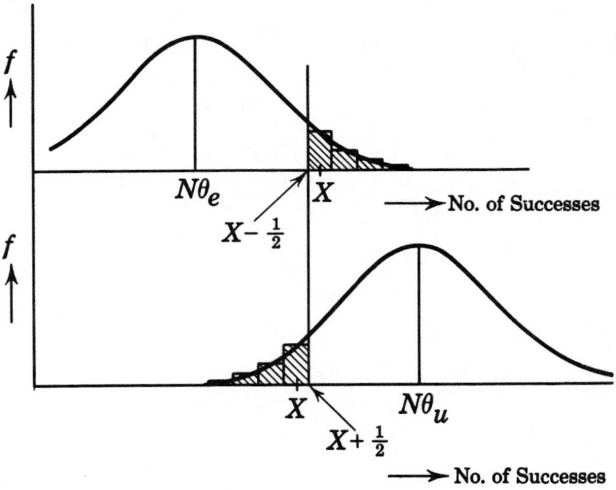

FIG. 31 CONFIDENCE LIMITS FOR THE PARAMETER OF A BINOMIAL DISTRIBUTION

5.14 **Confidence Limits for the Probability of Success in a Binomial Population** If X is the number of successes in N trials, the probability of success in each trial being θ, we know from § 3.3 that

$$E(X) = N\theta \tag{5.14.1}$$

and

$$V(X) = N\theta(1 - \theta) \tag{5.14.2}$$

If N is fairly large and θ not too near 0 or 1, the distribution of X is approximately normal, particularly if we make the correction for continuity mentioned in § 3.11. Thus if θ_l, θ_u are the lower and upper 95% confidence limits for θ, we have (see Figure 31)

$$X - \tfrac{1}{2} - N\theta_l = 1.96[N\theta_l(1 - \theta_l)]^{1/2} \tag{5.14.3}$$

and

$$N\theta_u - (X + \tfrac{1}{2}) = 1.96[N\theta_u(1 - \theta_u)]^{1/2} \tag{5.14.4}$$

These two equations give θ_l and θ_u, respectively, as the solution of a quadratic equation.

If N is quite large, it will often be sufficient to replace θ_l or θ_u on the right-hand side of Eqs. (3) and (4) by the sample proportion X/N, and to ignore the continuity correction. If we do so, the approximate confidence limits are given by

$$N\theta_l = X - 1.96\, X^{1/2}\left(1 - \frac{X}{N}\right)^{1/2} \tag{5.14.5}$$

and

$$N\theta_u = X + 1.96\, X^{1/2}\left(1 - \frac{X}{N}\right)^{1/2} \tag{5.14.6}$$

EXAMPLE 3 If in 400 binomial trials we find 280 successes, what are the 95% confidence limits for θ?

(a) The approximate limits given by Eqs. (5) and (6) are

$$400\theta_l = 280 - 1.96[280(0.30)]^{1/2} = 262$$

$$\theta_l = 0.655$$

and

$$400\theta_u = 280 + 1.96[280(0.30)]^{1/2} = 298$$

$$\theta_u = 0.745$$

(b) From Eq. (3), on squaring both sides, we obtain

$$(279.5 - 400\theta_l)^2 = (3.84)(400)\theta_l(1 - \theta_l)$$

which, on collecting terms and dividing by the coefficient of θ_l^2, becomes

$$\theta_l^2 - 1.3937\theta_l + 0.4836 = 0$$

The solution of this quadratic gives as the *smaller* root (the only one that satisfies the original equation before squaring) $\theta_l = 0.652$. Eq. (4) similarly gives the quadratic equation

$$\theta_u^2 - 1.3987\theta_u + 0.4871 = 0$$

the *larger* root of which is 0.744. In this example the approximate method gives almost as good results as the more exact one.

5.15 **Confidence Limits for the Difference of Probabilities in Two Binomial Populations** If we are given two fairly large samples which we suspect may come from two binomial populations with different parameters θ_1 and θ_2, we can similarly construct confidence limits for the *difference* of these parameters. If the confidence interval includes the value zero, the inference will be that the parameters are not significantly different (at the level of significance determined by the confidence coefficient).

Let us suppose that d is the difference of the two sample proportions of successes:

$$d = p_1 - p_2 = \frac{X_1}{N_1} - \frac{X_2}{N_2} \tag{5.15.1}$$

Then, by Theorem 1.16 and Bienaymé's Theorem (§ 2.14), we have

$$E(d) = E(p_1) - E(p_2) = \theta_1 - \theta_2 \tag{5.15.2}$$

$$\begin{aligned} V(d) &= V(p_1) + V(p_2) \\ &= N_1^{-1}\theta_1(1 - \theta_1) + N_2^{-1}\theta_2(1 - \theta_2) \end{aligned} \tag{5.15.3}$$

If the samples are large enough that we may use the normal approximation, d will also be approximately normal, and

$$\frac{d - (\theta_1 - \theta_2)}{[N_1^{-1}\theta_1(1 - \theta_1) + N_2^{-1}\theta_2(1 - \theta_2)]^{1/2}} \approx z \tag{5.15.4}$$

For the 95% limits we may put $z = \pm 1.96$, and solve for $\theta_1 - \theta_2$. Since we do not know θ_1 and θ_2 separately, we must replace them in the denominator of Eq. (4) by their estimators, p_1 and p_2. The 95% confidence limits are then given approximately by

$$\theta_1 - \theta_2 = d \pm 1.96[N_1^{-1}p_1(1 - p_1) + N_2^{-1}p_2(1 - p_2)]^{1/2} \tag{5.15.5}$$

EXAMPLE 4 A company selling "XX" tires conducted a survey among car owners in each of two districts, A and B. In district A, 750 persons said they planned to purchase tires shortly and 300 said they intended to buy the XX brand. In district B, 600 persons planned to purchase tires and 210 intended to get XX tires. Does there appear to be a significant difference between districts A and B with regard to the proportions of prospective XX purchasers?

Here $d = 0.40 - 0.35 = 0.05$. The approximate standard error of d is

$$\left[\frac{(0.40)(0.60)}{750} + \frac{(0.35)(0.65)}{600}\right]^{1/2} = 0.0264$$

so that

$$\begin{aligned} \theta_1 - \theta_2 &= 0.05 \pm 0.052 \\ &= -0.002 \text{ to } 0.102 \end{aligned}$$

Since the interval includes zero (although only just) we can say that at the 5% level the observed difference d is not significant. The result is, however, close to the borderline of significance.

* 5.16 **Sampling for Proportion of Successes from a Finite Population** If the sample of size N is drawn "without replacements" from a finite population of size M, the distribution of the sample proportion of successes p is not binomial but hypergeometric (see § 3.5). The expectation and variance of p are given by

$$(5.16.1)\qquad \begin{cases} E(p) = \theta \\ V(p) = \dfrac{M - N}{N(M - 1)}\,\theta(1 - \theta) \end{cases}$$

For large N (and of course still larger M), the distribution of p is approximately normal, and confidence limits for θ may be determined as in § 5.14, with the appropriate correction for the variance.

* 5.17 **Use of Binomial Probability Paper** A special graph paper, designed by Mosteller and Tukey [5], may be used to obtain quick approximate solutions of estimation problems involving binomial populations. The scheme is based on Fisher's angular transformation (see § 3.15), $p = \sin^2 A$, which has the effect of making the variance of A a function of the sample size only (proportional to $1/N$) and also of improving the approximation to normality. A specimen of this graph paper is shown in Figure 32.

The scales of x and y are square-root scales. The horizontal distance of a point marked x from the origin is proportional to $x^{1/2}$, and similarly for y. A quarter-circle is drawn through the points marked 100 on each axis, and on this circle $x + y = 100$. The angles A, in degrees, are marked on this circle and the abscissa of a point A is the corresponding p (multiplied by 100). At a distance of $\sqrt{N}$ from the origin, in a direction given by A, the variance of A on the circle of radius $\sqrt{N}$ is practically constant, independent both of N and of θ. Any straight line through the origin passes through points for which y/x is constant, and is called a *split*. A 40-60 split, for example, passes through the point $x = 40$, $y = 60$.

Suppose that in a sample of 10 we find 7 "successes," and therefore 3 "failures." We say that the *paired count* for the sample is (7,3) and plot it as a right-angled triangle with the right angle at (7,3) and the sides each one unit long, parallel to the axes. When one of the coordinates is larger than about 100, the one-unit length is scarcely more than the width of a pencil line.

In order to test whether the observed value of p (7/10) is significantly different from a hypothetical θ (say 1/2), we measure the perpendicular distance from the plotted triangle to the 50-50 split. When the numbers x and y are small, there are two distances, called the *short* and the *long distance*, measured from the two acute angles of the triangle, and these are interpreted by reference to the scale at the top of the paper (marked Full Scale). A distance of one unit on this scale corresponds to a standard normal deviate of 1, so that a distance of two units on the scale represents very nearly the 5% level of significance (when we are interested in the *magnitude* of the difference between p and θ rather than in the sign). The long and short distances each give a significance level and the observed result must be regarded as significant at some level in between. In the illustration above, the two distances are 1.6 and 1.0, so that the observed p is not significantly different from 1/2 at the 5% level of significance.

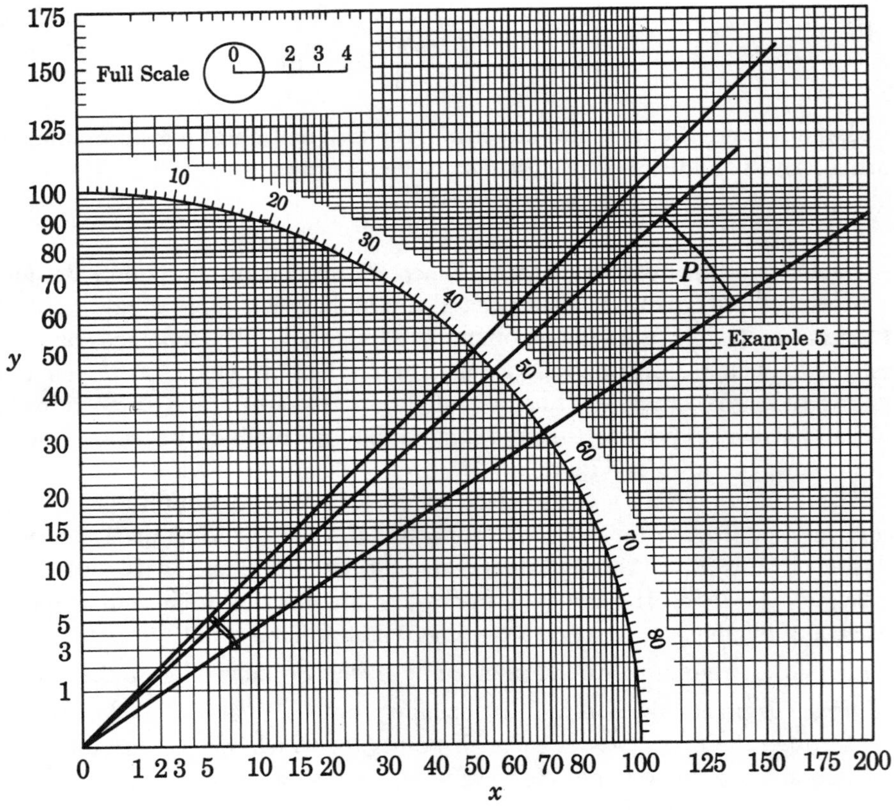

FIG. 32 BINOMIAL PROBABILITY GRAPH PAPER

EXAMPLE 5 In an opinion poll 124 "yes" answers were received to a certain question out of 200 replies. Find 95% confidence limits for the true proportion of persons who would answer "yes" in the population sampled.

The paired count is (124,76), and this is plotted as P in Figure 32 (the triangle is practically a point). Two splits are drawn such that they lie at perpendicular distances of two scale units from P. These splits cut the quarter-circle at (55, 45) and (69, 31) so that the 95% confidence limits for θ are 0.55 and 0.69.

5.18 **Confidence Limits for the Parameter of a Binomial Distribution with Small Samples** The normal approximation is not really justified for small samples, particularly when θ is not close to 0.5. By the use of cumulative binomial tables (such as those mentioned in References [3] and [4] of Chapter 3) it is possible to determine the parameter θ_l of a binomial such that, for example, the observed value of X cuts off the upper $2\frac{1}{2}\%$ tail of the distribution. In the same way, θ_u can be found such that the same X cuts off the lower $2\frac{1}{2}\%$, X itself being included in the tail (see Figure 31). These values θ_l and θ_u give the 95% confidence limits for θ.

Thus if $N = 20$ and $X = 6$, we find from the tables that $B(6, 20, 0.11) = 0.01755$ and $B(6, 20, 0.12) = 0.02602$. By interpolation, the value of θ_l corresponding to $B(6, 20, \theta_l) = 0.025$ is about 0.119. Also $B(7, 20, 0.55) = 0.97859$ and $B(7, 20, 0.54) = 0.97349$, giving θ_u, corresponding to 0.975, as 0.543. The 95% confidence limits for θ are therefore 0.119 and 0.543.

It may be noted for comparison that the approximate method of Eqs. (5.14.5) and (5.14.6) gives 0.099 and 0.511. The method of Eqs. (5.14.3) and (5.14.4) gives 0.128 and 0.543, so that even with an N as small as 20 the normal approximation, with a continuity correction, is fairly satisfactory.

Mention may be made of special tables [6] by Mainland and others, prepared for the Department of Medical Statistics at New York University College of Medicine. These give 95% and 99% confidence limits for θ for a considerable range of sample sizes and observed proportions, and include all cases that are likely to arise in practical statistics.

PROBLEMS

A. (§§ 5.1–5.6)

1. The variate X is distributed in a population with density $f(x|\theta) = 2(\theta - x)/\theta^2$, $0 < x < \theta$. It is desired to estimate θ from a single observation by using the statistic $T = 2X$. Write down the density function for T, integrate to find $F(t|\theta)$, and calculate the values of t_1 and t_2 from Eq. (5.3.1) when $\varepsilon = 0.05$. Plot the curves C_ε and $C_{1-\varepsilon}$, for $\theta \leq 1$. If the observed x is 0.02, find 90% confidence limits for θ. *Hint:* t_1 and t_2 are given by solutions of quadratic equations for t_1/θ and t_2/θ. In each case only one solution is possible, since $t \leq 2\theta$.

2. If X is uniformly distributed on an interval of unit length with centre at $x = \theta$, an estimator for θ is the mid-range of a sample, that is, the mean of the smallest and the largest observed values. If T is this estimator for a sample of size 4, the density function for T is $f(t) = 32(0.5 - |t - \theta|)^3$. If $\theta = 1$, calculate the values of t_1 and t_2 corresponding to $\varepsilon = 0.025$. Sketch the confidence belt for θ and find 95% confidence limits corresponding to an observed $t = 1.2$. *Hint:* If $\theta = 1$, $0.5 < t < 1.5$. Treat the cases $t < 1$ and $t > 1$ separately.

3. If X is normally distributed with mean μ and variance σ^2, the arithmetic mean $\bar{X}$ of a sample of size N is normally distributed with mean μ and variance σ^2/N. If $\bar{X}$ is used as an estimator of μ, find 99% confidence limits for μ corresponding to an observed value of $\bar{X}$. (The variance σ^2 is assumed to be known.)

4. Assuming the distribution of the arithmetic mean, as given in Problem 3, show that $\bar{X}$ is a consistent and unbiased estimator of μ.

5. The distribution of the mid-range of a sample of size N for a uniform distribution of X on $(\theta - \frac{1}{2}, \theta + \frac{1}{2})$ has the density function $f(t) = N2^{N-1}(\frac{1}{2} - |t - \theta|)^{N-1}$. (Compare Problem A-2, for $N = 4$.) Show that the mid-range is a consistent and unbiased estimator of θ. *Hint:* Prove that $E(T - \theta) = 0$ and $E(T - \theta)^2 = [2(N + 1)(N + 2)]^{-1}$. Treat separately the integrals from $\theta - \frac{1}{2}$ to θ and from θ to $\theta + \frac{1}{2}$.

6. If X has a rectangular (uniform) distribution on the interval $(0,\theta)$ and if R is the range for a sample of size N (the highest value x_N minus the lowest value x_1), the distribution function for R, for a given θ, is $F(r|\theta) = (r/\theta)^N(N\theta/r - N + 1)$. Show that the statistic $T = R/\theta$ has a distribution independent of θ and that fiducial limits for θ with confidence coefficient α are given by x_N and R/t_α, where α is the probability that $T > t_\alpha$. *Hint:* α is the probability that $t_\alpha < R/\theta \leq 1$, i.e., that $1 \leq \theta/R < t_\alpha^{-1}$. Write this as a fiducial probability for θ. For given α, t_α is the root of an equation of degree N. Note that θ must be at least equal to x_N.

7. For the rectangular distribution $f(x) = \theta^{-1}$, $0 < x < \theta$, prove that fiducial limits for θ with coefficient α, based on a sample of size two with values x_1 and x_2, are x_2 and $(x_2 - x_1)/(1 - \alpha^{1/2})$. *Hint:* Use Problem 6 with $N = 2$.

8. For the same distribution as in Problem 7, an estimator of θ, based on a sample of two, is $x_1 + x_2$. The density function is $f(t) = t/\theta^2$, for $t < \theta$ and $f(t) = (2\theta - t)/\theta^2$ for $t > \theta$. Show that confidence limits for θ, with confidence coefficient α, are given by $(x_1 + x_2)/[2 - (1 - \alpha)^{1/2}]$ and $(x_1 + x_2)/(1 - \alpha)^{1/2}$, except that when the lower limit is below x_2 it must be replaced by x_2.

Work out numerical values if $x_1 = 3$, $x_2 = 5$, and $\alpha = 0.9$. Compare these limits with those given by Problem 7 for the same data.

9. (a) A sample of N objects is taken from a large binomial population in which a proportion θ of the objects possess a certain attribute A. If p is the proportion of objects possessing this attribute in the sample, show that $pq/(N - 1)$ is an unbiased estimator of $\theta(1 - \theta)/N$, where $q = 1 - p$.

(b) Suppose that the sample is selected, one item at a time, until m of the selected items are A's. Calculate the probability that the size of the sample is N, and show that $(m - 1)/(N - 1)$ is an unbiased estimator of θ. *Hint:* Find the probability that in the first $N - 1$ items there are $m - 1$ A's and that the Nth item is an A. The distribution of $N - m$ is negative binomial.

B. (§§ 5.7–5.11)

1. Write out the proof of the statement in Eq. (5.7.8), that $E(\langle pq \rangle) = \mu'_{pq}$.

2. In the following table, X represents the number of defective items produced by a machine in one day's operation, and f is the frequency of occurrence of X over a period of 200 days. Compute the first four k-statistics for this empirical distribution, which is roughly Poisson. (Note that X is discrete. There is no occasion to use either an auxiliary variable or Sheppard's corrections.)

X	f
0	102
1	59
2	31
3	8
4 or more	0
	200

3. Find the standard errors of k_1 and k_2 and the estimated covariance of k_1 and k_2 for the data of Problem 2.

4. Find the standard errors of k_1 and k_2 as calculated for the u variable, for the data on weights used in §§ 5.9 and 5.10. (Use the corrected values of k_2 and k_4.)

C. (§§ 5.12–5.16)

1. A normal population has mean 20 and standard deviation 2. A sample of six items from the population has a mean 18.2. Can the sample be reasonably regarded as a random one, using the 5% level of significance? *Hint:* Calculate the probability that a *random* sample would have a mean differing from 20 by as much as 1.8. Alternatively, find 95% confidence limits for μ and see whether these include the true value 20.

2. A normal population of times has a standard deviation 0.104 sec. A random sample of 12 items from the population has a mean 12.33 sec. Calculate 90% confidence limits for the population mean. What is the smallest sample size we should use if we want to be 95% sure that the sample mean will not differ from the (unknown) population mean by more than 0.05 sec.?

3. A group of 120 freshmen in arts at a large university take an achievement test in mathematics and obtain a mean score 70 with a standard deviation 14. Another group of 80 in engineering take the same test and obtain a mean score 75 with a standard deviation 12. Is the difference in the means significant at the 1% level? *Hint:* The samples are large enough for the populations to be regarded as normal. The variance of the difference of the means is the sum of the variances of the two means separately. (Compare § 5.15.) As an estimate of the population variance take the weighted mean of the sample variances, weighted according to the sample size less one. Assume both populations of freshmen are large compared with the sample sizes.

4. If 400 eggs are selected at random from a large consignment and 50 are found to be bad, what are the approximate 99% confidence limits for the proportion of bad eggs in the whole consignment? Calculate also the more exact confidence limits for comparison.

5. A physician treats 20 patients suffering from a certain disease and 11 of them die. The mortality rate for this disease, based on thousands of cases, is 42%. Is the physician's sample significantly different from the population, at the 5% level? *Hint:* Calculate the probability that $X \geq 11$, assuming normality.

6. In order to test the efficacy of a drug said to prevent sea-sickness, 25 men who had always developed symptoms of sickness when subjected to the motion of a rocking machine were given the drug. On a further trial with the machine, 15 of these men were found to be immune to the motion. Find 95% confidence limits for the proportion of men liable to seasickness who would be rendered immune by taking this drug. (Assume approximate normality.)

7. In a poll of 148 men and 152 women the question was asked, "Do you approve of the practice of tipping, by and large?", and 89 of the men and 116 of the women answered "yes." Construct approximate 95% confidence limits for the *difference* between the proportion of "yes" answers in the male population sampled and that in the female population sampled. (Both populations may be taken as large compared with the samples.)

8. Random samples of 50 students each were taken from (a) a freshman class in arts and science numbering 248 and (b) a freshman class in engineering numbering 187. Both sample groups were given a mathematical aptitude test, and the numbers reaching a pass standard were 35 and 41 respectively. Test the hypothesis that the proportion of passes would be the same in both classes if all members were tested. *Hint:* The two populations are finite. Use Eq. (5.16.1). Calculate the probability of a difference as great numerically as that observed if the stated hypothesis were true.

9. A research worker wishes to estimate the mean of a population using a random sample so large that the probability will be at least 0.95 that the sample mean will not differ from the population mean by more than 25% of the population standard deviation. How large should the sample be?

10. Obtain some binomial probability paper and solve Problems C-5 and C-6 graphically.

REFERENCES

[1] Neyman, J., and Pearson, E. S., "On the Use and Interpretation of Certain Test Criteria for Purposes of Statistical Inference," *Biometrika*, **20A,** 1928, pp. 175–240 and 263–294.
See also Neyman, J., *Lectures and Conferences on Mathematical Statistics and Probability*, 2nd ed., U.S. Dept. of Agriculture, 1952.

[2] Fisher, R. A., "Inverse Probability," *Proc. Camb. Phil. Soc.*, **26,** 1930, pp. 528–535.
Fisher, R. A., "The Fiducial Argument in Statistical Inference," *Ann. Eugenics*, **6,** 1935, pp. 391–398.
See also his book *Statistical Methods and Scientific Inference*, Oliver and Boyd, 1956.

[3] Pitman, E. J. G., "Statistics and Science," *J. Amer. Stat. Ass.*, **52,** 1957, pp. 322–330.

[4] Tukey, J. W., "Some Sampling Simplified," *J. Amer. Stat. Ass.* **45,** 1950, pp. 501–519.

[5] Mosteller, F., and Tukey, J. W., "The Uses and Usefulness of Binomial Probability Paper," *J. Amer. Stat. Ass.*, **44,** 1949, pp. 174–212. This graph paper is obtainable from the Codex Book Co., Inc., Norwood, Mass.

[6] Mainland, D., Herrera, L., and Sutcliffe, M. I., *Statistical Tables for Use with Binomial Samples*, N.Y. Univ. College of Medicine, 1956.

Chapter 6

ESTIMATION, TESTING AND DECISION MAKING

6.1 **Maximum Likelihood Point Estimation** A very useful method of estimation, which has been vigorously promoted by Fisher, is the method of maximum likelihood. The general idea is to choose as estimator of a parameter θ that function of the sample observations which will, when substituted for θ, make the probability of the sample a maximum. In other words, for this value of θ the observed sample is also the most likely sample.

Consider, for instance, a binomial population with parameter θ. The variate observed is the number of successes X in N trials, and the probability that $X = x$ is

$$f(x|\theta) = \binom{N}{x} \theta^x(1-\theta)^{N-x}$$

As a function of θ, this is a maximum when $\partial f/\partial\theta = 0$ and $\partial^2 f/\partial\theta^2 < 0$. Since

$$\partial f/\partial\theta = \binom{N}{x}[x\theta^{x-1}(1-\theta)^{N-x} - (N-x)\theta^x(1-\theta)^{N-x-1}]$$

$$= f(x|\theta)[x\theta^{-1} - (N-x)(1-\theta)^{-1}]$$

the critical value $\hat{\theta}$ of θ is given by

$$x\hat{\theta}^{-1} - (N-x)(1-\hat{\theta})^{-1} = 0$$

or

$$\hat{\theta} = \frac{x}{N}$$

It is easy to verify that this value does indeed correspond to a maximum for f and not a minimum. The maximum likelihood estimator is therefore identical with the unbiased estimator used for θ in Chapter 5.

If the continuous variate X has a probability density $f(x|\theta, \theta_\alpha)$ which depends on a parameter θ and possibly on other parameters represented jointly by θ_α, the *likelihood* of a set of sample values $x_1, x_2 \ldots x_N$ is defined by

$$L = f(x_1|\theta, \theta_\alpha) \cdot f(x_2|\theta, \theta_\alpha) \ldots f(x_N|\theta, \theta_\alpha) \tag{6.1.1}$$

The likelihood is therefore a joint probability density for the whole sample, but not a probability. When X is discrete, L is just the joint probability for the observed sample (all the items being supposed independently selected from the same population). The principle of maximum likelihood states that we should choose as an estimator of the parameter θ that statistic T (if it exists) which maximizes L for variations in θ, whatever the values of the other parameters θ_α may be.

In practice, the logarithm of L is usually more convenient than L itself. Since $\log L$ is a monotone-increasing function of L, a value of θ which maximizes L also maximizes $\log L$. The maximum likelihood estimator is therefore given by solving for θ the equations

$$\frac{\partial}{\partial\theta}(\log L) = 0, \qquad \frac{\partial^2}{\partial\theta^2}(\log L) < 0 \tag{6.1.2}$$

EXAMPLE 1 For a normal population, with parameters μ, σ, the density function is

$$f(x|\mu, \sigma) = (2\pi\sigma^2)^{-1/2} \exp\left[-\frac{(x-\mu)^2}{2\sigma^2}\right]$$

Therefore, for a sample of size N with values $x_1, x_2 \ldots x_N$,

$$L = (2\pi\sigma^2)^{-N/2} \exp\left[-\sum_i \frac{(x_i-\mu)^2}{2\sigma^2}\right]$$

and

$$\log L = -\frac{N}{2}\log(2\pi) - N\log\sigma - \sum_i \frac{(x_i-\mu)^2}{2\sigma^2} \tag{6.1.3}$$

Differentiating with respect to μ, we find

$$\frac{\partial(\log L)}{\partial\mu} = \sum_i \frac{x_i-\mu}{\sigma^2}$$

and

$$\frac{\partial^2(\log L)}{\partial\mu^2} = \frac{-N}{\sigma^2}$$

The maximum likelihood estimator for μ is therefore $\hat{\mu}$, given by

$$\sum_i (x_i - \hat{\mu}) = 0$$

or

$$\hat{\mu} = \frac{1}{N}\sum_i x_i = m$$

the sample mean. This result is independent of the value of the other parameter σ.

If, however, we try to find an estimator for σ in the same way, we get

$$\frac{\partial(\log L)}{\partial\sigma} = -\frac{N}{\sigma} + \sum_i \frac{(x_i - \mu)^2}{\sigma^3}$$

and, on putting this equal to zero, the value of $\hat{\sigma}$ is not independent of μ. The extraneous parameter in such a case is often called a *nuisance parameter*. There is no maximum likelihood estimator for σ by itself, but it is possible to get joint maximum likelihood estimators for μ and σ by solving together the two equations,

$$\frac{\partial}{\partial\mu}(\log L) = 0 \qquad \text{and} \qquad \frac{\partial}{\partial\sigma}(\log L) = 0$$

These give $\hat{\mu} = m$, $\hat{\sigma}^2 = \dfrac{1}{N}\sum_i (x - m)^2$, so that the joint maximum likelihood estimators for μ and σ^2 are the sample mean (m) and the sample second moment (m_2). It may be noted that, although the former is unbiased, the latter is not, since, as we have already found,

$$E(m_2) = (N - 1)\frac{\sigma^2}{N}$$

and not σ^2 itself.

6.2 **Sufficient Estimators** Some characteristics of estimators were mentioned in § 5.6, but there are others which are also important. A statistic T is said to be a *sufficient* estimator of θ, or, in Fisher's terminology, *exhaustive*, if it uses all the relevant information in the sample. If the likelihood function is expressed in the form

(6.2.1) $$L = g(t|\theta)h(x_1, x_2 \ldots x_N|t, \theta)$$

or

$$\log L = \log g + \log h$$

where g is the density function for T and h is the conditional density function for $x_1 \ldots x_N$, given that $T = t$, then it may happen that h does not depend on θ. If so, T is a sufficient estimator of θ.

For suppose U is another statistic obtainable from the observations. The distribution of U for a given T will depend upon h, but since h does not involve θ, the statistic U can provide no information about θ which is not already given by T.

It is desirable to have a sufficient estimator where possible, since then we know that we are utilizing all the information about θ that we can get from the sample, but sufficiency alone does not define a statistic very precisely. If T is sufficient, so is a function of T.

Sufficient statistics exist in only a relatively few special cases. It is one of the merits of the maximum likelihood method of estimation that if a sufficient statistic does exist for a parameter the maximum likelihood estimator is sufficient.

In Example 1 above, the sum $\sum_i (x_i - \mu)^2$ occurs in the expression for $\log L$ and this can be split up into a part depending on m and μ and a part independent of μ. Thus

$$\begin{aligned}\sum_i (x_i - \mu)^2 &= \sum_i (x_i - m + m - \mu)^2 \\ &= \sum_i (x_i - m)^2 + N(m - \mu)^2 + 2(m - \mu)\sum_i (x_i - m) \\ &= Nm_2 + N(m - \mu)^2\end{aligned}$$

since $\sum_i (x_i - m) = 0$ from the definition of m. Equation (6.1.3) can therefore be written

$$\log L = -\frac{N}{2}\log(2\pi) - N\log\sigma - \frac{Nm_2}{2\sigma^2} - \frac{N(m-\mu)^2}{2\sigma^2}$$

so that, apart from constants,

$$\log g = -\frac{N}{2\sigma^2}(m - \mu)^2, \qquad \log h = -\frac{Nm_2}{2\sigma^2}$$

It is clear that h is a function of the sample values not depending on μ, while g is a function of the estimator m and of μ. Therefore, m is a sufficient estimator for μ.

6.3 **Properties of Maximum Likelihood Estimators** The following five properties are the main reasons for recommending the use of maximum likelihood (m.l.) estimators:

(a) The m.l. estimator is *consistent*. If $f(x|\theta)$ is continuous in x, and also continuous and monotonic in θ over an interval including the true value θ_0, and if T is the m.l. estimator of θ, then T converges in probability to θ_0 as the sample size increases. The proof holds also for a discrete variate if we replace each value by an interval over which we suppose the frequency distributed uniformly. Details may be found in [1].

(b) The m.l. estimator tends to *normality* as N increases. The conditions in (a) are supposed to hold, together with some further conditions on the continuity of $\partial f/\partial\theta$.

(c) The m.l. estimator is *most-efficient* (see § 5.6). The variance of the m.l. estimator T is given by

$$(6.3.1) \qquad [V(T)]^{-1} = N\int_{-\infty}^{\infty}\left(\frac{\partial \log f}{\partial\theta}\right)^2_{\theta=\theta_0} f\,dx = NE\left[\left(\frac{\partial \log f}{\partial\theta}\right)^2\right]_{\theta=\theta_0}$$

where, on the right-hand side, θ is to be put equal to θ_0. If the domain of f does not depend on θ, Eq. (1) is equivalent to

$$(6.3.2) \qquad [V(T)]^{-1} = -NE\left(\frac{\partial^2 \log f}{\partial\theta^2}\right)_{\theta=\theta_0}$$

Since $\log L = \sum_i \log f(x_i)$, and since the x_i all have the same distribution, Eq. (2) may be written

$$[V(T)]^{-1} = -E\left(\frac{\partial^2 \log L}{\partial \theta^2}\right)_{\theta=\theta_0} \tag{6.3.3}$$

This is a convenient way of finding the variance of a m.l. estimator.

It can be shown that if U is any estimator for θ, which in large samples is normally distributed about θ_0 with variance $V(U)$, and if the domain of f does not depend on θ, then

$$[V(U)]^{-1} \leq NE\left[\left(\frac{\partial \log f}{\partial \theta}\right)^2\right]_{\theta_0} \tag{6.3.4}$$

It follows from Eq. (1) and property (b) above that the m.l. estimator is most efficient in the sense described in § 5.6.

(d) If a sufficient estimator exists for θ, the m.l. estimator is *sufficient*.

For, by § 6.2, if T is sufficient and $g(t|\theta)$ is its density function,

$$L = g(t|\theta) \cdot h(x_1, x_2 \ldots x_N|t) \tag{6.3.5}$$

where h is a function of the sample values which does not depend on θ. Therefore,

$$\frac{\partial}{\partial \theta}(\log L) = \frac{1}{g}\frac{\partial g}{\partial \theta} = \psi(\theta, t) \tag{6.3.6}$$

The m.l. estimator is given by putting $\psi(\theta, t) = 0$ and solving for θ. The result is obviously a function of t, say $\phi(t)$. The estimator is therefore $\phi(T)$, and since T is sufficient, so is $\phi(T)$.

(e) The m.l. estimator is *invariant* under functional transformations. This means that if T is the m.l. estimator of θ, and if $u(\theta)$ is a function of θ, then $u(T)$ is the m.l. estimator for $u(\theta)$.

If, for example, we are dealing with a normal population (for which $\mu_4 = 3\sigma^4$) and we know that the m.l. estimator for σ^2 is the sample second moment m_2 (that is, $N^{-1}\sum(x_i - m)^2$), we conclude that the m.l. estimator for μ_4 is $3m_2{}^2$ and not m_4. Of course, m_4 might be used as an estimator, but it would not have as small a sampling variance as $3m_2{}^2$.

This property of invariance is not true of all estimators. If T is an unbiased estimator of θ, for example, it does not follow that T^2 is an unbiased estimator of θ^2.

EXAMPLE 2 If the population is normal, and if σ is supposed known, the m.l. estimator of μ is the sample mean m, as shown in Example 1.

Since $\log f = -\frac{1}{2}\log(2\pi\sigma^2) - \dfrac{1}{2\sigma^2}(x - \mu)^2$, we have

$$\frac{\partial \log f}{\partial \mu} = \frac{(x - \mu)}{\sigma^2}$$

$$\frac{\partial^2 \log f}{\partial \mu^2} = -\frac{1}{\sigma^2}$$

and

$$-NE\left(\frac{\partial^2 \log f}{\partial \mu^2}\right) = N \int_{-\infty}^{\infty} \frac{1}{\sigma^2} f\,dx = \frac{N}{\sigma^2}$$

The variance of m is therefore σ^2/N, in agreement with the result found in Chapter 5.

* 6.4 **The Cramér-Rao Inequality** If T is an estimator of θ, and if its *bias* is b, as defined by the relation

$$E(T) = \theta + b \tag{6.4.1}$$

where b may depend upon θ, then

$$E[(T-\theta)^2] \geq \frac{\left(1 + \dfrac{db}{d\theta}\right)^2}{N \displaystyle\int_{-\infty}^{\infty} \left(\frac{\partial \log f}{\partial \theta}\right)^2 f(x|\theta)\,dx} \tag{6.4.2}$$

If $g(t|\theta)$ is the frequency function for T, the equality sign in relation (2) holds if, and only if,

$$\begin{cases} \text{(a)} & T \text{ is sufficient} \\ (b) & \dfrac{\partial \log g}{\partial \theta} = k(T-\theta) \end{cases} \tag{6.4.3}$$

where k is independent of T but may depend on θ.

If T is an *unbiased* estimator, so that $b = 0$, $E[(T-\theta)^2]$ becomes the variance of T, and Eq. (2) is

$$V(T) \geq \left[N \int_{-\infty}^{\infty} \left(\frac{\partial \log f}{\partial \theta}\right)^2 f\,dx\right]^{-1} \tag{6.4.4}$$

which is formally the same as (6.3.4), although the conditions imposed on the estimator are different.

The relation (4) was proved independently by Cramér and Rao, although it was found earlier by Fisher for the special case of a normal population. For a proof see [2].

EXAMPLE 3 For the two-parameter gamma distribution of Eq. (4.4.6),

$$f(x) = e^{-x/\beta} x^{\alpha-1} [\beta^\alpha \Gamma(\alpha)]^{-1}, \qquad 0 \leq x < \infty$$

Suppose that α is known and we wish to find an estimator for β. We have

$$\log f = \frac{-x}{\beta} + (\alpha - 1)\log x - \alpha \log \beta - \log \Gamma(\alpha)$$

so that

$$\log L = -\sum \frac{x_i}{\beta} - N\alpha \log \beta + \text{terms independent of } \beta.$$

Therefore,

$$\frac{\partial(\log L)}{\partial \beta} = \sum \frac{x_i}{\beta^2} - \frac{N\alpha}{\beta} = 0$$

giving as the m.l. estimator

$$\hat{\beta} = \sum \frac{x_i}{N\alpha} = \frac{m}{\alpha}$$

where m is the sample mean.

Since, for this distribution $E(x) = \alpha\beta$, it follows that $E(\hat{\beta}) = \beta$, and the estimator $\hat{\beta}$ is therefore unbiased. The variance is given by

$$[V(\hat{\beta})]^{-1} = -N \int_0^\infty \frac{\partial^2(\log f)}{\partial \beta^2} f\, dx$$

$$= -N \int_0^\infty \left(\frac{-2x}{\beta^3} + \frac{\alpha}{\beta^2} \right) f\, dx$$

$$= \frac{2N\alpha}{\beta^2} - \frac{N\alpha}{\beta^2} = \frac{N\alpha}{\beta^2}$$

The variance of $\hat{\beta}$ is therefore $\beta^2/N\alpha$. The likelihood function may be written

$$\log L = -\frac{N\alpha}{\beta}\hat{\beta} - N\alpha \log \beta + (\alpha - 1) \sum \log x_i - N \log \Gamma(\alpha)$$

$$= \log g + \log h$$

where

$$\log g = -(N\alpha\,\hat{\beta})/\beta - N\alpha \log \beta + \text{terms independent of } \beta$$

and $\log h$ is independent of β. This shows that $\hat{\beta}$ is sufficient. Also,

$$\frac{\partial(\log g)}{\partial \beta} = \frac{N\alpha\hat{\beta}}{\beta^2} - \frac{N\alpha}{\beta}$$

$$= \frac{N\alpha}{\beta^2}(\hat{\beta} - \beta)$$

which is of the form of Eq. (6.4.3), condition (b). The two conditions for the sign of equality in (6.4.4) are therefore satisfied.

* 6.5 **Approximate Calculation of a Maximum Likelihood Estimator** It happens sometimes that the method of maximum likelihood leads to equations which are very troublesome to solve. In such a case it may be useful to find a

simpler, but less efficient, estimator and apply a correction to bring it nearer to the desired form.

If T is the m.l. estimator and U is an estimator which is not quite as efficient as T, we know that $(\partial \log L/\partial\theta)_T = 0$, and $E(\partial^2 \log L/\partial\theta^2)_T = -[V(T)]^{-1}$, at least for large N.

By Taylor's theorem,

$$\left(\frac{\partial \log L}{\partial\theta}\right)_U = \left(\frac{\partial \log L}{\partial\theta}\right)_T + (U - T)\left(\frac{\partial^2 \log L}{\partial\theta^2}\right)_T$$

$$+ \text{ terms of higher order.}$$

Since we suppose that the quantity $U - T$ is small and since we can approximately replace $(\partial^2 \log L/\partial\theta^2)_T$ by its expected value, we have, on neglecting the higher terms,

$$T \approx U + V(T)\left(\frac{\partial \log L}{\partial\theta}\right)_U \tag{6.5.1}$$

The last term on the right-hand side is a correction to be applied to U to bring it nearer to T. The value of $V(T)$ is obtained from Eq. (6.3.2).

EXAMPLE 4 For the Cauchy distribution—

$$f(x) = \frac{1}{\pi}\cdot\frac{1}{1 + (x - \theta)^2}, \qquad -\infty < x < \infty$$

—the sample mean is not a good estimator of θ, since it is no better than a single observation. The sample median may be used and in large samples has a variance $\pi^2/4N$. The m.l. estimator is given by

$$\log L = -\sum_i \log[1 + (x_i - \theta)^2] - N \log \pi,$$

$$\frac{\partial(\log L)}{\partial\theta} = 2\sum_i \frac{(x_i - \theta)}{1 + (x_i - \theta)^2} = 0$$

As an equation in θ, this gives a polynomial of degree $2N - 1$, which even for fairly small N is difficult to solve. From Eq. (6.3.2),

$$\begin{aligned} -[V(T)]^{-1} &= N\int_{-\infty}^{\infty}\left(\frac{\partial^2 \log f}{\partial\theta^2}\right) f\,dx \\ &= \frac{2N}{\pi}\int_{-\infty}^{\infty}\frac{(x - \theta)^2 - 1}{[1 + (x - \theta)^2]^3}\,dx \\ &= \frac{4N}{\pi}\int_0^{\infty}\frac{u^2 - 1}{(1 + u^2)^3}\,du \\ &= -\frac{N}{2} \end{aligned}$$

so that $V(T) = 2/N$.

The efficiency of the median is therefore $(2/N) \div (\pi^2/4N) = 8/\pi^2$, or about 0.8. The improved estimator will be the median U plus the correcting term $V(T)(\partial \log L/\partial\theta)_U$, i.e., $\dfrac{4}{N}\sum_i \dfrac{x_i - U}{1 + (x_i - U)^2}$.

6.6 **Tests of Hypotheses** There is a large class of statistical problems concerned with testing whether or not some hypothesis is true. For example, a machine is turning out thread, and we would like to be reasonably sure that the breaking strength is at least 100 lb weight; if not, the machine may have to be re-adjusted. We can take samples of the thread at intervals and test them, but because of the variability of the product (inherent in the process of manufacture) the samples will vary among themselves. We can, however, use them to test the hypothesis (H_0) that the mean breaking strength (μ) of the thread produced is at least 100 lb wt against the alternative hypothesis (H_1) that μ is less than 100 lb wt.

The hypothesis which we set up and proceed to test by experiment is called a null hypothesis. On the basis of the sample we can take various possible actions. We can (1) reject the null hypothesis, which in this example may mean dismantling the machine, (2) accept the null hypothesis, which means that we happily accept the product of the machine as up to standard, (3) declare that further experimentation is necessary before we can make a decision. If the size of the sample is fixed beforehand, this third procedure is not open to us, but in sequential sampling, as we shall see later, tests are continued until we feel justified in taking either action (1) or action (2).

In taking such an action on the basis of a sample we run a risk of doing the wrong thing. Obviously we may commit either of two kinds of error: we may reject a hypothesis which is really true (this will be called a *rejection error*, or an *error of the first kind*), or we may accept a hypothesis which is really false (this will be called an *acceptance error*, or an *error of the second kind*).

Tests are usually made by computing some statistic (e.g., an arithmetic mean or a variance) from the observations and noting whether or not this computed value lies in some particular interval, or set of intervals, previously chosen on the axis of real numbers. The part of the real axis so chosen is called the *region of rejection*, and the hypothesis H_0 is rejected if the computed value lies in this region. Thus, if the population is known to be normally distributed about a value μ with unit variance, and if H_0 is the hypothesis that μ is zero, we shall be inclined to reject this hypothesis if a sample of N observations gives a mean too far from zero. If H_0 were true, the sample mean would depart from zero by as much as $1.96\, N^{-1/2}$ in only 5% of random samples of size N. By taking as our region of rejection that part of the real axis outside the bounds $\pm 1.96N^{-1/2}$, we run a risk of wrongly rejecting H_0, but the chance of doing so is only 0.05. By suitably choosing the region of rejection we can make this chance what we like, depending on the circumstances of the problem and the consequences of making a wrong decision.

In some types of problem the test statistic may be a *pair* of numbers, and then the region of rejection may be represented by a part of the x-y plane. The concept can obviously be extended to three or more dimensions. However, in most of the day-to-day problems of practical statistics the statistic used is one-dimensional and the region of rejection is an interval or pair of intervals on the real axis.

6.7 **Simple and Composite Hypotheses** An hypothesis which is equivalent to a complete specification of the distribution is said to be *simple*. Otherwise, it is *composite*. Thus, if a population is known to be normal and to have variance σ^2, the hypothesis that the mean is μ_0 is a simple one, since the mean and variance together specify a normal distribution completely. The alternative hypothesis could be simple also—if it were known, for instance, that the population mean must be either μ_0 or μ_1 and could not have any other value. More usually, the alternative would be composite. It could be a *two-sided alternative* (namely, that μ is either less than or greater than μ_0), or it could be a *one-sided alternative* (that $\mu > \mu_0$, for instance, supposing that we have good reason to believe that it cannot possibly be less). We may, for example, want to know whether a new kind of fertilizer, applied in a particular way, will increase the yield of a crop, but feel quite certain that it will not actually diminish the yield. It would be reasonable in this case to use a one-sided alternative.

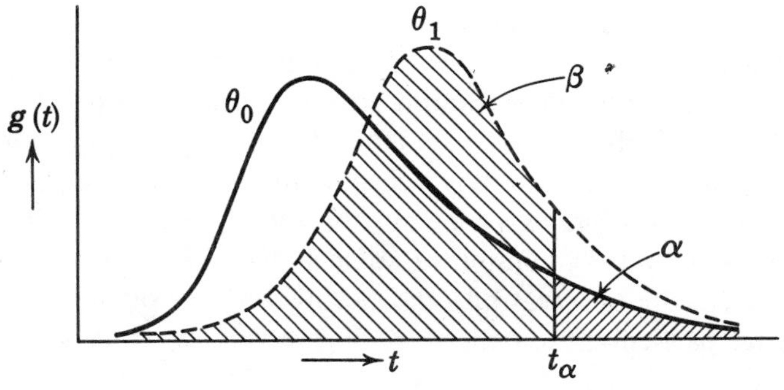

FIG. 33 ERRORS OF THE FIRST AND SECOND KIND

6.8 **The Size and Power of a Test** Suppose we want to test the simple null hypothesis (H_0), that $\theta = \theta_0$, against the simple alternative hypothesis (H_1), that $\theta = \theta_1$, by means of a test statistic T. In order to fix our region of rejection we shall need to know the density function of T when $\theta = \theta_0$, say $g(t|\theta_0)$. In general, g will depend on the sample size N, and the region of rejection (R) will be an interval on the t-axis (e.g., the interval $t > t_\alpha$ in Figure 33).

If the probability that T falls in R, when H_0 is true, is α, then α is the probability of committing a rejection error (that is, of rejecting H_0 when it is true).

This probability is often called the *size* of the test. It is given by

(6.8.1)
$$\alpha = \int_{(R)} g(t|\theta_0)\, dt$$

and is represented by the heavily-shaded area in Figure 33. In practice, R is usually chosen so that α is 0.05 or 0.01, although of course any convenient size can be used.

If the statistic T is discrete, the integral will be replaced by a sum, and it will not generally be possible to make the size *exactly* 0.05 or other preassigned value. The region of rejection will be a set of values of T, the probabilities of which add up to something near the required α.

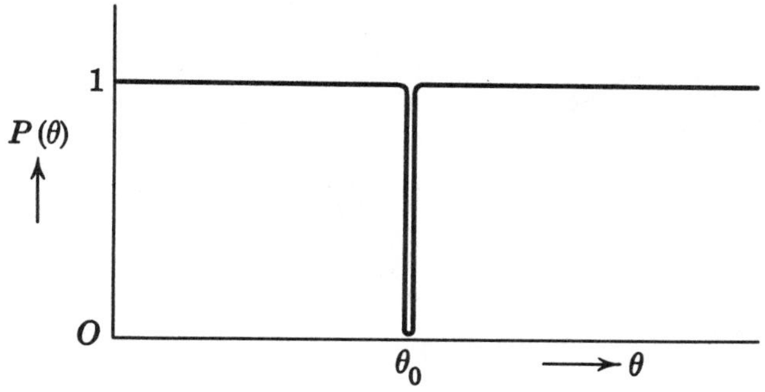

FIG. 34 POWER FUNCTION OF AN IDEAL TEST

If the alternative hypothesis H_1 is true, there will be a different distribution of T, with density $g(t|\theta_1)$. The probability of committing an acceptance error (that is, of accepting H_0 if H_1 is true) is given by

(6.8.2)
$$\beta = \int_{(A)} g(t|\theta_1)\, dt$$

where A is the region of acceptance (all possible values of t outside of R). This probability is represented by the lightly-shaded area in Figure 33. The *power* of the test is defined by

(6.8.3)
$$P(\theta_1) = 1 - \beta = \int_{(R)} g(t|\theta_1)\, dt$$

It is the probability of rejecting H_0 if H_1 is true (that is, if H_0 *should* be rejected). Obviously, we would like a test to be as powerful as possible for the same size.

The power depends, of course, on θ_1. If θ is any value of θ_1, $P(\theta)$ is called the *power function* of the test for θ_0 against θ. If θ is near to θ_0, the power will usually be small, and if $\theta = \theta_0$ it becomes equal to α. For θ far removed from θ_0 the power will usually be near to 1, since any reasonable test should be able to decide between very different hypotheses. The ideal power function would be something like the one sketched in Figure 34, in which $\alpha = 0$ and $P(\theta) = 1$ for all θ not equal to θ_0. Both kinds of error would then be zero, but such ideal tests are not available in practice.

An actual power function will be more like that depicted in Fig. 35. A low power can be tolerated for θ near θ_0, since no great harm is done if we do make the mistake of accepting θ instead of θ_0. Where θ differs considerably from θ_0, so that it might be a serious matter to mistake one for the other, the power is near 1 and the acceptance error β is small.

The function $\beta(\theta) = 1 - P(\theta)$ is often used instead of the power function, particularly in industrial practice, and is called the *operating characteristic* (O.C.) of the test. The graph of the O.C. is like that of the power function turned upside-down, with 0 and 1 interchanged.

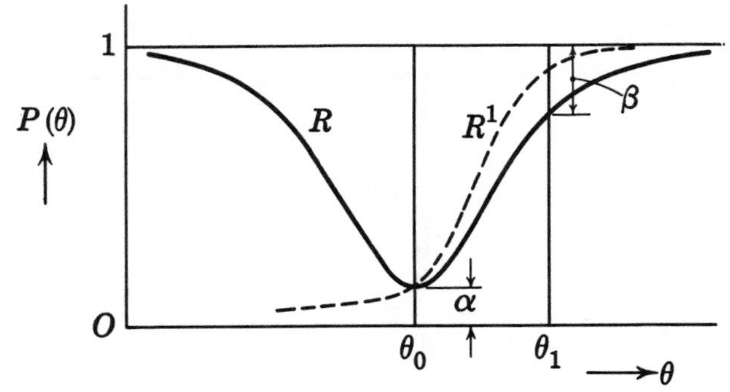

FIG. 35 POWER FUNCTIONS OF ALTERNATIVE TESTS

The two types of error that we have defined depend on *conditional* probabilities, the probability that T falls in R when H_0 is true (denoted by $T \in R|H_0$) or the probability that T falls in A when H_1 is true (denoted by $T \in A|H_1$). If we assert, on the basis of our observations, that H_0 is not true, the chance that we are wrong depends not only on these conditional probabilities but also on the prior probability (previous to our observations) that H_0 actually is true. If p_0 is the prior probability of H_0, it follows from the rules for probability calculations in Chapter 1 that the chance of being wrong in rejecting H_0 is given by

$$P(T \in R|H_0) \cdot p_0 = \alpha p_0$$

and the chance of being wrong in accepting H_0 is

$$P(T \in A|H_1) \cdot (1 - p_0) = \beta(1 - p_0)$$

where A is the region of acceptance (the whole domain of T outside of R). If the null hypothesis H_0 that we choose to test is one that has a small prior probability of being true, the chance of being wrong in rejecting it may be much less than the size of the test α.

It is often possible to choose the region of rejection R in different ways, even though the size α remains constant. Each choice of R will give rise to a different power function. Figure 35 illustrates the possibility that the test using a region

R' may be more powerful than the test using R for all values of $\theta_1 > \theta_0$, but may be less powerful when $\theta_1 < \theta_0$.

If for every value of θ, except θ_0, the power curve of R' is below that of R, the test using R is said to be *uniformly more powerful* than that using R'. If this holds for all possible choices of R' then the test using R is a *uniformly most powerful* test. In a few cases such tests have been found.

6.9 **The Neyman-Pearson Theorem** Let X be a random variable with density function $f(x)$. (If X is discrete, the necessary modifications in the proof can easily be made). We suppose that $f(x)$ depends on a parameter θ for which we would like to test the simple hypothesis H_0 (that $\theta = \theta_0$) against the simple hypothesis H_1 (that $\theta = \theta_1$). The test consists of rejecting H_0 if the observed value of X lies in a region R and accepting H_0 otherwise. The size of the test is

(6.9.1) $$\alpha = \int_{(R)} f(x|\theta_0)\,dx$$

and the power is

(6.9.2) $$P = \int_{(R)} f(x|\theta_1)\,dx$$

Suppose now that R' is any other region of the domain of X for which

(6.9.3) $$\int_{(R')} f(x|\theta_0)\,dx \le \alpha$$

If for *every* such R' it is true that

(6.9 4) $$\int_{(R')} f(x|\theta_1)\,dx \le \int_{(R)} f(x|\theta_1)\,dx$$

then R is a most-powerful test, of size not greater than α, for testing θ_1 against θ_0. Neyman and Pearson [3] proved that if a region R exists satisfying (1) and such that x belongs to R whenever

(6.9.5) $$\frac{f(x|\theta_0)}{f(x|\theta_1)} < c$$

where c is some constant, and does not belong to R whenever this ratio $>c$, then R is a most-powerful test of size not greater than α. The ratio in (5) is called the *likelihood ratio*, and will be denoted by $L(x)$.

As well as merely distinguishing between two fixed values θ_0 and θ_1, the likelihood ratio test applies more generally. Thus, suppose the possible values of θ form a set Ω (which may, for example, be the interval 0 to 1, or the interval $-\infty$ to ∞). The null hypothesis H_0 may specify that θ belongs to some subset ω of Ω (for instance, the single value 0.5, or the interval from 0.4 to 0.6) and H_1 is then the hypothesis that θ belong to $\Omega - \omega$. The likelihood ratio is defined as the ratio of the maximum likelihood under H_0 to the maximum likelihood under H_1:

$$L(x) = \frac{\max_{\theta \in \omega} f(x|\theta)}{\max_{\theta \in \Omega - \omega} f(x|\theta)} \tag{6.9.6}$$

If $L(x)$ is small, the observed x will be more likely under H_1 than under H_0, so that it would be unreasonable to maintain H_0. The test consists in rejecting H_0 when $L(x) < c$, c being such that

$$P(L(x) < c|H_0) = \alpha \tag{6.9.7}$$

Many useful tests in statistics are likelihood ratio tests. The statistic X may consist of a set of N independent observations, forming a random sample, and the likelihood will then be a joint probability density for the N observations. When N is large, the distribution of $-2 \log L(X)$ under H_0 is approximately a chi-square distribution with degrees of freedom depending on the number of parameters concerned (one, in the case discussed above). This was shown by Wilks [4]. If H_1 is true, the distribution of $-2 \log L(X)$ is approximately non-central chi-square (see Appendix A.13). Tables of the non-central chi-square distribution may be found in [5].

When the parent population is normal, as we shall see in the next section, the chi-square distribution of $-2 \log L(X)$ holds exactly, even for $N = 1$.

6.10 **The One-Sided Normal Test** Suppose the population is normal, with known variance σ^2, and suppose we wish to distinguish between two possible values of the mean, μ_0 and μ_1, where $\mu_1 > \mu_0$.

The test statistic is the sample mean m computed from N observations. Since m is normally distributed with mean μ (μ is either μ_0 or μ_1) and variance σ^2/N,

$$f(m|\mu) = \left(\frac{N}{2\pi\sigma^2}\right)^{1/2} e^{-N(m-\mu)^2/2\sigma^2} \tag{6.10.1}$$

The likelihood ratio is

$$L(m) = \frac{f(m|\mu_0)}{f(m|\mu_1)} = \exp\left\{\frac{N}{2\sigma^2}\left[(m-\mu_1)^2 - (m-\mu_0)^2\right]\right\} \tag{6.10.2}$$

$$= \exp\left[\frac{N}{2\sigma^2}(\mu_1 - \mu_0)(\mu_1 + \mu_0 - 2m)\right]$$

This will be less than some positive constant c if $\mu_1 + \mu_0 - 2m < c_1$, where c_1 is another constant depending on c and on the known quantities μ_1, μ_0, σ^2 and N. Actually $c_1 = \log c \Big/ \left[\frac{N}{2\sigma^2}(\mu_1 - \mu_0)\right]$. The relation $\mu_1 + \mu_0 - 2m < c_1$ implies

$$m > c_2, \qquad c_2 = \frac{\mu_1 + \mu_0 - c_1}{2}$$

so that the test reduces to rejecting H_0 (that $\mu = \mu_0$) if the sample mean is greater than a certain value, c_2. This value can be determined by deciding on the size of the test.

If α is the probability that $m > c_2$, given that $\mu = \mu_0$,

$$\alpha = \int_{c_2}^{\infty} f(m|\mu_0)\, dm \tag{6.10.3}$$

By the transformation $(m - \mu_0)N^{1/2}/\sigma = v$, this becomes

$$\begin{aligned} \alpha &= (2\pi)^{-1/2} \int_{v_0}^{\infty} e^{-v^2/2}\, dv \\ &= 1 - \Phi(v_0) \end{aligned} \tag{6.10.4}$$

where $v_0 = (c_2 - \mu_0)N^{1/2}/\sigma$ and $\Phi(v)$ is the cumulative distribution function for the normal law.

If we choose $\alpha = 0.05$, we find from the tables of the normal law that $v_0 = 1.645$, so that

$$c_2 = \mu_0 + 1.645\sigma N^{-1/2} \tag{6.10.5}$$

The test therefore consists in rejecting H_0 in favor of H_1 if $m > \mu_0 + 1.645\sigma N^{-1/2}$

The power of this test is given by

$$\begin{aligned} P &= \int_{c_2}^{\infty} f(m|\mu_1)\, dm \\ &= 1 - \Phi(v_1) \end{aligned} \tag{6.10.6}$$

where $v_1 = (c_2 - \mu_1)N^{1/2}/\sigma = 1.645 - (\mu_1 - \mu_0)N^{1/2}/\sigma$. Thus, if $\mu_1 - \mu_0 = 0.3\sigma$ and $N = 9$, the power is $1 - \Phi(0.745) = 0.228$. With a sample of size 9 there is therefore a probability equal to 0.228 of detecting a difference $\mu_1 - \mu_0$ as great as 0.3 of the standard deviation, if it is known that this difference is positive. This is a *one-sided test.*

It may be noted that this test can be regarded as a test of the simple hypothesis H_0 against the composite alternative H_1 (that $\mu > \mu_0$). The power is then a function of μ. The set Ω of possible values of μ is the set of real numbers $\geq \mu_0$, and the set ω consists of the single number μ_0.

The likelihood ratio is

$$L(m) = \frac{f(m|\mu_0)}{\max\limits_{\mu > \mu_0} f(m|\mu)} \tag{6.10.7}$$

where $f(m|\mu)$ is given by Eq. (1). This density, $f(m|\mu)$, is a maximum for variations in μ when the exponential factor is equal to 1, that is, when $\mu = m$. If the sample mean m should happen to be less than μ_0, the maximum would be when μ is arbitrarily near to μ_0. Therefore, $L(m) = 1$ if $m \leq \mu_0$, but if $m > \mu_0$,

$$L(m) = e^{-N(m-\mu_0)^2/2\sigma^2} \tag{6.10.8}$$

This is less than c if $m - \mu_0 > c_1$, and therefore if $m > c_2$. The number c_2 is determined by

$$\alpha = \int_{c_2}^{\infty} f(m|\mu_0)\,dm \tag{6.10.9}$$

$$= 1 - \Phi\,(v_0)$$

with $v_0 = (c_2 - \mu_0)N^{1/2}/\sigma$.

The test consists in rejecting H_0 if $m > c_2$. If m should be less than μ_0, H_0 will naturally be accepted. The power is given by

$$P(\mu) = \int_{c_2}^{\infty} f(m|\mu)\,dm \tag{6.10.10}$$

$$= 1 - \Phi(v)$$

with $v = (c_2 - \mu)N^{1/2}/\sigma$. Since $-2 \log L(m)$, from Eq. (8), is equal to $N(m - \mu_0)^2/\sigma^2$, which on hypothesis H_0 is the square of a standard normal variate, it follows that in this example $-2 \log L(m)$ has a chi-square distribution with one degree of freedom, regardless of the value of N.

6.11 **The Two-Sided Normal Test** If the null hypothesis H_0 is that $\mu = \mu_0$ (for a normal parent population with variance σ^2), and the two-sided alternative H_1 is that either $\mu > \mu_0$ or $\mu < \mu_0$, we shall have, for a test based on the sample mean,

$$L(m) = e^{-N(m-\mu_0)^2/2\sigma^2} \tag{6.11.1}$$

which is less than c if $|m - \mu_0| > c_1$.

The region of rejection therefore consists of two parts—from $-\infty$ to $\mu_0 - c_1$ and from $\mu_0 + c_1$ to ∞. For a given α, c_1 is fixed by the relation

$$\int_{-\infty}^{\mu_0 - c_1} f(m|\mu_0)\,dm + \int_{\mu_0 + c_1}^{\infty} f(m|\mu_0)\,dm = \alpha \tag{6.11.2}$$

Because of the symmetry of the normal distribution, both integrals are equal to $\alpha/2$, and as in § 6.10 we find

$$\alpha/2 = 1 - \Phi(v), \qquad v = c_1 N^{1/2}/\sigma \tag{6.11.3}$$

For $\alpha = 0.05$, $c_1 = 1.96\sigma N^{-1/2}$. The power is given by

$$P(\mu) = 1 - \int_{\mu_0 - c_1}^{\mu_0 + c_1} f(m|\mu)\,dm \tag{6.11.4}$$

$$= 1 - \Phi\,(v_1) + \Phi\,(v_0)$$

where

$$v_1 = \frac{N^{1/2}}{\sigma}(\mu_0 + c_1 - \mu) = 1.96 - N^{1/2}\sigma^{-1}(\mu - \mu_0)$$

and

$$v_0 = \frac{N^{1/2}}{\sigma}(\mu_0 - c_1 - \mu) = -1.96 - N^{1/2}\sigma^{-1}(\mu - \mu_0)$$

EXAMPLE 5 What size of sample is necessary in order to detect with probability 0.8 a difference between the population mean and the assumed value μ_0 amounting to as little as 0.2σ, given that the probability of rejection error (of stating that a difference exists when in fact it does not) is 0.05?

Here $c_1 = 1.96\sigma N^{-1/2}$,

$$P(\mu) = 1 - \Phi(v_1) + \Phi(v_0) = 0.8$$

with

$$v_1 = 1.96 \pm 0.2N^{1/2}$$

$$v_0 = -1.96 \pm 0.2N^{1/2}$$

(The two plus signs, or the two minus signs, go together.) For fairly large N (taking the plus signs), $\Phi(v_1) \approx 1$, $\Phi(v_0) \approx 0.8$, so that $v_0 \approx 0.8416$, giving $N = 196$. With this value, $v_1 = 4.76$, and $\Phi(v_1)$ is certainly close enough to 1. A sample size of 196 will therefore give the required power. The same result follows if we use the minus signs, with $\Phi(v_0) \approx 0$, $\Phi(v_1) \approx 0.2$, and $v_1 \approx -0.8416$.

* 6.12 **The Randomized Neyman-Pearson Theorem** It is possible to increase the power of a test, in certain circumstances, by allowing a randomized decision. The total domain of the statistic X is divided into three parts: R, A, and D. If the observed x falls in R, H_0 is rejected, and if it falls in A, H_0 is accepted, but there is also a doubtful region D. If x falls in D we toss a coin or draw a card or consult a table of random numbers—that is, we employ some randomizing procedure which gives us a known *probability* of rejecting H_0.

We can define a *test function* $\psi(x)$ by letting $\psi(x) = 1$ if $x \in R$, $\psi(x) = 0$ if $x \in A$ and $\psi(x) = \psi_0$ if $x \in D$, $\psi(x)$ being in all cases the probability of rejection of H_0 and ψ_0 being a number between 0 and 1.

The randomized Neyman-Pearson theorem states that if $L(x)$ is defined as in (6.9.5), and if

(6.12.1)
$$\begin{cases} \psi(x) = 1 & \text{when} \quad L(x) < c \\ \psi(x) = 0 & \text{when} \quad L(x) > c \\ \psi(x) = \psi_0 & \text{when} \quad L(x) = c \end{cases}$$

then the test with test-function $\psi(x)$ is most powerful of size α for testing H_0 against H_1. The value of c is determined by

(6.12.2)
$$P(L(X) < c | H_0) \le \alpha$$

and the value of ψ_0 by

(6.12.3)
$$P(L(X) < c | H_0) + \psi_0 P(L(X) = c | H_0) = \alpha$$

If the region D includes only a single value of x, ψ_0 is uniquely determined, but in other cases several values of ψ_0 may be found to satisfy this equation.

EXAMPLE 6 Suppose we want to test the hypothesis that the proportion of defectives in a large lot of manufactured articles is not more than 10%, and we decide to do so by taking a sample of four items and noting the number (X) of defectives. Clearly X can take only the values 0, 1, 2, 3 or 4, and the larger X is, the more readily we shall reject the hypothesis.

The probability of exactly x defectives is

$$f(x|\theta) = \binom{4}{x}\theta^x(1-\theta)^{4-x} \tag{6.12.4}$$

If $\theta = 0.10$, this expression, for $x = 2, 3, 4$, takes values 0.0486, 0.0036 and 0.0001, respectively. If $\theta < 0.10$, these values will be still smaller. We might therefore take as the region of rejection the set of values $x = 3$ and $x = 4$, and the rejection error will be

$$\sum_{(R)} f(x|\theta) \le 0.0037, \qquad \theta \le 0.10 \tag{6.12.5}$$

If, however, we include $x = 2$ in R, we have

$$\sum_{(R)} f(x|\theta) \le 0.0523, \qquad \theta \le 0.10$$

and, at least for some values of θ, the size of the test will be greater than 0.05. The non-randomized test would therefore tell us to reject H_0 if $X = 3$ or 4, and accept H_0 if $X = 0$, 1 or 2. The power of this test is

$$P(\theta) = \sum_{x=3}^{4} f(x|\theta) = \theta^4 + 4\theta^3(1-\theta) \tag{6.12.6}$$

For $\theta = 0.20$, this is 0.027.

Suppose now we use a randomized test, and decide to reject H_0 with probability ψ_0 when $X = 2$. The probability of this under H_0 is 0.0486, so that Eq. (3) gives, for $\alpha = 0.05$,

$$0.0037 + 0.0486\,\psi_0 = 0.05 \tag{6.12.7}$$

Therefore, $\psi_0 = 0.95$. The randomized test is:

$$\begin{cases} \text{reject } H_0 \text{ if } X = 3 \text{ or } 4 \\ \text{accept } H_0 \text{ if } X = 0 \text{ or } 1 \\ \text{reject } H_0 \text{ with probability } 0.95 \text{ if } X = 2 \end{cases}$$

A way of rejecting H_0 with probability 0.95 would be to use a table of random two-digit numbers. Before opening the table, decide arbitrarily on a particular page, a particular column, and a particular position in the column (say seventh from the top). Then look up the number. If it lies between 00 and 94, inclusively, reject H_0.

The power of this randomized test is

$$P(\theta) = \sum_{3}^{4} f(x|\theta) + 0.95f(2|\theta) \tag{6.12.8}$$
$$= \theta^4 + 4\theta^3(1-\theta) + 5.70\theta^2(1-\theta)^2$$

For $\theta = 0.20$ this is 0.173, greater than for the non-randomized test.

In the above example we did not need to find the likelihood ratio, but we can easily do so. The maximum of Eq. (4) under H_0 is $\binom{4}{x}(0.10)^x(0.90)^{4-x}$ if $x > 0$, or 1 if $x = 0$, and the maximum under H_1 is $\binom{4}{x}\left(\frac{x}{4}\right)^x\left(1-\frac{x}{4}\right)^{4-x}$ (given by $\theta = \frac{x}{4}$) if $x > 0$, or $(0.90)^4$ when $x = 0$. Therefore $L(x) = (10/9)^4$ when $x = 0$ and $L(x) = \left(\frac{0.40}{x}\right)^x\left(\frac{3.60}{4-x}\right)^{4-x}$ when $x > 0$. For $x = 2$, this is 0.1296, which is the c of Eqs. (2) and (3). The probability that $L(X) = c$ is the same as the probability that $X = 2$.

EXAMPLE 7 Suppose the null hypothesis H_0 is that X is a random variable with a rectangular distribution of mean 2 and range 2, and the alternative H_1 is that X has a rectangular distribution of mean 4 and range 4. It is clear from Figure 36 that H_0 must be accepted when $1 < x < 2$ and rejected when

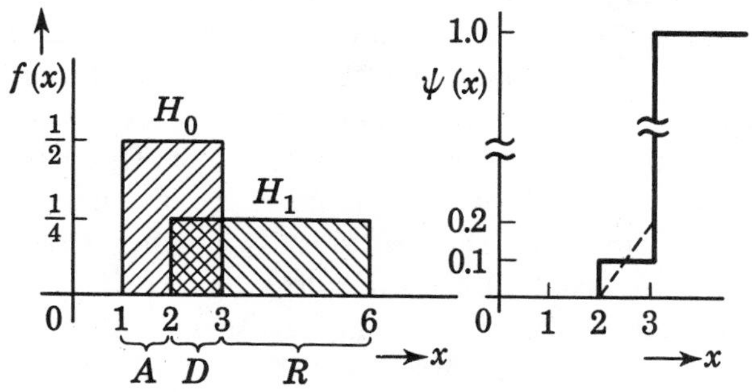

FIG. 36 RANDOMIZED TEST

$3 < x < 6$. The only doubt arises where the two distributions overlap, for $2 < x < 3$ (the region D). Evidently, $L(x) = \infty, 2, 0$ for the regions A, D and R. If in the region D the probability of rejection is ψ_0, we have

$$\alpha = P(3 < x < 6|H_0) + \psi_0 P(2 < x < 3|H_0)$$
$$= 0 + \psi_0(1/2)$$

If $\alpha = 0.05$, $\psi_0 = 0.10$. The power of the test is

$$P = P(3 < x < 6|H_1) + \psi_0 P(2 < x < 3|H_1)$$
$$= \tfrac{3}{4} + 0.1(\tfrac{1}{4}) = \frac{31}{40}$$

The value of ψ_0 is here not unique. If we take $\psi_0(x) = (x - 2)/5$, $2 < x < 3$, this will give the same size and power as $\psi_0 = 0.10$. The error α is now determined by

$$\alpha = P(3 < x < 6|H_0) + \int_2^3 \psi_0(x)P(2 < x < 3|H_1)\,dx$$
$$= \int_2^3 \frac{(x-2)}{10}\,dx = 0.05$$

6.13 **Statistical Decisions and Risk** The practical problem of the statistician is usually that of making a decision in the face of uncertainty. The problem may be one of deciding on the best value to use for some characteristic of a population, such as the variance, or it may involve deciding between alternative hypotheses. The decision may require *action*, for example, accepting or rejecting a lot of manufactured articles after having inspected a random sample, or recommending the use of a particular fertilizer for increasing the expected yield of a crop, after analysing some experimental results. We should like to have a sound guiding principle for use in making such decisions, but we must not expect too much from any single principle. The circumstances of a particular problem will be all-important.

There are two general decision principles which have been quite widely used, one associated with the names of Bayes and Laplace, the other due to Abraham Wald, although these are by no means the only possibilities. The Bayes rule is to choose that course of action which has the largest expectation of gain (or, which comes to the same thing, the smallest expectation of loss). This rule assumes that we know, or can estimate, the prior probabilities of the various possible situations with which we may be faced. The Wald, or minimax, principle is to choose that action which minimizes the maximum loss that could occur in the worst possible case. This is evidently a rather pessimistic attitude, but it does minimize the risk of a disastrous loss.

Both principles require the person making the decision to give numerical values to the gains, or losses, which will ensue from the various possible actions. Sometimes this is a fairly straightforward matter of cost accounting, and the values can be given in dollars and cents. If the problem is concerned with accepting or rejecting a lot of manufactured articles, on the basis of some sampling scheme, it will generally be possible to estimate fairly accurately the costs of sampling and inspection of individual items, and also the losses involved in accepting a poor lot or rejecting a good one. If, however, the problem is to decide between alternative medical treatments of a disease, the error of saying that a proposed new treatment is no better than the old one, when in fact it *is* better, may cost lives which might have been saved had the new treatment been adopted. Even when the gain is monetary, it may be argued that its value is different in different circumstances. A sum of $50 does not look the same to a millionaire and to a hobo. Economists have attempted to make a scale of "utility" to measure satisfactions and preferences, and where costs, or losses and gains, are mentioned later in this book the units may, if desired, be understood as units of utility.

The risk of making a wrong decision may often be very greatly reduced by taking a large number of observations, but sampling and experimentation cost money, or at least time and effort, and this should be reckoned in the total accounting. Wald therefore introduced a *risk function* which depends partly on the decision made or the action taken, and partly on the cost of experimenting so as to have a basis for decision.

Suppose we base our decision to take one of k possible actions $a_1, a_2 \ldots a_k$ on a single sample of N observations of a variate X. Let the cost be c_N, a bounded, non-negative number depending on N and possibly on the actual set of observations. (If all observations cost the same, c_N is proportional to N.) The probability of action a_i will depend on the decision rule d which is used, and may be denoted by $p(a_i|d)$, and d of course depends on the set of observations $x_1 \ldots x_N$. There will be a joint likelihood function $f(x_1 \ldots x_N)$ for any given set of values of X, and this function will generally depend on one or more parameters, the values of which represent the unknown state of Nature. For convenience we will suppose that there is only one parameter, θ, which can take a set of values symbolised by Ω.

If the statistician takes action a_i when θ is really equal to θ_j, we can suppose that his loss is $L(a_i, \theta_j)$. Wald regarded this loss as always non-negative, and equal to zero when the best possible decision in the circumstances is made. Any other decision involves a positive loss. The expected loss for the given set of observations is

(6.13.1) $$E[L(a_i, \theta_j)] = \sum_{i=1}^{k} L(a_i, \theta_j)p(a_i|d)$$

and the expected loss, whatever the sample observations may turn out to be, is

(6.13.2) $$r_1(\theta_j) = \sum_{i=1}^{k} \int L(a_i, \theta_j)p(a_i|d)f(x)\,dx$$

where $f(x)\,dx$ is written for $f(x_1 \ldots x_N)\,dx_1 \ldots dx_N$ and the integral is over the whole N-dimensional sample space.

The expected cost of the observations will be

(6.13.3) $$r_2(\theta_j) = \int c_N f(x)\,dx$$

since this cost will not depend on the subsequent action a_i. The risk function is the sum of r_1 and r_2, namely,

(6.13.4) $$r(\theta_j) = r_1(\theta_j) + r_2(\theta_j)$$

EXAMPLE 8 A zoologist wants to estimate the average number (μ) of a particular organism per unit volume in the water of a lake. He takes a sample of volume v and counts the number of such organisms (X) in this volume. He estimates μ by the ratio $x/v\,(= m)$, where x is the observed value of X.

The distribution of X may be assumed to be Poisson, so that

$$P(X = x|\mu) = (\mu v)^x \frac{e^{-\mu v}}{x!} \tag{6.13.5}$$

If m is equal to μ, the estimate is correct and there is presumably no loss. The loss will depend on the size of the error. It can hardly depend on the first power of the error, since if it did there would be a negative loss (a gain) when the error was in one direction. The simplest thing is to suppose that the loss is proportional to $(m - \mu)^2$. The loss due to estimating μ as m is, therefore,

$$L(m, \mu) = k(m - \mu)^2$$

The expected loss, whatever the observed x, is

$$\begin{aligned} r_1(\mu) &= \sum_{x=0}^{\infty} k\left(\frac{x}{v} - \mu\right)^2 \cdot (\mu v)^x \frac{e^{-\mu v}}{x!} \\ &= k\frac{\mu}{v} \end{aligned} \tag{6.13.6}$$

since this is a discrete distribution and the integral therefore becomes a sum.

The cost of the sample c may be added to r_1 to give the risk function.

6.14 **Bayes' Principle** This principle assumes that there is a prior probability distribution for the unknown parameter θ (which we may think of as a state of Nature). We will denote this probability density by p_θ. One hypothesis we might make regarding Nature is that θ belongs to the set ω (a subset of the set Ω of all possible values of θ). We investigate this hypothesis by taking a set of observations, which have values $x_1, x_2 \ldots x_N$ (collectively denoted by x). The probability of this set, given that θ belongs to ω, is $\int_{(\omega)} P(x|\theta)p_\theta \, d\theta$, the integration (or sum) being over all values of θ such that θ belongs to ω. The probability of the same set, whatever the value of θ, is $\int_{(\Omega)} P(x|\theta)p_\theta \, d\theta$. Therefore, the probability that θ belongs to ω, *given the observed set of values* x, is

$$P(\theta \in \omega|x) = \frac{\int_{(\omega)} P(x|\theta)p_\theta \, d\theta}{\int_{(\Omega)} P(x|\theta)p_\theta \, d\theta} \tag{6.14.1}$$

This rule was first clearly stated by Bayes [6], and used by him for reasoning back from the observed sample to the population sampled. Bayes recognized, however, that the use of this rule of inference depends upon knowing the prior probabilities p_θ, and except in artificial illustrations we seldom know much about these quantities. Bayes suggested, although apparently with some misgivings, that if we know nothing whatever about p_θ we should assume as a basis for action that all possible values of θ are equally likely. This suggestion was adopted, rather uncritically, by Laplace, but it was so vigorously attacked in recent times (mainly by Fisher) that the rule fell into disrepute. It is now beginning to be generally recognized that Bayes' approach may be very helpful in certain situations.

If the loss function for action a_i and state of Nature θ is $L(a_i, \theta)$, and if the prior probability density of θ is p_θ, the expected loss associated with a_i is

$$L(a_i) = \int_{(\Omega)} L(a_i, \theta) p_\theta \, d\theta \tag{6.14.2}$$

The Bayes principle is to choose that particular a_i for which $L(a_i)$ is a minimum, or, which comes to the same thing, that for which the utility is a maximum. If some information is available on p_θ, from other experiments or from intuition, this can be used, but if no information is available, p_θ is to be taken as uniform over all θ.

EXAMPLE 9 A dealer buys fuses in lots of 10,000 and sells them at 10 cents each, with a double-money-back guarantee if they prove defective. To protect himself he takes a sample of N for destructive testing, and refuses to buy the lot if n or more of the samples are defective. What value should he choose for n?

The probability of x defectives, if the proportion of defectives in the whole lot is θ, is

$$b(x, N, \theta) = \binom{N}{x} \theta^x (1 - \theta)^{N-x}$$

The probability of accepting a lot with proportion θ is $\sum_{x=0}^{n-1} b(x, N, \theta) = 1 - B(n, N, \theta)$ and the expected net income in dollars received from such a lot is

$$u(n, \theta) = (1 - 2\theta)\left(1000 - \frac{N}{10}\right)(1 - B(n, N, \theta))$$

since N of the 10,000 have been destroyed in sampling, and the dealer has to pay out 20 cents for each defective one he sells. Suppose he estimates the prior probability of θ as p_θ. His expected income from the decision rule he has adopted is

$$u(n) = \int_0^1 u(n, \theta) \cdot p_\theta \, d\theta$$

and he should choose n so as to make this as great as possible. If he feels that any value of θ is as likely as any other, he will put $p_\theta = 1$, and maximize the quantity

$$\frac{u(n)}{1000 - N/10} = \int_0^1 \sum_{x=0}^{n-1} \binom{N}{x} (1 - 2\theta)\theta^x (1 - \theta)^{N-x} \, d\theta \tag{6.14.3}$$

The integral can be evaluated in terms of beta functions and reduces to $\frac{n(N + 1 - n)}{(N + 1)(N + 2)}$, which is a maximum when $n = (N + 1)/2$. The lot should be rejected if there are k or more defectives in a sample of size $2k - 1$.

It may be considered unrealistic to suppose that p_θ is constant, and the dealer could, for example, suppose that θ is equally likely to be anywhere between 0.01 and 0.05, but is quite unlikely to be outside these bounds. That is, he could assume a rectangular distribution for θ. The integral in Eq. (3) will then be expressible in incomplete beta functions, and by the use of tables the maximizing value of n can be found.

6.15 **Wald's Principle (Minimax Principle)** To avoid having to estimate the prior probabilities, Wald suggested the principle of choosing the action which would minimize the maximum risk that could be feared whatever the state of Nature might be. In the example above, if θ could be as high as 1, or even a little more than 0.5, this principle would tell the dealer to refuse to accept the lot without troubling to sample it at all. If, however, he feels that the worst possible lot would have a θ equal to 0.05, say, he will choose n so as to maximize $u(n, 0.05)$. This will mean accepting the lot without sampling, since then he gets as much income as possible and avoids the loss of the fuses destroyed in testing.

EXAMPLE 10 This is a simplified betting problem, suggested by Sprowls [7]. The bettor has two possible actions in each case, to bet or not to bet. A bet is always to win and always at the same odds; if he wins he gains α, if he does not win he loses β. He has a system which gives a probability θ of picking a winner, and he decides whether or not to bet on any particular race by the number of wins, x, recorded in the N previous races on which he has bet. If $x \geq n$, he will bet; if $x < n$, he will not. The problem is to decide on n.

Assuming that the races can be treated as statistically independent events, the probability of exactly x wins is $b(x, N, \theta)$ so that the probability of betting is $B(n, N, \theta)$. If the bettor decides to bet, his expected loss per race is $\beta(1 - \theta) - \alpha\theta$, which is positive for $\theta < \theta_0$, where $\theta_0 = \beta/(\alpha + \beta)$.

If he decides not to bet at all, his gain will be zero, but he will lose what he might have won by betting if $\theta > \theta_0$. The risk function is

$$
\begin{aligned}
(6.15.1)\qquad r(n, \theta) &= [\beta(1 - \theta) - \alpha\theta]\cdot B(n, N, \theta), \qquad \theta \leq \theta_0 \\
r(n, \theta) &= [\alpha\theta - \beta(1 - \theta)]\cdot[1 - B(n, N, \theta)], \qquad \theta \geq \theta_0
\end{aligned}
$$

Using the normal approximation to the cumulative binomial, we have $B \approx 1 - \Phi(z)$, where

$$
(6.15.2)\qquad z = \frac{n - 1/2 - N\theta}{[N\theta(1 - \theta)]^{1/2}}
$$

The Wald principle is to pick n so as to minimize the maximum value of r over all possible θ. This minimum occurs when the maximum of r for $\theta \leq \theta_0$ is equal to the maximum of r for $\theta \geq \theta_0$. The actual calculation of these maxima can be done numerically with the help of good tables of the normal law (e.g., reference [8] of Chapter 3), and it turns out that the maxima occur at $\theta \approx \theta_0 \pm 0.752$ $[\theta_0(1 - \theta_0)/N]^{1/2}$. The approximate solution of the problem is to take n as $N\theta_0 = N\beta(\alpha + \beta)^{-1}$, and if $x \geq n$ to bet on the next race.

6.16 **Game Theory and Statistics** A subject which has come to the fore in recent years is the theory of games, which, besides its application to ordinary parlor games, has a very distinct relevance to economics and to military strategy. A good many statistical problems can be thought of as games played against Nature, although Nature of course is not a malevolent opponent, out to make things as bad as possible for the statistician. It is hardly surprising that in these circumstances Wald's principle is often unduly pessimistic, and several other decision criteria have been suggested. References [8], [9] and [10] may be consulted for more details.

PROBLEMS

A. (§§ 6.1–6.5)

1. Prove that the maximum likelihood estimator for the parameter μ of a Poisson distribution is the sample mean m, and that the variance of m is μ/N, where N is the sample size.

2. Show that, for the Poisson distribution, m is a sufficient estimator for μ. Also show that condition (b) of Eq. (6.4.3) is satisfied.

3. For a normal population with mean μ and variance σ^2, the mean and the median of a sample of N are both consistent estimators of μ. For large N the variance of the median is approximately $\pi\sigma^2/2N$. Show that as an estimator of μ the median is roughly 64% efficient.

4. Suppose that the mean μ of a normal population is known, but that the variance σ^2 is to be estimated. Show that the sample variance k_2, although unbiased, has an efficiency $(N - 1)/N$ and is therefore only asymptotically most efficient, while the sample second moment about μ is both unbiased and most efficient, for any N.

5. Show that for a binomial population with probability of success θ in each trial, the maximum likelihood estimator of θ is the proportion of successes p in a sample. Show also that the variance of p, as given by Eq. (6.3.3), agrees with that previously found, namely, $\theta(1 - \theta)/N$.

6. A one-parameter gamma variate has the density function $f(x) = x^{\alpha-1}e^{-x}/\Gamma(\alpha)$, $x \geq 0$. Write down the equations for determining from a sample of size N the maximum likelihood estimator for α and its variance. (In order to solve these equations, tables of the digamma function $d \log \Gamma(\alpha)/d\alpha$ and the trigamma function $d^2 \log \Gamma(\alpha)/d\alpha^2$ must be used. (See H. T. Davis, Tables of the Higher Mathematical Functions, Bloomington, Indiana, 1933–5.)

7. Prove that an unbiased estimator of α in Problem 6 is the arithmetic mean m of the sample. Show also that the efficiency of this estimator is $\{\alpha\, d^2[\log \Gamma(\alpha)]/d\alpha^2\}^{-1}$. (This quantity tends to zero as α decreases to 0. The nearer α is to zero the more skew is the distribution.)

8. The mean absolute deviation for a sample of size N and mean m is defined as $d = \sum_i |x_i - m|/N$. For samples from a normal population of mean μ and variance σ^2, the variance of d is given by

$$\frac{2\sigma^2(N-1)}{\pi N^2}\{\tfrac{1}{2}\pi + [N(N-2)]^{1/2} - N + \sin^{-1}[1/(N-1)]\}$$

Compare the asymptotic efficiency of the quantity $d\sqrt{\pi/2}$ with that of the sample standard deviation as estimators of σ. *Hint:* Prove that $V(d) = \sigma^2(1 - 2/\pi)/N + O(1/N^2)$, and note that $V(s) = \sigma^2/2N + O(1/N^2)$.

9. Show that for the continuous distribution with density $f(x) = \theta e^{-\theta x}, 0 \leq x < \infty$, confidence limits for θ for a large sample, with confidence coefficient 0.95, are given by $(1 \pm 1.96/\sqrt{N})/m$ where m is the sample mean. *Hint:* For large N, the m.l. estimator is approximately normal.

10. Show that for the rectangular population with density $f(x) = (\beta - \alpha)^{-1}$, $0 < \alpha < x < \beta$, joint maximum likelihood estimators for α and β are the smallest and largest members of the sample respectively. *Hint:* Show that these give the greatest possible value for L.

11. Suppose that the discrete variate X is binomially distributed except that it cannot take the value 0. The probability that $X = x (x = 1, 2 \ldots n)$ is given by $f(x) = \binom{n}{x}\theta^x(1 - \theta)^{n-x}[1 - (1 - \theta)^n]^{-1}$. If the numbers of successes in N repetitions of the sequence of n trials are $x_1, x_2, \ldots x_N$, show that the maximum likelihood estimator $\hat{\theta}$ of θ is given by the solution of the equation $n\hat{\theta} = m[1 - (1 - \hat{\theta})^n]$, where m is the arithmetic mean of the x_i. Find $\hat{\theta}$ for the case $n = 2$.

12. Obtain an equation for the maximum likelihood estimator of ρ derived from a sample of size N from the bivariate standard normal population with joint density function

$$f(x, y) = (2\pi)^{-1}(1 - \rho^2)^{-1/2} \exp\left(-\frac{1}{2}\frac{x^2 - 2\rho xy + y^2}{1 - \rho^2}\right)$$

Show that the variance of this estimator is $(1 - \rho^2)^2/[N(1 + \rho^2)]$. *Hint:* $E(x^2) = E(y^2) = 1$, $E(xy) = \rho$.

B. (§§ 6.6–6.12)

1. The yield in bushels of a certain type and size of potato plot is found to be normally distributed with a standard deviation of 2.36. It is hoped that the application of a certain fertilizer will increase the yield by at least 0.5 bushel. How large a sample of plots should be used to detect a difference of this amount, using the mean sample yield as a criterion, with a test of size 5% and power 90%?

2. An experimenter knows that a distribution is approximately normal with standard deviation 1.2. He wishes to test the hypothesis that the population mean is 75 against the alternative hypothesis that $\mu > 75$, using a sample of size N and a test of size 1%. What test should he use? Calculate the power for $N = 9$ and for $\mu = 75.5$, 76, 76.5. What size sample should he take if he wants to be 95% sure of detecting a difference as small as one unit from the assumed value 75, still using a test of size 1%?

3. A population has the Poisson distribution with parameter μ, which may have the values 1 or 2 but no others. Find the likelihood-ratio test for testing H_1(that $\mu = 1$) against H_2 (that $\mu = 2$), using the mean of 10 observations of X as the criterion. Assume that the probability of error of the first kind is not greater than 0.05, and calculate the power of the test. *Hint:* The distribution of the sum of N independent Poisson variates with parameter μ is also Poisson with parameter $N\mu$. Use a table of the cumulative Poisson function for numerical results.

4. Develop the likelihood-ratio test for a binomial population, for testing the simple hypothesis $\theta = \theta_0$ against the simple alternative $\theta = \theta_1$ (where $\theta_1 > \theta_0$), using as a criterion the number x of successes in the first n trials. If the size of the test is approximately α and the power approximately $1 - \beta$, find a relation to determine n. Give a numerical result for $\theta_0 = 0.5$, $\theta_1 = 0.7$, $\alpha = 0.05$, $\beta = 0.10$. *Hint:* Use the normal approximation to the binomial.

5. Find the likelihood-ratio test for testing the significance of the difference between the mean of a sample and an assumed population mean μ_0, the population being normal with unknown standard deviation σ. *Hint:* Find the ratio of the maximum likelihood over all σ for a given μ_0 to that over all σ and all μ. The test is equivalent to Student's t-test, which will be discussed more fully in Chapter 8.

6. Assume that the random variables Y_i, $i = 1, 2 \ldots N$, are independent and normally distributed with means $\eta = \alpha + \beta x_i$, and a common variance σ^2, for known values of x_i. Write down the likelihood function for the set of Y_i and hence obtain joint maximum likelihood estimators for the parameters α, β and σ. (These estimators will be discussed more fully in Chapter 11.)

7. It is desired to test the null hypothesis that a certain coin is fair (that is, that the probability θ of a head when the coin is tossed is 0.5) by counting the number of heads x in n tosses. Show that the likelihood-ratio test of H_0 against the alternative hypothesis H_1 (that θ is either less than or greater than 0.5) is equivalent to rejecting H_0 when $|x - n/2| > k$, where k is determined by the size of the test.

If the test is to be of size 0.05 and power 0.9 to detect a difference of 0.02 in θ from the assumed value 0.5, how large should n be? *Hint:* The likelihood-ratio test may be written: reject H_0 when $f(x) > c$, where $f(x) = x \log(x/n) + (n - x)\log(1 - x/n)$. Show that $f(x)$ has a minimum at $x = n/2$ and is symmetrical about this value. For the second part of the question use the normal approximation to the binomial.

C. (§§ 6.13–6.15)

1. Carry out the integration indicated in Eq. (6.14.3) and show that it reduces to the stated value. *Hint:* Use Eq. (4.5.3) and express the gamma functions as factorials.

2. A bag contains 10 balls, either black or white, but it is not known how many of each. A ball is drawn at random, looked at and replaced, and three times running the ball so selected is white. What is the probability that the bag contains at least five white balls? *Hint:* Use Bayes' rule, with sums instead of integrals. Obtain a numerical result by assuming a constant value for the prior probability of θ white balls ($\theta = 0, 1, 2 \ldots 10$).

3. Instead of the assumption at the end of Problem 2, suppose that the bag was filled by picking 10 balls at random from a very large number of black and white balls mixed in equal proportions. What is now the probability of at least five white balls, after seeing the three white balls drawn? *Hint:* The prior probabilities are binomial.

4. A set of 100 independent observations is made on a variate X which is normally distributed with unknown mean μ and known variance 25 units. The null hypothesis H_0 is that $\mu = 0$, and the alternative hypothesis H_1 is that $\mu = 2$ (these are the only possibilities). The decision whether to accept H_0 or H_1 is made on the basis of the mean (m) of X for the 100 observations. If H_0 is true, the losses corresponding to these two decisions (d_0 and d_1) are 0 and 25, respectively; if H_1 is true the losses are 10 and 0, respectively.

Given that ξ is the prior probability of H_0, show that, on the Bayes principle of minimizing the expected loss, the decision d_0 should be taken if $m < c$ where $c = 1 + (1/8) \log_e\{5\xi/[2(1 - \xi)]\}$. *Hint:* The mean m is normally distributed with variance $\frac{1}{4}$. Use Bayes' rule to find the probability of H_0 after the sample has been examined.

5. In Problem 4 above, find the probability $\alpha(\xi)$ of rejecting H_0 if true and the probability $\beta(\xi)$ of accepting H_0 if false.

The expected loss is $25\xi\alpha(\xi) + 10(1 - \xi)\beta(\xi)$. Compute this quantity for various values of ξ between 0 and 1 and find approximately for what value the expected loss is a maximum.

6. Solve Problem 4 above using the minimax principle. With ξ unknown, the maximum risk is $25\alpha(\xi)$ if H_0 is true and $10\beta(\xi)$ if H_1 is true. The minimum of this maximum risk occurs when $25\alpha(\xi) = 10\beta(\xi)$. Show that c is then 1.096, and that the corresponding value of ξ is about 0.46. (This is the least favorable value of ξ.)

7. A buyer of manufactured articles in large lots is willing to accept a lot if the proportion θ of defective articles is less than θ_0, but will wish to reject it if $\theta > \theta_0$. If he accepts a lot, his loss $L_1(\theta)$ will be zero if $\theta \leq \theta_0$ but positive if $\theta > \theta_0$. If he rejects a lot his loss $L_2(\theta)$ will be zero if $\theta \geq \theta_0$ but positive if $\theta < \theta_0$. He bases his decision on the number of defectives r in a sample of N (assumed binomial). What will be his decision rule on the Bayes principle if the prior probability of θ is $\xi(\theta)$? Show that this is equivalent to the rule: accept the lot if $r < c$, where c is some fixed number. *Hint:* He will accept with r defectives if the expected loss in accepting is less than the expected loss in rejecting. Show that if this is true for $r = c$ it is also true for $r = c - 1$ and so for all $r < c$.

REFERENCES

[1] Kendall, M. G., *Advanced Theory of Statistics, Vol. II*, Chas. Griffin, 1951.

[2] Cramér, H., *Mathematical Methods of Statistics*, Princeton Univ. Press, 1946.

[3] Neyman, J., and Pearson, E. S., "On the Problem of the Most Efficient Tests of Statistical Hypotheses," *Phil. Trans. Roy. Soc.* **A231,** 1933, pp. 289–337.

[4] Wilks, S. S., "The Large Sample Distribution of the Likelihood Ratio for Testing Composite Hypotheses," *Ann. Math. Stat.* **9,** 1938, pp. 60–62.

[5] Fix, Evelyn, *Tables of Non-Central Chi-Square*, Univ. of Calif. Press, 1949.

[6] Bayes, Thomas, "An Essay Towards Solving a Problem in the Doctrine of Chances," *Phil. Trans. Roy. Soc.*, **53,** 1763, pp. 370–418. This essay is reprinted, with biographical material regarding Bayes, in *Biometrika*, **45,** 1958, pp. 293–315.

[7] Sprowls, R. C., "Statistical Decisions by the Method of Minimum Risk: an Application," *J. Amer. Stat. Ass.*, **45,** 1950, pp. 238–248.

[8] Keeping, E. S., "Statistical Decisions," *Amer. Math. Monthly*, **63,** 1956, pp. 147–159.

[9] Bross, I. D. J., *Design for Decision*, Macmillan, 1953. This is written for popular consumption and contains little mathematics.

[10] Luce, R. D., and Raiffa, H., *Games and Decisions*, Wiley, 1957. This gives a very full discussion of general principles.

Chapter 7

SOME SAMPLING PROCEDURES

7.1 **Random and Less Random Samples** Sampling is undertaken in order to find out something about a population without having to examine every item in it. By "a population," we mean any collection (usually large) of elements such as people, pigs, farms, coin-tosses, incomes, or whatever it may be, about which we want some information. Often an important decision must be made on the basis of knowledge obtained from the sample, so that it is useful to be able to estimate how reliable this knowledge actually is. Sampling theory is concerned with ways of estimating, and perhaps improving, the precision of the information obtainable from a sample about a population.

Any procedure for making such an estimate must be based on the theory of probability. That is, it must suppose that the sample is random. Most of the theory of estimation, hypothesis-testing and decision-making that we have been considering in the last two chapters is based on the concept of a random sample. A sample of given size is said to be random if every possible sample of this size in the population (supposedly finite) has a calculable probability of being chosen, but this probability need not be the same for all items. If, however, the sample (of size N) is selected in such a way that every combination of N elements in the population has an *equal* probability of being chosen, the process is called *simple random sampling*. This is the usual assumption in theoretical statistics, although in actual sample surveys simple random sampling is rarely used. For reasons of cost and administrative convenience, as well as in order to improve precision, some modification of the simple random design is generally adopted.

For a sample of N from a finite population of size M the probability that any individual item will be drawn is N/M. If the population is infinite, this probability is zero, but it still makes sense in many cases to assume that one item is as likely to be drawn as another (see § 5.1). When a number of tosses, made with a particular coin, is considered as a sample of the practically infinite number of tosses that might conceivably be made with this same coin, the sample is obviously not random in the strict sense, since it consists of the first N items of the population in order of time. However, we make the physical assumption that the order in time is quite irrelevant as far as the characteristic of any toss (heads or tails) is concerned, so that the first N tosses form effectively a random sample.

If some prior information is available, it may be possible to use *stratified sampling* and so gain in precision over simple random sampling. In this procedure, the population is divided into groups, the elements within a group being more alike than those in the population as a whole. If a simple random sample is drawn from each group, we still have a probability sampling procedure, but we have insured that each group is represented in the total sample. The groups are called *strata*, and the process of dividing the population into groups is called *stratification*. This procedure normally reduces the sampling variance of the variate measured. It is particularly effective when there are extreme values in the population—stratification with regard to income levels, for example, is a common practice. The costs of sampling may differ considerably from one stratum to another (as between urban and rural households, for instance) and these costs may be important in setting up the strata.

The people who conduct sample surveys are usually much concerned with questions of cost. They want the maximum precision per dollar spent, and therefore tend to favor *cluster sampling*. This is a method of reducing costs by first taking a random sample of groups or clusters and then taking sub-samples from the clusters selected. To take a sample of 3000 households from the population of the United States, we might first draw a sample of, say, 50 counties and then sample these proportionately to their total populations. A simple random sample would probably be spread over many more than 50 counties, and would need much more travel and supervision.

Cluster sampling may not be very efficient as far as precision of the estimate is concerned. The best results occur when the clusters each contain very diversified elements—just the opposite from the requirements for stratified sampling.

Another common procedure in some types of sampling surveys is *systematic sampling*. To draw 500 cards from a file containing 10,000 cards, we could select a random number between 1 and 20 (say 13) and then take every 20th card, beginning with the 13th. That is, we could pick the cards numbered 13, 33, 53, and so on. If the order of the cards has nothing whatever to do with the variate for which we are sampling, this gives us effectively a random sample and it is easy to apply. To sample housing units in a city one might, for instance, take every 12th block and every fifth house in the block, but it would be well to make sure that the procedure adopted did not lead to picking out an undue proportion of corner houses—at least if the object of the survey has any connection with economic status. Corner houses often pay higher taxes and are generally occupied by people with higher incomes than non-corner houses.

A method of selecting a sample often employed in public-opinion polls is that of *quota* sampling, in which an interviewer is instructed to fill a specified quota by finding as best he can persons satisfying certain restrictions—he may be asked to contact a specified number of persons of a particular sex, age-group, and income-group, for example, but no attempt is made to make the sample random. This method is apt to introduce a completely unknown bias into the estimates made.

Any method of sampling which is not random, but tries to pick out a typical or representative sample, may be called *purposive sampling*. This may be useful if only a very small sample can be taken, and if the person picking the sample has good judgment and expert knowledge, but there is no statistical theory available for measuring the reliability of the results obtained. Sometimes, of course, a random sample is from the nature of things practically unobtainable—a sample of fish from the sea, for instance—and we are forced to use any kind of sample we can get. Nevertheless a probability sample should be obtained whenever possible, and only then is the theory of sampling strictly applicable. For a fuller discussion of sampling procedures see [1] and [2].

* 7.2 **Stratified Sampling** Suppose the population of size M is divided into strata of sizes $M_1, M_2 \ldots M_k$, and a simple random sample of size N_i is taken from the ith stratum. If $X_{i\alpha}$ is the measured characteristic for the αth item in the ith stratum, the ith stratum mean for the population is

$$\mu_i = \frac{1}{M_i} \sum_{\alpha=1}^{M_i} X_{i\alpha}, \qquad \sum M_i = M, \tag{7.2.1}$$

and the over-all mean for the population is

$$\mu = \frac{1}{M} \sum_{i=1}^{k} M_i \mu_i \tag{7.2.2}$$

The estimator of μ, based on the stratified sample, is $\bar{X}$, where

$$\bar{X} = \frac{1}{M} \sum M_i \bar{X}_i, \qquad \bar{X}_i = \frac{1}{N_i} \sum_{j=1}^{N_i} X_{ij} \tag{7.2.3}$$

and X_{ij} is the value of X for the jth item in the sample from the ith stratum. Note that $\bar{X}$ is a weighted mean of the sample stratum means, with weights depending on the sizes of the strata in the *population*, so that these sizes must be known fairly accurately before the method can be used.

From Eq. (5.7.3), $E(\bar{X}_i) = \mu_i$, and therefore, by Eq. (3) above,

$$E(\bar{X}) = \frac{1}{M} \sum M_i E(\bar{X}_i) = \mu \tag{7.2.4}$$

by Eq. (2). That is, $\bar{X}$ is an unbiased estimator of μ. We will now show that the variance of $\bar{X}$ is given by

$$V(\bar{X}) = \frac{1}{M^2} \sum_i \frac{M_i(M_i - N_i)}{N_i} \sigma_i^2 \tag{7.2.5}$$

where $\sigma_i^2 = (M_i - 1)^{-1} \sum_\alpha (X_{i\alpha} - \mu_i)^2$, the population variance in the ith stratum.

Since $\bar{X}$ is a linear combination of the $\bar{X}_i$ with coefficients M_i/M, and since the strata are sampled independently, we can use Bienaymé's Theorem (§ 2.14).

$$\begin{aligned} V(\bar{X}) &= \sum_i \left(\frac{M_i}{M}\right)^2 \cdot V(\bar{X}_i) \\ &= \sum_i \left(\frac{M_i}{M}\right)^2 \kappa'_{2i}(N_i^{-1} - M_i^{-1}) \end{aligned}$$

by Eq. (5.11.7), where $\kappa'_{2i} = \sigma_i^2$. If f_i is the sampling fraction N_i/M_i,

$$V(\bar{X}) = \frac{1}{M^2} \sum_i \frac{M_i^2}{N_i} (1 - f_i)\sigma_i^2 \tag{7.2.6}$$

which is equivalent to Eq. (5).

Since $k_{2i}[= (N_i - 1)^{-1} \sum_j (X_{ij} - \bar{X}_i)^2]$ is an unbiased estimator of κ'_{2i}, we may estimate the variance of $\bar{X}$ by means of

$$\hat{V}(\bar{X}) = \frac{1}{M^2} \sum_i \frac{M_i^2}{N_i} (1 - f_i)k_{2i} \tag{7.2.7}$$

and k_{2i} is the sample variance of the sample from the ith stratum.

If f_i is the same for each stratum, and equal to f, say, the sampling is said to be *proportionate* (the N_i are proportional to the M_i). Then

$$\begin{aligned} V(\bar{X}) &= \frac{1 - f}{M^2} \sum \frac{M_i}{f} \sigma_i^2 \\ &= \frac{1 - f}{MN} \sum_i M_i \sigma_i^2 \end{aligned} \tag{7.2.8}$$

since $N \doteq f M$.

If X and Y are two variates, both measured for each item in the sample, the covariance of $\bar{X}$ and $\bar{Y}$ is similarly given by

$$C(\bar{X}, \bar{Y}) = \frac{1}{M^2} \sum_i \frac{M_i^2}{N_i} (1 - f_i)\pi_i \tag{7.2.9}$$

where $\pi_i = (M_i - 1)^{-1} \sum_\alpha (X_{i\alpha} - \mu_i)(Y_{i\alpha} - \nu_i)$, the population covariance in the ith stratum, ν_i being the stratum mean for Y. As before, π_i may be estimated from the sample covariance in this stratum.

The gain in precision due to using proportionate sampling, compared with simple random sampling from the whole population, may be found by comparing Eq. (7.2.8) with the expression for the variance of the mean of a random sample, namely,

$$V_r(\bar{X}) = \sigma^2(1 - f)/N \tag{7.2.10}$$

where $\sigma^2 = (M - 1)^{-1} \sum_\alpha (X_\alpha - \mu)^2$, and α takes any value from 1 to M. Therefore,

$$V_r(\bar{X}) - V(\bar{X}) = \frac{1 - f}{N} \left(\sigma^2 - \frac{1}{M} \sum M_i \sigma_i^2 \right) \tag{7.2.11}$$

In practice, the M_i are usually so large that the distinction between M_i and $M_i - 1$ is unimportant, and we can put

$$M\sigma^2 \approx \sum_{\alpha=1}^{M} (X_\alpha - \mu)^2, \qquad M_i\sigma_i^2 \approx \sum_{\alpha=1}^{M_i} (X_\alpha - \mu_i)^2$$

Since

$$(X_\alpha - \mu_i)^2 = (X_\alpha - \mu)^2 + (\mu - \mu_i)^2 + 2(\mu - \mu_i)(X_\alpha - \mu)$$

we have

$$M_i\sigma_i^2 \approx \sum_1^{M_i} (X_\alpha - \mu)^2 + M_i(\mu - \mu_i)^2 + 2M_i(\mu - \mu_i)(\mu_i - \mu)$$
$$= \sum_1^{M_i} (X_\alpha - \mu)^2 - M_i(\mu - \mu_i)^2$$

so that

(7.2.12) $$V_r(\bar{X}) - V(\bar{X}) \approx \frac{1-f}{MN} \sum_i M_i(\mu - \mu_i)^2$$

and this expression is never negative. The gain from proportionate sampling is greater, the greater the differences between the stratum means.

The question arises as to the *optimum choice* of the N_i. It was proved by Neyman [3] that for a fixed N the variance of $\bar{X}$ is least when N_i is proportional to $M_i\sigma_i$. That is, we should choose N_i so that

(7.2.13) $$N_i/N = M_i\sigma_i/\sum(M_i\sigma_i).$$

This, of course, supposes that some information, from a preliminary survey or from previous experience, is available about the σ_i.

If the cost c_i per unit of sampling also varies from one stratum to another, and if the total cost c of the whole survey is fixed, it may be shown that the optimum sampling number for the ith stratum is proportional to $M_i\sigma_i/c_i^{1/2}$.

* 7.3 **Cluster Sampling** In simple cluster sampling, the elements of the population are grouped in clusters which themselves are the primary sampling units. In a one-stage plan all the elements in the selected clusters (picked by simple random sampling) are included in the sample. In a two-stage plan a random sub-sample is selected from each primary sampling unit, and of course further stages of sampling can be introduced.

We will suppose that there are k clusters in the population, with sizes $M_i (i = 1, 2 \ldots k)$, and l of these are selected in the first stage. From the jth selected cluster $(j = 1, 2 \ldots l)$, the number of second-stage items picked is N_j. This is represented diagrammatically in Figure 37, where samples are indicated from some, but not all, of the clusters. Let $X_{i\alpha}$ be the value of X for the αth item in the ith cluster, and X_{jh} the value for the hth item picked from the jth selected cluster $(\alpha = 1, 2 \ldots M_i,\ h = 1, 2 \ldots N_j)$.

With a notation similar to that used before,

$$M_i\mu_i = \sum_\alpha X_{i\alpha} \qquad M\mu = \sum_i M_i\mu_i$$
$$N_j\bar{X}_j = \sum_h X_{jh} \qquad N\bar{X} = \sum_j N_j\bar{X}_j$$
$$M = \sum M_i \qquad N = \sum N_j$$

We take as the estimator of μ the quantity

$$\overline{X} = \frac{k}{Ml}\sum_j M_j\overline{X}_j \tag{7.3.1}$$

and it can be shown that

$$E(\overline{X}) = \mu \tag{7.3.2}$$

and

$$V(\overline{X}) = \frac{k}{M^2 l}\left[(k-l)\tau^2 + \sum_i \frac{M_i(M_i - N_i)}{N_i}\sigma_i^2\right] \tag{7.3.3}$$

where

$$\tau^2 = (k-1)^{-1}\sum_i (M_i\mu_i - M\mu/k)^2 \tag{7.3.4}$$

and σ_i^2 has the same meaning as in Eq. (7.2.5).

From Eq. (1), $E(\overline{X}) = \frac{k}{Ml}\sum_j E(M_j\overline{X}_j)$. Now the actual value of $M_j\overline{X}_j$ depends upon two random events, the selection of the jth cluster and the selection of the N_j items from this cluster. It is shown in Appendix A.14 that if X is a random variable depending on Y which is itself a random event, then $E(X) = E[E(X|Y)]$. Here the event Y is the choice of the jth cluster. Given this choice, the expectation of $M_j\overline{X}_j$ is $M_j\mu_j$, where μ_j is the mean for the whole cluster. Therefore

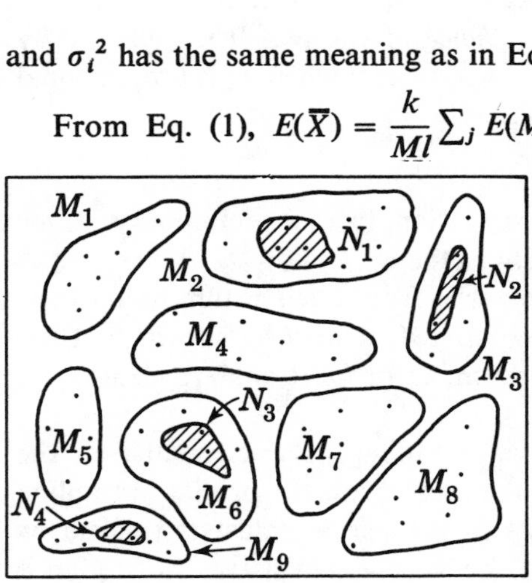

FIG. 37 CLUSTER SAMPLING

$$E(M_j\overline{X}_j) = E(M_j\mu_j) = \frac{1}{k}\sum_i M_i\mu_i \tag{7.3.5}$$

since this cluster is one of k clusters, all with an equal chance of being picked. It follows that

$$\begin{aligned} E(\overline{X}) &= \frac{1}{Ml}\sum_j\sum_i M_i\mu_i \\ &= \frac{1}{M}M\mu = \mu. \end{aligned} \tag{7.3.6}$$

To find the variance of $\overline{X}$ we need Theorem 2 of Appendix A.14, according to which $V(X) = E[V(X|Y)] + V[E(X|Y)]$, where X is replaced by $\overline{X}$ and Y means the choice of a particular set of l clusters from the set of all k clusters in the population. When this set is fixed, the problem becomes one of stratified sampling with l strata. The variance of the jth mean $\bar{X}_j$ is, by Eq. (5.11.7)

$$V(\bar{X}_j) = \sigma_j^2\left(\frac{1}{N_j} - \frac{1}{M_j}\right)$$

where $\sigma_j^2 = (M_j - 1)^{-1} \sum_\alpha (X_{j\alpha} - \mu_j)^2$, so that

$$V(\bar{X}|Y) = \frac{k^2}{l^2M^2} \sum_j M_j^2\sigma_j^2\left(\frac{1}{N_j} - \frac{1}{M_j}\right)$$

The expectation of this for any choice of the l clusters is

$$\text{(7.3.7)} \qquad E[V(\bar{X}|Y)] = \frac{k}{lM^2} \sum_i M_i^2\sigma_i^2\left(\frac{1}{N_i} - \frac{1}{M_i}\right)$$

which is the second term in Eq. (3). The first term represents the part of the variance due to first-stage sampling. Since $E(M_j\bar{X}_j|\,Y) = M_j\mu_j$,

$$\text{(7.3.8)} \qquad E(\bar{X}|Y) = \frac{k}{Ml} \sum_j M_j\mu_j = \frac{k}{M}\,\overline{M_j\mu_j}$$

By Eq. (5.11.7),

$$V(\overline{M_j\mu_j}) = \kappa_2' \,(1/l \quad 1/k)$$

where

$$\kappa_2' = [\textstyle\sum_i (M_i\mu_i - M\mu/k)^2]/(k - 1),$$

so that

$$\text{(7.3.9)} \qquad V[E(\bar{X}|Y)] = \frac{k^2}{M^2}\left(\frac{1}{l} - \frac{1}{k}\right)\frac{1}{k - 1} \sum_i \left(M_i\mu_i - \frac{M\mu}{k}\right)^2.$$

This gives the first term in Eq. (3). This term is small when the clusters are very much alike in size and composition.

* 7.4 **Systematic Sampling** Suppose the population consists of the elements $E_1, E_2 \ldots E_m$, arranged in some fixed order. Any systematic sample consists of the elements $E_i, E_{k+i}, E_{2k+i} \ldots E_{(N-1)k+i}$, where i is one of the numbers $1, 2 \ldots k$, and $Nk \leq M$. Usually, for a sample of size N, k is chosen so that Nk is as near to M as possible.

Systematic sampling divides the population in effect into strata, each consisting of k successive units, and chooses one sampling unit per stratum. The choice is not random, however, since the unit chosen occupies the same ordinal position in each stratum. Since a systematic sample is spread evenly over the population, it often gives a good estimate of the mean, although the variance of the estimator is in most cases greater than that for a simple random sample.

If X_α is the variate measured on the αth element the sample mean for the ith systematic sample is

$$m_i = \frac{X_i + X_{k+i} + \dots + X_{(N-1)k+i}}{N} \tag{7.4.1}$$

Since there are k values of i, all equally likely,

$$E(m_i) = \sum_i \frac{m_i}{k} = \frac{1}{Nk} \sum_{\alpha=1}^{Nk} X_\alpha \tag{7.4.2}$$

If $Nk = M$, this expression is the population mean μ, so that m_i is an unbiased estimator for μ. Also,

$$\begin{aligned} V(m_i) &= E(m_i - \mu)^2 \\ &= \frac{1}{k}\sum_i (m_i - \mu)^2 = \frac{1}{N^2k}\sum_i (Nm_i - N\mu)^2 \end{aligned} \tag{7.4.3}$$

Now by Eq. (1), $Nm_i - N\mu = (X_i - \mu) + (X_{k+i} - \mu) + \dots + (X_{(N-1)k+i} - \mu)$ so that

$$\sum_{i=1}^{k} (Nm_i - N\mu)^2 = \sum_{\alpha=1}^{Nk} (X_\alpha - \mu)^2 + 2\sum_{j=1}^{N-1} \sum_{\beta=1}^{k(N-j)} (X_\beta - \mu)(X_{\beta+jk} - \mu) \tag{7.4.4}$$

The first term on the right-hand side is $(M - 1)\sigma^2$. The second term vanishes if there is no correlation between pairs such as X_β and $X_{\beta+jk}$, separated by jk items. Correlation of this type is called *serial correlation.*

If the items are serially uncorrelated,

$$V(m_i) = \frac{(M-1)\sigma^2}{N^2k} = \frac{M-1}{M}\frac{\sigma^2}{N} \tag{7.4.5}$$

The corresponding value for the variance of the mean m_r of a random sample of size N is $(N^{-1} - M^{-1})\sigma^2$, so that

$$\frac{V(m_i)}{V(m_r)} = \frac{M-1}{M-N}$$

This is greater than 1 for any $N > 1$. However, if there is a sufficiently large negative serial correlation, $V(m_i)$ may be less than $V(m_r)$. See [4].

* 7.5 **Double Sampling** It is sometimes useful to take a preliminary large sample in order to get some information which will serve as a basis for drawing a sub-sample for further investigation, particularly if the large sample can be obtained rather cheaply. The information is used for stratification, or in other ways, in order to increase the precision of estimates from the smaller and more costly sub-sample.

For instance, suppose we are concerned with a variable T, such as total sales in all retail stores of a certain type, which may be rather hard to obtain. The large stores of this type will be much more important in providing an estimate of T than the small ones, and we would like to have a much larger sampling fraction of the larger stores. As a preliminary to the sampling design we might make a survey of a large fraction of the population of stores, obtaining only simple information on size (say the number of employees), and use this to decide on the sub-sample which will be investigated to determine T.

Suppose we classify the stores in the original sample of N (from a population of M) as large (N_1) or small (N_2). The corresponding numbers in the population are M_1 and M_2. If the sub-sample consists of all the large stores and n_2 of the small ones (sampling fraction $n_2/N_2 = 1/k$), an unbiased estimator of T is

(7.5.1) $$\hat{T} = \frac{1}{f}(T_1 + kT_2)$$

where T_1 is the total sales for the N_1 large stores and T_2 is the total for the n_2 small stores. Here f is the primary sampling fraction N/M. The total sub-sample size is $N_1 + n_2$. Equation (1) may be written

(7.5.2) $$\hat{T} = \frac{M}{N}\left[\sum_{i=1}^{N_1} X_{1i} + \frac{N_2}{n_2}\sum_{j=1}^{n_2} X_{2j}\right]$$

where X_{1i} is the sales figure for the ith store in the large group and X_{2j} that for the jth store in the sub-sample of the small group.

The expectation of $\hat{T}$ for a fixed n_2, and for a fixed set of N_2 units from which the sub-sample of size n_2 is picked, is given by

(7.5.3) $$E(\hat{T}|n_2, N_2) = \frac{M}{N}\left[\sum_{i=1}^{N_1} X_{1i} + \sum_{i=1}^{N_2} X_{2i}\right] = \frac{M}{N}\sum_{i=1}^{N} X_i$$

where X_i is now the value for the ith store in the sample, regardless of whether it belongs to the one group or the other. Then

(7.5.4) $$E(\hat{T}) = E[E(\hat{T}|n_2, N_2)] = \sum_{i=1}^{M} X_i = T$$

The variance of $\hat{T}$, as shown below, is given by

(7.5.5) $$V(\hat{T}) = \frac{M}{N}[(M - N)\sigma^2 + (k - 1)M_2\sigma_2{}^2]$$

where

$$\sigma^2 = (M - 1)^{-1}\sum_{i=1}^{M}(X_i - \mu)^2, \quad \mu = \frac{T}{M}$$

and

$$\sigma_2{}^2 = (M_2 - 1)^{-1} \sum_{i=1}^{M_2} (X_{2i} - \mu_2)^2, \quad \mu_2 = \sum_{i=1}^{M_2} \frac{X_{2i}}{M_2}$$

To prove this we need Eq. (A.14.6) of the Appendix, namely,

$$V(X) = E[V(X|Y)] + V[E(X|Y)]$$

where Y stands for a fixed set of N_2 small stores and the fixed number n_2. The second term on the right is just the variance of the right-hand side of Eq. (3), that is of $M\bar{X}$. This is $M^2\left(\frac{1}{N} - \frac{1}{M}\right)\sigma^2$, which is the first term of Eq. (5).

The conditional variance of $\hat{T}$ is

$$(7.5.6) \qquad V(\hat{T}|n_2, N_2) = \frac{M^2N_2{}^2}{N^2}\left(\frac{1}{n_2} - \frac{1}{N_2}\right)s_2{}^2$$
$$= \frac{N_2}{f^2}(k-1)s_2{}^2$$

where $s_2{}^2 = (N_2 - 1)^{-1} \sum_{j=1}^{N2} (X_{2j} - \bar{X}_2)^2$. This follows from Eq. (2), since the first term is constant (under the stated condition) and the second term is $(MN_2)/N$ times the sub-sample mean $\bar{X}_2$.

The expectation of Eq. (6) for a given n_2 and given k (i.e., for given number N_2 of small units although not for a fixed set of N_2) is $(N_2/f^2)(k - 1)\sigma_2{}^2$, and the expectation of this for given k is

$$\frac{k-1}{f^2}\sigma_2{}^2E(N_2) = \frac{k-1}{f^2}\sigma_2{}^2\frac{NM_2}{M} = (k-1)\frac{M}{N}M_2\sigma_2{}^2$$

which is the second term of Eq. (5).

The optimum allocation of sample sizes will depend on the relative costs of the first and second sample. These often differ quite considerably. The original large sample may, for instance, be obtained by a mailed questionnaire, and a sub-sample of the non-responders may be followed up with personal interviews, which are considerably more expensive. As a cost function (apart from fixed overhead costs) for the whole survey, we might assume

$$(7.5.7) \qquad C = NC_0 + N_1C_1 + n_2C_2$$

where C_0 is the cost of selecting and examining a unit in the large sample, C_1 is the additional unit cost for the large units, and C_2 is the additional unit cost for sub-sampling the small units. We want to find the optimum values of N and k $(= N_2/n_2)$ for fixed variance and minimum cost. From Eq. (7),

$$(7.5.8) \qquad E(C) = NC_0 + \frac{N}{M}M_1C_1 + \frac{1}{k}\frac{NM_2}{M}C_2$$

We wish to minimize this, subject to a fixed value, say ε^2, for the variance of the estimate $\hat{T}$. The method is to use a Lagrange undetermined multiplier λ (see Appendix A.15) and form the function

$$F(N, k, \lambda) = E(C) + \lambda[V(\hat{T}) - \varepsilon^2] \tag{7.5.9}$$

Setting $\partial F/\partial N$ and $\partial F/\partial k$ equal to zero, we get

$$C_0 + \frac{M_1}{M} C_1 + \frac{M_2}{kM} C_2 + \lambda\left[-\frac{M^2}{N^2}\sigma^2 - \frac{MM_2(k-1){\sigma_2}^2}{N^2}\right] = 0 \tag{7.5.10}$$

and

$$-\frac{NM_2}{k^2M} C_2 + \lambda\frac{MM_2{\sigma_2}^2}{N} = 0$$

Eliminating λ/N^2 from these equations, we get

$$k^2 = \frac{C_2M\sigma^2 - C_2M_2{\sigma_2}^2}{C_0M{\sigma_2}^2 + C_1M_1{\sigma_2}^2} = \frac{MC_2}{M_1}\frac{\sigma^2/{\sigma_2}^2 - M_2/M}{C_1 + C_0M/M_1} \tag{7.5.11}$$

If we put $V(\hat{T}) = \varepsilon^2$ in Eq. (5), we get an expression for N, namely,

$$N = \frac{M^2\sigma^2}{M\sigma^2 + \varepsilon^2}\left[1 + (k-1)\frac{M_2{\sigma_2}^2}{M\sigma^2}\right] \tag{7.5.12}$$

where k is given by Eq. (11).

If we took a simple random sample of N' from the population of M, large enough to give the same variance for $\hat{T}$ (now $M\bar{X}$), we should have

$$\varepsilon^2 = V(M\bar{X}) = M^2V(\bar{X}) = M^2\sigma^2\left(\frac{1}{N'} - \frac{1}{M}\right)$$

so that

$$N' = \frac{M^2\sigma^2}{M\sigma^2 + \varepsilon^2} \tag{7.5.13}$$

This is the first factor in the expression for N, Eq. (12). However, although N is larger than N', the cost of the double sample is less than that of the proportionate single sample.

EXAMPLE 1 Suppose that

$$M = 20{,}000, \qquad M_1 = 1{,}000, \qquad (M_2 = 19{,}000)$$
$${\sigma_1}^2 = 500, \qquad {\sigma_2}^2 = 5, \qquad \sigma^2 = 34$$

(the variances are estimates from a preliminary investigation), and

$$C_0 = 0.25, \qquad C_1 = 2, \qquad C_2 = 1 \quad \text{(dollars)}$$

Suppose the preliminary estimate of T is 29,000 units, and we want ε to be not more than 0.04 of this, or 1160. Then by Eq. (13),

$$N' = 6710$$

The cost of a single proportionate sample of this size would be $6710C_0 + 335C_1 + 6375C_2 = \$8723$.

From Eqs. (11) and (12) we find $k^2 = 16.7$, so that $k \approx 4$, and $N = 9520$. This gives $N_1 = 476$, $n_2 = \frac{1}{4}(9520 - 476) = 2261$. The cost of the double sampling method would therefore be $9520C_0 + 476C_1 + 2261C_2 = \$5593$.

This is considerably cheaper than the cost of a single sample to give the same precision.

7.6 **Sequential Sampling** In any fixed-size sampling procedure the total number of items in the sample is decided beforehand and this number of items is drawn and examined. However, it is sometimes practicable and economical to draw the sample items one at a time and examine them as they are drawn. This type of sampling is called *sequential.*

Suppose a certain hypothesis H_0 regarding the parent population is to be tested (for instance, the hypothesis that the proportion of defectives in a large batch of machine parts is not greater than p_0). On the basis of the first m sample items tested we may make one of three decisions: (a) to accept H_0, (b) to reject H_0, (c) to test one more item. The process is terminated when our decision rule leads us to either (a) or (b). The *expected* number of observations required to reach one of these two decisions is less than we would need in order to make the same decision on the basis of a single fixed-size sample. Of course, it may happen that the sequential procedure will take more observations than the fixed-size one (although this is unlikely) and it may not always be convenient in practice to take the samples one at a time, but, by and large, sequential sampling is a definitely economical procedure. For full details, Wald's book [5] should be consulted.

Sequential testing may be illustrated by the theory of the *random walk.* Suppose B, O, A are three points on a straight line, where A is a paces to the right of O and B is b paces to the left. If I start at O and take one pace per second in a random direction (backwards or forwards) along the line, how long will it take me to reach either A or B? This is the random walk problem in a simple form. It can be proved that the walk will eventually terminate. The probability of oscillating back and forth without ever reaching either A or B is zero. In the sequential decision process each new item tested is like a pace in the random walk—it leads towards decision (a) or decision (b). Eventually one of these two decisions will be actually reached.

The type of sequential test suggested by Wald is a *likelihood ratio test.* Let us suppose that we are testing a simple null hypothesis H_0 against a simple alternative H_1 (see § 6.7). Let $f_0(x_1)$ be the probability (or probability density) that the variable X takes the value x_1 when H_0 is true, and similarly $f_1(x_1) = P(X = x_1 | H_1)$. The joint likelihood of the given set of m observations $x_1\ x_2, \ldots x_m$ under H_0 is

$$p_{0m} = f_0(x_1) \cdot f_0(x_2) \ldots f_0(x_m)$$

and the joint likelihood under H_1 is

$$p_{1m} = f_1(x_1) \cdot f_1(x_2) \ldots f_1(x_m)$$

The test suggested is to calculate p_{1m}/p_{0m} for each successive value of m and to continue testing as long as the ratio lies between two specified limits A and B ($A > B$). The process is terminated when for the first time either $p_{1m}/p_{0m} \geq A$ or $p_{1m}/p_{0m} \leq B$. In the first case H_0 is rejected, and in the second case it is accepted.

If we put $z_i = \log f_1(x_i) - \log f_0(x_i)$, we have

$$\log\left(\frac{p_{1m}}{p_{0m}}\right) = \sum_{i=1}^{m} \log\left[\frac{f_1(x_i)}{f_0(x_i)}\right] = \sum_{i=1}^{m} z_i = Z_m \tag{7.6.1}$$

and the test is terminated as soon as $Z_m \geq \log A$ or $Z_m \leq \log B$. Since Z_m is a sum of the m random variables z_i, the analogy with a random walk is clear.

The values of A and B are determined by the risks we are prepared to take in coming to the one decision or the other. If the probability of a rejection error (an error of the first kind) is α and the probability of an acceptance error (second kind) is β, and if n observations lead to the rejection of H_0, then $p_{1n}/p_{0n} = (1 - \beta)/\alpha$, since this is the ratio of the probabilities of H_1 and H_0 for a sample which leads to the rejection of H_0. Therefore,

$$\frac{1-\beta}{\alpha} \geq A \tag{7.6.2}$$

Similarly,

$$\frac{\beta}{1-\alpha} \leq B \tag{7.6.3}$$

In practice we usually take $A = (1 - \beta)/\alpha$ and $B = \beta/(1 - \alpha)$. If the distribution is such that one extra observation will make little difference to the value of p_{1n}/p_{0n}, there will be no appreciable error in doing this.

*** 7.7 Number of Observations Required for a Final Decision in Sequential Sampling** Let n be the smallest integer for which $Z_n \geq \log A$ or $Z_n \leq \log B$. We would like to find the expected value of n and compare it with the fixed sample size N which would give the same probabilities α and β of error.

Since $Z_n = z_1 + z_2 + \ldots + z_n$, and n is a random variable,

$$\begin{aligned} E(Z_n) &= E[E(Z_n|n)] = E[nE(z)] \\ &= E(n) \cdot E(z) \end{aligned} \tag{7.7.1}$$

where $E(z)$ is the expected value of any of the z_i.

If the test leads to the rejection of H_0, $E(Z_n)$ will be nearly $\log A$, and, if the test leads to the acceptance of H_0, $E(Z_n)$ will be nearly $\log B$, so that

$$E(Z_n) \approx \gamma \log A + (1 - \gamma) \log B \tag{7.7.2}$$

where γ is the probability of rejecting H_0 (γ will be α if H_0 is true or $1 - \beta$ if H_1 is true). From Eqs. (1) and (2),

$$E(n) \approx \frac{\gamma \log A + (1 - \gamma)\log B}{E(z)} \tag{7.7.3}$$

EXAMPLE 2 Suppose the variate X is normally distributed with unit variance, and that under H_0 the mean is μ_0 and under H_1 it is $\mu_1(>\mu_0)$. Then

$$\begin{cases} f_0(x) = (2\pi)^{-1/2}e^{-(x-\mu_0)^2/2} \\ f_1(x) = (2\pi)^{-1/2}e^{-(x-\mu_1)^2/2} \end{cases} \tag{7.7.4}$$

Let $E_0(n)$, $E_1(n)$ be the expected values of n under H_0 and H_1 respectively. Then

$$\begin{cases} E_0(n) = \dfrac{\alpha \log[(1-\beta)/\alpha] + (1-\alpha)\log[\beta/(1-\alpha)]}{E_0(z)} \\[2ex] E_1(n) = \dfrac{(1-\beta)\log[(1-\beta)/\alpha] + \beta \log[\beta/(1-\alpha)]}{E_1(z)} \end{cases} \tag{7.7.5}$$

From Eq. (4),

$$\begin{aligned} z &= \log f_1(x) - \log f_0(x) \\ &= -\frac{\mu_1{}^2 - \mu_0{}^2}{2} + x(\mu_1 - \mu_0) \end{aligned}$$

Under H_0, $E(x) = \mu_0$, so that

$$\begin{aligned} E_0(z) &= -\frac{\mu_1{}^2 - \mu_0{}^2}{2} + \mu_0(\mu_1 - \mu_0) \\ &= -\tfrac{1}{2}(\mu_1 - \mu_0)^2 \end{aligned} \tag{7.7.6}$$

and similarly, $E_1(z) = \frac{1}{2}(\mu_1 - \mu_0)^2$. Therefore, from Eq. (5), $E_0(n)$ and $E_1(n)$ may be found.

Now if N is the fixed sample size corresponding to the same values of α and β, and we use the statistic $\bar{X}$ (which is normally distributed with mean μ_0 and variance $1/N$ when H_0 is true),

$$\alpha = (2\pi)^{-1/2} \int_{\lambda_0}^{\infty} e^{-t^2/2}\, dt, \qquad \lambda_0 = \sqrt{N}(c - \mu_0)$$

where c is the critical value for $\bar{X}$. Similarly,

$$1 - \beta = (2\pi)^{-1/2} \int_{\lambda_1}^{\infty} e^{-t^2/2}\, dt, \qquad \lambda_1 = \sqrt{N}(c - \mu_1)$$

Therefore $\alpha = 1 - \Phi(\lambda_0)$, $\beta = \Phi(\lambda_1)$ and

$$N = \frac{(\lambda_0 - \lambda_1)^2}{(\mu_1 - \mu_0)^2} \tag{7.7.7}$$

From Eqs. (5), (6) and (7) it appears that $E_0(n)/N$ and $E_1(n)/N$ are independent of $(\mu_1 - \mu_0)^2$ and so may be calculated for any given α and β. Thus for $\alpha = 0.05$, $\beta = 0.1$, we have

$$\lambda_0 = 1.645, \qquad \lambda_1 = -1.282, \qquad (\lambda_0 - \lambda_1)^2 = 8.57,$$

$$\frac{E_0(n)}{N} = \frac{0.05 \log 18 + 0.95 \log 0.1053}{-\frac{1}{2}(8.57)}$$

$$= 0.465$$

$$\frac{E_1(n)}{N} = \frac{0.9 \log 18 + 0.1 \log 0.1053}{\frac{1}{2}(8.57)}$$

$$= 0.555$$

There is an expected saving of 53.5% if H_0 is true or 44.5% if H_1 is true, in the number of observations required.

A lower limit may be calculated for the probability that the sequential process will terminate before n reaches some preassigned number n_0. If this probability is $P_0(n_0)$ under H_0 and $P_1(n_0)$ under H_1, then it may be shown that

$$P_0(n_0) \geq \Phi(\delta_0), \qquad P_1(n_0) \geq 1 - \Phi(\delta_1)$$

where

$$\delta_0 = \frac{\log B - n_0 E_0(z)}{\sqrt{n_0}\,\sigma_0(z)}, \qquad \delta_1 = \frac{\log A - n_0 E_1(z)}{\sqrt{n_0}\,\sigma_1(z)}$$

and $\sigma_0(z)$, $\sigma_1(z)$ are the standard deviations of z under hypotheses H_0 and H_1.

In Example 2 above, $\sigma_0(z) = \sigma_1(z) = \mu_1 - \mu_0$ since the standard deviation of X is 1. Therefore,

$$\delta_0 = \frac{\log 0.1053 + \frac{n_0}{2}(\mu_1 - \mu_0)^2}{\sqrt{n_0}(\mu_1 - \mu_0)}$$

$$\delta_1 = \frac{\log 18 - \frac{n_0}{2}(\mu_1 - \mu_0)^2}{\sqrt{n_0}(\mu_1 - \mu_0)}$$

With $\alpha = 0.05$ and $\beta = 0.1$ and a fixed sample size of 1000, we could detect a difference $\mu_1 - \mu_0$ amounting to $2.927/\sqrt{1000} = 0.0926$ (see §6.10).

With $n_0 = 1000$, $\delta_0 = 0.694$, $\delta_1 = -0.476$, so that $P_0(n_0) \geq 0.756$, and $P_1(n_0) \geq 0.683$. The probability is therefore at least 0.68 that a sequential test of this kind will terminate in a decision for acceptance or rejection of H_0 before the sample size reaches 1000.

7.8 **The Truncated Sequential Test** If it happens that the test is still not terminated for some n_0 beyond which it is not convenient to continue testing, a reasonable decision rule is the following:

If $Z_{n_0} \leq 0$, accept H_0; if $Z_{n_0} > 0$, accept H_1.

The probabilities of the two kinds of error, $\alpha(n_0)$ and $\beta(n_0)$ under this rule are slightly different from α and β. It may be shown that

$$\alpha(n_0) \leq \alpha - \Phi(v_1) + \Phi(v_2) \tag{7.8.1}$$
$$\beta(n_0) \leq \beta + \Phi(v_3) - \Phi(v_4)$$

where

$$v_1 = \delta_0 - \frac{\log B}{\sqrt{n_0}\sigma_0(z)} = -\sqrt{n_0}E_0(z)/\sigma_0(z)$$
$$v_2 = v_1 + \frac{\log A}{\sqrt{n_0}\sigma_0(z)}$$
$$v_3 = \delta_1 - \frac{\log A}{\sqrt{n_0}\sigma_1(z)} = -\sqrt{n_0}E_1(z)/\sigma_1(z)$$
$$v_4 = v_3 + \frac{\log B}{\sqrt{n_0}\sigma_1(z)}$$

These are upper bounds and probably higher than necessary. For $n_0 = 1000$ and the data of Example 2, $v_1 = -v_3 = 1.464$, $v_2 = 2.451$, $v_4 = -2.233$. These values give $\alpha(n_0) \leq 0.114$, $\beta(n_0) \leq 0.159$.

If $N = 100$, and we decide to stop at $n_0 = 300$, whatever happens, the upper bounds for $\alpha(n_0)$ and $\beta(n_0)$ are 0.052 when $\alpha = \beta = 0.05$, so that truncating in this way would make very little difference to the probabilities of error.

7.9 **The Sequential Test for a Binomial Distribution** We assume that the objects in a large group (a "lot" in the language of sampling) can be classified as either "defective" or "satisfactory". A lot will be acceptable if the proportion of defectives $\theta \leq \theta'$, but otherwise the buyer will want to refuse it. It is supposed that the buyer has to make a decision on the basis of a sample, and therefore he may make either of the two kinds of error we have discussed previously. He may refuse a good lot or accept a bad one, and must decide what risks he is prepared to run of making either of these mistakes. Suppose he decides that it would be a serious matter to refuse a lot with $\theta < \theta_0$ and that it would be unfortunate to accept a lot with $\theta > \theta_1$, where of course $\theta_0 < \theta' < \theta_1$. He will want to keep the probability of committing these serious errors down below say α and β, respectively, where both these numbers are fairly small compared with 1. Having decided on θ_0, θ_1, α and β, he can construct a sequential test.

Randomly selected items from the lot are taken one at a time and inspected. Suppose the number of defectives in the first m units tested is d_m. Then

$$\frac{p_{1m}}{p_{0m}} = \frac{\theta_1^{d_m}(1-\theta_1)^{m-d_m}}{\theta_0^{d_m}(1-\theta_0)^{m-d_m}} \tag{7.9.1}$$

will give a likelihood ratio test for hypothesis H_0 (that $\theta = \theta_0$) against hypothesis

H_1 (that $\theta = \theta_1$). If $\theta < \theta_0$ the probability of rejecting the lot will be even less than for $\theta = \theta_0$, and if $\theta > \theta_1$ the probability of accepting the lot will be less than for $\theta = \theta_1$. The same test may therefore be regarded as a test for the composite hypothesis $\theta \leq \theta_0$ against the composite hypothesis $\theta \geq \theta_1$.

The test consists in rejecting H_0 if $Z_m = \log\left(\frac{p_{1m}}{p_{0m}}\right) \geq \log\frac{1-\beta}{\alpha}$, accepting H_0 if $Z_m \leq \log\frac{\beta}{1-\alpha}$, and continuing the test if neither of these is true. This is equivalent to setting up an acceptance number A_m and a rejection number R_m for each value of m and continuing the test as long as $A_m < d_m < R_m$. The numbers A_m and R_m are given by

$$A_m = \frac{\log\dfrac{\beta}{1-\alpha} + m\log\dfrac{1-\theta_0}{1-\theta_1}}{\log\dfrac{\theta_1}{\theta_0} + \log\dfrac{1-\theta_0}{1-\theta_1}} \tag{7.9.2}$$

$$R_m = \frac{\log\dfrac{1-\beta}{\alpha} + m\log\dfrac{1-\theta_0}{1-\theta_1}}{\log\dfrac{\theta_1}{\theta_0} + \log\dfrac{1-\theta_0}{1-\theta_1}} \tag{7.9.3}$$

Since A_m and R_m depend linearly on m, they define a sloping band of constant width on a diagram with d_m plotted against m. For $\alpha = \beta = 0.05$, $\theta_0 = 0.001$ and $\theta_1 = 0.03$, the lines representing A_m and R_m as functions of m are

$$\begin{cases} A_m = 0.00859m - 0.858 \\ R_m = 0.00859m + 0.858 \end{cases} \tag{7.9.4}$$

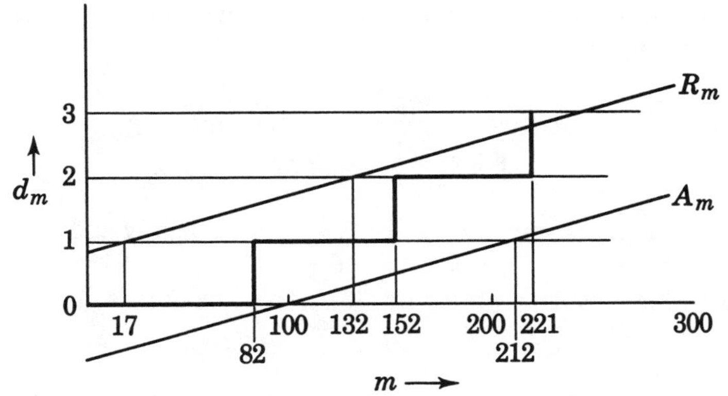

FIG. 38 SEQUENTIAL BINOMIAL TEST

In Figure 38 an imaginary sampling experiment is represented by the stepped line. In the first 81 items tested there were no defectives, the 82nd was defective, the second defective turned up at the 152nd test, the third at the 221st. This last test took the cumulative polygon outside the rejection line and therefore the lot was refused.

A lot under this scheme will be accepted if the first 100 items tested show no defectives and will be refused if a defective appears in the first 17. If only one defective has appeared in 212, the lot will be accepted; if two appear in the first 132 it will be refused; and so on.

The probability P_θ of accepting the lot for any given θ can be expressed as a function of θ. It decreases from 1, when $\theta = 0$, to 0 when $\theta = 1$. When $\theta = \theta_0$, $P_\theta = 1 - \alpha$ and when $\theta = \theta_1$, $P_\theta = \beta$. If θ_0 and θ_1 are not too far apart, the approximate value of P_θ is given by

$$P_\theta \approx (A^h - 1)/(A^h - B^h) \tag{7.9.5}$$

where $A = (1 - \beta)/\alpha$, $B = \beta/(1 - \alpha)$ and h is the non-zero root of

$$\theta\left(\frac{\theta_1}{\theta_0}\right)^h + (1 - \theta)\left(\frac{1 - \theta_1}{1 - \theta_0}\right)^h = 1 \tag{7.9.6}$$

that is, of

$$\theta = \frac{1 - \left(\dfrac{1 - \theta_1}{1 - \theta_0}\right)^h}{\left(\dfrac{\theta_1}{\theta_0}\right)^h - \left(\dfrac{1 - \theta_1}{1 - \theta_0}\right)^h} \tag{7.9.7}$$

By choosing various values of h, we can calculate corresponding values of θ and P_θ and plot the curve. This is a sort of *operating characteristic* or *power curve* of the test. It indicates the probability of accepting a lot with any given proportion of defectives. With the data assumed above there is an even chance of accepting a lot with $\theta = 0.009$.

The expected number of observations n necessary to reach a final decision, one way or the other, is given approximately by

$$E(n) \approx \frac{P_\theta \log B + (1 - P_\theta)\log A}{\theta \log\left(\dfrac{\theta_1}{\theta_0}\right) - (1 - \theta)\log\left(\dfrac{1 - \theta_0}{1 - \theta_1}\right)} \tag{7.9.8}$$

As a function of θ, this starts at 100 for $\theta = 0$, rises slightly and then decreases as θ increases, becoming 1 for $\theta = 1$. For $\theta = 0.02$, $E(n) = 53$ and for $\theta = 0.03$, $E(n) = 36$ (using the data of Figure 38). The function is indeterminate at $P_\theta = 0.5$ ($\theta = 0.0086$).

7.10 **Tolerance Limits** Tolerance limits are limits within which we are confident that at least a *specified proportion* of the population will lie (with of course a specified degree of confidence). We may for instance claim with 95% confidence that at least 90% of a particular population will have values of X between given limits. If these limits are the smallest and the greatest values observed in a sample of N, we may ask how large N should be to justify the claim.

If x_1 and x_N are the least and greatest values of X for the sample, and if $f(x)$ is the density function for X, the proportion of the population lying between x_1 and x_N is

$$v = \int_{x_1}^{x_N} f(x)\,dx \tag{7.10.1}$$

The density function for v is

$$g(v) = N(N-1)v^{N-2}(1-v), \qquad 0 \le v \le 1 \tag{7.10.2}$$

The probability that $v \ge \beta$ is therefore

$$\int_{\beta}^{1} g(v)\,dv$$

and for a confidence coefficient of $1 - \alpha$ this probability is $1 - \alpha$. Integrating Eq. (2) we obtain

$$\alpha = N\beta^{N-1} - (N-1)\beta^N \tag{7.10.3}$$

from which N can be obtained for given α and β. For $\alpha = 0.05$ and $\beta = 0.99$, we find $N = 473$. This means that if we take a random sample of size 473 from a population in which X is distributed continuously, there is a probability 0.95 that at least 99% of the population will have X values between the least and the greatest values found in the sample. This result is independent of the form of the distribution.

PROBLEMS

A. (§ 7.2)

1. Households in a town are stratified into a high-rent stratum (4,000 items) and a low-rent stratum (20,000 items). The variate X, of which the average is to be estimated, is thought to have a standard deviation in the first stratum about three times that in the second. How should a total sample of 1000 be divided between the two strata?

2. The farms in a certain county are stratified according to size in seven strata, as shown in the table below. For the variate X (the number of acres in corn) the stratum means μ_i and the stratum standard deviations σ_i are as given. If it is required to take a sample of 100 farms for estimating some quantity closely related to X, how should these farms be allocated among the strata (a) with proportionate sampling, (b) with optimum sampling? Compare the precision of each of these methods with that of simple random sampling.

Farm Size (Acres)	No. of Farms	μ_i	σ_i
0 - 40	394	5.4	8.3
41 - 80	461	16.3	13.3
81 - 120	391	24.3	15.1
121 - 160	334	34.5	19.8
161 - 200	169	42.1	24.5
201 - 240	113	50.1	26.0
241 -	148	63.8	35.2

Hint: The precision varies inversely as the variance. The variance of the mean of a simple random sample is $\sigma^2(M - N)/(MN)$, where σ^2 is the overall variance of X. This can be found from the σ_i and μ_i by the formula: $(M - 1)\sigma^2 = \sum_i (M_i - 1)\sigma_i^2 + \sum_i M_i(\mu_i - \mu)^2$, $\mu = \sum_i M_i\mu_i/M$.

3. Prove that if $V_r(\bar{X})$ is the variance for a random sample and $V_p(\bar{X})$ that for a proportionate sample, then

$$V_r(\bar{X}) - V_p(\bar{X}) = \frac{M - N}{MN(M - 1)}\left[\sum M_i(\mu_i - \mu)^2 - \frac{1}{M}\sum (M - M_i)\sigma_i^2\right]$$

4. A variate X is distributed in the population with density e^{-x}, $x \geq 0$. The population is divided into two strata at the point x_0 and a stratified sample of size N is taken with proportionate sampling. Show that the variance of the sample estimator $\bar{X}$ of the population mean is $N^{-1}[1 - (x_0^2e^{-x_0})/(1 - e^{-x_0})]$, and find for what value of x_0 this is least. *Hint:* The population is infinite, so that the sampling fraction f is zero. The *ratios* M_1/M and M_2/M are given by the integral of e^{-x} from 0 to x_0 and from x_0 to ∞, respectively.

B. (§ 7.3)

1. From the following artificial population with three clusters, suppose that two clusters are selected and two units are selected from each cluster. Find the variance of the unbiased estimator for μ. If the sampling is proportional to cluster size (one item from cluster 1, two from 2, three from 3) what is the variance?

Cluster No. (i)	$X_{i\alpha}$	M_i
1	0, 1	2
2	1, 2, 2, 3	4
3	3, 3, 4, 4, 5, 5	6

Note that the clusters are widely dissimilar, so that the first term in Eq. (7.3.3) is much the larger of the two.

2. If the sub-sample number for each cluster sampled is proportional to the size of that cluster, show that the estimator of μ reduces to $kT/(Mlf)$ where f is the common sampling fraction N_j/M_j and T is the total of the X_{jh} for all the items in the combined sample.

3. An alternative method of selecting the l clusters is to sample with probabilities proportional to cluster size. That is, the probability z_i of selecting the ith cluster is M_i/M. (Note that this means sampling with replacement. The same cluster may appear more than once in the sample.)

If the sub-sample size is the same (N) for each cluster sampled, show that T/Nl is an unbiased estimator of μ, where T is defined as in Problem 2. *Hint:* Let T_j be the total for the jth cluster. Show that, for a given value of j, $E(T_j) = N\mu_j$ and that, over all j, $E(\mu_j) = \mu$.

4. The variance of the estimator in Problem 3 is

$$\sum_i [M_i(\mu_i - \mu)^2 + (M_i - N)\sigma_i^2/N]/(lM)$$

Show that with the data of Problem 1, this estimator has a considerably smaller variance than either of those in Problem 1.

C. (§ 7.4)

1. Show that the variance of the mean of a systematic sample may be written

$$V(m_i) = (M - 1)\frac{\sigma^2}{M} - (N - 1)\frac{s_w^2}{N}$$

where

$$s_w^2 = \sum_{ij} \frac{(X_{ij} - m_i)^2}{k(N - 1)}$$

which is the average of the variances within the separate systematic samples. *Hint:* $\sum_{ij} (X_{ij} - \mu)^2 = \sum_{ij} (X_{ij} - m_i)^2 + \sum_{ij} (m_i - \mu)^2$, and this last term is $N \sum_i (m_i - \mu)^2$.

2. Prove from the result of Problem 1 that $V(m_i) < V(m_r)$ if and only if $s_w^2 > \sigma^2$, where m_r is the mean of a random sample of size N. This result indicates that systematic sampling is more precise than random sampling when the variance within a sample tends to be larger than that in the whole population, that is, when the sample is markedly heterogeneous.

3. The following table exhibits an artificial population with a fairly steady rising trend; here $M = 40$, $N = 4$, $k = 10$, and each column is a separate systematic sample. Calculate the estimator for μ from each of these ten samples. Find the variance of these estimators and compare with the average variance within samples and the variance of the mean of a random sample from the given population.

Systematic Sample Numbers									
1	2	3	4	5	6	7	8	9	10
0	1	1	2	5	4	7	7	8	6
6	8	9	10	13	12	15	16	16	17
18	19	20	20	24	23	25	28	29	27
26	30	31	31	33	32	35	37	38	38

4. Calculate the serial correlation coefficient ρ_k for a lag of k, defined by

$$k(N - 1)\sigma^2\rho_k = \sum_{\beta=1}^{k(N-1)} (X_\beta - \mu)(X_{\beta+k} - \mu)$$

for the data of Problem 3. *Hint:*

$$\sum_\beta (X_\beta - \mu)(X_{\beta+k} - \mu) = \sum_\beta (X_\beta X_{\beta+k}) - \mu(\sum X_\beta + \sum X_{\beta+k}) + k(N - 1)\mu^2$$

5. If the serial correlation coefficient ρ_{jk} for a lag of jk is defined by

$$k(N - j)\sigma^2 \cdot \rho_{jk} = \sum_{\beta=1}^{k(N-j)} (X_\beta - \mu)(X_{\beta+jk} - \mu)$$

and if $\rho_{jk} = (\rho_k)^j$, where ρ_k is the coefficient for a lag of k, show that when terms of order $1/N^2$ are neglected, the ratio of the variances of the two estimators m_i and m_r is given approximately, for large M, by $V(m_i)/V(m_r) = (1 + \rho_k)/(1 - \rho_k)$.

D. (§ 7.5)

1. In a double sampling scheme where the items are stratified in two classes (e.g., large and small), the first sample of N produces N_1 large and N_2 small items. A random sub-sample of n_1 is drawn from the N_1 items and an independent random sample of n_2 from the N_2 items, and the variate X is measured on these sub-samples. If T_1, T_2, are the totals of X for the n_1 and n_2 items, and T is the total for the whole population, show that an unbiased estimator of T is $\hat{T} = (hT_1 + kT_2)/f$ and that its variance is

$$V(\hat{T}) = \frac{(M - N)\sigma^2 + (h - 1)M_1\sigma_1^2 + (k - 1)M_2\sigma_2^2}{f}$$

where $h = N_1/n_1$; $k = N_2/n_2$, $f = N/M$.

2. Show that the expression for the variance in Problem 1 may be written, if we ignore the differences between M and $M - 1$, M_1 and $M_1 - 1$, M_2 and $M_2 - 1$,

$$V(\hat{T}) = \frac{1 - f}{f}[M_1(\mu_1 - \mu)^2 + M_2(\mu_2 - \mu)^2] + M_1\sigma_1^2\frac{(h - f)}{f} + M_2\sigma_2^2\frac{(k - f)}{f}$$

(See the Hint following Problem A-2).

3. Suppose the farms in Problem A-2 are divided into two classes, called large (over 160 acres) and small (160 acres or under). A first, comparatively cheap, sample of 200 is taken and this gives 40 large and 160 small farms. The variate X is measured on a subsample of 30 of the large farms and 50 of the small ones. Calculate the variance of T for the double sample and compare it with the variance measured on a random sample of 100 from the original population.

Farm Size	M_i	μ_i	σ_i^2	$M_i\sigma_i^2$
Large	430	51.6	922	396,500
Small	1580	19.4	312	493,000
Population	2010	26.3	617	1,239,300

4. Find expressions for the optimum values of N, h and k, using the method of § 7.5 and supposing that only n_1 of the N_1 large units obtained in the first sample are sub-sampled (with additional cost C_1) and that $h = N_1/n_1$. *Hint:* Differentiate F partially with respect to N, h and k.

5. Apply the results of Problem 4 to the data of Problem 3, assuming that $C_0 = 0.1$, $C_1 = 0.8$, $C = 1.0$, and calculate the values of h and k for optimum sampling. If the standard deviation of the estimator $\hat{T}$ is not to exceed 2650 acres, calculate the size of the primary sample and the expected cost of getting the required information with double sampling. Calculate also the size and expected cost of a simple random sample from the population to give the same precision of estimation.

E. (§§ 7.6–7.10)

1. Suppose that in a certain population the probability θ that an individual is defective is either 0.1 or 0.3, but cannot have any other value. We wish to test the hypothesis H_0 that $\theta = 0.1$ against the alternative H_1 that $\theta = 0.3$, on the basis of a *fixed-sample* test. The test consists in accepting H_0 if the number of defectives d_N in a sample of size N is less than k; otherwise we accept H_1. Find N and k if the risks of error are $\alpha = 0.02$ and $\beta = 0.03$.

2. Construct a sequential acceptance-and-rejection chart for Problem 1 above.

3. Calculate the approximate expected number of trials before a decision is reached by the method of Problem 2. *Hint:* Use Eq. (7.9.8), first assuming H_0 and then assuming H_1.

4. Perform an imaginary sampling experiment from the population of Problem 1, as follows: read off a set of one-digit random numbers (say a column from the table in Appendix B.1), regarding each digit as a sample item. Count each zero as a defective (this corresponds to hypothesis H_0). Continue until a decision is reached, using the chart constructed in Problem 2, and note the number of trials necessary to reach the decision. Repeat 20 times, using different sets of random numbers, and note the average number of trials required. Compare with the result of Problem 3 for H_0.

5. Suppose the proportion of defectives in a population can vary from 0 to 1, but that acceptance limits are fixed at 0.1 and 0.3. Construct the operating characteristic (or power curve) of the binomial sequential test, with $\alpha = 0.02$, $\beta = 0.03$.

6. Construct a sequential acceptance-and-rejection chart, for testing the binomial probability $\theta = 0.5$ against the alternative $\theta = 0.7$, given the risks of error as $\alpha = 0.1$, $\beta = 0.2$. If the following table represents the results of a sequence of trials, x_m being the number of successes in m trials, show graphically that the sampling terminates with a decision in favour of $\theta = 0.5$ at the 10th trial.

m	1	2	3	4	5	6	7	8	9	10
x_m	0	0	1	1	2	3	3	4	4	4

7. Construct a sequential test for the mean μ of a Poisson distribution, to test $\mu = \mu_0$ against $\mu = \mu_1 (\mu_1 > \mu_0)$. Find the expected sample size and the power function of this test for given α and β. *Hint:* The power function for any value of μ is given by $P_\mu = (A^h - 1)/(A^h - B^h)$, where h is the non-zero number for which

$$\sum_{x=0}^{\infty} [p(x, \mu_1)/p(x, \mu_0)]^h p(x, \mu) = 1.$$

Here $p(x, \mu)$ is the Poisson probability for x successes with parameter μ. Show that h is given by the relation $\mu + (\mu_1 - \mu_0)h = (\mu_1/\mu_0)^h \mu$.

REFERENCES

[1] Hansen, M. H., Hurwitz, W. N., and Madow, W. G., *Sampling Survey Methods and Theory* (two vols.), Wiley, 1953.

[2] Cochran, W. G., *Sampling Techniques*, Wiley, 1953.

[3] Neyman, J., "On the two different aspects of the Representative Method: the Method of Stratified Sampling and the Method of Purposive Selection," *J. Roy. Stat. Soc.*, **97**, 1934, pp. 558–606.

[4] Madow, W. G., and Madow, L. H., "On the Theory of Systematic Sampling," *Ann. Math. Stat.*, **15**, 1944, pp. 1–24.

[5] Wald, A., *Sequential Analysis*, Wiley, 1947.

Chapter 8

EXACT TESTS ON SAMPLES FROM A NORMAL POPULATION

8.1 **The Assumption of Normality** We have already obtained in Chapter 5 some moments of the distribution of the sample mean and variance in samples of size N from a finite population, and we have noted the simplication in these results when the parent population is supposed to be infinite in size and normal in distribution. Thus the expectation and variance of k_1 are given by

$$(8.1.1)\qquad \begin{cases} E(k_1) = \mu \\ V(k_1) = \dfrac{\sigma^2}{N} \end{cases}$$

where μ and σ are the parameters of the normal parent distribution. Also,

$$(8.1.2)\qquad \begin{cases} E(k_2) = \kappa_2 = \sigma^2 \\ V(k_2) = \dfrac{2\sigma^4}{(N-1)} \end{cases}$$

furthermore, k_1 and k_2 are uncorrelated, i.e.,

$$(8.1.3)\qquad C(k_1, k_2) = 0$$

The skewness and kurtosis of the distribution of k_1 were shown to be zero, which suggested that the distribution of k_1 might be normal, but the methods of Chapter 5 were unsuitable for finding exact distributions. Straightforward methods of finding density functions for statistics often lead to intractable mathematical expressions, but when the parent population is normal, the calculations are relatively simple, and a good deal of work has been done using the basic assumption of normality. Throughout this chapter we assume, unless otherwise stated, a normal parent population.

The most popular tests among practical statisticians are tests which depend on normality in the parent population, and these tests are often used in situations where the assumption of normality is decidedly dubious. Fortunately, however, the tests are usually quite *robust*, which means that considerable departures from normality will not affect them very much. When there is grave doubt about the assumption, non-parametric (or distribution-free) tests should be used, even at some sacrifice of power.

8.2 **The Distribution of the Sample Mean** The fact that the distribution of the sample mean is normal when the population is normal is easily demonstrated. We saw in § 2.8 that the cumulant generating function for a linear function $L(=\sum_j C_jX_j)$ of independent variates X_j is given by

$$K_L(h) = \sum_j K_j(C_jh) \tag{8.2.1}$$

where $K_j(h)$ is the c.g.f. for X_j. If all the X_j are normal with mean μ and variance σ^2, and if $L = \bar{X} = \sum X_j/N$,

$$\begin{aligned} K_L(h) &= \sum_{j=1}^{N}\left(\mu\frac{h}{N} + \frac{\sigma^2h^2}{2N^2}\right) \\ &= \mu h + \frac{\sigma^2}{N}\cdot\frac{h^2}{2} \end{aligned} \tag{8.2.2}$$

and this is the c.g.f. for a normal distribution with mean μ and variance σ^2/N. On the assumption that a distribution is uniquely determined by its c.g.f., this proves the normality of L. The distribution of the variance and higher moments is not, however, so easily obtained.

8.3 **The Distribution of the Sample Variance** One way of arriving at the distribution of the variance is to find the joint distribution of the mean and variance, and then integrate over all possible values of the mean. For simplicity we will choose the origin for X in such a way that the population mean is zero. This will have no effect at all on the distribution of the variance. If the N observed sample values of X are $x_1, x_2 \ldots x_N$, the joint density function for the sample is

$$f(x_1, x_2 \ldots x_N) = (2\pi\sigma^2)^{-N/2}e^{-\Sigma x_i^2/2\sigma^2} \tag{8.3.1}$$

Now $\sum x_i^2 = \sum(x_i - \bar{x} + \bar{x})^2 = \sum(x_i - \bar{x})^2 + N\bar{x}^2 + 2\bar{x}\sum(x_i - \bar{x})$ where $\bar{x}$ is the sample mean. Since $\sum(x_i - \bar{x}) = 0$ and $\sum(x_i - \bar{x})^2 = Nm_2$, where m_2 is the second sample moment about the mean, we have

$$\sum x_i^2 = N(\bar{x}^2 + m_2) \tag{8.3.2}$$

Therefore (1) may be written

$$f(x_1, x_2 \ldots x_N) = (2\pi\sigma^2)^{-N/2}e^{-N(\bar{x}^2+m_2)/2\sigma^2} \tag{8.3.3}$$

Since m_2 is proportional to the sample variance k_2, the relation being

$$Nm_2 = (N-1)k_2 \tag{8.3.4}$$

the distribution of k_2 is easily obtainable from that of m_2.

There are two methods of proceeding in a problem like this—one is the analytical method, which involves a good deal of algebraic manipulation; the other method is geometrical and requires considerable spatial intuition. Fisher's original approach was geometric, but various writers since have given analytical ofs. Both promethods are explained in detail in [1].

In the analytical method we change the set of variables $x_1, x_2 \ldots x_N$ to a new set, of which two will be $\bar{x}$ and m_2, the variables which appear in Eq. (3). We therefore need $N - 2$ new variables, $w_1, w_2 \ldots w_{N-2}$, which can be chosen in the most convenient way and which will later disappear. Then

$$f(x_1, x_2 \ldots x_N)\, dx_1\, dx_2 \ldots dx_N = g(w_1, w_2 \ldots w_{N-2}, \bar{x}, m_2)\, dw_1 \ldots d\bar{x}\, dm_2 \tag{8.3.5}$$

Since, by Eq. (3), $f(x_1, \ldots x_N)$ is already expressed in terms of the new set, we merely have to work out the relation between the differentials. This is

$$dx_1 \ldots dx_N = |J|\, dw_1 \ldots d\bar{x}\, dm_2 \tag{8.3.6}$$

where J is the Jacobian of the old variables with respect to the new (see Appendix A.4). For a certain particular choice of the w's, we find

$$J = \tfrac{1}{2} N^{(N-1)/2} m_2^{(N-3)/2} D \tag{8.3.7}$$

where D is an expression, in the form of a determinant, depending only on the w's. It follows that

$$g(w_1, w_2 \ldots \bar{x}, m_2) = \frac{N^{(N-1)/2}}{2(2\pi\sigma^2)^{N/2}} |D| m_2^{(N-3)/2} \exp\left[-\frac{N}{2\sigma^2}(\bar{x}^2 + m_2)\right] \tag{8.3.8}$$

If we now integrate over all possible values of the w's (this integration does not actually have to be carried out), we know that the result must be of the form

$$h(\bar{x}, m_2) = C m_2^{(N-3)/2} \exp\left[-\frac{N}{2\sigma^2}(\bar{x}^2 + m_2)\right] \tag{8.3.9}$$

where C is some constant depending on the bounds of integration of the w's and on the constant factors in Eq. (8). We do not need, at this stage, to know exactly what it is. Since this joint distribution is of the form $f_1(\bar{x})$ multiplied by $f_2(m_2)$, it is clear that $\bar{x}$ and m_2 are *independent* variates. To obtain the distribution of m_2 we simply have to integrate over all possible values of $\bar{x}$ ($-\infty$ to $+\infty$). This gives

$$\begin{aligned} f(m_2) &= \int_{-\infty}^{\infty} h(\bar{x}, m_2)\, d\bar{x} \\ &= C m_2^{(N-3)/2} e^{-Nm_2/2\sigma^2} \int_{-\infty}^{\infty} e^{-N\bar{x}^2/2\sigma^2}\, d\bar{x} \\ &= C_1 m_2^{(N-3)/2} e^{-Nm_2/2\sigma^2} \end{aligned} \tag{8.3.10}$$

where

$$C_1 = C\left(\frac{2\pi\sigma^2}{N}\right)^{1/2} \tag{8.3.11}$$

The constant C_1 can now be found, since

$$\int_0^{\infty} f(m_2)\, dm_2 = 1 \tag{8.3.12}$$

By the substitution $u = Nm_2/2\sigma^2$, this becomes

$$C_1\left(\frac{2\sigma^2}{N}\right)^{(N-1)/2}\int_0^\infty u^{(N-3)/2}e^{-u}\,du = 1$$

from which, since the integral is $\Gamma\left(\dfrac{N-1}{2}\right)$,

$$C_1 = \frac{\left(\dfrac{N}{2\sigma^2}\right)^{(N-1)/2}}{\Gamma\left(\dfrac{N-1}{2}\right)} \tag{8.3.13}$$

This gives, with Eq. (10), the distribution of m_2. That of k_2 is found from Eq. (4), since

$$\begin{aligned} g(k_2)\,dk_2 &= f(m_2)\,dm_2 \\ &= f(m_2)\frac{N-1}{N}\,dk_2 \end{aligned}$$

so that

$$\begin{aligned} g(k_2) &= \frac{N-1}{N}C_1\left(\frac{N-1}{N}k_2\right)^{(N-3)/2} e^{-(N-1)k_2/2\sigma^2} \\ &= \left(\frac{N-1}{2\sigma^2}\right)^{(N-1)/2} k_2^{(N-3)/2}\frac{e^{-(N-1)k_2/2\sigma^2}}{\Gamma\left(\dfrac{N-1}{2}\right)} \end{aligned} \tag{8.3.14}$$

This is more concisely expressed in terms of $n = N - 1$, which is the number of degrees of freedom in the expression for the variance. With this notation,

$$g(k_2) = \left(\frac{n}{2\sigma^2}\right)^{n/2}\frac{1}{\Gamma(n/2)}k_2^{(n-2)/2}e^{-nk_2/2\sigma^2} \tag{8.3.15}$$

If we put $nk_2/\sigma^2 = \chi^2$, this becomes the ordinary χ^2 distribution with n degrees of freedom. Thus,

$$g(k_2)\,dk_2 = f(\chi^2)\,d\chi^2 = \frac{n}{\sigma^2}\cdot f(\chi^2)\,dk_2$$

so that

$$f(\chi^2) = \frac{1}{2}\left(\frac{\chi^2}{2}\right)^{(n-2)/2}\frac{e^{-\chi^2/2}}{\Gamma(n/2)} \tag{8.3.16}$$

the same as Eq. (4.6.4).

If we put $u = \dfrac{\chi^2}{2} = \dfrac{nk_2}{2\sigma^2}$, u is a gamma variate with parameter $n/2$. Its density function is

$$f(u) = u^{(n-2)/2}\frac{e^{-u}}{\Gamma(n/2)}$$

and the moments and cumulants can be found from § 4.4. The rth cumulant of u is $(n/2)(r-1)!$, so that the rth cumulant of k_2 is

$$\kappa_r = \left(\frac{2\sigma^2}{n}\right)^r \frac{n}{2}(r-1)! \qquad (8.3.17)$$

$$= \sigma^{2r}(r-1)!\left(\frac{2}{n}\right)^{r-1}$$

* 8.4 **The Geometrical Approach to the Joint Distribution of the Mean and Variance** Because one cannot visualize space of more than three dimensions, we will carry through the discussion for $N = 3$. The argument is similar for larger samples.

We suppose as before that the population mean is zero. The observed sample values x_1, x_2, x_3 are considered as the coordinates of a point in the sample space of 3 dimensions. The sample mean $\bar{x}$ is given by

$$\bar{x} = \tfrac{1}{3}(x_1 + x_2 + x_3) \qquad (8.4.1)$$

For a given value of $\bar{x}$ this equation represents a plane equally inclined to all three axes.

The sample second moment m_2 is given by

$$m_2 = \tfrac{1}{3}[(x_1 - \bar{x})^2 + (x_2 - \bar{x})^2 + (x_3 - \bar{x})^2] \qquad (8.4.2)$$

and for given $\bar{x}$ and m_2 this represents a sphere of radius $(3m_2)^{1/2}$ with its centre at the point $(\bar{x}, \bar{x}, \bar{x})$.

The sphere and plane intersect in a circle of center M, where $OM = \sqrt{3}\bar{x}$ and $MP = (3m_2)^{1/2}$ (see Figure 39).

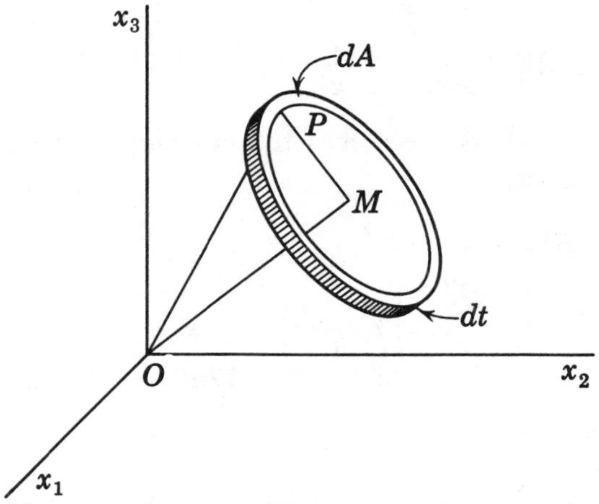

FIG. 39 ELEMENT OF THREE-DIMENSIONAL SAMPLE SPACE

If $\bar{x}$ increases slightly, the plane of Eq. (1) moves parallel to itself in the direction OM a distance $d(OM) = dt = \sqrt{3}d\bar{x}$.

If m_2 also increases, the circle of centre M enlarges, the radius increasing by an amount $d(MP) = (\sqrt{3}/2)m_2^{-1/2}\, dm_2$, which produces an increase of area

$$\begin{aligned} dA &= 2\pi r\, dr = 2\pi(3m_2)^{1/2}(\sqrt{3}/2)m_2^{-1/2}\, dm_2 \\ &= 3\pi\, dm_2 \end{aligned}$$

The volume of the ring-shaped element so formed is

$$dV = dA \cdot dt = 3\sqrt{3}\pi\, dm_2\, d\bar{x} \tag{8.4.3}$$

The probability that a sample point will lie in this element of volume is, by Eq. (8.3.3),

$$\begin{aligned} dP &= (2\pi\sigma^2)^{-3/2} \exp[-3(\bar{x}^2 + m_2)/2\sigma^2]\, dV \\ &= \frac{3\sqrt{3}}{2\sqrt{2\pi}}\sigma^{-3} \exp[-3(\bar{x}^2 + m_2)/2\sigma^2]\, dm_2\, d\bar{x} \end{aligned} \tag{8.4.4}$$

and this is the same as Eq. (8.3.9) when we put $N = 3$. By Eqs. (8.3.11) and (8.3.13)

$$C = \frac{\left(\dfrac{N}{2\pi\sigma^2}\right)^{1/2}\left(\dfrac{N}{2\sigma^2}\right)^{(N-1)/2}}{\Gamma\left(\dfrac{N-1}{2}\right)} = \frac{1}{\sqrt{\pi}}\frac{\left(\dfrac{N}{2\sigma^2}\right)^{N/2}}{\Gamma\left(\dfrac{N-1}{2}\right)}$$

and when $N = 3$ this becomes

$$\frac{1}{\sqrt{\pi}\sigma^3}\left(\frac{3}{2}\right)^{3/2} = \frac{3\sqrt{3}}{2\sqrt{2\pi}}\sigma^{-3}$$

as in Eq. (4).

The rest of the argument is just as before. In the N-dimensional case, the hypersphere of N dimensions intersects the hyperplane in a hypersphere of $N - 1$ dimensions. The radius is $\sqrt{Nm_2}$ and the "area" is $K(Nm_2)^{(N-1)/2}$, where $K = \pi^{(N-1)/2}/\Gamma\left(\dfrac{N+1}{2}\right)$. The element of "volume" is $\sqrt{N}\, d\bar{x}\, dA$ $= KN^{N/2}\left(\dfrac{N-1}{2}\right)m_2^{(N-3)/2}\, dm_2\, d\bar{x}$ and this, with Eq. (8.3.3), gives the same result as Eq. (8.3.9).

8.5 The Distribution of Student's *t* If $\bar{x}$ is the mean of a sample of N from a normal distribution with mean μ and variance σ^2, then, as we have already seen, the variate

$$z = N^{1/2}\frac{(\bar{x} - \mu)}{\sigma} \tag{8.5.1}$$

is a standard normal variate. If, however, σ is replaced by its estimator $s = k_2^{1/2}$, the quantity

$$t = N^{1/2}\frac{(\bar{x} - \mu)}{s} \tag{8.5.2}$$

is not normally distributed, except asymptotically for large N. The distribution of t (actually of $tN^{-1/2}$) was originally obtained by W. S. Gosset, writing under the pen name of "Student" [2], and the importance of this distribution in a variety of practical situations was later emphasized by Fisher.

The joint density function for $\bar{x}$ and s may be obtained from Eq. (8.3.9) by noting that $m_2 = (N-1)s^2/N$, so that

$$(8.5.3)\quad \begin{aligned} f(\bar{x}, s)\,d\bar{x}\,ds &= h(\bar{x}, m_2)\,d\bar{x}\,dm_2 \\ &= C\left(\frac{N-1}{N}s^2\right)^{(N-3)/2}\exp\left[-\frac{N\bar{x}^2}{2\sigma^2}\right]\exp\left[-\frac{(N-1)s^2}{2\sigma^2}\right]\left(\frac{N-1}{N}\right)2s\,ds\,d\bar{x} \\ &= As^{N-2}\exp\left[-\frac{N\bar{x}^2}{2\sigma^2}\right]\exp\left[-\frac{(N-1)s^2}{2\sigma^2}\right]d\bar{x}\,ds \end{aligned}$$

where

$$A = \left(\frac{2N}{\pi}\right)^{1/2}\frac{\left(\frac{N-1}{2}\right)^{(N-1)/2}}{\sigma^N\Gamma\left(\frac{N-1}{2}\right)}$$

and where we are assuming as before that $\mu = 0$.

If we change the variables from $\bar{x}$ and s to t and s, we obtain as the joint density function for t and s

$$(8.5.4)\qquad g(t, s) = As^{N-1}N^{-1/2}\exp\left[-\frac{s^2t^2}{2\sigma^2}\right]\exp\left[-\frac{(N-1)s^2}{2\sigma^2}\right]$$

since, for fixed s, $s\,dt = N^{1/2}\,d\bar{x}$ and since $g(t, s)\,dt\,ds = f(\bar{x}, s)\,d\bar{x}\,ds$.

By integrating over all values of s, we obtain the desired density function for t, namely,

$$(8.5.5)\qquad \begin{aligned} f(t) &= \int_0^\infty g(t, s)\,ds \\ &= n^{-1/2}\left[B\left(\frac{1}{2}, \frac{n}{2}\right)\right]^{-1}\left(1 + \frac{t^2}{n}\right)^{(n+1)/2} \end{aligned}$$

(see Appendices A.6 and A.7). Here $n = N - 1$ and is called the number of *degrees of freedom* for t.

The important characteristic of $f(t)$ is that it is independent of σ, and therefore tables of this function can be used to test hypotheses about the mean of a population, irrespective of what the variance may be.

The graph of $f(t)$ is a symmetrical unimodal curve, tailing off towards zero at both ends. The tails are higher, and the central peak is higher, than for a normal curve of the same mean, variance and total area (see Figure 40, drawn for $n = 4$). As n increases, the curve becomes more and more nearly normal. To show this we note that

$$(8.5.6)\qquad \begin{aligned} \lim_{n\to\infty}\left(1 + \frac{t^2}{n}\right)^{-(n+1)/2} &= \lim_{n\to\infty}\left(1 + \frac{t^2}{n}\right)^{-1/2}\cdot\lim_{n\to\infty}\left(1 + \frac{t^2}{n}\right)^{-n/2} \\ &= (1)\cdot(e^{-t^2/2}) \end{aligned}$$

(see Appendix A.1), and by using Stirling's approximation (Appendix A.2) it is easy to prove that

$$\lim_{n\to\infty}\left\{n^{-1/2}\left[B\left(\frac{1}{2},\frac{n}{2}\right)\right]^{-1}\right\} = \lim_{n\to\infty}\frac{\Gamma\left(\frac{n+1}{2}\right)}{(\pi n)^{1/2}\Gamma\left(\frac{n}{2}\right)} = (2\pi)^{-1/2}$$

The limit of $f(t)$ is therefore $(2\pi)^{-1/2}e^{-t^2/2}$, which is the density function for a standard normal variate.

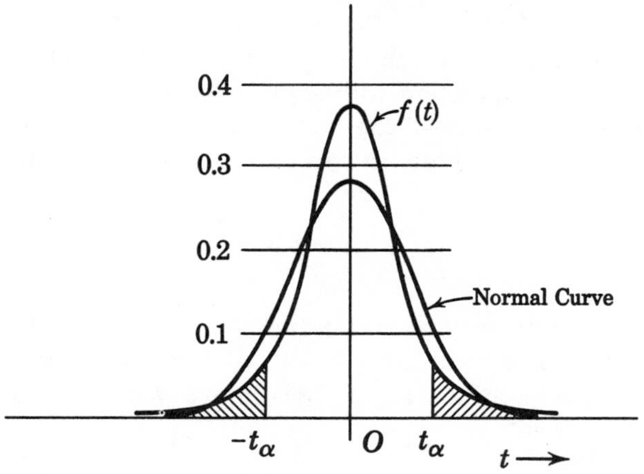

FIG. 40 GRAPHS OF THE T-DISTRIBUTION AND NORMAL DISTRIBUTION

From symmetry, the odd-order moments of $f(t)$ are all zero, but the $(2r)^{\text{th}}$ moment is

(8.5.7)
$$\mu_{2r} = 2n^{-1/2}\left[B\left(\frac{1}{2},\frac{n}{2}\right)\right]^{-1}\int_0^\infty t^{2r}\left(1+\frac{t^2}{n}\right)^{-(n+1)/2} dt$$
$$= \frac{n^r(2r-1)(2r-3)\ldots 1}{(n-2)(n-4)\ldots(n-2r)}$$

Thus $\mu_2 = \dfrac{n}{n-2}$, $\mu_4 = \dfrac{3n^2}{(n-2)(n-4)}$, so that $\kappa_4/\kappa_2{}^2 = \mu_4/\mu_2{}^2 - 3 = 6/(n-4)$. For $n > 4$ this is always positive.

From Eq. (2), $\dfrac{t^2}{n} = \dfrac{N(\bar{x}-\mu)^2}{ns^2} = \dfrac{[N^{1/2}(\bar{x}-\mu)/\sigma]^2}{ns^2/\sigma^2}$. The numerator of this fraction is the square of a standard normal variate (and therefore a χ^2 variate with one d.f.) and the denominator (by § 8.3) is an independent χ^2 variate with n d.f. The fraction itself is therefore the ratio of two gamma variates with parameters 1/2 and $n/2$, respectively (see § 4.6), and this ratio may be shown to have a beta-prime distribution with parameters 1/2, $n/2$. It is often useful to know that any statistic t has the Student-t distribution if t^2/n is the ratio of two independent variates distributed respectively as χ^2 with 1 and n degrees of freedom.

8.6 **Tables of t and Approximations to t** In Appendix B.4 there is a table of the integral

$$\alpha = P(t \geq t_\alpha) = \int_{t_\alpha}^{\infty} f(t)\,dt \tag{8.6.1}$$

where $f(t)$ is given by Eq. (8.5.5). This table gives, for all values of n from 1 to 30 and for selected values of the probability α, the values of t_α satisfying Eq. (1); that is, it gives those values of t which in a random sample of size $n + 1$ from a normal population will be exceeded with probability α.

Tables of t often give instead the values which will be exceeded *numerically* with probability α. This procedure corresponds to the equation

$$\alpha = P(|t| \geq t_\alpha) = 1 - \int_{-t_\alpha}^{t_\alpha} f(t)\,dt \tag{8.6.2}$$

Because of the symmetry of t, this probability is just twice that given by Eq. (1). If the two-tailed probability is wanted from the table in the Appendix, the probabilities given at the head of the columns should be doubled.

The variance of t is $n/(n - 2)$, so that $t\left(\frac{n-2}{n}\right)^{1/2}$ is a standardized variate and is approximately normal for fairly large n (say $n > 30$). Thus, the approximate value of t from Eq. (1) for $n = 30$ and $\alpha = 0.05$ is given by multiplying the corresponding normal variate (1.645) by $(30/28)^{1/2}$. This gives 1.703, whereas the correct value is 1.697

A better approximation, given by Hendricks [3], is

$$z \approx 2\frac{\Gamma\left(\frac{n+1}{2}\right)t}{\Gamma\left(\frac{n}{2}\right)(t^2 + 2n)^{1/2}} \tag{8.6.3}$$

where z is the normal variate giving the same probability as the actual t. Thus, if $n = 30$ and $t = 1.697$, the corresponding z would be 1.644, which is quite close to the true value 1.645.

If n is even, the factor $\Gamma\left(\frac{n+1}{2}\right)\Big/\Gamma\left(\frac{n}{2}\right)$ in Eq. (3) is equivalent to $\frac{\pi^{1/2}}{2^{n-1}}\cdot\frac{(n-1)!}{[(n/2-1)!]^2}$ and if n is odd it is equivalent to $\frac{2^{n-1}}{\pi^{1/2}}\cdot\frac{\left[\left(\frac{n-1}{2}\right)!\right]^2}{(n-1)!}$.

In either case a good approximation for large n is

$$\frac{\Gamma\left(\frac{n+1}{2}\right)}{\Gamma\left(\frac{n}{2}\right)} \approx \left(\frac{n}{2}\right)^{1/2}\left[1 - \frac{1}{4n} + \frac{1}{32n^2} - \dots\right] \tag{8.6.4}$$

An excellent approximation to t, given by Cornish and Fisher, is the following:

Let t_α and z_α be defined by

$$\alpha = \int_{t_\alpha}^{\infty} f(t)\, dt = \int_{z_\alpha}^{\infty} \phi(z) dz \tag{8.6.5}$$

where $\phi(z) = (2\pi)^{-1/2} e^{-z^2/2}$. Then

$$t_\alpha \approx z_\alpha + \frac{z_\alpha^{\ 3} + z_\alpha}{4n} + \frac{5z_\alpha^{\ 5} + 16z_\alpha^{\ 3} + 3z_\alpha}{96n^2} \tag{8.6.6}$$

For $n = 30$, and $\alpha = 0.05$, $z_\alpha = 1.645$, the second term is 0.0508 and the third is 0.00158, so that $t_\alpha \approx 1.697$, which is correct to four figures.

8.7 **Confidence Limits for the Population Mean** Equation (8.6.2) may be written

$$1 - \alpha = 2\int_0^{t_\alpha} f(t)\, dt \tag{8.7.1}$$

where $t = N^{1/2}(\bar{x} - \mu)/s$. There is a probability $1 - \alpha$ that the observed value of the statistic t, for a sample of size N from a normal population with mean μ, will lie within the limits $\pm t_\alpha$ (see Figure 40). This is equivalent to the statement that the $100(1 - \alpha)\%$ confidence limits for μ, corresponding to observed values of $\bar{x}$ and s for the sample, are given by

$$\mu = \bar{x} \pm s t_\alpha N^{-1/2}. \tag{8.7.2}$$

Example 1 [4] Electric meters are adjusted to work synchronously with a standard meter. After adjustment, a sample of 10 meters was tested by means of precision instruments. If the standard meter is rated at 1000, the observed ratings for the sample were as given under x in the following table. The question to be answered is whether the meters tested can reasonably be regarded as a random sample from a normal population with mean 1000, or whether there is a systematic deviation from this standard. From the data, $\bar{x} = 994$, $s^2 = (744 - 160)/9 = 64.9$, and if $\alpha \doteq 0.05$, the value of t_α for nine degrees of freedom is 2.262. The 95% confidence limits for μ, therefore, are

$$994 \pm \frac{2.262}{\sqrt{10}}\sqrt{64.9} = 994 \pm 5.8$$

or 998.2 to 999.8. Since these limits do not quite include 1000, there is a barely significant deviation (at the 5% level) from the standard value assumed.

The null hypothesis here is that $\mu = 1000$ and the alternative hypothesis (which we are led to accept) is that $\mu < 1000$. If μ were greater than 1000, the probability of the observed t value (or less) would be less than 2.5%.

TABLE 8.1

x	$u = x - 990$	u^2
983	−7	49
1002	12	144
998	8	64
996	6	36
1002	12	144
983	−7	49
994	4	16
991	1	1
1005	15	225
986	−4	16
	40	744

8.8 **Confidence Limits for the Difference of Means in Two Populations** If we suppose that two independent samples come from two different normal populations with means μ_1 and μ_2 but with a common variance σ^2, we can form confidence limits for the difference $\mu_1 - \mu_2$. If these limits include zero, there is no significant difference (at the chosen level) between the means.

Suppose the samples are of sizes N_1 and N_2, with means $\bar{x}_1$, $\bar{x}_2$, and variances s_1^2, s_2^2, respectively. An unbiased estimate of σ^2, based on both samples, is

$$\hat{\sigma}^2 = \frac{n_1 s_1^2 + n_2 s_2^2}{n_1 + n_2} \tag{8.8.1}$$

where $n_1 = N_1 - 1$, $n_2 = N_2 - 1$. This follows at once from the fact that $E(s_1^2) = \sigma^2$ and $E(s_2^2) = \sigma^2$, so that $E(\hat{\sigma}^2) = \sigma^2$.

By hypothesis, $\bar{x}_1$ and $\bar{x}_2$ are both normal with means μ_1 and μ_2 and variances σ^2/N_1 and σ^2/N_2, respectively. Therefore, $\bar{x}_1 - \bar{x}_2$ is normal with mean $\mu_1 - \mu_2$ and variance $\sigma^2(1/N_1 + 1/N_2)$. If we substitute for σ^2 the unbiased estimate of Eq. (1) we obtain the statistic

$$[\bar{x}_1 - \bar{x}_2 - (\mu_1 - \mu_2)]\left[\hat{\sigma}^2\left(\frac{1}{N_1} + \frac{1}{N_2}\right)\right]^{-1/2}$$

which has the Student-t distribution with $n_1 + n_2$ degrees of freedom. The $100(1 - \alpha)\%$ confidence limits for $\mu_1 - \mu_2$ are therefore given by

$$\mu_1 - \mu_2 = \bar{x}_1 - \bar{x}_2 \pm t_\alpha\left[\frac{n_1 s_1^2 + n_2 s_2^2}{n_1 + n_2}\cdot\frac{N_1 + N_2}{N_1 N_2}\right]^{1/2} \tag{8.8.2}$$

EXAMPLE 2 Two batches of concrete were made with slightly different qualities of sand. From each batch four cylinders were made up and tested for compressive strength (lb/in.^{2}), with the results shown:

Batch No.	*Values of X*	$\bar{X}$
1	1690, 1580, 1745, 1685	1675
2	1550, 1445, 1645, 1545	1546

The variances for the two samples are 4750 and 6673, respectively. These values are close enough to justify an assumption that the two batches do not differ in variance (a test for this will be given later, § 8.13). The question is whether the *means* differ significantly. We have $\bar{x}_1 - \bar{x}_2 = 129$, $n_1 = n_2 = 3$, $N_1 = N_2 = 4$, and the 95% confidence limits for $\mu_1 - \mu_2$ are

$$129 \pm 2.447\left[\frac{11423}{2}\cdot\frac{1}{2}\right]^{1/2}$$

$$= 129 \pm 131 \text{ or } -2 \text{ to } 260.$$

These limits include zero; therefore, at the level of significance chosen, the hypothesis that there is no difference between the means for the two batches must be accepted. The observed difference is, however, almost significant at this level.

8.9 **Confidence Limits for the Difference of Means in Paired Samples** In some types of experimental work, the two samples which are compared are not independent random samples but are deliberately paired in such a way as to reduce as much as possible all accidental differences other than those due to the particular effect which is being investigated. Thus in testing the effect of a drug on some property of the blood, the *same* group of experimental animals might be examined before and after administration of the drug. This procedure renders the experiment more precise, since it eliminates the random variability between one group of animals and another; this variability might, or might not, affect the particular property under investigation. If the sample size is N, the number of degrees of freedom is $n(= N - 1)$ instead of $2n$ as it would be if two independent random samples of size N had been used.

Again, in comparing the yields of two varieties of apple, one can imagine an experiment in which pairs of trees of the two varieties are grown side by side, in a dozen different locations. Differences of soil fertility, drainage, chemical composition of the soil, etc, will then be almost entirely eliminated from the comparison of yields, since each variety in any one pair is growing under almost the same conditions as the other variety, and the method of paired samples would be applicable.

The method of analysis in such a case is to obtain the N differences between members of a pair, $d_i = x_{1i} - x_{2i}$, where the subscript 1 refers to one sample and the subscript 2 to the other. (Thus 1 might refer to an animal before treatment and 2 to the same animal after treatment). On the null hypothesis that the treatment has really no effect (in the case of the apple trees, that the varieties do not really differ in mean yield), the expectation of d_i is zero. If s^2 is the variance of the differences, i.e.,

$$ns^2 = \sum d_i^2 - N\bar{d}^2 \tag{8.9.1}$$

the quantity $\bar{d}N^{1/2}s^{-1}$ has the Student-t distribution with n degrees of freedom.

On the hypothesis of a true difference δ between the means,

$$(\bar{d} - \delta)N^{1/2}s^{-1} = \pm t_\alpha \tag{8.9.2}$$

where t_α is the value of t which is exceeded numerically with probability α. Then Eq. (2) can be written

$$\delta = \bar{d} \pm st_\alpha N^{-1/2} \tag{8.9.3}$$

which gives $100(1 - \alpha)\%$ confidence limits for δ.

EXAMPLE 3 The following table gives pH values for the arterial blood of dogs (a) breathing normally, (b) after a period of breathing air containing 5% carbon dioxide.

TABLE 8.2

Dog Number	(a) x_1	(b) x_2	$d = x_1 - x_2$
1	7.42	7.26	0.16
2	7.53	7.30	0.23
3	7.36	7.26	0.10
4	7.43	7.39	0.04
5	7.43	7.38	0.05
6	7.15	6.69	0.46
7	7.50	7.32	0.18
8	7.34	7.26	0.08
9	7.45	7.23	0.22
10	7.42	7.06	0.36
11	7.53	7.34	0.19
12	7.48	7.28	0.20
13	7.42	7.29	0.13

The mean value of d is $\bar{d} = 0.1846$, and $s^2 = (0.6149 - 0.4430)/12 = 0.0143$, so that $\bar{d}N^{1/2}s^{-1} = 5.57$. For 12 d.f., the 1% value of t (for a two-tailed test) is 3.055, so that the probability is considerably less than 0.01 that a random sample of 13 animals would exhibit a mean difference as great numerically as that found if there were really no effect of the treatment. The hypothesis that breathing 5% of CO_2 has no effect on the pH value for the blood is decisively rejected.

If one could feel quite confident, *before the experiment is performed*, that if there is any effect it could be only one way (could result only in a *lowered pH* value) one would be justified in using a one-tailed test. The 1% value is then 2.681 and the probability of the observed result on the null hypothesis is even lower than before.

In terms of confidence limits, the 99% confidence limits for δ would be 0.1846 + 0.0332 (3.055), or 0.083 to 0.286. This expresses in another way the fact that there is a highly significant difference produced by the treatment.

*** 8.10 The *t*-Test and Maximum Likelihood** Suppose the null hypothesis H_0 is that our sample of N values of X comes from a normal population with mean μ_0 and variance σ^2, and the alternative hypothesis H_1 is that the sample comes from a normal population with mean $\mu_1(> \mu_0)$ and the same variance σ^2. For convenience, we can take μ_0 as zero (this simply means subtracting a constant amount μ_0 from all the observed values of X).

Under H_0, the likelihood for the particular set of observed values $x_1, x_2 \ldots x_N$ is

$$L_0 = (2\pi\sigma^2)^{-N/2} \exp\left(-\sum \frac{x_i^2}{2\sigma^2}\right) \tag{8.10.1}$$

while under H_1 it is

$$L_1 = (2\pi\sigma^2)^{-N/2} \exp\left(-\sum \frac{(x_i - \mu_1)^2}{2\sigma^2}\right) \tag{8.10.2}$$

The region of rejection R must satisfy the condition

$$\int_{(R)} L_0 \, dx_1 \ldots dx_N = \alpha \tag{8.10.3}$$

for a fixed value α of the probability of wrongly rejecting H_0. However, R can be chosen in many ways. To make the test as powerful as possible we should choose R to maximize the probability of rejecting H_0 when H_1 is true. That is, the probability

$$P = \int_{(R)} L_1 \, dx_1 \ldots dx_N \tag{8.10.4}$$

is to be a maximum subject to the condition of (3). This implies that we should maximize $\int_{(R)} (L_1 - \lambda L_0) \, dx_1 \ldots dx_N$ without restriction, λ being a Lagrange multiplier (see Appendix A.15). If we include in R all the points for which $L_1 - \lambda L_0 > 0$ and exclude all points for which $L_1 - \lambda L_0 < 0$ we shall make the integral as large as possible. The boundary of R is given by $L_1 - \lambda L_0 = 0$.

This equation is equivalent to

$$\log L_1 = \log \lambda + \log L_0$$

which reduces to

$$\sum (x_i - \mu_1)^2 = \sum x_i^2 - c, \qquad c = 2\sigma^2 \log \lambda$$

This is again equivalent to $\bar{x} = c_1$, where c_1 is another constant depending on σ, μ_1, N and λ. The region of rejection is defined by $\bar{x} > c_1$.

The situation can perhaps be appreciated geometrically for the special case $N = 3$, as in § 8.4. The more general case requires a familiarity with N-dimensional geometry. The likelihood L_0 is constant on the sphere $\sum x_i^2 =$ constant (with center at the origin). Equation (3) implies that on any such sphere we can pick a region of rejection R equal in area to a fixed fraction α of the surface, and we can agree to reject H_0 when the sample point lies in this region. The maximum likelihood condition implies that this region is the "cap" cut off on the sphere by the plane $\bar{x} = c_1$, this plane being perpendicular to a line equally inclined to all three axes. (See Fig. 41, where the shaded area represents the cap cut off the sphere by the plane.) The boundaries of all the caps for different spheres lie on a circular cone, with vertex at the origin.

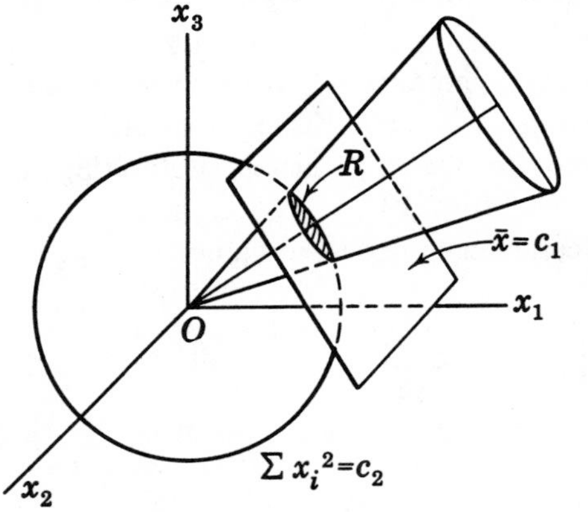

FIG. 41 REGION OF REJECTION OF H_0 WHEN $N = 3$

The plane $\bar{x} = c_1$ lies at a distance $\sqrt{3}c_1$ from the origin. The sphere $\sum x_i^2 = c_2$ is of radius $\sqrt{c_2}$ and the fractional area of the cap is given by

(8.10.5) $$\alpha = \frac{\sqrt{c_2} - \sqrt{3}c_1}{2\sqrt{c_2}}$$

Now the sample variance for a sample with representative point (x_1, x_2, x_3), lying on the intersection of the sphere and the plane, is

(8.10.6) $$\begin{aligned} s^2 &= \tfrac{1}{2}[(x_1 - \bar{x})^2 + (x_2 - \bar{x})^2 + (x_3 - \bar{x})^2] \\ &= \tfrac{1}{2}(x_1^2 + x_2^2 + x_3^2 - 3\bar{x}^2) \\ &= \tfrac{1}{2}(c_2 - 3c_1^2) \end{aligned}$$

and t for this sample is

(8.10.7) $$t = \sqrt{3}\,\frac{\bar{x}}{s} = \sqrt{6}\,\frac{c_1}{(c_2 - 3c_1^2)^{1/2}}$$

For $N = 3$ (and therefore $n = 2$) the distribution of Student's t reduces to

(8.10.8) $$f(t) = (2\sqrt{2})^{-1}(1 + t^2/2)^{-3/2}$$

and the integral of this from t_α to ∞ is

(8.10.9) $$\int_{t_\alpha}^{\infty} f(t)\,dt = \tfrac{1}{2}[1 - t_\alpha(2 + t_\alpha^2)^{-1/2}]$$

Substituting for t_α the value given by Eq. (7) we find precisely the α of Eq. (5), so that the one-tailed t-test (namely, reject H_0 when $t > t_\alpha$) is the same as the maximum likelihood test described above. It follows from Eqs. (5) and (7) that $t = (1 - 2\alpha)(2\alpha - 2\alpha^2)^{-1/2}$ so that the value of t for such a sample is independent of the value of σ^2 and of the particular value of μ_1 chosen for hypothesis H_1. The test is therefore uniformly most powerful against any H_1 with $\mu_1 > 0$.

Similarly, if $\mu_1 < 0$ the test $-t > t_\alpha$ is uniformly most powerful. However, if μ_1 may be greater or less than zero, no uniformly most powerful test exists, except in the special class of *unbiased* tests. A test is said to be unbiased if its power function for testing the hypothesis that a parameter θ is equal to θ_0 has a minimum at the value θ_0. The ordinary two-tailed t test ($|t| > t_\alpha$) provides a uniformly most powerful unbiased test of H_0 against H_1, where H_1 is the hypothesis $|\mu_1| > 0$.

* 8.11 **The Power of the *t*-Test** The power is the value of P given by Eq. (8.10.4) subject to the condition of (8.10.3). As in § 8.5, the probability $L_1\, dx_1 \ldots dx_N$ can be expressed as $f(\bar{x}, s)\, d\bar{x}\, ds$, except that now the population mean is to be taken as μ_1 instead of zero. We have

(8.11.1) $$f(\bar{x}, s) = As^{n-1} \exp\left(-\frac{ns^2}{2\sigma^2}\right)\exp\left[-\frac{N(\bar{x} - \mu_1)^2}{2\sigma^2}\right]$$

where $A = (2N/\pi)^{1/2}(n/2)^{n/2} \Big/ \left[\sigma^N \Gamma\left(\frac{n}{2}\right)\right]$ with $n = N - 1$ as usual.

If we define t as $t = N^{1/2}\bar{x}/s$, we can find P by integrating Eq. (1) over all $\bar{x} > N^{-1/2}st_\alpha$ and over all s from 0 to ∞. For any point in the region so delimited, t will be greater than t_α. Therefore,

(8.11.2) $$P = A\int_0^\infty s^{n-1}e^{-ns^2/2\sigma^2} \int_{N^{-1/2}st_\alpha}^\infty \exp\left[-\frac{N(\bar{x} - \mu_1)^2}{2\sigma^2}\right] d\bar{x}\, ds$$

On putting $z = N^{1/2}(\bar{x} - \mu_1)/\sigma$ and $\chi^2 = ns^2/\sigma^2$, so that z is a standard normal variate and χ^2 is the ordinary χ^2 variate with n degrees of freedom, Eq. (2) becomes

(8.11.3) $$P = \frac{1}{2\Gamma(n/2)}\int_0^\infty \left(\frac{\chi^2}{2}\right)^{(n-2)/2} e^{-\chi^2/2} \int_\gamma^\infty \phi(z)\, dz\, d(\chi^2)$$

where

$$\phi(z) = (2\pi)^{-1/2}e^{-z^2/2}$$

and

(8.11.4) $$\gamma = t_\alpha \chi n^{-1/2} - \mu_1 N^{1/2}\sigma^{-1}$$

This integral can be evaluated numerically for given values of n, α, and μ_1/σ. The power function depends on σ (through the quantity γ) and therefore some preliminary information about σ is necessary if we want to use the power function. We could, for example, calculate the size of sample necessary to detect, with a given probability, a given deviation of μ_1/σ from zero but without the information about σ itself we cannot say anything about μ_1.

* 8.12 **The Non-Central *t*-Distribution** If the distribution of t on hypothesis H_1 is calculated as in § 8.5 from Eq. (8.11.1), we obtain, after some reduction,

$$f_1(t) = C\left(1 + \frac{t^2}{n}\right)^{-(n+1)/2} e^{-nk^2/t^2} Hh(k) \tag{8.12.1}$$

where

$$C = \frac{n!}{2^{(n-1)/2}\Gamma(n/2)(\pi n)^{1/2}}, \quad k = -\frac{N^{1/2}\mu_1 t}{\sigma\sqrt{t^2 + n}}$$

and

$$Hh(k) = \int_0^\infty \frac{v^n}{n!} e^{-\frac{1}{2}(v+k)^2} \, dv \tag{8.12.2}$$

This function is known as Airey's function. The quantity δ, defined by

$$\delta = N^{1/2} \frac{\mu_1}{\sigma} \tag{8.12.3}$$

is called the *non-centrality parameter*. Any statistic of the form

$$t = (z + \delta)w^{-1/2} \tag{8.12.4}$$

where z is a standard normal variate and nw is an independent variate distributed as χ^2 with n degrees of freedom, has the non-central t-distribution.

When $\delta = 0$, $k = 0$ and $Hh(k) = \dfrac{2^{(n-1)/2}\Gamma\left(\dfrac{n+1}{2}\right)}{n!}$.

The density function then reduces to that for the ordinary Student-t, Eq. (8.5.5).

The power of the t test is the integral of $f_1(t)$ from t_α to ∞,

$$P = \int_{t_\alpha}^\infty f_1(t) \, dt \tag{8.12.3}$$

where t_α is given by

$$\alpha = \int_{t_\alpha}^\infty f(t) \, dt \tag{8.12.4}$$

This gives the same result as Eq. (8.11.3).

Extensive tables of non-central t have recently been provided by Resnikoff and Lieberman [5]. These give the density function $f_1(t)$, the cumulative distribution function $F_1(t)$, and certain percentage points of the distribution. Since $f_1(t)$ depends on two parameters, n and δ, the tables are of triple entry. Values of n go from 2 to 49, 2(1)24(5)49.* The argument used is $t/\sqrt{n}$ instead of t, this arrangement being more convenient for tabulation. The parameter δ is expressed in terms of z_α, where z_α is the standard normal variate exceeded with probability

*This notation means that n goes by steps of 1 to $n = 24$ and then by steps of 5 to $n = 49$.

α—as in Eq. (8.6.5)—the quantity tabulated being $\delta = N^{1/2}z_\alpha$, for ten selected values of α from 0.001 to 0.25.

A rough approximation to the non-centrality parameter for moderate-sized samples is

$$\delta = t_\alpha - z_P\left(\frac{1 + t_\alpha^2}{2n}\right)^{1/2} \tag{8.12.5}$$

where z_P is the standard normal variate exceeded with probability P. If β is the probability of error of the second kind, $P = 1 - \beta$. This approximation is useful in estimating the difference in the mean which can be detected with given probabilities of error.

Example 4 Suppose we are using a sample size $N = 17$, and are willing to allow errors $\alpha = 0.05$, $\beta = 0.2$. That is, we will accept a risk 0.05 of wrongly rejecting the hypothesis that $\mu = 0$ and a risk 0.2 of wrongly accepting the hypothesis that $\mu = \mu_1$. How large must μ_1 be?

Using the approximation Eq. (5), with $n = 16$,

$$t_\alpha = 1.746, \qquad z_P = -0.842$$
$$\delta = 1.746 + 0.842(1.0466) = 2.627$$

and therefore $\mu_1/\sigma = \delta/\sqrt{17} = 0.637$. This means that we should have about an 80% chance of detecting a real difference in the mean, from the assumed value zero, equal to about 0.64 times the standard deviation.

Some tables compiled by Neyman and Tokarska [6] are rather more convenient than the larger tables [5] for this particular type of problem. They give for $\alpha = 0.05$ and 0.01 and for $n = 1(1)30$, the values of δ corresponding to selected values of β. From these tables, with $\alpha = 0.05$, $\beta = 0.2$, and $n = 16$, we find $\delta = 2.60$, giving $\mu_1/\sigma = 0.631$.

* 8.13 **Sampling Inspection by Variables** A procedure which depends upon non-central t is that of accepting or rejecting a lot according to the percentage of defectives p, where "defective" is defined as meaning that a measured random variable X has a value above some fixed standard u. This variate X is supposed to be normal with unknown parameters μ and σ. The method is to measure the mean m and the standard deviation s for a sample of N and accept the lot if

$$m + ks \le u \tag{8.13.1}$$

where k is a constant. This criterion can be written as

$$\sqrt{N}\,\frac{u - m}{s} \ge \sqrt{N}k$$

or

$$\frac{\dfrac{\sqrt{N}(u - \mu)}{\sigma} - \dfrac{\sqrt{N}(m - \mu)}{\sigma}}{s/\sigma} \ge \sqrt{N}k \tag{8.13.2}$$

Now ns^2/σ^2 is distributed as χ^2 with n degrees of freedom, and $\sqrt{N}(m - \mu)/\sigma$ is a standard normal variate, so that the left-hand side of (2) has a non-central t distribution* with $n(= N - 1)$ d.f. and non-centrality parameter given by

(8.13.3) $$\delta = \frac{\sqrt{N}(u - \mu)}{\sigma} = \sqrt{N}z_p$$

where z_p is the standard normal variate exceeded with probability p. The acceptance criterion is therefore

(8.13.4) $$t \geq \sqrt{N}k, \qquad t = \sqrt{N}\,\frac{u - m}{s}$$

The power function (the probability of accepting the lot) is

(8.13.5) $$P = \int_{\sqrt{N}k}^{\infty} f_1(t)\,dt$$

and several values can be found by interpolation in the tables of Resnikoff and Lieberman for given N and k and different p. Thus if $N = 10$ and $k = 1.72$, we find that when $p = 0.10$ (corresponding to $\delta = 4.053$), $P = 0.224$, and when $p = 0.004$ (corresponding to $\delta = 8.386$), $P = 0.969$.

Conversely, if we fix two points on the power curve, we can find N and k and so set up a sampling acceptance plan. Thus suppose we want the values of P corresponding to $p_1 = 0.01$ and $p_2 = 0.15$ to be $1 - \alpha(= 0.99)$ and $\beta(= 0.10)$, respectively. That is, if p is as low as 0.01, we shall be almost certain (probability 0.99) to accept the lot. If p is as high as 0.15, we shall be very likely (probability 0.9) to reject it. The corresponding values of N and k are found by trial and error, using the tables. We want to find two consecutive values of n, say $n - 1$ and n, such that for $n - 1$ there is a t' for which simultaneously

$$\int_{t'}^{\infty} f_1(t)\,dt \leq 0.99, \qquad \delta = \sqrt{N - 1}z_{0.01} = 2.326\sqrt{N - 1}$$

and

$$\int_{t'}^{\infty} f_1(t)\,dt \geq 0.10, \qquad \delta = \sqrt{N - 1}z_{0.15} = 1.036\sqrt{N - 1}$$

while for n there is a t'' for which simultaneously

$$\int_{t''}^{\infty} f_1(t) \geq 0.99, \qquad \delta = \sqrt{N}z_{0.01} = 2.326\sqrt{N}$$

and

$$\int_{t''}^{\infty} f_1(t)\,dt \leq 0.10, \qquad \delta = \sqrt{N}z_{0.15} = 1.036\sqrt{N}$$

***The negative of this side has the non-central t distribution with parameter $-\delta$, but this is the same as saying that the side itself is non-central t with parameter δ.**

We find that for 16 d.f., with $p = 0.01$ and $t'/4 = 1.55$, P is 0.9896, while for 17 d.f. and the same p and t'', P is 0.9942. Also, for 16 d.f. and $t'/4 = 1.55$, with $p = 0.15$, $P = 0.1064$ while for 17 d.f. with the same p and t'', $P = 0.079$. We can therefore choose $n = 17$ ($N = 18$). Here it happened that t' and t'' are identical. With $n = 17$, we find by interpolation that $t' = 6.43$ corresponds to $P = 0.99$; k is therefore $6.43/\sqrt{18} = 1.516$. The sampling plan is to take a sample of size 18 and accept the lot if $t \geq 6.43$. This will be a little stricter than desired but near enough for practical purposes.

8.14 **Confidence Limits for the Variance of a Population** Since the quantity ns^2/σ^2 is distributed as χ^2 with n d.f. for samples from a normal parent population, it is easy to construct confidence limits for σ^2 corresponding to a given sample variance. It is merely necessary to find from a table of χ^2 the values χ_1^2 and χ_2^2 such that

(8.14.1)
$$\int_{\chi_1^2}^{\infty} f(\chi^2)\, d\chi^2 = \frac{\alpha}{2}, \qquad \int_0^{\chi_2^2} f(\chi^2)\, d\chi^2 = \frac{\alpha}{2}.$$

Then the lower and upper confidence limits for σ^2 are given respectively by

(8.14.2)
$$\sigma_1^2 = \frac{ns^2}{\chi_1^2}, \qquad \sigma_2^2 = \frac{ns^2}{\chi_2^2}$$

and the confidence coefficient is $100(1 - \alpha)\%$. It is not, of course, necessary that the two tails should each be $\alpha/2$ in area, provided that the sum is equal to α. However, it is usual to take them as equal.

EXAMPLE 5 The variance of a sample of size 10 is 0.064. What are the 95% confidence limits for σ^2?

The values of χ_1^2 and χ_2^2 corresponding to $\alpha = 0.05$ are 19.023 and 2.700, for 9 d.f. The confidence limits are therefore $\sigma_1^2 = \dfrac{0.576}{19.023} = 0.030$ and $\sigma_2^2 = \dfrac{0.576}{2.700} = 0.213$.

Confidence limits for the standard deviation σ should, strictly, be obtained from the distribution of s. Nevertheless, it is customary to obtain the limits for σ^2 and use the positive square roots, in spite of the fact that the square root of s^2 is not an unbiased estimator for σ.

8.15 **Distribution of the Variance Ratio** Suppose s_1^2 and s_2^2 are the observed variances for two samples, of sizes N_1 and N_2, drawn from normal populations with variances σ_1^2 and σ_2^2 respectively. We can test the null hypothesis H_0 that $\sigma_1^2 = \sigma_2^2$ by calculating the distribution of the *variance ratio*

(8.15.1)
$$F = s_1^2/s_2^2$$

This ratio is generally denoted by F in honour of Sir R. A. Fisher, although Fisher originally used the related statistic

(8.15.2) $$z = \tfrac{1}{2}\log_e(s_1^2/s_2^2) = \tfrac{1}{2}\log_e F,$$

which has a more nearly symmetrical distribution than F.

On the null hypothesis, with $\sigma_2^2 = \sigma_1^2$,

(8.15.3) $$\frac{n_1 F}{n_2} = \frac{n_1 s_1^2/\sigma_1^2}{n_2 s_2^2/\sigma_1^2} = \frac{\chi_{n_1}^2}{\chi_{n_2}^2},$$

both the numerator and denominator being χ^2 variates with n_1 and n_2 d.f. ($n_1 = N_1 - 1$ and $n_2 = N_2 - 1$). The ratio is therefore a beta-prime variate (see § 4.5) with parameters $\alpha = n_1/2$, $\beta = n_2/2$, and its density function is

(8.15.4) $$f(x) = x^{\alpha-1}(1+x)^{-\alpha-\beta}/B(\alpha, \beta)$$

with $x = n_1F/n_2$.

The density function for F is given by

$$g(F)\,dF = g(F)\frac{n_2}{n_1}\,dx = f(x)\,dx$$

so that

(8.15.5) $$g(F) = \frac{\left(\dfrac{n_1}{n_2}\right)^{n_1/2}}{B\left(\dfrac{n_1}{2}, \dfrac{n_2}{2}\right)} \frac{F^{(n_1/2)-1}}{\left(1 + \dfrac{n_1 F}{n_2}\right)^{(n_1+n_2)/2}}, \qquad 0 \le F < \infty$$

The numbers n_1 and n_2 are called the degrees of freedom for F. This is a positively skew distribution. The mode (the value of F corresponding to a maximum of $g(F)$) is at $F = \dfrac{n_2}{n_1}\cdot\dfrac{n_1 - 2}{n_2 + 2}$ which is always less than 1. The expectation of F is given by

(8.15.6) $$E(F) = n_2/(n_2 - 2), \qquad n_2 > 2$$

which is independent of n_1 and is always greater than 1.

The distribution of Fisher's z is found by writing $F = e^{2z}$, $dF = 2F\,dz$, and its density function is

(8.15.7) $$f(z) = \frac{2n_1^{n_1/2}n_2^{n_2/2}}{B\left(\dfrac{n_1}{2}, \dfrac{n_2}{2}\right)} \cdot \frac{e^{n_1 z}}{(n_1 e^{2z} + n_2)^{(n_1+n_2)/2}}$$

8.16 **Tables of the Distributions of *F* and *z*** Table B.5 in the Appendix gives for various values of n_1 and n_2 the upper 5% and the upper 1% points of the distribution of F. A complete table of the probability integral of F would be quite bulky, being a triple-entry table, but since we usually merely want to know whether an observed F value is significant or not, it is sufficient to have, for each pair of values of n_1 and n_2, a few values of F corresponding to common levels of significance. In the tables of Fisher and Yates [7], 20%, 10%, 5%, 1% and 0.1% points are given for both F and z.

Interpolation for n_1 and n_2, when necessary, should be harmonic instead of linear. Thus, suppose the 1% point is required for $n_1 = 60$ and $n_2 = 55$. The table gives the 1% points for 50 and 55 and for 75 and 55 as 1.90 and 1.82, respectively. If x is the value for 60 and 55, harmonic interpolation gives $\dfrac{1.90 - x}{1.90 - 1.82} = \dfrac{1/50 - 1/60}{1/50 - 1/75}$, so that $x = 1.86$.

EXAMPLE 6 For two samples, of sizes 8 and 12, the observed variances are 0.064 and 0.024, respectively. Since the table refers only to the *upper* points of the distribution, we will take the subscript 1 to refer to the sample with the larger variance. Then $n_1 = 7$, $n_2 = 11$, and $F = \dfrac{0.064}{0.024} = 2.67$, with 7 and 11 degrees of freedom.

The 5% point is 3.01 and the 1% point is 4.88. The probability of a value of F at least as great as 2.67 is therefore more than 0.05, and the two samples are not significantly different at this level. From the Fisher and Yates tables we note that the 10% point is 2.34, so that the difference *is* significant at the 10% level.

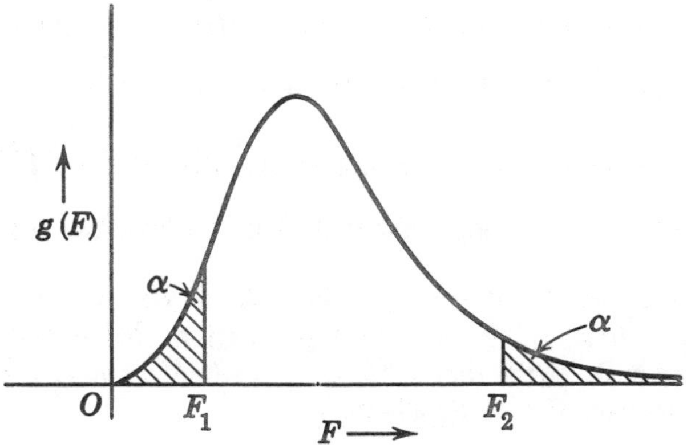

FIG. 42 THE F-DISTRIBUTION

A departure from the null hypothesis of equality of variances could just as well give a value of F less than 1 as a value greater than 1. Corresponding to any F_2 such that $\int_{F_2}^{\infty} g(F)\,dF = \alpha$ there is a value of F_1 such that $\int_0^{F_1} g(F)\,dF = \alpha$ (see Figure 42, where it is assumed that $\alpha < 0.5$).

If we put $u = 1/F$ in the equation

$$\alpha = \int_{F_2}^{\infty} g(F)\,dF \tag{8.16.1}$$

we obtain

$$\alpha = \int_0^{1/F_2} \frac{\left(\dfrac{n_2}{n_1}\right)^{n_2/2}}{B\left(\dfrac{n_2}{2}, \dfrac{n_1}{2}\right)} \cdot \frac{u^{(n_2/2)-1}\,du}{\left(1 + \dfrac{n_2 u}{n_1}\right)^{(n_1+n_2)/2}} \tag{8.16.2}$$

The integrand here is the same as $g(u)$, in Eq. (8.15.5) but with n_1 and n_2 interchanged, and $1/F_2$ is the same as F_1. Therefore to find the *lower* 5% point, say, for a given n_1 and n_2 we take the reciprocal of the *upper* 5% point, after interchanging n_1 and n_2. This makes it unnecessary to have tables for both ends of the distribution. It should be noted, however, that when we use a two-tailed test (supposing that the departures from equality of variance may be in either direction) the probabilities given in the table must be doubled. The 5% point becomes a 10% point, for example.

If we write $F = \dfrac{n_2}{n_1}\cdot\dfrac{1-x}{x}$, or, equivalently,

$$x = n_2(n_2 + n_1 F)^{-1} \tag{8.16.3}$$

it is a straightforward matter to show that x is a beta-variate with parameters $n_2/2, n_1/2$. The distribution function of x is therefore an incomplete beta function and the tables of this function can be used to calculate the probability that x is less than some observed value. Thus in Example 6, we should have $x = \dfrac{11}{11 + 7F}$ $= 0.371$. The probability of a value not greater than this is $I_x\left(\dfrac{11}{2}, \dfrac{7}{2}\right) = 0.071$, which, as previously noted, is greater than 0.05 but less than 0.1.

***8.17 The Power of the *F*-Test** As in § 8.15, we assume that the null hypothesis H_0 is that $\sigma_1{}^2 = \sigma_2{}^2$ (both populations being normal). Let the alternative hypothesis H_1 be that $\sigma_1{}^2/\sigma_2{}^2 = \lambda$, which we may take greater than 1. Then H_0 is rejected if $F > F_2$ where

$$\int_{F_2}^{\infty} g(F)\,dF = \alpha \tag{8.17.1}$$

The power of this test is the probability that $F > F_2$, under H_1,

$$P = Pr(F > F_2 | H_1) \tag{8.17.2}$$

Now the statistic $F\sigma_2{}^2/\sigma_1{}^2 = \dfrac{s_1{}^2/\sigma_1{}^2}{s_2{}^2/\sigma_2{}^2}$ on hypothesis H_1 has the F distribution with n_1 and n_2 d.f. This follows because $(n_1/n_2)\cdot(F\sigma_2{}^2/\sigma_1{}^2)$ is the ratio of two

χ^2 variates with n_1 and n_2 d.f., respectively. If, therefore, F_P is a value of F which is exceeded with probability P, under H_0,

$$P = Pr\left(F\frac{\sigma_2^2}{\sigma_1^2} > F_P \middle| H_1\right) \tag{8.17.3}$$

From Eqs. (2) and (3), $F_2 = \sigma_1^2 F_P/\sigma_2^2 = \lambda F_P$.

We can use tables of F (preferably those by Merrington and Thompson, see [7]) to calculate λ for selected values of P. For example if $N_1 = N_2 = 10$, $\alpha = 0.05$ and $P = 1 - \beta = 0.5$, we find $F_2 = 3.18$ and $F_P = 1.00$, so that $\lambda = 3.18$. This means that we have an even chance of recognizing a difference between the variances when one is 3.18 times the other if we agree to accept a 5% chance of rejecting the null hypothesis when the variances are really equal.

The tables can also be used to estimate the size of sample necessary to have a given chance of observing a stated difference in variance. Thus suppose that a suggested new process of manufacturing some metal part might be expected to reduce the standard deviation of tensile strength by a factor of 1.41 (which means halving the variance). We would need, in order to have an even chance of detecting such an effect, samples of size 25, and to have a 95% chance we would need samples of nearly 100, the rejection error remaining at 5%.

*** 8.18 The Variance of Sample Skewness and Kurtosis** It is possible by means of long and rather tedious algebra to work out the moments of k_3, k_4, etc. in samples from a population with known cumulants. Even for a normal parent population the expressions are long, for any moments above the second. Here we shall simply state a few results for the normal case.

For the third k-statistic,

$$\begin{cases} E(k_3) = 0 \\ V(k_3) = 6\kappa_2^3 \dfrac{N}{(N-1)(N-2)} \end{cases} \tag{8.18.1}$$

and an unbiased estimator of this variance is

$$\hat{V}(k_3) = \frac{6k_2^3 N(N-1)}{(N-2)(N+1)(N+3)} \tag{8.18.2}$$

For the fourth k-statistic,

$$\begin{cases} E(k_4) = 0 \\ \hat{V}(k_4) = \dfrac{24k_2^4 N(N-1)^2}{(N-3)(N-2)(N+3)(N+5)} \end{cases} \tag{8.18.3}$$

The variances of $g_1 = k_3/k_2^{3/2}$ and of $g_2 = k_4/k_2^2$ (the sample skewness and kurtosis respectively) were worked out by R. A. Fisher, who found that

$$V(g_1) = \frac{6N(N-1)}{(N-2)(N+1)(N+3)} \tag{8.18.4}$$

(8.18.5)
$$V(g_2) = \frac{24N(N-1)^2}{(N-3)(N-2)(N+3)(N+5)}$$

For large N these approximate to the values $6/N$ and $24/N$, respectively.

8.19 **The Distribution of Extreme Values** It is sometimes convenient to judge a sample by the largest (or smallest) item in it. If the observed values of X for the sample are arranged in ascending order,

$$x_1 \leq x_2 \leq \ldots \leq x_N$$

and if $F(x)$ is the distribution function for the parent population, the probability that $x_N \leq x$ is $F(x)^N$. This is true because if $x_N \leq x$, the same inequality must hold for all the other values in the sample, which are all supposed to be independent. The probability that the largest item in a sample of size N from a standard normal population is less than x is given by

(8.19.1)
$$P(x_N \leq x) = [\Phi(x)]^N$$

where

$$\Phi(x) = (2\pi)^{-1/2} \int_{-\infty}^{x} e^{-u^2/2}\, du$$

Values of the lower and upper percentage points of the largest value x_N have been calculated by Tippett and Pearson [8]. The same table applies to the smallest value x_1 with a change of sign and a reversal of the terms "upper" and "lower." A brief extract from this table is appended.

TABLE 8.3

Sample Size	Upper Percentage Points	
N	5%	1%
5	2.319	2.877
10	2.568	3.089
15	2.705	3.207
20	2.799	3.289
30	2.929	3.402
50	3.082	3.539
100	3.283	3.718
1000	3.884	4.264

Such a table is useful in some types of quality control problems. If, for example, a manufacturer is producing a certain article for which the average breaking strength should be 180 lb with a standard deviation of not more than 12 lb, and if routine samples of size 10 are tested, the *lowest* value in a sample should not be below 180 − 12(3.089) = 142.9 lb more than once in 100 times. If such a low value is observed it might be worth while to look into possible causes.

* 8.20 **The Rejection of Extreme Observations** The question of whether to reject an extreme value (often called a "straggler") in a set of observations is one that sometimes poses a difficulty in experimental work. If we can assume that the sample comes from a normal population of known mean and variance, the distribution of the extreme value as given in § 8.19 will enable us to calculate the risk we run in rejecting the straggler. In practice, however, the mean and variance are not generally known and we must substitute estimates derived from the sample itself, but the distribution is then not precisely that of § 8.19.

For a sample of size N with mean $\bar{x}$ and standard deviation s, the distribution of

$$u_i = (x_i - \bar{x})/s \tag{8.20.1}$$

where x_i is the value of X for the straggler suspected, was worked out by W. R. Thompson [9]. He found that the quantity

$$t_i = u_i \left[\frac{N - 2}{\dfrac{(N-1)^2}{N} - u_i^2} \right]^{1/2} \tag{8.20.2}$$

has the Student-t distribution with $N - 2$ d.f., so that the probability of such a value arising by chance in a normal parent population can be calculated. This, however, refers to a *single* observation and not to the smallest or largest in a sample, and care should therefore be used in interpretation. In a sample of 20 from the same normal population we could expect that one, by pure chance, would reach a t-value corresponding to probability 0.05. As a rough rule-of-thumb, one might agree to require a probability of less than 0.01 for a sample of size less than 10 and a probability of less than 0.005 for one of size 10 to 20, before rejecting the extreme value.

W. J. Dixon [10] has suggested the use of a simple ratio criterion for the rejection of x_N, and this requires very little computation. For samples of size 8 to 12 we compute $r_{11} = (x_N - x_{N-1})/(x_N - x_2)$, and for larger samples $r_{22} = (x_N - x_{N-2})/(x_N - x_3)$. If the ratio exceeds a critical value R_α, the probability is less than α that the extreme value x_N comes from the same normal population as the rest of the observations. When the *lowest* value in the set is the one suspected, the observations should be placed in reverse order so that x_N is

TABLE 8.4

	r_{11}					r_{22}			
N	8	9	10	11	12	13	14	15	16
$R_{0.05}$	.608	.564	.530	.502	.479	.611	.586	.565	.546
$R_{0.01}$	.717	.672	.635	.605	.579	.697	.670	.647	.627

still the straggler. Table 8.4 is a brief extract from the tables [11] giving values of R_α for $\alpha = 0.05$ and 0.01, and N from 8 to 16. If r_{11} (or r_{22}) is greater than $R_{0\cdot 05}$, the observation x_N may reasonably be rejected. This table applies when we do not know before seeing the data whether we shall want to test the highest or the lowest value, and so corresponds to $\alpha = 0.025$ and 0.005 in Dixon's tables.

EXAMPLE 7 The following 15 observations were made of the vertical semi-diameter of the planet Venus in seconds of arc (43.00″ have been subtracted from each reading and the readings have been rearranged in ascending order):

$$-1.40,\ -0.44,\ -0.30,\ -0.24,\ -0.22,\ -0.13,\ -0.05,$$
$$0.06,\ 0.10,\ 0.18,\ 0.20,\ 0.39,\ 0.48,\ 0.63,\ 1.01$$

The observation -1.40 is rather suspiciously low. The mean of all the readings is 0.018 and the standard deviation is 0.551 so that, from Table 8.3, the lowest value in the sample should not be below $0.018 - 2.705\,(0.551) = -1.47$ more than once in 20 times, if the sample mean and variance apply to the population. The observed -1.40 is therefore not too unreasonable, on this supposition. According to Thompson's criterion, $u_i = -2.574$ and $t = -3.55$, with 13 d.f., and the probability of a single value as low as this is less than 0.005. We might, on this criterion, reasonably reject the straggler. If we do so, and then test the remaining 14 observations, the largest has a t of 2.874 with 12 d.f. Since the probability is between 0.05 and 0.01, we should be chary of rejecting it.

Applying Dixon's criterion, we get $r_{22} = 0.585$, which is greater than the value 0.565 in the table, and so would lead to rejection. After rejecting the lowest value, the ratio for the highest remaining value is 0.424, and this suggests retention of the straggler.

8.21 **The Distribution of the Range** The range of a sample, with the observed values placed in ascending order of size, is given by

$$R = x_N - x_1 \tag{8.21.1}$$

If $F(x)$ is the distribution function for the parent population, the probability that $N - 2$ values lie between x_1 and x_N is $[F(x_N) - F(x_1)]^{N-2}$. The probability that one specified observation has the value x_1 (to $x_1 + dx_1$) is $f(x_1)\,dx$, and similarly the probability for one specified observation to be equal to x_N is $f(x_N)\,dx_N$. Since, however, there are $N(N - 1)$ ways in which these two extreme observations may appear in the original order of the observations (x_1 could be in any one of N places and x_N in any of the remaining $N - 1$ places), the probability of a sample with lowest value x_1 and highest value x_N is

$$P(x_1, x_N) = N(N-1)[F(x_N) - F(x_1)]^{N-2} f(x_1) f(x_N)\,dx_1\,dx_N \tag{8.21.2}$$

Putting $x_N = x_1 + R$, we find as the joint probability density for x_1 and R

$$f(x_1, R) = N(N-1)[F(x_1 + R) - F(x_1)]^{N-2} f(x_1) f(x_1 + R) \tag{8.21.3}$$

Integrating over all values of x_1, we obtain the probability density for R, namely,

$$g(R) = N(N-1)\int_{-\infty}^{\infty} [F(x_1+R) - F(x_1)]^{N-2} f(x_1) f(x_1+R)\, dx_1 \tag{8.21.4}$$

The distribution function for R is

$$G(R) = \int_0^R g(u)\, du. \tag{8.21.5}$$

If we write $F(x_1 + u) - F(x_1) = y$, then, for a fixed x_1, $dy = f((x_1 + u)\, du$. Substituting from Eq. (4) in Eq. (5) and reversing the order of integration (which is legitimate here), we obtain

$$\begin{aligned} G(R) &= N(N-1)\int_{-\infty}^{\infty} f(x_1) \int_{u=0}^{u=R} y^{N-2}\, dy\, dx_1 \\ &= N \int_{-\infty}^{\infty} f(x_1)[F(x_1+R) - F(x_1)]^{N-1}\, dx_1 \end{aligned} \tag{8.21.6}$$

The expected value of R, $E(R)$, has been calculated by Tippett (see, e.g., [12], page 338) as

$$E(R) = \int_{-\infty}^{\infty} [1 - F^N - (1-F)^N]\, dx \tag{8.21.7}$$

where F stands for $F(x)$. For a standard normal parent population ($\mu = 0$, $\sigma = 1$), $F(x) = \Phi(x) = (2\pi)^{-1/2}\int_{-\infty}^{x} e^{-u^2/2}\, du$. If the range in a sample from this population is denoted by w, Tippett's values of $E(w)$ for $N = 2(1)500(10)1000$ are given in *Biometrika Tables for Statisticians, Vol. I*, Table 27. Examination of this table shows that for N between 350 and 550 the value of $E(w)$ is close to 6. This is the reason for the common practice of estimating roughly the standard deviation from a sample of several hundred items as one-sixth of the range.

Values of $G(w)$ for the standard normal population have been calculated for $N = 2$ to 20 by E. S. Pearson and H. O. Hartley and may be found in the *Biometrika Tables*, Table 23. More complete tables are given in reference [13]. The expression for $G(w)$ may be reduced to

$$G(w) = \left[2\Phi\left(\frac{w}{2}\right) - 1\right]^N - 2N\int_{w/2}^{\infty} [\Phi(u) - \Phi(u-w)]^{N-1}\phi(u)\, du \tag{8.21.8}$$

but the evaluation must be carried out by numerical methods. The distribution does not approach normality as N increases.

In practice the range is used mainly for small samples, of size 5 or 10, say, such as commonly occur in the applications of quality control in industry. The range is certainly a very convenient measure of dispersion because of the simplicity of its calculation. For large samples the distribution becomes very sensitive to departures from normality in the parent population, particularly as regards kurtosis.

* 8.22 **Tests of Hypotheses concerning the Variance of a Normal Population by use of the Sample Range** By comparing the observed range with the tabulated value of $E(w)$ for a standard normal population, we may estimate the standard deviation of the actual normal population. For samples of size ≤ 10 the efficiency of this method is at least 85%, and the calculation is very easy.

If the observed range is R, an unbiased estimator of σ is given by

$$\hat{\sigma} = R/E(w) = kR \tag{8.22.1}$$

The values of k for a few sample sizes are given in the following table, which also includes upper and lower 0.5 percentage points for $w = R/\sigma$. These may be used in establishing 99% confidence limits for σ based on the range of a single sample.

TABLE 8.5[a]

N	k	Lower 0.5%	Upper 0.5%
2	0.886	0.01	3.97
3	.591	.13	4.42
4	.486	.34	4.69
5	.430	.55	4.89
6	.395	.75	5.03
7	.370	.92	5.15
8	.351	1.08	5.26
9	.337	1.21	5.34
10	.325	1.33	5.42
15	.288	1.80	5.70
20	.268	2.12	5.89

[a]Extracted from Table 22, reference [8], by kind permission of Professor E. S. Pearson and the publishers of *Biometrika*.

Thus if the observed range in a sample of five items is $R = 8$, the estimate of σ would be $8(0.430) = 3.44$. The 99% confidence limits for σ would be $8/(4.89)$ and $8/(0.55)$, that is, 1.64 and 14.5.

If we wish to test the hypothesis H_0 (that $\sigma = 1$) against the alternative H_1 (that $\sigma > 1$) and if we use a test of size α, the critical value of w will be the upper $100\,\alpha$ % point. Thus for $\alpha = 0.005$, the critical value for a sample of size 10 is 5.42 (see Table 8.5). If $\alpha = 0.05$ we find from a larger table that the critical value is 4.47. The following sample of random normal numbers,

$$-2.015,\ -0.623,\ -0.699,\ 0.481,\ -0.586,$$
$$-0.579,\ -0.120,\ 0.191,\ 0.071,\ -3.001$$

has a range $w = 3.482$, and hence the hypothesis H_0 would not be rejected by this test. The test is less powerful than that based on the sample variance (using a table of chi-square) but is easier to apply.

*** 8.23 The Range in Samples from a Rectangular Distribution** If the parent population is rectangular, so that (with suitable units) $F(x) = x, 0 \leq x \leq 1$, the distribution function for the range can easily be obtained explicitly. We have

$$G(R) = N \int_0^1 [F(x_1 + R) - F(x_1)]^{N-1} \, dx_1 \tag{8.23.1}$$

Now $F(x_1 + R) = x_1 + R$ as long as $x_1 + R \leq 1$, but, when $x_1 > 1 - R$, $F(x_1 + R)$ remains equal to 1. The region of integration for a given R must therefore be split into two parts, from 0 to $1 - R$ and from $1 - R$ to 1 (see Figure 43). Then,

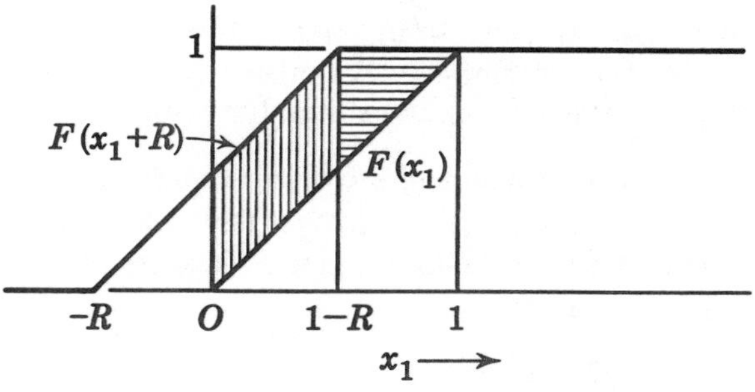

FIG. 43 DISTRIBUTION FUNCTION FOR RECTANGULAR DISTRIBUTION

$$\begin{aligned} G(R) &= N \int_0^{1-R} (x_1 + R - x_1)^{N-1} \, dx_1 \\ &\quad + N \int_{1-R}^1 (1 - x_1)^{N-1} \, dx_1 \\ &= NR^{N-1}(1 - R) + R^N = NR^{N-1} - (N - 1)R^N \end{aligned} \tag{8.23.2}$$

The probability density is

$$g(R) = N(N - 1)R^{N-2}(1 - R) \tag{8.23.3}$$

which has a maximum at $R = (N - 2)/(N - 1)$. The expected value of R is $(N - 1)/(N + 1)$.

A rectangular population is not quite as artificial as it may appear. In the production of machine parts in a factory to rather narrow specification limits, when only those articles which comply with the specification are included in the population, the hypothesis of a rectangular distribution seems not unreasonable.

*** 8.24 A Test for Equality of Two Rectangular Populations, Based on the Range** If R_1 and R_2 are the ranges in two random samples of sizes N_1 and N_2, assumed to come from rectangular populations of widths C_1 and C_2, respectively, the distribution of the quotient of ranges R_1/R_2, under the null hypothesis that $C_1 = C_2 = C$, has been worked out by Rider [14].

The probability-density for $U = R_1/R_2$ turns out to be independent of C and is given by

$$(8.24.1)\quad f(u) = \frac{[N_1N_2(N_1 - 1)(N_2 - 1)] \times [(N_1 + N_2)u^{N_1-2} - (N_1 + N_2 - 2)u^{N_1-1}]}{(N_1 + N_2)(N_1 + N_2 - 1)(N_1 + N_2 - 2)}, \quad \text{if } 0 \leq u \leq 1$$

and

$$(8.24.2)\quad f(u) = \frac{[N_1N_2(N_1 - 1)(N_2 - 1)] \times [(N_1 + N_2)u^{-N_2} - (N_1 + N_2 - 2)u^{-N_2-1}]}{(N_1 + N_2)(N_1 + N_2 - 1)(N_1 + N_2 - 2)}, \quad \text{if } 1 \leq u \leq \infty.$$

The expected value of u is $(N_1 - 1)N_2/[(N_1 + 1)(N_2 - 2)]$.

Rider gives a table for the quotient of ranges which will be exceeded in 5% of random samples, and this table may be used for testing the null hypothesis.

EXAMPLE 8 The width of a slot in a certain airplane part was measured to the thousandth of an inch in a sample of five parts on the first day of production and again in a sample of 10 parts two days later. The results (in thousandths of an inch in excess of 0.800 in.) were

(1) 77, 80, 78, 72, 78
(2) 75, 77, 75, 76, 77, 79, 75, 78, 77, 76

We see that $N_1 = 5$, $R_1 = 8$, $N_2 = 10$, $R_2 = 4$. Then $u = 2$. The probability of a value as great as this can be obtained by integrating Eq. (2) from $u = 2$ to ∞ and is 0.0013. It appears, therefore, that the second sample is pretty definitely more uniform than the first. The 5% critical value of u is actually 1.27.

This test is analogous to the F-test, discussed in §§ 8.15 and 8.16. The null hypothesis that the quotient of population ranges is 1 is tested against the alternative hypothesis that the quotient is greater than 1. We can always make $u \geq 1$ by choosing for sample (1) that with the greater range.

8.25 The Distribution of Order Statistics The rth order-statistic of a sample of size N is the rth smallest variate-value in the sample. If the values are arranged in ascending order of size, $x_1 \leq x_2 \leq x_3 \ldots \leq x_N$, the rth order-statistic is x_r. For a sample of size $2r + 1$, the $(r + 1)^{\text{th}}$ order-statistic is the *median* (the middle value).

In a sample of size N from a population with a continuous distribution function $F(x)$, the probability that $x_r = x$ (to $x + dx$) is

$$(8.25.1)\quad g(x)\,dx = C[F(x)]^{r-1}[1 - F(x)]^{N-r}f(x)\,dx$$

since there are $r - 1$ observations smaller than x and $N - r$ observations larger than x. The constant C is found from the condition

$$\int_{-\infty}^{\infty} g(x)\,dx = 1 \tag{8.25.2}$$

Writing $F(x) = F, f(x)\,dx = dF$, and noting that F goes from 0 to 1 as x goes from $-\infty$ to ∞, we see that Eq. (2) gives

$$C\int_{0}^{1} F^{r-1}(1-F)^{N-r}\,dF = 1 \tag{8.25.3}$$

whence

$$C^{-1} = B(r, N-r+1) \tag{8.25.4}$$

For a *rectangular* parent population, $F(x) = x$ and $g(x)$ is simply the ordinary beta-distribution.

For a *normal* population, the study of the distribution involves the numerical calculation of certain integrals. Thus for the *median*, with $N = 2r + 1$, and with $F(x) = \Phi(x)$, we have

$$g(x)\,dx = C[\Phi(x)]^r\,[1-\Phi(x)]^r\,d\Phi(x) \tag{8.25.5}$$

where

$$C^{-1} = B(r+1, r+1) = (r!)^2/N! \tag{8.25.6}$$

Note that the r of Eq. (1) is now replaced by $r + 1$, since the median is x_{r+1} when $N = 2r + 1$. For $N = 2r$, the median is taken as $(x_r + x_{r+1})/2$.

It is obvious from the symmetry of the parent population (which we have taken as standardized) that the expected value of the median will be zero. The variance is

$$\begin{aligned} V(x_{r+1}) &= \int_{-\infty}^{\infty} x^2\,g(x)\,dx \\ &= \frac{N!}{(r!)^2}\int_{-\infty}^{\infty} x^2\Phi^r(1-\Phi)^r\phi(x)\,dx \end{aligned} \tag{8.25.7}$$

where $\phi(x) = (2\pi)^{-1/2}e^{-x^2/2}$, and Φ is written for $\Phi(x)$. If $(1 - \Phi)^r$ is expanded binomially as

$$(1-\Phi)^r = \sum_{j=0}^{r}(-1)^j\binom{r}{j}\Phi^j \tag{8.25.8}$$

integration by parts, with $x\,e^{-x^2/2}$ as one part, yields the result

$$V(x_{r+1}) = 1 + \frac{N!}{4\pi(r!)^2}\sum_j(-1)^j\binom{r}{j}(r+j)(r+j-1)T_{r+j-2} \tag{8.25.9}$$

where

$$T_{r+j-2} = \int_{-\infty}^{\infty}\Phi^{r+j-2}(2\pi)^{-1/2}\exp(-3x^2/2)\,dx \tag{8.25.10}$$

By calculating the different integrals of this type, the variance, and also higher moments, may be obtained. An approximate expression for the variance of the median in odd samples from a standard normal population is

$$V(x_{r+1}) = \frac{\pi}{2(N+2)} + \frac{\pi^2}{4(N+2)(N+4)} \tag{8.25.11}$$

For a sample of size 11 this gives 0.133, the correct value being 0.137.

The corresponding variance of the mean is 0.091, so that the efficiency of the median in a sample of this size is 66.4%. As N increases, the efficiency tends to the value $2/\pi = 0.637$. On the average, therefore, we can get about as good a value of the population mean from the median of 100 observations as from the mean of 64.

If $\tilde{\mu}$ is the population median, for a population with density $f(x)$, so that

$$\int_{-\infty}^{\tilde{\mu}} f(x)\,dx = 1/2 \tag{8.25.12}$$

then as $N \to \infty$ the sample median is approximately normally distributed with mean $\tilde{\mu}$ and variance $[4(N + 2)f^2(\tilde{\mu})]^{-1}$.

For the rectangular distribution, for which $f(\tilde{\mu}) = 1$, this value is exact. For the normal distribution, $f(\tilde{\mu}) = (2\pi)^{-1/2}$, and the approximation is $\pi/[2(N + 2)]$.

* 8.26 **The Asymptotic Distribution of the Extreme Value** The distribution of the largest value in a sample is of interest in certain applications, as for instance in predicting the occurrence of exceptional floods in river flow. If in Eq. (8.25.1) we put $r = N$, the probability density for the largest value is found, as in § 8.19, to be

$$g(x) = N[F(x)]^{N-1}f(x)\,dx \tag{8.26.1}$$

and the distribution function is

$$G(x) = [F(x)]^N \tag{8.26.2}$$

Let x_0 be defined by the relation

$$N[1 - F(x_0)] = 1 \tag{8.26.3}$$

Since $N[1 - F(x_0)]$ is the expected number of values exceeding x_0 in a sample of size N, Eq. (3) states that in such a sample we may expect x_0 to be exceeded just once.

We will first suppose that the distribution in the parent population is *exponential*, so that

$$F(x) = 1 - e^{-\alpha x}, \qquad f(x) = \alpha e^{-\alpha x}, \qquad \text{and } e^{\alpha x_0} = N$$

Therefore,

$$[F(x)]^N = [1 - e^{-\alpha x}]^N = \left[1 - \frac{1}{N}e^{-\alpha(x-x_0)}\right]^N$$

and the limit of this as $N \to \infty$ is $\exp[-e^{-\alpha(x-x_0)}]$. We have as the limiting distribution, therefore,

$$\lim G(x) = \exp[-e^{-\alpha(x-x_0)}] \tag{8.26.4}$$

and, consequently,

$$\lim g(x) = \alpha e^{-\alpha(x-x_0)} \exp[-e^{-\alpha(x-x_0)}] \tag{8.26.5}$$

If $y = \alpha(x - x_0)$ the density function for y is

$$h(y) = e^{-y} \exp(-e^{-y}) \tag{8.26.6}$$

which is a form used by Gumbel in a study of floods [15].

For the *normal* distribution,

$$f(x) = (2\pi)^{-1/2} e^{-x^2/2}$$

and

$$1 - F(x_0) = (2\pi)^{-1/2} \int_{x_0}^{\infty} e^{-u^2/2}\, du \tag{8.26.7}$$

The integral on the right of Eq. (7) is asymptotically equivalent to $f(x_0)(1/x_0 - 1/x_0^3 + \dots)$, so that from Eq. (3)

$$\frac{1}{N} = (2\pi)^{-1/2} e^{-(1/2)x_0^2} \left(\frac{1}{x_0} - \frac{1}{x_0^3} + \dots\right)$$

or

$$-\tfrac{1}{2}x_0^2 \approx \log x_0 + \tfrac{1}{2}\log(2\pi) - \log N$$

Using only the leading terms for large N,

$$x_0^2 \approx 2 \log N \tag{8.26.8}$$

In the exponential distribution of Eq. (4), $\alpha^{-1} = [1 - F(x_0)]/f(x_0)$. The corresponding expression for the normal law is asymptotically equal to x_0^{-1}, and in fact, as proved by Fisher and Tippett [16] the limiting distribution of the extreme value is the same as that of Eq. (4) with $\alpha = x_0$, where x_0 is given by (8).

8.27 The Effect of Non-Normality Since we do not usually know whether a sample comes from a normal universe or not, it is natural to ask what difference it would make in the t-test or the F-test if the universe were not normal. Bartlett [17] and others have shown that the t-test gives quite good results even for considerable departures from normality, although the one-tailed test is more vulnerable in this respect than the two-tailed test. For a skew parent population the true significance level may be considerably under- or overestimated by using the ordinary tables. The effects of skewness and kurtosis in the parent population on the power function of the t-test have been discussed by Srivastava [18]. A positive skewness tends to reduce the power when the power is low and increase it when it is high; a negative skewness has the opposite effect. Kurtosis, unless quite marked, seems to have comparatively little effect. With increase in sample size, of course, the effect of non-normality diminishes.

Some experimental work on the F-test for samples from non-normal populations was carried out by E. S. Pearson, and this suggested that the test may be used with populations differing quite considerably from normal without serious error. W. G. Cochran [19] has discussed the effect of non-normality on the t-test and F-test, and concludes that a tabular 5% may perhaps mean anything from 4% to 7% and a tabular 1% anything between $\frac{1}{2}$% and 2%. The effect of non-normality is usually to increase the apparent significance of results, which suggests caution when interpreting results near the borderline of significance.

Unless data are very extensive, it is seldom possible to demonstrate that they are not normal. The standard errors of skewness and kurtosis are so large with samples of moderate size that only very marked non-normality could be detected. If there is reason to suspect non-normality, from the nature of the data, it is advisable to try a transformation. The logarithm of the variate, or the square root, or the inverse sine, may be more nearly normal (see Chapters 3 and 4, §§ 3.15, 3.16 and 4.8, and also reference [20]).

PROBLEMS

A. (§§ 8.1–8.4)

1. If X is normally distributed with mean 0 and variance 1, and if m is the mean of a random sample of 16 items, show that the odds are about 370 to 1 against obtaining an m numerically greater than $\frac{3}{4}$.

2. Assume that the mean age at death of men who are alive at age 20 is 59.13 years, with a standard deviation of 10.2 years. An insurance company would like to feel fairly sure (probability at least 0.99) that the mean age at death in its own group of men aged 20 will not differ from 59.13 years by more than 1 year. Assuming a normal distribution, how large should the group be?

3. The mean of a particular normal distribution is equal to the standard deviation of the mean of samples of 100 from the same distribution. Find the probability that the mean of a sample of 25 will be negative.

4. How large a sample should be taken from a normal population if the probability is to be 0.95 that the sample mean will not differ from the population mean by more than one-quarter of the population standard deviation?

5. Prove that the density function of the statistic k_2, for given σ, has a maximum at $k_2 = (n - 2)\sigma^2/n$, where $n = N - 1$. (Hence the estimator $nk_2/(n - 2)$ has the property that its most likely value is the true value σ^2.)

6. Show that the moment generating function of the distribution of k_2 is $M(h) = (1 - 2h\sigma^2/n)^{-n/2}$. Hence obtain the c.g.f., and the expectation and variance of k_2.

7. For what value of α is the expectation of $(\alpha k_2 - \sigma^2)^2$ a minimum? (The quantity αk_2 is a "least squares" estimator of σ^2, different from the unbiased estimator and the one in Problem 5).

8. If s is the positive square root of k_2, prove that

$$E(s) = \sigma\left(\frac{2}{n}\right)^{1/2} \Gamma\left(\frac{n+1}{2}\right) \Big/ \Gamma\left(\frac{n}{2}\right)$$

and

$$V(s) = \sigma^2\left[1 - \frac{2}{n}\left\{\Gamma\left(\frac{n+1}{2}\right) \Big/ \Gamma\left(\frac{n}{2}\right)\right\}^2\right]$$

Hence show that

$$E(s) = \sigma\left(1 - \frac{1}{4n} + O\frac{1}{n^2}\right)$$

and

$$V(s) = \left(\frac{\sigma^2}{2n}\right)\left(1 - \frac{1}{4n} + O\frac{1}{n^2}\right)$$

Hint: Use the Stirling formula,

$$\log \Gamma(x) = \frac{1}{2}\log 2\pi + \left(x - \frac{1}{2}\right)\log x - x + \frac{1}{12x} - \ldots.$$

to evaluate $\log\left\{\Gamma\left(\frac{n+1}{2}\right) \Big/ \Gamma\left(\frac{n}{2}\right)\right\}$. See (8.6.4).

B. (§§ 8.5–8.13)

1. Four different boxes of Eddy's matches, from the same carton, contained 55, 58, 53 and 57 matches. Obtain 95% confidence limits for the mean number of matches in boxes of the same kind.

2. The tensile strength (X), in pounds, of a certain type of cable was measured for 12 samples. The results were: 182, 178, 185, 184, 180, 179, 177, 185, 174, 179, 183, 186. Calculate 90% confidence limits for the mean of X in this type of cable. *Hint:* Use the auxiliary variate $U = X - 180$.

3. A machine producing mica insulating washers is supposed to turn them out with a mean thickness of 10 mils (1 mil = 0.001 in.). A random sample of nine washers from the output of this machine has a mean thickness 9.5 mils with a standard deviation 0.60 mil. Is the output significantly different from standard with respect to thickness?

4. In the course of archaeological investigations at a certain site, 16 lower first molars were found with mean length 13.57 mm and standard deviation 0.72 mm. From a near-by site, nine lower first molars were taken with mean 13.06 mm and standard deviation 0.62 mm. Can the two finds be reasonably regarded as samples of the same population?

5. Two samples of herring were measured for length (mm) with the following results:

(1) 192, 179, 181, 193, 215, 181, 178

(2) 173, 194, 194, 187, 168, 186, 176, 191, 191, 178, 185, 160

Find 95% confidence limits for the difference in the mean lengths for the two populations sampled.

6. Twelve hogs were fed on diet A, fifteen others on diet B. The gains in weight for the individual hogs in pounds, over the same period, were as follows:

A 25, 30, 28, 34, 24, 35, 13, 32, 24, 30, 31, 35.

B 44, 34, 22, 18, 47, 31, 40, 30, 32, 35, 18, 21, 35, 29, 22.

On the assumption that the diet may affect the mean gain without affecting the variance of gains, obtain 90% confidence limits for the average *increase* in gain with diet *B* over that with diet *A*.

7. A physiological experiment was carried out to test the effect of an injection of secretin on the percentage of reticulocytes in the blood of rabbits. Seventeen rabbits were tested, before and after injection, and the mean increase was 0.0635. The standard deviation of the increases was 0.168. Was there a significant effect at the 5% level?

8. A paired feeding experiment on pigs was conducted to determine the relative value of limestone and bone meal for bone development. The variable is the percentage ash content in the shoulder-blade.

Pair Number	Limestone	Bone Meal
1	49.2	57.5
2	53.3	54.9
3	50.6	52.2
4	52.0	53.3
5	46.8	57.6
6	50.5	54.1
7	52.1	54.2
8	53.0	54.3

Is the difference between the two sets of values significant at the 5% level?

9. (*Snedecor*) An agronomist, interested in the effect of superphosphate on the yield of corn, added the fertilizer to a mixture of manure and lime. Five pairs of adjacent plots were used for the trial, the plots in each pair being as alike as possible except that one was treated with the old fertilizer (without superphosphate) and the other with the new. The plots with the new fertilizer yielded, respectively, 20, 6, 4, 3 and 2 bushels per acre more than the corresponding controls. Was the value of the superphosphate demonstrated?

If the increased yields had been 5, 6, 4, 3 and 2 bushels per acre, would the verdict have been different? Explain the apparent paradox.

10. Complete the proof of the statement in § 8.5 that Student's t-distribution tends to normal as $n \to \infty$. *Hint:* See Eq. (8.6.4).

11. If $t = n^{1/2} \cot \phi$, show that the density function for ϕ is $C \sin^{n-1}\phi (0 \leq \phi \leq \pi)$, where $C = 1/B(1/2, n/2)$.

12. A sample of size 20 is used for testing the hypothesis that $\mu = 0$ against the alternative hypothesis that $\mu = 0.5\sigma$, where σ is the population standard deviation. If Student's t is used as the criterion, and the size of the test is 0.05, what is the power? *Hint:* Use the approximation of Eq. (8.12.5).

13. It is desired to test the hypothesis that $\mu = 0$ against the alternative hypothesis that $\mu = \mu_1 (\mu_1 > 0)$. If the standard deviation of the population is 10 units and a sample of size 17 is used, find the least value of μ_1 that could be detected by the t-test, assuming that the risks for both kinds of error are not more than 0.05.

14. Two samples, each of size 10, come from populations with means μ_1 and μ_2 and a common variance σ^2. The null hypothesis H_0 is that $\mu_2 - \mu_1 \leq 0$ and the alternative H_1 is that $\mu_2 - \mu_1 = k\sigma (k > 0)$. Show that if α is not greater than 0.05, a value of k at least 1.37 could be detected by the t-test, with power at least 0.9. *Hint:* If the two samples have means m_1 and m_2 and variances s_1^2 and s_2^2, the quantity $(m_2 - m_1)/\{(s_1^2 + s_2^2)/10\}^{1/2}$ has the t distribution with 18 d.f. if $\mu_1 = \mu_2$. Under H_1 it has a non-central t-distribution with parameter $k\sqrt{5}$, since the variance of $m_1 - m_2 = \sigma^2/5$.

15. A large lot of manufactured articles is rated on the percentage p of defective items, an item being reckoned defective if the value of a normal variate X is at least 3.

A prospective purchaser will want to reject a lot with probability 0.95 if $p > 2.5$. A sample of 10 items is used for a non-central t-test. What criterion should be used for accepting the lot? *Hint:* Find δ from Eq. (8.13.3); k can be obtained from the tables of non-central t, using Eq. (8.13.5) with $P = 0.05$, or approximately from Eq. (8.12.5) with $z_P = 1.645$ and $t_\alpha = k\sqrt{10}$.

C (§§ 8.14–8.17)

1. The variance of a random sample of size 5 is 29.83. Calculate 90% confidence limits for the population variance.

2. Two chemists, A and B, each repeat a protein analysis 20 times. If the sets of values obtained are denoted by X_i, Y_i, respectively, it is found that $\Sigma X_i = 196.40$, $\Sigma X_i^2 = 1928.6560$, $\Sigma Y_i = 205.16$, $\Sigma Y_i^2 = 2104.7152$. Determine whether there is a significant difference in precision between the two sets of analyses, precision being inversely proportional to the variance.

3. In two series of hauls to determine the number of plankton organisms inhabiting the waters of a lake, the following results were found:

Series I: 80, 96, 102, 77, 97, 110, 99, 88, 103, 108

Series II: 74, 122, 92, 81, 104, 92, 90.

In Series I the hauls were made in succession at the same place; in Series II they were made at different points scattered over the lake. Does there appear to be a greater variability between different places than exists at different times at the same place?

4. For the data on feeding of hogs in Problem B-6, determine whether the assumption of a common variance under both diets is justified.

5. If $x = n_2/(n_2 + n_1F)$ prove that x is a beta-variate with parameters $n_2/2$, $n_1/2$.

6. When $n_1 = 2$, show that the upper significance level of F corresponding to probability p is $n_2(p^{-2/n_2} - 1)/2$. *Hint:* The integral of $g(F)$ from F_1 to ∞ is p, where F_1 is the required level.

7. Find the upper 5% point for F with 2 and 4 degrees of freedom by direct integration of $g(F)$. Compare with the value in Table B.5 in the appendix.

8. What is the smallest ratio λ of two variances ($\lambda > 1$) that can be detected by an F-test with two samples of size 10, the size of the test being 0.05 and the power 0.95? *Hint:* When $n_1 = n_2$, the 95% point for F is the reciprocal of the 5% point.

9. An approximation to Fisher's z for given P, n_1 and n_2 has been devised by A. H. Carter, namely,

$$z_F = z_P \frac{(h+k)^{1/2}}{h} - \left(\frac{1}{n_1} - \frac{1}{n_2}\right)\left(k + \frac{5}{6} - \frac{s}{3}\right)$$

where $s = 1/n_1 + 1/n_2$, $h = 2/s$, $k = (z_P^2 - 3)/6$ and z_P is the *normal* standard variate exceeded with probability P. Use this approximation for $n_1 = n_2 = 19$ to find z_F, and hence F, when $P = 0.25$. Then determine the smallest variance ratio detectable with two samples of size 20 and a test of size 0.05 and power 0.25. *Hint:* $z_F = \frac{1}{2}\log_e F$.

D (§§ 8.18–8.26)

1. A sample of 10 observations is taken from a normal population with mean 250 and standard deviation 10. What value for the largest member of the sample would be exceeded only once in 20 samples; what value only once in 100 samples?

2. A quantity is measured 10 times with the following results: 236, 251, 249, 252, 248, 254, 246, 257, 243, 274. Should the largest of these observations be rejected according to Thompson's criterion?

3. In the following measurements of an angle (degrees and minutes omitted, values in seconds of arc) would it be reasonable to reject the lowest reading? 51.75, 47.85, 47.40, 48.90, 44.45, 48.45, 51.05, 48.85, 50.95, 50.60, 47.75, 49.20, 50.55.

4. Apply Dixon's criterion to the data of Problems 2 and 3.

5. The following frequency distribution was found for the range R in 200 samples of size 10 from an artificial, approximately normal, population with mean 20 and standard deviation 4:

R	5	6	7	8	9	10	11	12	13	14	15	16	17	18	19	20	21	22	23
f	2	4	4	14	11	20	25	28	25	17	13	13	5	9	3	3	3	0	1

Calculate the mean range and estimate the standard deviation of the population, assuming that the true value 4 is unknown. *Hint:* The estimator of σ is $k\bar{R}$, where k is found from Table 8.5

6. Let u denote the mid-range of a sample, that is, $u = (x_1 + x_N)/2$. For the rectangular population with density $f(x) = \frac{1}{2}$, $-1 < x < 1$, the density for the mid-range is given by $g(u) = \frac{1}{2}N(1 - |u|)^{N-1}$. Prove that the variance of the mid-range is $2/[(N + 1)(N + 2)]$. Hence show that the mid-range is, for all $N > 2$, more efficient than the arithmetic mean as an estimator of the population mean. *Hint:* Separate the interval of integration into two parts, -1 to 0, and 0 to 1.

7. A sample of size N is taken from the exponential population with density e^{-x}, $x \geq 0$. Find the density function of the range and show that its expected value is $1 + \frac{1}{2} + \frac{1}{3} + \ldots + 1/(N - 1)$. *Hint:* In the integral for $E(R)$ put $u = 1 - e^{-R}$, and expand $\log(1 - u)$ in a series.

8. Samples of size 4 are taken from the population of Problem 7. Find 95% confidence limits for the range in such samples.

9. For a sample of size N from the rectangular distribution $f(x) = 1/b$, $0 < x < b$, show that R/b is a beta-variate with parameters $N - 1$ and 2. Hence obtain the mean and variance of the distribution of R.

10. Numbers are drawn at random from the interval (0,1). How many are required before the probability will exceed 0.95 that the range of the sample will be at least 0.5? *Hint:* Show that N is given by the inequality $2^{N-2} > 5(N + 1)$. Solve by trial for small values of N.

11. A sample of odd size $N(= 2r + 1)$ is taken from the rectangular population with density 1, $0 < x < 1$. The median is the $(r + 1)^{\text{th}}$ member of the sample when arranged in ascending order. Prove that the expectation of the median is $\frac{1}{2}$ and its variance is $1/(4N + 2)$.

12. Prove that the density function for the range w in samples of size 3 from a standard normal population is $g(w) = (3/\pi^{1/2})e^{-w^2/4}[\Phi(w/\sqrt{6}) - \frac{1}{2}]$. *Hint:* Use Eq. (8.21.4) with $F(x) = \Phi(x)$, and w for R. Put $x_1 = (v - w)/2$ and obtain

$$g(w) = 3(2\pi)^{-3/2}\, e^{-w^2/4} \int_{-\infty}^{\infty} e^{-v^2/4} \int_{(v-w)/2}^{(v+w)/2} e^{-u^2/2}\, du\, dv$$

Change to oblique coordinates $x = u - v/2$, $z = v\sqrt{5}/2$ and integrate over the strip of horizontal width w, x going from $-w/2$ to $w/2$ and z from $-\infty$ to ∞.

REFERENCES

[1] Kenney, J. F., and Keeping, E. S., *Mathematics of Statistics, Part II,* Van Nostrand, 1951.

[2] Student, "On the Probable Error of a Mean," *Biometrika,* **6,** 1908, pp. 1–25. A biographical sketch of W. S. Gosset will be found in *J. Amer. Stat. Assoc.,* **33,** 1938, pp. 226–228.

[3] Hendricks, W. A., "An Approximation to Student's Distribution," *Ann. Math. Stat.*, **7**, 1936, pp. 210–221.

[4] Hald, A., *Statistical Theory with Engineering Applications*, Wiley, 1952.

[5] Resnikoff, G. J., and Lieberman, G. J., *Tables of the Non-Central t-Distribution*, Stanford Univ. Press, 1957.

[6] Neyman, J., and Tokarska, B., "Errors of the Second Kind in Testing Student's Hypothesis," *J. Amer. Stat. Ass.*, **31**, 1936, pp. 318–326.

[7] Fisher, R. A., and Yates, F., *Statistical Tables for Biological, Agricultural and Medical Research*, Oliver and Boyd, 3rd ed., 1948. See also tables by Merrington, M., and Thompson, C. M., in *Biometrika*, **33**, 1943, pp. 73–88.

[8] Pearson, E. S., and Hartley, H. O., *Biometrika Tables for Statisticians*, Table 24, Cambridge Univ. Press, 1954.

[9] Thompson, W. R., On a Criterion for the Rejection of Observations and the Distribution of the Ratio of Deviation to Sample Standard Deviation, *Ann. Math. Stat.*, **6**, 1935, pp. 214–219.

[10] Dixon, W. J., "Analysis of Extreme Values," *Ann. Math. Stat.*, **21**, 1950, pp. 488–506.

[11] Dixon, W. J., "Ratios involving Extreme Values," *Ann. Math. Stat.*, **22**, 1951, pp. 68–78. See also Crow, E. L., Davis, F. A., and Maxfield, M. W., *Statistics Manual*, Dover, 1961.

[12] Kendall, M. G., and Stuart, A., *The Advanced Theory of Statistics, Vol. I.* Chas. Griffin, 1958.

[13] Harter, H. L., and Clemm, D. S., *Probability Integrals of the Range and of the Studentized Range, Vols. I and II*, W.A.D.C. Technical Report 58–484, Wright Air Development Center, 1959.

[14] Rider, P. R., "The Distribution of the Quotient of Ranges in Samples from a Rectangular Population," *J. Amer. Stat. Assn.*, **46**, 1951, pp. 502–507.

[15] Gumbel, E. J., *Statistics of Extremes*, Columbia Univ. Press, 1958.

[16] Fisher, R. A., and Tippett, L. H. C., "Limiting forms of the Frequency-Distribution of the Largest or Smallest Member of a Sample," *Proc. Camb. Phil. Soc.*, **24**, 1928, pp. 180–190.

[17] Bartlett, M. S., "The Effect of Non-Normality on the *t*-Distribution," *Proc. Camb. Phil. Soc.*, **31**, 1935, pp. 223–231.

[18] Srivastava, A. B. L., "Effect of Non-Normality on the Power Function of the *t*-Test," *Biometrika*, **45**, 1958, pp. 421–430.

[19] Cochran, W. G., "Some Consequences when the Assumptions for the Analysis of Variance Are Not Satisfied, *Biometrics*, **3**, 1947, pp. 22–38.

[20] Bartlett, M. S., The Use of Transformations, *Biometrics*, **3**, 1947, pp. 39–52.

Chapter 9

ANALYSIS OF VARIANCE

9.1 **Tests of Homogeneity of Variance** The analysis of variance is a widely used technique for separating the observed variance in a group of samples into portions which are traceable to different sources. Thus the different samples may have undergone different treatments which possibly affect the general level of the measured variate X. If all the samples are lumped together into one grand sample, the observed variance will be partly due to differences between the individual members of the same original sample and partly due to the effects of the different treatments. The method of analysis of variance enables us to estimate how much of the variance is attributable to the one cause and how much to the other, and so to decide whether or not the treatments have produced any significant effects.

Much more elaborate experimental designs than this can be analysed by comparable methods, and several of the more usual designs will be considered in this chapter. All the common analysis-of-variance tests rest on certain assumptions, such as normality of the distribution of X and additivity of treatment effects, and among these assumptions is one on the constancy of variance as between samples. It is supposed that, apart from possibly affecting the average value of X, the different treatments (for instance) do not change the sample variances. Methods which do not depend on these assumptions will be mentioned later on, but meanwhile a test for the homogeneity of variance as between a group of samples will be considered.

For *two* samples, the technique of § 8.15 may be used. We suppose therefore that we have k samples ($k > 2$), and that for the ith sample, of size N_i, the observed values are x_{ij}, $j = 1, 2 \ldots N_i$, $i = 1, 2 \ldots k$. We assume that all the samples are independent and come from normal populations with means μ_1, $\mu_2 \ldots \mu_k$ and variances σ_1^2, $\sigma_2^2 \ldots \sigma_k^2$. The null hypothesis H_0 is that $\sigma_1^2 = \sigma_2^2 = \ldots = \sigma_k^2$ ($= \sigma^2$, say). The alternative hypothesis H_1 is that these variances are not all equal.

Under H_0, the likelihood function is

$$(9.1.1) \qquad L_0 = \frac{1}{(2\pi\sigma^2)^{N/2}} \exp\left[-\frac{1}{2}\sum_{i,j}\left(\frac{x_{ij}-\mu_i}{\sigma}\right)^2\right]$$

where $N = \sum N_i$. Under H_1, the likelihood function is

$$(9.1.2) \qquad L_1 = \frac{1}{(2\pi)^{N/2}\sigma_1^{N_1}\ldots\sigma_k^{N_k}} \exp\left[-\tfrac{1}{2}\sum_{i,j}\left(\frac{x_{ij}-\mu_i}{\sigma_i}\right)^2\right]$$

The joint maximum likelihood estimators of μ_i and σ under H_0 are

(9.1.3)
$$\begin{cases} \hat{\mu}_i = \dfrac{1}{N_i}\sum_j x_{ij} = \bar{x}_i \\ \hat{\sigma}^2 = \dfrac{1}{N}\sum_{i,j}(x_{ij}-\bar{x}_i)^2 = \sum_i \dfrac{S_i}{N} \end{cases}$$

where

(9.1.4)
$$S_i = \sum_j (x_{ij}-\bar{x}_i)^2$$

which is the sum of squares of deviations from the mean for the ith sample. The sample variance is $v_i = S_i/n_i$, where $n_i = N_i - 1$.

The joint maximum likelihood estimators of μ_i and σ_i under H_1 are

(9.1.5)
$$\begin{cases} \hat{\mu}_i = \bar{x}_i \\ \hat{\sigma}_i^{\,2} = S_i/N_i \end{cases}$$

The likelihood ratio is, therefore,

(9.1.6)
$$\frac{(L_0)_{\max}}{(L_1)_{\max}} = \frac{(S_1/N_1)^{N_1/2}\ldots(S_k/N_k)^{N_k/2}}{(S/N)^{N/2}} = L$$

where $S = \sum S_i$. Then H_0 will be rejected if $L < c$. The constant c is so chosen that

(9.1.7)
$$P(L < c|H_0) \le \alpha$$

As mentioned in § 6.9, the distribution of $-2\log L$ for large N is approximately χ^2 with $k - 1$ degrees of freedom. This number is the number of parameters under H_1, namely, $2k$, less the number under H_0, namely, $k + 1$. In this case,

(9.1.8)
$$-2\log L = N\log\frac{S}{N} - \sum N_i \log\frac{S_i}{N_i}$$

It was shown by Bartlett [1], that the approximation to χ^2 may be improved by replacing each N_i by n_i ($= N_i - 1$) and therefore N by n ($= N - k$). In effect this replaces the maximum likelihood estimators by unbiased estimators. Furthermore, the approximation will hold reasonably well down to values of n_i as small as 4 or 5 if a correcting factor is introduced in Eq. (8). We can therefore in most cases assume that the quantity

(9.1.9)
$$M = C^{-1}\left[n\log\frac{S}{n} - \sum n_i\log\frac{S_i}{n_i}\right]$$

is distributed like χ^2 with $k - 1$ d.f., where

9.1.10)
$$C = 1 + \frac{\sum\dfrac{1}{n_i} - \dfrac{1}{n}}{3(k-1)}$$

A small value of L will mean a large value of M, since $M \approx -2 \log L$, and this will lead to rejection of the null hypothesis. It should be noted that the logarithms in Eq. (9) are to base e.

For even smaller values of n_i (down to 2, say) an improved approximation was given by Hartley [2], and tables based on this approximation have been compiled by Catherine Thompson and Maxine Merrington [3].

EXAMPLE 1 To test the effect of a small proportion of coal in the sand used for making concrete, several batches were mixed under practically identical conditions except for the variation in the percentage of coal. From each batch, four cylinders were made and tested for breaking strength in lb/in^2. One cylinder in the third sample was defective, so there were only three items in this sample. The results are given in Table 9.1.

TABLE 9.1

Sample No.	1	2	3	4	5
Percentage coal	0	0.05	0.1	0.5	1.0
Breaking strengths	1690 1580 1745 1685	1550 1445 1645 1545	1625 1450 1510	1725 1550 1430 1445	1530 1545 1565 1520
Mean	1675	1546	1528	1538	1540
S_i/n_i	4750	6673	7908	18,475	383
$n_i \log_{10} S_i/n_i$	11.03	11.47	7.80	12.80	7.75

From this table, we obtain the values $n = 14$, $S/n = 7619$, $n \log(S/n) - \sum n_i \log(S_i/n_i) = 2.303\,[54.35 - 50.85] = 8.06$. Also $\sum (1/n_i) = 4/3 - 1/2 = 11/6$, so that $C = 1.15$ and $M = 8.06/1.15 = 7.02$. With 4 d.f., this value of χ^2 corresponds to a P of 0.13, so that even the rather large differences in the estimates of variance, given by S_i/n_i in the above table, are not really significant in view of the small sample sizes.

Thompson and Merrington's tables should preferably be used for values of n_i as small as those appearing in the above table. These give the 5% point for the distribution of $-2 \log L$ as about 10.7 and the 1% point as about 14.9. The observed value, 8.06, is therefore not significant at the 5% level.

9.2 A Test for Difference of Means in k Samples The simplest application of analysis of variance occurs in the problem of deciding whether a group of samples come from populations which differ from one another in respect of their *mean* values of some measured variate X. It is assumed that they do not differ as regards the *variance* of X, and this homogeneity of variance may be tested by the methods given in § 9.1. An example is the measurement of breaking strength on the five samples of concrete cylinders in Table 9.1, in which the test of homogeneity has been shown to be satisfied.

If x_{ij} is the measured value of X on the jth member of the ith sample, $j = 1, 2 \dots N_i$, $i = 1, 2 \dots k$, the mathematical model we assume is

$$x_{ij} = \mu + \alpha_i + \varepsilon_{ij} \tag{9.2.1}$$

where for every i and j, ε_{ij} is normally distributed with expectation 0 and variance σ^2. This means that the measured x_{ij} for any individual item is made up of three parts which are added together, an over-all average value denoted by μ, an effect due to the particular treatment undergone by the ith sample, denoted by α_i (and supposed to be the same for all members of this sample), and a random or error term ε_{ij} due to many unspecified causes. These ε_{ij} are supposed to be uncorrelated with each other.

Adding the x_{ij} for all items in all k samples, we get

$$\sum_{i,j} x_{ij} = N\mu + \sum_i N_i\alpha_i + \sum_{i,j} \varepsilon_{ij} \tag{9.2.2}$$

where $N = \sum N_i$. We can suppose that μ and the α_i are so adjusted that $\sum N_i \alpha_i = 0$. If this does not happen to be the case at first, and if $\sum N_i \alpha_i = h$, we simply have to subtract $h/(k\, N_i)$ from each α_i and add h/N to μ. Then if $\bar{x}$ is the over-all mean of the x_{ij} ($= N^{-1} \sum_{ij} x_{ij}$), we see that $\bar{x}$ is an unbiased estimator of μ. The total sum of squares for all the x_{ij} may be defined by

$$\begin{aligned} S_t &= \sum_{i,j} (x_{ij} - \bar{x})^2 \\ &= \sum_{i,j} x_{ij}{}^2 - G \end{aligned} \tag{9.2.3}$$

where $G = N\bar{x}^2 = (\sum_{i,j} x_{ij})^2/N$. Now

$$x_{ij} - \bar{x} = x_{ij} - \bar{x}_i + \bar{x}_i - \bar{x}$$

where $\bar{x}_i$ is the mean of X for the ith sample, and therefore

$$(x_{ij} - \bar{x})^2 = (x_{ij} - \bar{x}_i)^2 + (\bar{x}_i - \bar{x})^2 + 2(\bar{x}_i - \bar{x})(x_{ij} - \bar{x}_i).$$

The first expression for S_t above can then be written

$$\begin{aligned} S_t &= \sum_{i,j} (x_{ij} - \bar{x}_i)^2 + \sum_i N_i(\bar{x}_i - \bar{x})^2 \\ &\quad + 2\sum_i \left[(\bar{x}_i - \bar{x}) \sum_j (x_{ij} - \bar{x}_i)\right] \end{aligned} \tag{9.2.4}$$

The last term in this equation vanishes since $\sum_j (x_{ij} - \bar{x}_i) = 0$. Also $\sum_j (x_{ij} - \bar{x}_i)^2$ is the sum of squares for the x_{ij} belonging to the ith sample, which we may denote by S_i, so that the first term on the right-hand side of Eq. (4) is $\sum_i S_i$. This is generally called the sum of squares *within samples*, denoted by S_w. The remaining term depends on the means of the various samples and their relation to the over-all mean. It is called the sum of squares *between samples*, denoted by S_b. Then

$$S_t = S_w + S_b \tag{9.2.5}$$

where

$$\begin{aligned} S_w &= \sum_{i,j} (x_{ij} - \bar{x}_i)^2 \\ &= \sum_{i,j} x_{ij}^2 - \sum_i N_i \bar{x}_i^2 \\ &= \sum_{i,j} x_{ij}^2 - \sum_i \frac{\left(\sum_j x_{ij}\right)^2}{N_i} \end{aligned} \tag{9.2.6}$$

and

$$\begin{aligned} S_b &= \sum_i N_i(\bar{x}_i - \bar{x})^2 \\ &= \sum_i N_i \bar{x}_i^2 - N\bar{x}^2 \\ &= \sum_i \frac{\left(\sum_j x_{ij}\right)^2}{N_i - G} \end{aligned} \tag{9.2.7}$$

It should be noted that the splitting up of the total sum of squares into the two parts S_w and S_b is a matter of algebra and does not depend on any assumptions about the normality of the distribution or the constancy of variance between samples. However, on the null hypothesis that all the x_{ij} come from a single normal population with variance σ^2 (this is equivalent to assuming that all the α_i in Eq. (1) are separately zero), it follows that S_t/σ^2 is a χ^2-variate with $N - 1$ degrees of freedom. Similarly, within the ith sample, S_i/σ^2 is a χ^2-variate with $N_i - 1$ d.f., so that, by the addition theorem for independent χ^2-variates, S_w/σ^2 is distributed like χ^2 with $\sum_i (N_i - 1) = N - k$ d.f. Theorem 4.3 then tells us that S_b/σ^2 is a χ^2-variate with $N - 1 - (N - k) = k - 1$ d.f., and is independent of S_w.

Since the expectation of a χ^2-variate is equal to the number of degrees of freedom, it follows that

$$\begin{cases} E(S_t/\sigma^2) = N - 1 \\ E(S_w/\sigma^2) = N - k \\ E(S_b/\sigma^2) = k - 1 \end{cases} \tag{9.2.8}$$

so that

$$\begin{cases} E[S_t/(N - 1)] = \sigma^2 \\ E[S_w/(N - k)] = \sigma^2 \\ E[S_b/(k - 1)] = \sigma^2 \end{cases} \tag{9.2.9}$$

Furthermore, the ratio of the two independent unbiased estimators of σ^2, namely $S_b/(k-1)$ and $S_w/(N-k)$ has the F distribution with $k-1$ and $N-k$ degrees of freedom. These estimators are usually called "mean squares." The result is set out in an Analysis of Variance table, such as Table 9.2.

TABLE 9.2

Variation	Sum of Squares (S.S.)	Degrees of Freedom (D.F.)	Mean Square (M.S.)
Between samples	S_b	$k-1$	$S_b/(k-1)$
Within samples	S_w	$N-k$	$S_w/(N-k)$
Total	S_t	$N-1$	$S_t/(N-1)$

If the null hypothesis is not true, the α_i will not all be zero, and the mean square between samples will tend to be greater than the mean square within samples. The F-test will therefore be a one-tailed test, and the probabilities given in the table (Appendix B.5) are correct as stated there.

As an illustration we may consider the data of Example 1, § 9.1, in which $k=5$ and all the N_i are 4 (except N_3, which is 3). We find $\sum_{i,j} x_{ij}{}^2 = 46,842,150$, $\sum_{i,j} x_{ij} = 29{,}780$, whence $G = 46, 676, 232$, and $S_t = 165, 918$. Also the five values of $(\sum_j x_{ij})^2$ are $(6700)^2$, $(6185)^2$, $(4585)^2$, $(6150)^2$ and $(6160)^2$, whence $S_w = 106{,}661$. The value of S_b, by difference, is 59,257. The results of the analysis, set out in the form of Table 9.2, are as follows:

TABLE 9.3

Variance	S.S.	D.F.	M.S.
Between samples	59,257	4	14,814
Within samples	106,661	14	7,619
Total	165,918	18	9,218

The value of F is $(14814)/(7619) = 1.94$, with 4 and 14 d.f. Since the 5% point is 3.11 and the 1% point 5.03, it is clear that the observed value is not significant. We can therefore, as far as this test is concerned, accept the null hypothesis that the strength of the concrete was not affected by the different amounts of coal in the sand used in making it.

When the k samples are all of the same size, say r, $N = rk$ and $\sum \alpha_i = 0$. Eq. (7) becomes

$$S_b = \frac{1}{r}\sum_i \left(\sum_j x_{ij}\right)^2 - G \tag{9.2.10}$$

9.3 **Two-Way Classification (Complete Blocks)** In a somewhat more complicated experimental design, the attempt is made to estimate two effects simultaneously. In the example above of the concrete cylinders, we might have allowed the concrete to set for different periods of time before testing its strength, In a field experiment on the yield of a certain crop, under different treatments. we might wish to estimate the effect of different locations of the experimental plots. (There might for instance be appreciable effects due to differences of soil moisture, drainage, slope, shade, etc., as well as natural differences of soil fertility). The experimental procedure is then to set out the plots in distinct blocks, the same number of plots in each block, arranged so that as far as possible the plots in any one block are relatively similar. The experimental treatments are applied randomly to the plots within each block. Figure 44

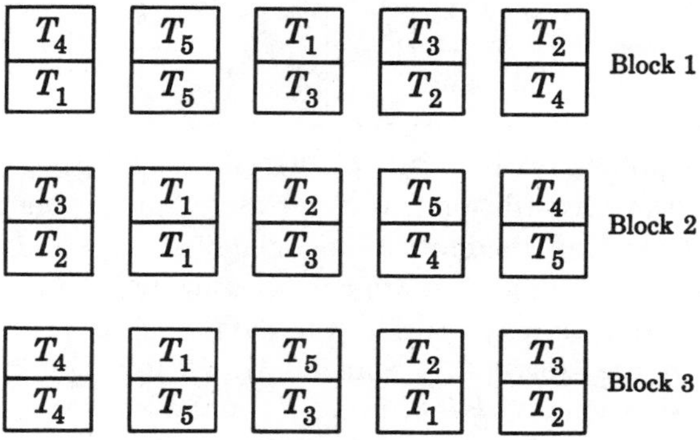

FIG. 44 COMPLETE RANDOMIZED BLOCKS

suggests a possible arrangement in which five treatments are used in each block, and each treatment is replicated on two plots. This is an illustration of a "complete block design." The purpose of randomization is to reduce as far as possible any systematic effects of the uncontrolled factors in the experiment, and to give increased justification for applying statistical theory.

In general we will suppose that we have a treatments and b blocks, and that each treatment in each block is replicated r times. The total number of individual items (plots) in the experiment is $N = abr$. A variate X is measured on each item, and we will denote by x_{ijk} the value of X for the ith treatment in the jth block, on the kth replicate. We suppose that the ith treatment has an effect on X measured by α_i, and that the blocks also have their effect, the jth block contributing β_j to X. There may also be a differential effect of treatments in different blocks, known as "interaction." (The ith treatment may not contribute the same amount to X in each block.) Assuming that these various effects can be added together, we have as our mathematical model:

$$x_{ijk} = \mu + \alpha_i + \beta_j + \gamma_{ij} + \varepsilon_{ijk} \tag{9.3.1}$$

where μ is the over-all average effect, γ_{ij} is the interaction (the extra contribution of the ith treatment in the jth block over and above the general effect of this treatment), and ε_{ijk} is the random effect shown by the kth replicate in the jth

block under the ith treatment. This random part of x_{ijk} is due to all the miscellaneous causes which may produce an effect but which are not specifically allowed for in the design of the experiment. It is usually considered as experimental error.

We can always adjust the origins from which α_i, β_j and γ_{ij} are measured, so as to satisfy the conditions:

(9.3.2) $$\sum_i \alpha_i = 0, \quad \sum_j \beta_j = 0, \quad \sum_i \gamma_{ij} = 0, \quad j = 1, 2 \ldots b$$

$$\sum_j \gamma_{ij} = 0, \quad i = 1, 2 \ldots a$$

We may suppose also that $E(\varepsilon_{ijk}) = 0$ and $V(\varepsilon_{ijk}) = \sigma^2$.

The total sum of squares may be split into four constituent parts, namely,

(9.3.3) $$S_t = S_a + S_b + S_{ab} + S_r$$

where

(9.3.4) $$S_t = \sum_{ijk} x_{ijk}{}^2 - G$$

(9.3.5) $$G = N\bar{x}^2 = N^{-1}\Big(\sum_{ijk} x_{ijk}\Big)^2$$

(9.3.6) $$S_a = (br)^{-1} \sum_i \Big(\sum_{jk} x_{ijk}\Big)^2 - G$$

(9.3.7) $$S_b = (ar)^{-1} \sum_j \Big(\sum_{ik} x_{ijk}\Big)^2 - G$$

(9.3.8) $$S_{ab} = r^{-1} \sum_{ij} \Big(\sum_k x_{ijk}\Big)^2 - S_a - S_b - G$$

(9.3.9) $$S_r = \sum_{ijk} x_{ijk}{}^2 - r^{-1} \sum_{ij} \Big(\sum_k x_{ijk}\Big)^2$$

The four terms on the right-hand side of Eq. (3) are, respectively, the sum of squares (S.S.) between treatments, the S.S. between blocks, the S.S. due to interaction and the S.S. between replicates. The degrees of freedom are $a - 1$, $b - 1$, $(a - 1)(b - 1)$ and $ab(r - 1)$.

If $\bar{x}_{i\cdot\cdot}$ is the mean of x_{ijk} taken over all blocks and replicates for the ith treatment, and if $\bar{x}$ is the mean of all the observed x_{ijk}, then

(9.3.10) $$S_a = \sum_{ijk} (\bar{x}_{i\cdot\cdot} - \bar{x})^2$$

and this is algebraically equivalent to Eq. (6). Similarly,

(9.3.11) $$S_b = \sum_{ijk} (\bar{x}_{\cdot j \cdot} - \bar{x})^2$$

where $\bar{x}_{\cdot j \cdot}$ is the mean for the jth block, over all treatments and replicates;

(9.3.12) $$S_r = \sum_{ijk} (x_{ijk} - \bar{x}_{ij\cdot})^2$$

where $\bar{x}_{ij}.$ is the mean over the k replicates for the ith treatment in the jth block; and

$$S_{ab} = \sum_{ijk} (\bar{x}_{ij}. - \bar{x}_{i}.. - \bar{x}._{j}. + \bar{x})^2 \tag{9.3.13}$$

The mean squares may be calculated as before. On the null hypothesis that all the α_i, β_j and γ_{ij} are zero, and on the assumption that the ε_{ijk} are normally distributed, these mean squares are all unbiased estimators of σ^2. Moreover, the mean squares for treatments, blocks and interactions are independent of the mean square for replicates, so that the ratios

$$\frac{S_a}{S_r}\cdot\frac{ab(r-1)}{a-1}, \quad \frac{S_b}{S_r}\cdot\frac{ab(r-1)}{b-1} \quad \text{and} \quad \frac{S_{ab}}{S_r}\cdot\frac{ab(r-1)}{(a-1)(b-1)}$$

all have the F distribution with the appropriate degrees of freedom. These ratios can therefore be used to test whether there are significant treatment effects, block effects, or interaction effects.

Unless r is greater than 1 there is no possibility, with this design, of testing for interaction by the ordinary F-test. If $r = 1$, the interaction effect is generally ignored or treated as part of the error. If it is assumed, however, that r_{ij} is of the form $C\,\alpha_i\,\beta_j$, where C is constant, an F-test of the hypothesis $r_{ij} = 0$ is possible. See [7], p. 130.

EXAMPLE 2 Tests were carried out on sheets of building material for permeability [4]. Specimens were selected from the output of each of three machines on each of nine days, and for each machine on each day three sheets were examined. The raw materials all came from a common store, but it was thought that the machines might vary in their quality of output and might also vary from day to day. The machines may be regarded as "treatments" and the days as "blocks," and there were three replicates. The randomizing within blocks was done by varying the order of sampling from the machines on the different days.

TABLE 9.4.

Variation	S.S.	D.F.	M.S.
Between machines	0.9168	2	0.4584
Between days	0.5534	8	0.0692
Interaction	0.8657	16	0.0541
Between replicates	2.0150	54	0.0373
Total	4.3509	80	

In this experiment the measured variate was the permeability (an average of eight measurements on each sheet). Since it appeared that the logarithm of the variate was more nearly normally distributed than the variate itself, the values in the above table all relate to the *common log* of the permeability.

This is denoted by x_{ijk}. From the experimental data, we find

$$N = 3 \cdot 9 \cdot 3 = 81$$

$$\sum_{ijk} x_{ijk} = 127.093, \qquad G = 199.4152$$

$$\sum_{ijk} x_{ijk}^2 = 203.7661, \qquad S_t = 4.3509$$

$$\sum_i \left(\sum_{jk} x_{ijk}\right)^2 = 5408.9627, \qquad S_a = 0.9168$$

$$\sum_j \left(\sum_{ik} x_{ijk}\right)^2 = 1799.7177, \qquad S_b = 0.5534$$

$$\sum_{ij} \left(\sum_{k} x_{ijk}\right)^2 = 605.2533, \qquad S_r = 2.0150$$

and therefore $S_{ab} = 0.8657$. The analysis of variance is given in Table 9.4. The F-ratio for interaction is 1.45, with 16 and 54 d.f. Since the 5% point is 1.83, the hypothesis of zero interaction is not rejected.

The F-ratio for days is 1.85, with 8 and 54 d.f. This is also non-significant at the 5% level.

The F-ratio for machines is 12.3, with 2 and 54 d.f. The 1% point is 5.02, so that there is a highly significant effect of the differences between machines.

If there are no interactions, the conclusions about the main effects are much simplified. If there seems to be a real difference between two treatments, for example, we can conclude that this difference persists in all blocks. But if there is appreciable interaction, a significant difference between the two treatments merely means that, *on the average over all blocks*, there is a difference. In some particular block this difference might not exist or might even be reversed in sign.

It may happen that the hypothesis of no interactions will be rejected by the ordinary statistical test while at the same time the hypothesis of zero main effects will be accepted. This means that there certainly are non-zero differences between blocks or treatments, but that when the block differences are averaged over the treatments, or the treatment differences over the blocks, the averages are not significantly different from zero.

9.4 **Estimation of Fixed Treatment Effects (Model I)** There are two ways of looking at the treatment effects. They may be regarded as fixed effects or as random variables, and in different situations either the one way or the other may be more appropriate. In Example 1 of § 9.1, the "treatments" were fixed percentages of coal in the sand used for making concrete, and any conclusions drawn from the experiment would presumably refer to these percentages and these only. However, it is conceivable that the specimens of sand used might have contained variable amounts of coal, drawn at random from some parent distribution, and in this case we could estimate the variance of the effect of added coal and apply the results of the experiment to percentages of coal outside the values actually observed.

In Example 2, above, the machine effects should probably be regarded as fixed, and the results applied to the particular machines used. The days, however, might well be considered a random sample of days, unless there was a special reason for selecting particular days for the experiment.

The mathematical model with fixed effects, which is the one we have been using, is sometimes called Model I. For the one-way classification of § 9.1, we have

$$x_{ij} = \mu + \alpha_i + \varepsilon_{ij}, \quad i = 1, 2 \ldots k, \quad j = 1, 2 \ldots N_i \tag{9.4.1}$$

and the mean of the ith sample is

$$\bar{x}_i. = \mu + \alpha_i + \bar{\varepsilon}_i. \tag{9.4.2}$$

The over-all weighted mean of the $\bar{x}_i.$, with weights N_i, is

$$\bar{x} = \mu + \bar{\varepsilon} \tag{9.4.3}$$

since $\sum N_i\alpha_i = 0$.

The expectation of $\bar{x}_i.$ is, therefore, $\mu + \alpha_i$, and its variance is σ^2/N_i. The sums of squares between samples is given by

$$\begin{aligned} S_b &= \sum_i N_i(\bar{x}_i. - \bar{x})^2 \\ &= \sum_i N_i[\bar{x}_i. - \mu - \alpha_i - (\bar{x} - \mu) + \alpha_i]^2 \\ &= \sum_i N_i[\bar{x}_i. - \mu - \alpha_i - (\bar{x} - \mu)]^2 + \sum_i N_i\alpha_i^2 \\ &\quad + 2\sum_i N_i\alpha_i[\bar{x}_i. - \mu - \alpha_i - (\bar{x} - \mu)] \end{aligned} \tag{9.4.4}$$

Now $\bar{x}_i. - \mu - \alpha_i$ is normal with expectation 0 and variance σ^2/N_i, and its weighted mean is $\bar{x} - \mu$. It therefore follows that $\sum_i(N_i/\sigma^2)[\bar{x}_i. - \mu - \alpha_i - (\bar{x} - \mu)]^2$ has the χ^2 distribution with $k - 1$ d.f., and hence

$$E\sum N_i[\bar{x}_i. - \mu - \alpha_i - (\bar{x} - \mu)]^2 = (k - 1)\sigma^2 \tag{9.4.5}$$

Also $E[\bar{x}_i. - \mu - \alpha_i - (\bar{x} - \mu)] = 0$, so that

$$E(S_b) = (k - 1)\sigma^2 + \sum_i N_i\alpha_i^2 \tag{9.4.6}$$

This shows that if the α_i are not all zero, the expectation of $S_b/(k - 1)$ is *greater* than σ^2, which justifies the use of the one-tailed F-test for treatment effects between samples. In the same way,

$$E(S_t) = (N - 1)\sigma^2 + \sum_i N_i\alpha_i^2 \tag{9.4.7}$$

and therefore,

$$E(S_w) = (N - k)\sigma^2 \tag{9.4.8}$$

The fixed treatment effects do not enter into the mean square *within* samples. From Eq. (3),

$$E(\bar{x}) = \mu \tag{9.4.9}$$

so that $\bar{x}$ is an unbiased estimator of μ. Also, from Eq. (1),

$$E\left(\sum_j x_{ij}\right) = N_i\mu + N_i\alpha_i \tag{9.4.10}$$

so that $\alpha_i = E(\bar{x}_i.) - \mu$. The quantity $\bar{x}_i. - \bar{x}$ is therefore an unbiased estimator of α_i. In Example 1, the estimates of α_i are $\hat{\alpha}_1 = 108$, $\hat{\alpha}_2 = -21$, $\hat{\alpha}_3 = -39$, $\hat{\alpha}_4 = -30$, $\hat{\alpha}_5 = -27$.

For the two-way classification of § 9.3, with fixed effects, a similar argument leads to the following results:

$$\begin{cases} E(S_a) = (a-1)\sigma^2 + br\sum_i \alpha_i^2 \\ E(S_b) = (b-1)\sigma^2 + ar\sum_j \beta_j^2 \\ E(S_{ab}) = (a-1)(b-1)\sigma^2 + r\sum_{ij} \gamma_{ij}^2 \\ E(S_r) = ab(r-1)\sigma^2 \\ E(S_t) = (abr-1)\sigma^2 + br\sum_i \alpha_i^2 + ar\sum_j \beta_j^2 + r\sum_{ij}\gamma_{ij}^2 \end{cases} \tag{9.4.11}$$

The estimators of α_i, β_j, γ_{ij} and μ are

$$\begin{cases} \hat{\alpha}_i = \bar{x}_{i}.. - \bar{x}, & i = 1, 2 \ldots a \\ \hat{\beta}_j = \bar{x}._j. - \bar{x}, & j = 1, 2 \ldots b \\ \hat{\gamma}_{ij} = \bar{x}_{ij}. - \bar{x}_i.. - \bar{x}._j. + \bar{x} \\ \hat{\mu} = \bar{x} \end{cases} \tag{9.4.12}$$

9.5 **Estimation of Variable Treatment Effects (Components of Variance) Model II** In Model II, the effects (even including the interaction) are treated as random variates which are normally and independently distributed. Thus in Eq. (9.2.1), α_i is regarded as a value of a random variate which has expectation zero and variance σ_α^2. We must suppose that the k samples actually examined are a random selection from a large population of possible samples. The members of this population may be denoted by the subscript u, and for each there is a "true" or expected value of X which we may denote by m_u. This quantity m_u is a random variable with a certain probability distribution over the population, and its expected value is μ. The difference between m_u and μ is the true effect of sample u, which we have denoted by α_u. The variance of α_u is the quantity σ_α^2. We have, therefore, for an actually selected sample i,

$$x_{ij} = m_i + \varepsilon_{ij} = \mu + \alpha_i + \varepsilon_{ij} \tag{9.5.1}$$

where ε_{ij} is the error term, namely, the difference between the true value m_i and the measured value on the jth replicate. It is assumed that the set of α_i and the set of ε_{ij} are completely independent and that the ε_{ij} have the same variance σ^2 for all i, and therefore,

(9.5.2) $$V(x_{ij}) = \sigma_\alpha^2 + \sigma^2$$

The quantities σ_α^2 and σ^2 are the *components of variance.*

It should be noted that two observations in the *same* sample will be correlated. Thus the covariance of x_{ij} and $x_{ij'}$ is given by

$$\begin{aligned} C(x_{ij}, x_{ij'}) &= E[(x_{ij} - \mu)(x_{ij'} - \mu)] \\ &= E[(\alpha_i + \varepsilon_{ij})(\alpha_i + \varepsilon_{ij'})] \\ &= E(\alpha_i^2) = \sigma_\alpha^2 \end{aligned}$$

since all the other terms have zero expectations by hypothesis. The quantity $\sigma_\alpha^2/(\sigma_\alpha^2 + \sigma^2)$ which is the correlation coefficient between two observations in the same sample, is called the *intra-class correlation coefficient.*

The usual null hypothesis to be tested is that $\sigma_\alpha^2 = 0$, which implies that $m_u = \mu$ for all values of u. If we suppose that all the samples are the same size (r), the sums of squares between samples and within samples are, as before,

(9.5.3) $$\begin{aligned} S_b &= r \sum_i (\bar{x}_{i\cdot} - \bar{x})^2 \\ &= r \sum_i [(\mu + \alpha_i + \bar{\varepsilon}_{i\cdot}) - (\mu + \bar{\alpha} + \bar{\varepsilon})]^2 \end{aligned}$$

and

(9.5.4) $$\begin{aligned} S_w &= \sum_{ij} (x_{ij} - \bar{x}_{i\cdot})^2 \\ &= \sum_{ij} (\varepsilon_{ij} - \bar{\varepsilon}_{i\cdot})^2 \end{aligned}$$

If the ε_{ij} are normal with expectation 0 and variance σ^2, S_w/σ^2 has the χ^2 distribution with $k(r - 1)$ d.f., so that

(9.5.5) $$E\frac{S_w}{k(r-1)} = \sigma^2$$

If also the α_i are normal with variance σ_α^2, and if we write η_i for the variable $\alpha_i + \bar{\varepsilon}_{i\cdot}$, then

(9.5.6) $$S_b = r \sum_i (\eta_i - \bar{\eta})^2$$

and the η_i are independently normal with expectation 0 and variance $\sigma_\alpha^2 + \sigma^2/r$. It follows that $S_b/[(\sigma_\alpha^2 + \sigma^2/r)r]$ is a chi-square variate with $k - 1$ d.f., so that

(9.5.7) $$E\frac{S_b}{k-1} = r\sigma_\alpha^2 + \sigma^2$$

The sums of squares S_b and S_w are statistically independent. The null hypothesis is therefore tested by the ordinary F-test of the ratio of mean squares between and within samples.

The *power* of this test is a function of σ_α^2/σ^2. If F has the ordinary F-distribution with $k - 1$ and $k(r - 1)$ d.f., and if F_α is the value of F exceeded with probability α, then the power is given by

(9.5.8)
$$P = Pr\left(F \geq F_\alpha \frac{\sigma^2}{\sigma^2 + \sigma_\alpha^2}\right)$$

For the two-way layout of § 9.3, a similar set of assumptions leads to the model

(9.5.9)
$$x_{ijk} = \mu + \alpha_i + \beta_j + \gamma_{ij} + \varepsilon_{ijk}$$

where the α_i, the β_j, the γ_{ij} and the ε_{ijk} are independently and normally distributed with zero expectations and variances $\sigma_\alpha^2, \sigma_\beta^2$, σ_γ^2 and σ^2, respectively. The variance of x_{ijk} is then given by

(9.5.10)
$$V(x_{ijk}) = \sigma_\alpha^2 + \sigma_\beta^2 + \sigma_\gamma^2 + \sigma^2$$

The sum of squares for A-effects is

(9.5.11)
$$\begin{aligned} S_a &= br \sum_i (\bar{x}_{i\cdot\cdot} - \bar{x})^2 \\ &= br \sum_i (\alpha_i - \bar{\alpha} + \bar{\gamma}_{i\cdot} - \bar{\gamma} + \bar{\varepsilon}_{i\cdot\cdot} - \bar{\varepsilon})^2 \end{aligned}$$

If $\eta_i = \alpha_i + \bar{\gamma}_{i\cdot} + \bar{\varepsilon}_{i\cdot\cdot}$, then η_i is normal with expectation 0 and variance $\sigma_\alpha^2 + \sigma_\gamma^2/b + \sigma^2/(br)$. Also its mean over the a values of i is $\bar{\eta} = \bar{\alpha} + \bar{\gamma} + \bar{\varepsilon}$. Therefore, $\dfrac{\sum_i(\eta_i - \bar{\eta})^2}{\sigma_\alpha^2 + \sigma_\gamma^2/b + \sigma^2/(br)}$ is χ^2 with $a - 1$ d.f., and

(9.5.12)
$$\begin{aligned} E[S_a/(a - 1)] &= br(\sigma_\alpha^2 + \sigma_\gamma^2/b + \sigma^2/(br)) \\ &= br\sigma_\alpha^2 + r\sigma_\gamma^2 + \sigma^2. \end{aligned}$$

A similar argument leads to the results:

(9.5.13)
$$E\frac{S_b}{b - 1} = ar\sigma_\beta^2 + r\sigma_\gamma^2 + \sigma^2$$

(9.5.14)
$$E\frac{S_{ab}}{(a - 1)(b - 1)} = r\sigma_\gamma^2 + \sigma^2$$

(9.5.15)
$$E\frac{S_r}{ab(r - 1)} = \sigma^2$$

The four sums of squares are statistically independent, and therefore ordinary F-tests of the hypotheses $\sigma_\alpha^2 = 0$, $\sigma_\beta^2 = 0$, $\sigma_\gamma^2 = 0$ may be carried out.

It should be noted that, contrary to the conclusions from the fixed-effects model, the interaction component of variance appears in the mean squares for A-effects and B-effects. The F-test for the null hypothesis $\sigma_\alpha^2 = 0$ must therefore be carried out by comparing the mean squares for A-effects and *interaction*, and similarly for $\sigma_\beta^2 = 0$. If $r > 1$, we can test for interaction by an F-test of Eq. (14) against Eq. (15).

Similar but more complicated models can be used when there are three main effects, with *three* second-order interactions and one third-order interaction. With Model II a complication arises when the interactions are not negligible, because it turns out to be impossible to apply the F-test directly in order to test for the main effects. An approximate method suggested by Satterthwaite may be tried in such cases—[5], [6].

* 9.6 **Mixed Models (Model III)** A layout in which it seems reasonable to regard one effect as fixed and another as random is said to be *mixed*. In a problem concerned with the daily output of workers in a factory using certain machines [7], we might be inclined to regard the workers as a random sample from a large population, but we might be interested in the performance of individual machines, perhaps of different makes.

Let x_{ijk} be the output, say, for the jth worker on the kth day that he is assigned to the ith machine, ($i = 1, 2 \ldots a, j = 1, 2 \ldots b, k = 1, 2 \ldots r$). The days will be regarded merely as replicates, the effects in which we are interested being the fixed effects α_i of machines and the random effects β_j of workers as well as their possible interactions. We assume that

(9.6.1) $$x_{ijk} = m_{ij} + \varepsilon_{ijk}$$

where m_{ij} is the "true" mean output of the jth worker on the ith machine, and the ε_{ijk} are independent normal variates with mean zero and variance σ^2. Since the jth worker is regarded as a random selection from a large population of workers, we can think of m_{ij} as a particular value of a random variable M_i which represents the mean output of a worker selected at random on the machine numbered i.

Let the expected value of M_i over the population of workers be denoted by μ_i and let the arithmetic mean of the μ_i over the i machines be denoted by μ. Then

(9.6.2) $$\mu_i = E(M_i), \qquad \mu = \bar{\mu}. = \sum \frac{\mu_i}{a}$$

The main effect of the ith machine is

(9.6.3) $$\alpha_i = \mu_i - \mu, \qquad \sum \alpha_i = 0$$

Suppose m_{iw} is the value of M_i for any worker labelled w in the whole population of workers. The true mean for this worker is the average of m_{iw} over the i machines and the main effect β_w of worker w is the excess of this over the general mean.

(9.6.4) $$\begin{aligned}\beta_w &= \overline{m}_{.w} - E(\overline{M}.) \\ &= \overline{m}_{.w} - \mu\end{aligned}$$

This is a random variable with expectation 0, and variance, say, σ_β^2.

The main effect of worker w, specific to the ith machine, may be defined as $m_{iw} - E(M_i) = m_{iw} - \mu_i$, and the excess of this above its average over the machines is called the *interaction* of the ith machine and the particular worker w, namely,

$$\gamma_{iw} = m_{iw} - \mu_i - \overline{m}_{.w} + \mu \tag{9.6.5}$$

For each i, this is a random variable with expectation zero and variance $\sigma_{\gamma i}^2$. Also, $\sum_i \gamma_{iw} = 0$ for each value of w.

From Eqs. (3), (4) and (5) we obtain

$$m_{iw} = \mu + \alpha_i + \beta_w + \gamma_{iw} \tag{9.6.6}$$

and, for any worker w and any day d,

$$x_{iwd} = \mu + \alpha_i + \beta_w + \gamma_{iw} + \varepsilon_{iwd} \tag{9.6.7}$$

where the ε_{iwd} are independent of each other and of β_w and γ_{iw}, and have a common variance σ^2. The γ_{iw}, for different values of i, are not necessarily independent of each other or of β_w, but have covariances depending on those of the random variables M_i.

The b workers actually used in the experiment may be regarded as a random sample from the whole population of workers, and the r days similarly form a sample of all possible days. The x_{ijk} of Eq. (1) has therefore the same form as the x_{iwd} of Eq. (7), but j takes only the values $1, 2 \ldots b$, and k takes only the values $1, 2 \ldots r$. That is,

$$x_{ijk} = \mu + \alpha_i + \beta_j + \gamma_{ij} + \varepsilon_{ijk} \tag{9.6.8}$$

The β_j for different values of j may be looked on as independent variates all having the same distribution as β_w, and the γ_{ij} are similarly independent with the same distribution as γ_{iw}, for any i. The ε_{ijk} can be regarded as independent of each other and of the β_j and the γ_{ij}.

For convenience of notation we may define σ_α^2 by the relation

$$(a-1)\sigma_\alpha^2 = \sum_i \alpha_i^2 \tag{9.6.9}$$

but it must be remembered that we are treating the α_i as fixed effects, so that σ_α^2 is *not* the variance of a random variable. Also we will define σ_γ^2 by

$$(a-1)\sigma_\gamma^2 = \sum_i \sigma_{\gamma i}^2 \tag{9.6.10}$$

The division of the total sum of squares into four parts may be carried out just as in Model I or Model II. We get

$$S_t = S_a + S_b + S_{ab} + S_r \tag{9.6.11}$$

where

(9.6.12)
$$\begin{cases} S_a = br\sum_i (\bar{x}_{i\cdot\cdot} - \bar{x})^2 \\ S_b = ar\sum_j (\bar{x}_{\cdot j\cdot} - \bar{x})^2 \\ S_{ab} = r\sum_{ij} (\bar{x}_{ij\cdot} - \bar{x}_{i\cdot\cdot} - \bar{x}_{\cdot j\cdot} + \bar{x})^2 \\ S_r = \sum_{ijk} (x_{ijk} - \bar{x}_{ij\cdot})^2 \end{cases}$$

It is straightforward to show, as in § 9.4, that if the ε_{ijk}, β_j and γ_{ij} are normally distributed

(9.6.13)
$$E\,\frac{S_a}{a-1} = br\sigma_\alpha^{\,2} + r\sigma_\gamma^{\,2} + \sigma^2$$

Also, by using Eq. (8) in the second equation of (12) and noting that $\bar{\gamma}_{\cdot j} = 0$ for all j, since $\sum_i \gamma_{ij} = 0$, we find

(9.6.14)
$$\begin{aligned} S_b &= ar\sum_j [(\beta_j - \bar{\beta}_\cdot) + (\bar{\varepsilon}_{\cdot j\cdot} - \bar{\varepsilon})]^2 \\ &= ar\sum_j (\beta_j - \bar{\beta}_\cdot)^2 + ar\sum_j (\bar{\varepsilon}_{\cdot j\cdot} - \bar{\varepsilon})^2 \\ &\qquad + 2ar\sum_j (\beta_j - \bar{\beta}_\cdot)(\bar{\varepsilon}_{\cdot j\cdot} - \bar{\varepsilon}) \end{aligned}$$

The expectation of $\sum_j (\beta_j - \bar{\beta}_\cdot)^2$ is $(b-1)\,\sigma_\beta^{\,2}$. Since $\bar{\varepsilon}_{\cdot j\cdot}$ is normal with variance σ^2/ar, the expectation of $\sum_j (\bar{\varepsilon}_{\cdot j\cdot} - \bar{\varepsilon})^2$ is similarly $(b-1)\sigma^2/ar$. The expectation of the product term vanishes because of the independence of β_j and ε_{ijk}. Therefore,

(9.6.15)
$$E(S_b) = ar(b-1)\sigma_\beta^{\,2} + (b-1)\sigma^2$$

so that

(9.6.16)
$$E\,\frac{S_b}{b-1} = ar\sigma_\beta^{\,2} + \sigma^2$$

In the same way we can write

(9.6.17)
$$S_{ab} = r\sum_{ij} (\gamma_{ij} - \bar{\gamma}_{i\cdot} + \bar{\varepsilon}_{ij\cdot} - \bar{\varepsilon}_{i\cdot\cdot} - \bar{\varepsilon}_{\cdot j\cdot} + \bar{\varepsilon})^2$$

Now $E\sum_j (\gamma_{ij} - \bar{\gamma}_{i\cdot})^2 = (b-1)\,\sigma_{\gamma i}^{\,2}$ since the γ_{ij} for given i are independent variates with variance $\sigma_{\gamma i}^{\,2}$. It follows that $E\sum_{ij} (\gamma_{ij} - \bar{\gamma}_{i\cdot})^2 = (a-1)(b-1)\,\sigma_\gamma^{\,2}$.

Let us define a variate η_{ij} by

$$\eta_{ij} = \bar{\varepsilon}_{ij\cdot} - \bar{\varepsilon}_{i\cdot\cdot}$$

Then $\eta_{ij} - \bar{\eta}_{\cdot j} = \bar{\varepsilon}_{ij\cdot} - \bar{\varepsilon}_{i\cdot\cdot} - \bar{\varepsilon}_{\cdot j\cdot} + \bar{\varepsilon}$.

Since $\bar{\varepsilon}_{ij.}$ has variance σ^2/r, the variance of η_{ij} is $(b - 1)\sigma^2/br$ (see § 2.14). The expectation of $\sum_i (\eta_{ij} - \bar{\eta}_{.j})^2$ is therefore $(a - 1)(b - 1)\sigma^2/br$, and that of $\sum_{ij} (\eta_{ij} - \bar{\eta}_{.j})^2$ is b times as great. From Eq. (17),

$$S_{ab} = r\sum_{ij} [(\gamma_{ij} - \bar{\gamma}_{i.})^2 + (\eta_{ij} - \bar{\eta}_{.j})^2 + 2(\gamma_{ij} - \bar{\gamma}_{i.})(\eta_{ij} - \bar{\eta}_{.j})]$$

The expectation of the product term is zero, because of the independence of γ_{ij} and ε_{ijk}, and therefore we have

$$E(S_{ab}) = r(a - 1)(b - 1)\sigma_\gamma^2 + (a - 1)(b - 1)\sigma^2$$

Dividing by $(a - 1)(b - 1)$, we obtain the expectation of the mean square,

$$E\frac{S_{ab}}{(a - 1)(b - 1)} = r\sigma_\gamma^2 + \sigma^2 \tag{9.6.18}$$

Finally, $S_r = ab\sum_k (\varepsilon_{ijk} - \bar{\varepsilon}_{ij.})^2$ and $E(S_r) = ab(r - 1)\sigma^2$, so that

$$E\frac{S_r}{ab(r - 1)} = \sigma^2 \tag{9.6.19}$$

The analysis of variance for the mixed Model III is set out in Table 9.5.

TABLE 9.5

Source of Variation	D.F.	M.S.	E(M.S.)
A-effects (fixed)	$a - 1$	$S_a/(a - 1)$	$br\sigma_\alpha^2 + r\sigma_\gamma^2 + \sigma^2$
B-effects (random)	$b - 1$	$S_b/(b - 1)$	$ar\sigma_\beta^2 + \sigma^2$
$A \times B$-effects (interaction)	$(a - 1)(b - 1)$	$S_{ab}/[(a - 1)(b - 1)]$	$r\sigma_\gamma^2 + \sigma^2$
Error	$ab(r - 1)$	$S_r/[ab(r - 1)]$	σ^2
Total	$abr - 1$		

The four sums of squares are pairwise independent, except for the pair S_b and S_{ab}. We can therefore test for interaction by comparing the mean squares for interaction and error, test for B-effects by comparing the mean squares for B-effects and error, and test for A-effects by comparing the mean squares for A-effects and interaction.

Estimates of σ_α^2, σ_β^2, σ_γ^2 and σ^2 can be calculated from the last column of Table 9.5. The estimator of μ is the over-all mean $\bar{x}$, and estimators of α_i, β_j and γ_{ij} are, respectively,

$$\hat{\alpha}_i = \bar{x}_{i..} - \bar{x} \tag{9.6.20}$$

$$\hat{\beta}_j = \bar{x}_{.j.} - \bar{x} \tag{9.6.21}$$

$$\hat{\gamma}_{ij} = \bar{x}_{ij.} - \bar{x}_{i..} - \bar{x}_{.j.} + \bar{x} \tag{9.6.22}$$

*** 9.7 Nested or (Hierarchal Models)** One type of incomplete design is that in which a factor B is "nested" within another factor A. This means that for each level of A there is a set of levels of B, and each B-level occurs in just one A-level. In the type of design considered earlier in this chapter, each B-level occurs in each A-level, and A and B are said to be *completely crossed.* If this is not so but if some B-level occurs in at least two A-levels, the factors are said to be *partly crossed.*

As an example, we may consider an experiment on the determination of the protein content of wheat. From many suitable laboratories in Canada, three were selected at random and, in each of these, five samples of wheat were analyzed on each of two days. All the 30 wheat samples were parts of one carefully selected master sample and were thoroughly mixed and randomized. There is presumably a main effect between the different laboratories, and also between days in each laboratory, but day number 1 in one laboratory has nothing whatever necessarily in common with day number 1 in another laboratory. In fact, we might not even use the same number of days in the different laboratories. The day-effect is said to be "nested" within the laboratory effect, or, to put it another way, the day-effect has a lower rank in the hierarchy of effects than the laboratory effect. This is why the nested model is sometimes called "hierarchal." There is no need for any interaction term in this model, since no B-level occurs with more than one A-level.

We will suppose that the A-effect occurs at a levels, denoted by i ($i = 1, 2 \dots a$), and that, nested within the ith level, there are b_i B-levels, denoted by j ($j = 1, 2 \dots b_i$). In the illustration above, the b_i were all equal to 2, but this is not necessary. We suppose also that corresponding to any given i and j there are r_{ij} replicates of the measurement of a variate X. The r_{ij} can be different for each pair of i and j, but it will be convenient to assume that they are all equal to r. The measurement on the kth replicate will then be denoted by x_{ijk} and is given by

$$x_{ijk} = m_{ij} + \varepsilon_{ijk} \tag{9.7.1}$$

where the ε_{ijk} can be regarded as "errors," which are independently and normally distributed with expectation 0 and variance σ^2, and are independent of the m_{ij}. The m_{ij} are the expectations (or true values) for the ith A-level and the jth B-level.

We may use either Model I, II or III for the effects. If we suppose, in the illustration mentioned earlier, that the laboratories are chosen at random from a large number, and if the days are also (as far as possible) random, we should use Model II, and this model will be assumed in the rest of this section. There is then a large population of A-levels, denoted in general by u, and for the uth level there is a large population of B-levels nested within it and denoted in general by v. The expectation of X, measured for the vth sub-level of the uth level, will be the random variable m_{uv}. The m_{ij} of Eq. (1) is the value of m_{uv} for $u = i$ and $v = j$.

If the conditional expectation of m_{uv} for fixed u is denoted by $m_{u\cdot}$, and the expectation of $m_{u.}$ is denoted by $m_{..}$, we may write

$$m_{uv} = \mu + \alpha_u + \beta_{uv} \tag{9.7.2}$$

where

$$\begin{cases} \mu = m_{..} \\ \alpha_u = m_{u\cdot} - m_{..} \\ \beta_{uv} = m_{uv} - m_{u\cdot} \end{cases} \tag{9.7.3}$$

As usual, μ is the over-all expectation of X. The random variable α_u has expectation 0 and variance $\sigma_\alpha{}^2$. The random variable β_{uv} has a conditional expectation (for given u) zero and conditional variance $\sigma_{\beta u}{}^2$. These variables are uncorrelated, as may be shown with the help of a theorem on conditional expectations in Appendix A.14. The proof goes as follows:

$$E(\alpha_u\beta_{uv}|u) = \alpha_u E(\beta_{uv}|u)$$
$$= 0$$

and therefore, by Eq. (A.14.4), $E(\alpha_u\beta_{uv}) = 0$, which shows that the correlation is zero, both variables having zero expectations.

For the variance of β_{uv}, by Eq. (A.14.6), we have

$$V(\beta_{uv}) = V[E(\beta_{uv}|u)] + E[V(\beta_{uv}|u)]$$

The first term on the right-hand side is zero; hence,

$$V(\beta_{uv}) = E(\sigma_{\beta u}{}^2) = \sigma_\beta{}^2 \tag{9.7.4}$$

The A-levels actually selected in the experiment may be denoted by u_i $(i = 1, 2 \ldots a)$, and the B-levels corresponding to u_i by v_j $(j = 1, 2 \ldots b_i)$. Then

$$m_{ij} = \mu + \alpha_i + \beta_{ij} \tag{9.7.5}$$

where α_i is the value of α_u for $u = u_i$ and β_{ij} the value of β_{uv} for $u = u_i$ and $v = v_j$. These are uncorrelated and are both independent of ε_{ijk}. We now suppose that they are also normally distributed.

The total sum of squares S_t is divided into three parts:

$$S_t = S_a + S_b + S_r \tag{9.7.6}$$

where

$$\begin{cases} S_a = r\sum_i b_i(\bar{x}_{i..} - \bar{x})^2 \\ S_b = r\sum_i\sum_j (\bar{x}_{ij.} - \bar{x}_{i..})^2 \\ S_r = \sum_{ijk} (x_{ijk} - \bar{x}_{ij.})^2 \end{cases} \tag{9.7.7}$$

It should be noted that the mean over i is a weighted mean when the b_i are not all equal. Thus,

$$\bar{x} = \frac{\sum_i b_i \bar{x}_{i\cdot\cdot}}{\sum_i b_i} = N^{-1} \sum_{ijk} x_{ijk}$$

where $N = r \sum b_i$.

These three sums of squares are statistically independent. Details of the proof of this statement may be found in Scheffé's book [7].

The quantity S_r/σ^2, under the normality assumption, is a χ^2-variate with $(r - 1) \sum b_i$ degrees of freedom. Also $S_b/(\sigma^2 + r\sigma_\beta^2)$ is a χ^2-variate with $\sum_i (b_i - 1)$ d.f. However, S_a is not in general equal to a constant times a chi-square variate. If it happens that $\sigma_\alpha^2 = 0$ (which means that there are no A-effects), it may be shown that $S_a/(\sigma^2 + r\sigma_\beta^2)$ is distributed like χ^2 with $a - 1$ d.f. Also, if all the b_i are equal to b (as in the example of the laboratory analyses described above), $S_a/(\sigma^2 + r\sigma_\beta^2 + br\sigma_\alpha^2)$ is distributed like χ^2 with $a - 1$ d.f. In the more general case, the expectation of S_a is given by

(9.7.8) $$E(S_a) = (a - 1)(\sigma^2 + r\sigma_\beta^2) + (a - 1)A\sigma_\alpha^2$$

where

(9.7.9) $$(a - 1)A = r\left(\sum_i b_i - \frac{\sum_i b_i^2}{\sum b_i}\right)$$

which, when $b_i = b$, reduces to $br(a - 1)$.

The analysis of variance table is, therefore, as given in Table 9.6.

TABLE 9.6

Variation	S.S.	D.F.	M.S.	E(M.S.)
A-effects	S_a	$a - 1$	$S_a/(a - 1)$	$\sigma^2 + r\sigma_\beta^2 + A\sigma_\alpha^2$
B-effects (nested in A)	S_b	$\sum_i b_i - a$	$S_b/(\sum b_i - a)$	$\sigma^2 + r\sigma_\beta^2$
Error	S_r	$(r - 1)\sum_i b_i$	$S_r/[(r - 1) \sum b_i]$	σ^2
Total	S_t	$r \sum_i b_i - 1$		

If $b_i = b$ for all i, $A = br$, $\sum b_i = ab$.

The hypothesis of no B-effects ($\sigma_\beta^2 = 0$) may be tested by an exact F-test of the mean square for B-effects divided by the mean square for error. A test of the hypothesis $\sigma_\alpha^2 = 0$ may be made by dividing the mean square for A-effects by the mean square for B-effects. The power of these tests can be calculated as in § 8.16, except for the test of $\sigma_\alpha^2 = 0$ when the b_i are unequal. In this case an approximate method must be used.

9.8 **Latin Square Designs** In a *complete* experimental design there is at least one observation for every possible combination of levels. The nested design is incomplete because each B-level appears in combination with one and only one A-level (that level within which it is nested). The Latin square is another type of incomplete design in which there are three factors, all with the same number of levels m, but in which observations are made on only m^2 instead of the possible m^3 combinations. This is obviously economical of effort, but the design is rather restrictive (enforcing the same number of levels for each factor) and the usual analysis does not allow for interactions.

A Latin square is a square array of symbols, m to a row or column, and such that each symbol appears once and only once in each row and each column. Thus in the figure below there are 5 letters arranged in this way. In an agricultural experiment these might represent 5 varieties of wheat planted in 25 plots arranged in 5 rows and 5 columns across a field. The purpose of the arrangement is to allow for a possible row-effect or column-effect on the yield, such as might be produced if there is a definite fertility gradient across the field in the direction of the columns or the rows. In an animal-feeding experiment there might be 5 types of ration used on 25 animals. If the animals belonged to 5 different litters (5 to each litter) and could be kept in 5 types of pens, 5 in each type, we could arrange a Latin square experiment in which each type of pen contained just 1 animal from each litter. The 5 rations would be allocated to the animals according to a Latin square, and we could then eliminate the effect of litters and of pens on the gains in weight.

A Latin square remains a Latin square when the rows, or the columns, are permuted among themselves. A *standard m* by *m* square is one in which the first row and the first column contain the m symbols in their natural order, as in the right-hand 5 by 5 square below.

E	*A*	*C*	*B*	*D*		*A*	*B*	*C*	*D*	*E*
B	*C*	*E*	*D*	*A*		*B*	*E*	*A*	*C*	*D*
A	*B*	*D*	*E*	*C*		*C*	*D*	*B*	*E*	*A*
C	*D*	*B*	*A*	*E*		*D*	*C*	*E*	*A*	*B*
D	*E*	*A*	*C*	*B*		*E*	*A*	*D*	*B*	*C*

With 5 symbols there are 56 different standard squares, and from each of these we may obtain 2880 Latin squares by permuting rows and columns. The number of possible Latin squares increases enormously as m increases. In an experimental design we should choose a Latin square at random from the whole set of possibilities. We could, for instance, first choose a standard square from a complete set of standard squares, such as is found in Fisher and Yates' tables [8] for the smaller values of m, and then permute at random the columns and the rows (except the first row). For this purpose a table of random permutations of the numbers 1 to 9 may be used—see, for example, [9].

The mathematical model for a Latin square design is

$$x_{ijk} = \mu + \alpha_i + \beta_j + \gamma_k + \varepsilon_{ijk} \qquad (9.8.1)$$

where i, j, k take values from 1 to m, but where only m^2 sets of triples (i, j, k) are permissible, these being dictated by the particular Latin square used. The α_i are the main effects for the first factor (say treatments), the β_j are those for the second factor (say rows) and the γ_k those for the third factor (say columns). The ε_{ijk} are assumed to be independently normal with expectation 0 and variance σ^2. The mean values for α_i, β_j and γ_k are adjusted to zero as usual. This model ignores any interaction between the three main effects. The null hypotheses to be tested are (a) all $\alpha_i = 0$, (b) all $\beta_j = 0$, (c) all $\gamma_k = 0$.

We will assume that the effects are fixed, as in Model I. The total sum of squares S_t may be divided into four parts:

$$S_t = S_a + S_b + S_c + S_e \tag{9.8.2}$$

where

$$S_t = \sum_{(i,j,k)\in D} (x_{ijk} - \bar{x})^2 = \sum_D x_{ijk}{}^2 - G \tag{9.8.3}$$

D being the actual set of triples (i, j, k) appearing in the square.

$$\begin{cases} S_a = m \sum_i (\bar{x}_{i\cdot\cdot} - \bar{x})^2 = m \sum_i \bar{x}_{i\cdot\cdot}{}^2 - G \\ S_b = m \sum_j (\bar{x}_{\cdot j\cdot} - \bar{x})^2 = m \sum_j \bar{x}_{\cdot j\cdot}{}^2 - G \\ S_c = m \sum_k (\bar{x}_{\cdot\cdot k} - \bar{x})^2 = m \sum_k \bar{x}_{\cdot\cdot k}{}^2 - G \end{cases} \tag{9.8.4}$$

Here $\bar{x}$ is the arithmetic mean of the m^2 observations and is an unbiased estimator for μ. G is the quantity $m^2\bar{x}^2$. Also $\bar{x}_{i\cdot\cdot}$ is the average of the m observations on the ith treatment, $\bar{x}_{\cdot j\cdot}$ is the average of the m observations in the jth row, and $\bar{x}_{\cdot\cdot k}$ is the average of the m observations in the kth column. The sum of squares for error S_e is calculated by difference from Eqs. (2), (3) and (4).

The number of degrees of freedom for treatments, rows and columns is $m - 1$ in each case. Since the total number is $m^2 - 1$, there are $m^2 - 3m + 2$ degrees of freedom for error. The expectations of the mean squares are given in Table 9.7, where $\sigma_\alpha{}^2$ is a symbol for $\sum_i \alpha_i{}^2/(m - 1)$ and similarly for $\sigma_\beta{}^2$ and $\sigma_\gamma{}^2$.

TABLE 9.7
Analysis of Variance for m × m Latin Square

Variation	S.S.	D.F.	M.S.	E(M.S.)
Treatments (A)	S_a	$m - 1$	$S_a/(m - 1)$	$\sigma^2 + m\sigma_\alpha{}^2$
Rows (B)	S_b	$m - 1$	$S_b/(m - 1)$	$\sigma^2 + m\sigma_\beta{}^2$
Columns (C)	S_c	$m - 1$	$S_c/(m - 1)$	$\sigma^2 + m\sigma_\gamma{}^2$
Error	S_e	$m^2 - 3m + 2$	$S_e/(m^2 - 3m + 2)$	σ^2
Total	S_t	$m^2 - 1$		

The four first sums of squares in the above table are statistically independent, so that F-tests of the three null hypotheses mentioned above may be applied to test for significant effects of treatments, rows or columns.

If there are interactions between these effects the interpretation of the results of the experiment is complicated and difficult. Usually the expectation of the error mean square will be increased, but the effect of interaction on the other mean squares may be to increase or reduce them.

* 9.9 **Balanced Incomplete Block Designs** In the complete randomized block design of § 9.3, every treatment appears in every block, possibly with several replications as well. But it is sometimes advisable to have smaller blocks, the size of which is dictated by special circumstances. If one were carrying out tests on the life of automobile tires of various makes, the four wheels of a car would form a natural "block."

We shall consider a fixed-effects model, even though it might well be more reasonable in some cases to think of the blocks as randomly chosen from a large population of blocks (this would be true for the car tires, for example, in the illustration above). We suppose that the blocks are all of the same size, that each treatment occurs r times, and that no treatment appears twice in the same block. Then if there are k "plots" (or items) in a block, b blocks, and a treatments, we obviously have

(9.9.1) $$N = ar = bk$$

where $k < a$.

A design is said to be *balanced* if each *pair* of treatments occurs in the same number of blocks. If this number is denoted by λ, it follows that in a balanced design

(9.9.2) $$(a - 1)\lambda = (k - 1)r$$

In the example illustrated in Table 9.8 there are seven treatments (denoted by letters A to G), and seven blocks, each of size 4, so that $\lambda = 2$. The pair BD, for example, occurs in blocks 5 and 7 only, and similarly for every other pair.

TABLE 9.8

Block Number						
1	2	3	4	5	6	7
C	*A*	*G*	*F*	*B*	*E*	*D*
G	*F*	*E*	*A*	*D*	*C*	*B*
E	*G*	*A*	*B*	*C*	*D*	*F*
F	*D*	*B*	*C*	*G*	*A*	*E*

The analysis is considerably simplified when the design is balanced. A further necessary condition for such a design is

(9.9.3) $$a \leq b$$

which implies $r \geq k$. Many designs satisfying conditions (1), (2) and (3) may be found in reference [9], Chapter 11. Once a suitable design has been selected, the numbering of the treatments and of the blocks, and the positions within the blocks, may all be randomized.

If we want to balance out the positions by ensuring that each treatment occurs just m times in each of the k positions in a block, we must add a fourth condition

$$b = ma \tag{9.9.4}$$

or $r = km$. This is satisfied in Table 9.8, with $m = 1$. Treatment A, for instance, occurs just once in each of the four positions in a block.

The mathematical model is

$$x_{ij} = \mu + \alpha_i + \beta_j + \varepsilon_{ij} \tag{9.9.5}$$

where the α_i are treatment effects, and the β_j are block effects and, as usual, $\sum \alpha_i = \sum \beta_j = 0$. It is assumed that there is no interaction. The ε_{ij} are independently normal with expectation zero and variance σ^2, but it must be observed that only some of the possible pairs i, j correspond to actual observations. If K_{ij} is the number of times the ith treatment occurs in the jth block, K_{ij} is either 0 or 1, and the observed values of x_{ij} correspond to those pairs for which $K_{ij} = 1$ (there are ar of these, out of ab pairs altogether).

If we define the ith treatment total g_i and the jth block total h_j by

$$g_i = \sum_j x_{ij}, \qquad h_j = \sum_i x_{ij} \tag{9.9.6}$$

the ith *adjusted treatment total* is

$$\begin{aligned} G_i &= g_i - k^{-1} \sum_j K_{ij} h_j \\ &= g_i - T_i/k \end{aligned} \tag{9.9.7}$$

where T_i is the sum of the block totals in which the ith treatment appears. The adjustment therefore consists in subtracting the sum of the block averages for those blocks in which the ith treatment occurs.

The total sums of squares S_t is given by

$$S_t = \sum_D x_{ij}^2 - G \tag{9.9.8}$$

where D refers to the set of pairs (i, j) for which $K_{ij} = 1$ and G is the correction for the grand mean,

$$G = \frac{\left(\sum_i g_i\right)^2}{N} \tag{9.9.9}$$

This total sum of squares is split into three parts:

$$S_t = S_{\text{t.e.b.}} + S_{\text{b.i.t.}} + S_e \tag{9.9.10}$$

where $S_{\text{t.e.b.}}$ is the sum of squares for *treatments, eliminating blocks*, and $S_{\text{b.i.t.}}$ is the sum of squares for *blocks, ignoring treatments*. These are given respectively by

(9.9.11) $$S_{\text{t.e.b.}} = \sum_i \frac{G_i^2}{rE}$$

(9.9.12) $$S_{\text{b.i.t.}} = \sum_j \frac{h_j^2}{k} - G$$

The number E, called the *efficiency factor* of the design, is defined by

(9.9.13) $$E = \frac{a(k-1)}{k(a-1)}$$

and is less than 1 because $k < a$. The sum of squares for error, S_e, is obtained by subtraction. The number of degrees of freedom for error is $N - 1 - (a-1) - (b-1) = N - a - b + 1$. The analysis of variance is outlined in the Table 9.9.

TABLE 9.9

Variation	S.S.	D.F.	E(M.S.)
Treatments (eliminating blocks)	$\sum_i G_i^2/(rE)$	$a - 1$	$\sigma^2 + r\sigma_\alpha^2 E$
Blocks (ignoring treatments)	$\sum_j h_j^2/k - G$	$b - 1$	
Error	(by subtraction)	$N - a - b + 1$	σ^2
Total	$\sum_D x_{ij}^2 - G$	$N - 1$	

In the last column of Table 9.9, σ_α^2 is defined as usual (for fixed effects) by

(9.9.14) $$\sigma_\alpha^2 = \sum_i \frac{\alpha_i^2}{a-1}$$

Treatment effects may be tested for significance by the ordinary F-test, and estimated by

(9.9.15) $$\hat{\alpha}_i = \frac{G_i}{rE}$$

If it is desired to test the hypothesis that all the block effects are zero, the numerator of the F-statistic is the mean square for "blocks eliminating treatments," given by $S_{\text{b.e.t.}}/(b-1)$, where

(9.9.16) $$S_{\text{b.e.t.}} = \sum_j \frac{h_j^2}{k} + \sum_i \frac{G_i^2}{rE} - \sum_i \frac{g_i^2}{r}$$

However, the block effects are usually regarded as less important than the treatment effects.

For details of the many other types of experimental design used in practice see reference [9], and for a full treatment of the theoretical considerations involved, reference [7] is invaluable.

9.10 **Departures from the Assumptions Underlying Analysis of Variance Techniques** The assumptions usually made fall under four heads: (a) additivity of effects, which implies an absence of interaction, (b) independence of the error terms, (c) normality of the distribution of error terms, (d) constancy of the variance of error terms, whatever the magnitude of the main effects. Taken together, these constitute a severe restriction on the type of data to which the techniques of analysis of variance are strictly relevant. It is highly desirable to know whether these assumptions can be relaxed appreciably in practice.

As regards *non-normality*, some empirical data on sampling from artificial non-normal populations suggest that the ordinary F-test is fairly robust and can be used without serious error even for considerable variations from normality. Care should be used in claiming significance when the probability is near the border-line, since, on the whole, non-normality tends to make results look more significant than they are.

Non-normality does not introduce any bias into point estimates of parameters (or linear combinations of parameters) or of the components of variance. It does, however, affect the validity of the F-tests, since without the assumption of normality it is not in general true that the mean squares have independent chi-square distributions.

Inferences about the mean μ which are valid for a normal population will also hold for almost any non-normal population as long as the sample is very large. This, however, is not true for the population variance σ^2. If the population has a kurtosis γ_2, the variance of s^2/σ^2, where s^2 is the sample variance, is increased for large N by a factor $1 + \frac{1}{2}\gamma_2$, and this may seriously affect any significance levels or calculations of power obtained from normal theory. Most inferences about variances, including F-tests, are subject to uncertainty due to non-normality, but these effects are less when the design is such that equal numbers of observations occur in each cell of the layout.

As regards *independence*, the simplest reasonable alternative is probably to assume that the observations are *serially correlated.* That is, if the observations, taken successively, are denoted by $x_1, x_2 \ldots$, there is a constant correlation ρ between x_i and x_{i+1}, but all other correlation coefficients are zero. It may be shown that under these conditions

$$(9.10.1) \qquad \begin{cases} E(\bar{x}) = \mu \\ V(\bar{x}) = \dfrac{\sigma^2}{N}\left[1 + 2\rho\left(1 - \dfrac{1}{N}\right)\right] \\ E(s^2) = \sigma^2(1 - 2\rho/N) \end{cases}$$

with ρ taking possible values from $-\frac{1}{2}$ to $\frac{1}{2}$. The probability that a confidence interval with nominal confidence coefficient $1 - \alpha$ does not cover the true value μ is, for large N, not α but $2[1 - \Phi(A)]$ where $A = z_{\alpha/2}(1 + 2\rho)^{-1/2}$, $z_{\alpha/2}$ being the standard normal variate exceeded with probability $\alpha/2$. For $\rho = -\frac{1}{2}$ and $\alpha = 0.05$, the probability $2(1 - \Phi(A))$ is zero and for $\rho = \frac{1}{2}$ it is 0.166, so that there is a chance of rather serious error in ignoring ρ.

The effect of *inequality of variances* in the one-way layout with fixed effects (§ 9.2) depends upon the sample sizes. If these are all equal, the effect is slight, but in general the variance of the ratio of mean squares (between and within samples) is increased, even for large values of N_i, by this inequality. The result is to increase the true probabilities for Type I errors, in the usual F-test for equality of means, beyond the nominal value α. The effect may be quite serious, leading even to a doubling or trebling of the error.

The most common procedure to reduce inequality of variance is to make use of transformations such as those mentioned in §§ 3.15, 3.16 and 4.8. The logarithmic transformation (using the logs of the observations instead of the observations themselves) is appropriate when their *percentage* error is approximately constant. Transformations to reduce inequality of variance often reduce non-normality also, but they may destroy the additivity of effects which existed before the transformation.

9.11 **Estimation of Contrasts** The ordinary F-test in an analysis of variance determines whether there is a significant difference between a group of means, but the investigator really wants to know more than this—which of the means differ significantly from which others. One should not, of course, pick out the two means which differ most (out of k samples, say) and apply the ordinary t-test to these, since the two selected in this way are clearly not chosen at random. There are $k(k - 1)/2$ pairs that might be chosen, of which the selected pair is one.

Any linear combination of treatment means (or other parameters) with coefficients adding up to zero is called a *contrast*. A difference of two means such as $\alpha_1 - \alpha_2$ is a contrast and so is an interaction: $\gamma_{ij} = m_{ij} - m_{i\cdot} - m_{\cdot j} + m_{\cdot\cdot}$.

Fisher has pointed out [10] that when one wishes to test a particular contrast (picked out *after* the results of the experiment are known), the F-test having failed to demonstrate a significant differentiation, one should be very cautious in claiming significance. He suggests that if the contrast is one out of $k(k - 1)/2$ possibilities, we should require significance at the level $\alpha/[k(k - 1)/2]$ instead of α.

As an illustration we may consider the data of Example 1, § 9.1. The F-test gives a non-significant difference between means, but let us nevertheless make a t-test for the two samples 1 and 3, for which the estimated difference is 147. As an independent estimate of the population variance, σ^2, we may use the combined mean square within samples, which is 7619, with 14 d.f. The t-value is $147\ [7619(\frac{1}{5} + \frac{1}{4})]^{-\frac{1}{2}} = 2.20$, and this is significant at the 5% level. The probability of a value numerically as great is about 0.045. If, however, we consider this difference as simply one out of 10, we should require a t-value of 3.33 (corresponding to a probability 0.005) before claiming significance. This would mean a difference of means of at least 222.

In general terms, for k samples each of size N, Fisher's test requires us to calculate a significant difference D given by

(9.11.1) $$D = t_{\varepsilon,\nu}(2M_e/N)^{1/2}$$

where M_e is the mean square for error, with ν degrees of freedom, $\varepsilon = \alpha/[k(k - 1)/2]$, and $t_{\varepsilon,\nu}$ is the upper $\varepsilon/2$ point of the t-distribution for ν degrees of freedom. Then D is the difference which should be regarded as significant at level α. The $100(1 - \alpha)\%$ confidence limits for the difference between two population means are $\bar{x}_1 - \bar{x}_2 \pm D$, where $\bar{x}_1$ and $\bar{x}_2$ are the observed sample means. If only differences as great as D are reported as significant, the expected number of wrong statements per experiment will be α. Thus for $\alpha = 0.05$ we should make only one wrong statement in 20 experiments of the same type.

If the F-test has shown significance and one desires to know which means differ significantly from which others, the *Newman-Keuls procedure* is perhaps the best. Let a be the number of levels of the factor under consideration (say treatments), and let M_e be the mean square used as the denominator of F in testing for this factor and ν the number of degrees of freedom corresponding to M_e. Also let r be the number of observations at each level of the factor. The observed means are arranged in order from the smallest to the greatest, and the test is carried out on sub-groups of p successive means beginning with the whole set (for which $p = a$). For any such group the significance is tested by comparing the observed range with the *critical range*, the latter being obtained from the distribution of the studentized range.

The *studentized range* of p observations, having an actual range R and coming from a population with variance σ^2, is R/s_ν where s_ν^2 is an independent estimate of σ^2 based on ν degrees of freedom. (Note that the *standardized* range is R/σ.) In the analysis of variance the estimate is usually based on the mean square for error.

The probability integral (distribution function) of the studentized range is given by

(9.11.2) $$F(q) = P(R/s_\nu \leq q) = \frac{\nu^{\nu/2}}{\Gamma(\nu/2)2^{(\nu-2)/2}} \int_0^\infty x^{\nu-1}e^{-\nu x^2/2}G(qx)\,dx$$

where $G(w)$ is the probability integral for the standardized range, § 8.21. Tables of percentage points of the studentized range may be found in [11], for the appropriate values of p and ν. If q is the upper 5% point, the critical range is found by multiplying q by s_ν. In the Newman-Keuls procedure the observations are each a mean of r original observations, and the estimated s_ν^2 is therefore M_e/r.

EXAMPLE 3 Table 9.10 gives measurements of tensile strength (kg/cm^2) on specimens of rubber. There were five batches, each batch affording six specimens. The mean, range and variance for each batch are given.

TABLE 9.10

	Batch Number				
	1	2	3	4	5
	177	116	170	181	177
	172	179	156	190	186
	137	182	188	210	199
	196	143	212	173	202
	145	156	164	172	204
	168	174	184	187	198
Mean	165.8	158.3	179.0	185.5	194.3
Range	59	66	56	38	27
Variance	468.6	653.1	406.0	196.3	111.5

The within-batch estimate of variance is $M_e = (1835.5)/5 = 367.1$, with 25 d.f., and $r = 6$, so that $s_v = 7.82$. The value of q for $v = 25$ and $p = 5$ is 4.16, giving a critical range of 32.5.

Arranging the actual means in order, we get

158.3, 165.8, 179.0, 185.5, 194.3

For the whole set, $R = 36.0$, which exceeds the critical range, and we may therefore conclude that there is a significant difference at the 5% level for the whole group of means. This test can therefore be regarded as a substitute for the analysis of variance test. In the present example the mean square between batches is 1273.5 with 4 d.f., and $F = 3.47$. Since the 5% point of F, with 4 and 25 d.f., is 2.76 and the 1% point is 4.18, the observed value is significant at the 5% level, which agrees with the result of the Newman-Keuls test.

We next proceed to omit the largest (or smallest) mean and find the critical range for both the remaining sets of 4 means. If a significant difference is found we test groups of 3 means, and so on. Any time that a group of means is found not to differ significantly it is underlined. Any two means not underlined by the same line differ significantly.

The q values for $p = 2, 3, 4$, with $v = 25$, are as follows (upper 5% points), giving the critical ranges shown:

p	2	3	4
q	2.92	3.52	3.89
c	22.8	27.5	30.4

Omitting 194.3, the range of the means is 27.2, and omitting 158.3, it is 28.5, neither of which is significant. It is not therefore necessary to go any further.

The set of means, underlined, is as shown below:

158.3 165.8 179.0 185.5 194.3

Only the contrast of the first and last is significant at the 5% level.

Applying the same method to the data of Example 1 (and assuming for convenience that the samples are all of the same size 4), we find that the critical range for $p = 5$ is 192 while the actual range is 147. There is therefore no significant difference between the means, as was found previously by the F-test, and the contrast of even the least and greatest is non-significant at the 5% level.

* 9.12 **The Power of Analysis of Variance Tests** In calculating the power of the F-test, it is convenient to make the change of variable

(9.12.1) $$x = \frac{n_1 F}{n_1 F + n_2}$$

where n_1 and n_2 are the degrees of freedom for F. Under the null hypothesis H_0 x is an ordinary beta-variate with parameters $\frac{1}{2}n_1$, $\frac{1}{2}n_2$. The distribution under the alternative hypothesis H_1 was worked out by Tang [12].

If we consider the one-way layout, with k samples each of size r, and if S_b, S_w are the sums of squares between treatments and within samples respectively, we have

(9.12.2) $$x = \frac{S_b}{S_b + S_w}, \qquad n_1 = k - 1, \qquad n_2 = k(r - 1)$$

With the usual notation for fixed treatment effects α_i (with $\sum \alpha_i = 0$), the null hypothesis is that $\sigma_\alpha^2 = 0$, where

(9.12.3) $$\sigma_\alpha^2 = \sum_i \frac{\alpha_i^2}{k - 1}$$

The alternative hypothesis H_1 is that not all the α_i are 0, so that σ_α^2 is not 0. The population variance is assumed constant and equal to σ^2.

Under H_0, S_b/σ^2 has the χ^2 distribution with n_1 d.f. Under H_1 it has the *non-central* χ^2 distribution (§ A.13) with non-centrality parameter

(9.12.4) $$\lambda = r(k - 1)\frac{\sigma_\alpha^2}{2\sigma^2}$$

On the other hand, S_w/σ^2 still has the ordinary χ^2 distribution with n_2 d.f. The density function for x under H_1 is

(9.12.5) $$f(x) = \frac{e^{-\lambda} x^{\frac{1}{2}n_1 - 1}(1 - x)^{\frac{1}{2}n_2 - 1} H(\lambda x)}{B(\frac{1}{2}n_1, \frac{1}{2}n_2)}$$

where

(9.12.6) $$H(x) = 1 + \frac{n}{n_1}\frac{x}{1!} + \frac{n(n - 2)}{n_1(n_1 + 2)}\frac{x^2}{2!} + \ldots$$

with $n = n_1 + n_2$. This function is called the *confluent hypergeometric function.* When $\lambda = 0$, $H(0) = 1$ and x is a beta-variate.

The null hypothesis is rejected when $x > x_\alpha$, x_α being determined for a given Type I error α by the relation

$$\alpha = \int_{x_\alpha}^{1} f(x|\lambda = 0)\, dx \tag{9.12.7}$$
$$= 1 - I_{x_\alpha}(\tfrac{1}{2}n_1, \tfrac{1}{2}n_2)$$

where I is the incomplete beta-function ratio tabulated by K. Pearson (§ 4.5). The power of the test is given by

$$P = 1 - \beta = \int_{x_\alpha}^{1} f(x)\, dx \tag{9.12.8}$$

where $f(x)$ is the function defined in Eq. (5). Tang's tables give values of x_α corresponding to $\alpha = 0.05$ and 0.01 and also the power P for selected values of λ. (Tang denotes our x by E^2, and our λ by $k\phi^2/2$.)

EXAMPLE 4 Suppose we have four treatments, each with five replicates. Then $n_1 = 3$, $n_2 = 16$. (If the experiment were done in randomized complete blocks, we should have to subtract the four degrees of freedom between blocks and so obtain $n_2 = 12$.) Suppose the true treatment effects α_i, expressed as percentages of $\hat{\mu}$, are $-5, -4, 3, 6$, and let an estimate of σ be 10% of $\hat{\mu}$ (obtained perhaps from previous experience). Then $\sigma_\alpha^2 = 86/3 = 28.7$, $\sigma^2 = 100$, $\lambda = 2.15$, $\phi = 1.04$, $n_1 = 3$, $n_2 = 16$. The tables show that for $\alpha = 0.05$, the power P when $\phi = 1.04$ is about 0.31, so that the chance of detecting differences as great as those just mentioned is only about 3 out of 10.

The same method will apply to a Latin square experiment. If the square is of side m, the degrees of freedom are $n_1 = m - 1$, $n_2 = (m - 1)(m - 2)$.

If we use Model II in which the treatment effects are random with variance σ_α^2, the power of the test is a function of $1 + \sigma_\alpha^2/\sigma^2$ (see § 9.5).

Recent investigations on the effect of non-normality suggest that moderate amounts of skewness and kurtosis have comparatively little effect on the power, kurtosis being more important than skewness in this respect.

PROBLEMS

A. (§§ 9.1–9.2)

1. Five samples, each of four seasoned mine-props, were tested for maximum load. The means and standard deviations of the maximum load (in units of 1000 lb wt) were as follows:

Sample No.	$\bar{x}_i$	$s_i = (S_i/n_i)^{1/2}$
1	42.0	10.10
2	52.0	12.06
3	65.5	5.45
4	51.8	9.54
5	73.5	19.26

Test approximately the homogeneity of the variances, using Bartlett's method.

2. Prove that when $k = 2$, the sum of squares between samples reduces to

$$S_b = \frac{N_1 N_2}{N_1 + N_2}(\bar{x}_1 - \bar{x}_2)^2$$

3. Show that the t-test for the significance of a difference between *two* sample means is a special case (when $k = 2$) of the F-test given in § 9.2 for k samples. *Hint:* S_b reduces to the expression given in Problem 2, and S_w becomes $n_1 s_1^2 + n_2 s_2^2$. Then $F = (N - 2)S_b/S_w$ and this is the square of t in § 8.8, when $\mu_1 = \mu_2$.

4. The following table represents the yield of wheat, in bushels per acre, for trial plots of land treated with four different levels of fertilizer. Each level was applied to five plots randomly chosen over a field.

Plot Number	Treatments			
	1	2	3	4
1	21	24	34	40
2	25	33	26	47
3	31	34	38	39
4	17	39	32	41
5	26	35	35	33

Determine whether there is a significant treatment effect.

5. Seed yields in hundreds of pounds per acre from three replicates of four varieties of alfalfa were as follows:

Plots	Varieties			
	1	2	3	4
1	4.7	5.1	9.1	4.9
2	5.2	4.8	6.3	5.2
3	2.4	3.2	5.8	5.3

Does there appear to be a significant difference between varieties?

B. (§§ 9.3–95)

1. The data represent sugar yields (tons/acre) for nine varieties of sugar beet. The design consists of five blocks each of nine plots, with no replications. The varieties were randomized among the plots in each block. Test for significant differences between varieties, and between blocks. *Hint:* The interaction, if any, is included with the error.

Block	Variety								
	A	*B*	*C*	*D*	*E*	*F*	*G*	*H*	*J*
1	1.94	1.70	2.23	2.14	1.80	1.82	1.91	1.90	1.98
2	2.08	1.96	2.26	2.08	2.23	2.06	2.06	2.25	2.03
3	1.86	1.83	2.22	2.16	1.67	2.03	2.22	1.92	1.81
4	2.21	1.60	2.08	2.16	2.11	1.96	2.14	1.99	1.77
5	2.03	2.13	2.02	2.17	2.01	2.28	2.28	2.02	1.88

2. In a greenhouse experiment on wheat, four fertilizer treatments of the soil and four chemical treatments of the seed were used (including in each case a control with no treatment). Each combination was applied to three plots which were placed at random in the available space. Show that there is negligible interaction between chemical treatments and fertilizers, but a large effect due to fertilizers.

Fertilizer	Chemical Treatment			
	1	2	3	4
1	21.4, 21.2, 20.1	20.9, 20.3, 19.8	19.6, 18.8, 16.6	17.6, 16.6, 17.5
2	12.0, 14.2, 12.1	13.6, 13.3, 11.6	13.0, 13.7, 12.0	13.3, 14.0, 13.9
3	13.5, 11.9, 13.4	14.0, 15.6, 13.8	12.7, 12.9, 13.1	12.4, 13.7, 13.0
4	12.8, 13.8, 13.7	14.1, 13.2, 15.3	14.2, 13.6, 13.3	12.0, 14.6, 14.0

3. In an experiment to determine whether five makes of automobile average the same number of miles per gallon, three cars of each make were selected at random in each of three cities and given a test run on one gallon of a standard gasoline. The table gives the number of miles travelled. Make an analysis of variance and determine whether there is a significant effect (a) of makes, (b) of cities.

Make	Los Angeles	San Francisco	Portland
A	20.3, 19.8, 21.4	21.6, 22.4, 21.3	19.8, 18.6, 21.0
B	19.5, 18.6, 18.9	20.1, 19.9, 20.5	19.6, 18.3, 19.8
C	22.1, 23.0, 22.4	20.1, 21.0, 19.8	22.3, 22.0, 21.6
D	17.6, 18.3, 18.2	19.5, 19.2, 20.3	19.4, 18.5, 19.1
E	23.6, 24.5, 25.1	17.6, 18.3, 18.1	22.1, 24.3, 23.8

4. Estimate the separate treatment effects in Problem A-4 and the separate variety effects in Problem A-5, using Model I.

5. Estimate the variety effects and block effects in Problem B-1, using Model I.

7. Prove the following identities:

(a) $\sum_{ij} (\bar{x}_{i..} - \bar{x})(\bar{x}_{ij.} - \bar{x}_{i..} - \bar{x}_{.j.} + \bar{x}) = 0$

(b) $\sum_{ij} (\bar{x}_{.j.} - \bar{x})(\bar{x}_{ij.} - \bar{x}_{i..} - \bar{x}_{.j.} + \bar{x}) = 0$

8. In Problem A-4, assume that the treatments are random selections from a normal population (e.g., samples of fertilizer might differ in the proportions of an active ingredient, and these samples might be chosen at random and applied in equal amounts in each treatment). Calculate the components of variance, the intra-class correlation, and the power of the *F*-test corresponding to a size of 0.05.

9. In an experiment on yield of sugar beets (tons/acre) there were two levels of irrigation treatment and three of fertilizer treatment, and each combination of treatments was carried out in five replications. The analysis of variance table was as follows:

Variation	S.S.	D.F.	M.S.
Irrigation	120.0	1	120.0
Fertilizer	221.7	2	110.9
Interaction	35.0	2	17.5
Error	108.0	24	4.5
Total	484.7	29	

Assuming that it makes sense to regard the irrigation and fertilizer effects as random (Model II), estimate the components of variance.

C. (§§ 9.6–9.8)

1. Regard Problem B-3 as one of mixed type (Model III), with the makes of cars fixed and the cities random. (That is, the results will apply only to the particular makes selected for the experiment, but there may be a wide variety of possible cities chosen for the trials.) Estimate the components of variance, and show that because of the high interaction there is not, on this model, a significant effect as between makes of cars.

2. Show that the sums of squares in Eq. (9.7.7) may be written:

$$S_a = r^{-1} \sum_i (b_i)^{-1} (\sum_{jk} x_{ijk})^2 - N\bar{x}^2$$

$$S_b = r^{-1} \sum_{ij} (\sum_k x_{ijk})^2 - r^{-1} \sum_i (b_i)^{-1} (\sum_{jk} x_{ijk})^2$$

$$S_r = \sum_{ijk} (x_{ijk}^2) - r^{-1} \sum_{ij} (\sum_k x_{ijk})^2$$

3. The following data, from Scheffé [7], purport to represent breaking strengths of tissues from different boxes of the same brand, purchased in three different cities. From each box six tissues were measured. Calculate the mean squares for cities, for boxes within cities, and for tissues within boxes, and test for the reality of a city effect and an effect of boxes within cities.

City	1		2			3			
Boxes	1	2	1	2	3	1	2	3	4
	1.59	1.72	2.44	2.27	2.46	1.36	1.59	1.73	1.53
	1.80	1.40	2.11	2.70	2.21	1.43	1.50	1.74	1.41
	1.72	2.02	2.41	2.36	2.50	1.48	1.50	1.65	1.64
	1.69	1.75	2.48	2.36	2.37	1.55	1.49	1.58	1.51
	1.71	1.95	2.36	2.16	2.24	1.53	1.47	1.49	1.52
	1.83	1.61	2.36	2.04	2.25	1.39	1.63	1.70	1.36

The boxes may be supposed chosen at random from a large number in each city. Treat the cities as random also and estimate the components of variance. *Hint:* Use the computation formulas of Problem 2.

4. In a Latin square layout, let T_c be the sum of squares of the column totals, T_r the S.S. of the row totals and T_t the S.S. of the treatment totals. Prove that S_a, S_b, S_c in Eq. (9.8.4) may be written: $S_a = T_c/m - G$, $S_b = T_r/m - G$, $S_c = T_t/m - G$.

5. The following table gives wheat yields (bushels/acre) for five fertilizer-treatments of plots arranged in a Latin square. Test for significance of row, column and treatment effects, using Model I.

Rows	Columns				
	1	2	3	4	5
1	34(*C*)	21(*A*)	52(*E*)	24(*B*)	40(*D*)
2	33(*B*)	45(*E*)	47(*D*)	26(*C*)	25(*A*)
3	31(*A*)	38(*C*)	34(*B*)	39(*D*)	38(*E*)
4	44(*E*)	41(*D*)	32(*C*)	17(*A*)	39(*B*)
5	33(*D*)	35(*B*)	26(*A*)	46(*E*)	35(*C*)

Hint: Use the computation method suggested in Problem 4.

D. (§§ 9.9–9.12)

1. Five detergents, lettered *A* to *E*, were compared as to number of soiled dinner plates washed in a basin before the foam disappeared from the basin. In each block there were three basins containing three different detergents, and the three dish-washers rotated after washing each plate. Analyze this balanced incomplete block experiment. (Data from Scheffé [7].) Test for differences between blocks as well as between detergents.

Detergent	Block									
	1	2	3	4	5	6	7	8	9	10
A	27	28	30	31	29	30				
B	26	26	29				30	21	26	
C	30			34	32		34	31		33
D		29		33		34	31		33	31
E			26		24	25		23	24	26

2. In Problem B-1, calculate a significant difference between pairs of varieties by Fisher's method, using 0.05 as the value of α. Which pairs appear to be significantly different at this level? *Hint:* To find the $\varepsilon/2$ point for t, use one of the approximations in § 8.6.

3. Apply the Newman-Keuls procedure to the data of Problem B-1, using the 5% level of significance. Find which pairs are significantly different at this level. *Hint:* For $\nu = 32$, the upper 5% values of q for $p = 9, 8, 7$ and 6 are 4.70, 4.58, 4.45 and 4.29 respectively.

4. A group of eight different kinds of alloy steels is to be tested for tensile strength. It is expected that the strengths will be of the order of 150,000 lb/in^2 and that the standard deviation for any one kind will be of the order of 3000 lb/in^2. How many specimens should be used for each kind of alloy if we want the error of the first kind not to exceed

5% and if we would like the probability of rejecting the null hypothesis to be at least 0.9 if in fact two of the alloys differ by 10,000 lb/in² or more? (Scheffé [7]). *Hint:* The minimum value of σ_α^2 satisfying the condition that two α_i differ by 10,000 is given by putting $\alpha_1 = -5{,}000$, $\alpha_8 = 5{,}000$, and all the other $\alpha_i = 0$. This makes $\sigma_\alpha^2 \geq (50{,}000)/7$. By means of charts prepared by Pearson and Hartley [13], suitable values of r and ϕ can be read off. Tang's tables are not so convenient for this purpose.

REFERENCES

[1] Bartlett, M. S., "Properties of Sufficiency and Statistical Tests," *Proc. Roy. Soc. A*, **160,** 1937, 268–282.

[2] Hartley, H. O., "Testing the Homogeneity of a Set of Variances," *Biometrika*, **31,** 1939, 249–255.

[3] Thompson, C. M., and Merrington, M., "Tables for Testing the Homogeneity of a Set of Estimated Variances," *Biometrika*, **33,** 1946, 296–304. These are reprinted in [11], pp. 180–182.

[4] Hald, A., *Statistical Theory with Engineering Applications*, Wiley, 1952. p. 472.

[5] Satterthwaite, F. E., "An Approximate Distribution of Estimates of Variance Components," *Biometrics Bulletin*, **2,** 1946, 110–114.

[6] Johnson, L. P. V., and Keeping, E. S., "Composite Mean Squares and their Degrees of Freedom," *Applied Statistics*, **1,** 1952, 202–205.

[7] Scheffé, H., *The Analysis of Variance*, Wiley, 1959, 238, 261.

[8] Fisher, R. A., and Yates, F., *Statistical Tables for Biological, Agricultural and Medical Research*, 3rd ed., Oliver and Boyd, 1948, Tables XV and XVI.

[9] Cochran, W. G., and Cox, G. M., *Experimental Designs*, 2nd ed., Wiley, 1957, p. 577.

[10] Fisher, Sir R. A., *The Design of Experiments*, 5th ed., Oliver and Boyd, 1949, pp. 57–58.

[11] Pearson, E. S., and Hartley, H. O., *Biometrika Tables for Statisticians*, Cambridge Univ. Press, 1954, pp. 176–177.

[12] Tang, P. C., "The Power Function of the Analysis of Variance Tests, with Tables and Illustrations of Their Use," *Statistical Research Memoirs*, **2,** 1938, pp. 126–157 (University College, London).

[13] Pearson, E. S., and Hartley, H. O., *Biometrika*, **38,** 1951, pp. 115–122.

Chapter 10

NON-PARAMETRIC STATISTICAL TESTS

10.1 **Non-Parametric or Distribution-Free Tests** Many common statistical tests are concerned with estimating parameters in a distribution function of known or assumed form (for the population), or with testing the significance of differences between samples on the hypothesis that they come from such a population. Thus the population may be supposed to be normal, with parameters μ and σ^2, and these may be estimated by the sample statistics m and s^2, or we may use an F-test to decide whether two samples which differ in variance may reasonably be presumed to come from normal populations with the same variance σ^2. Such tests, since they deal with population parameters, are called *parametric*.

There are, however, other tests which do not require assumptions about the parameters of the population from which the sample is drawn. These are called *non-parametric*, or *distribution-free*, since they are free of specific assumptions about the distribution in the parent population. Some assumptions, of course, have to be made—for example, that the observations constituting the sample are independent—but these assumptions are considerably weaker than those required for the usual parametric tests. The most obvious danger in using such tests as the t-test and the F-test is that the underlying assumption of *normality* may not be justified. It is true that these tests appear to be fairly insensitive to considerable departures from normality (they are said to be *robust tests*) but nevertheless for some kinds of data it would be rash to assume anything like a normal distribution and in such cases non-parametric tests should be used.

Non-parametric tests are generally simple to apply, not involving much computation. In those cases where a parametric test would also be applicable, a non-parametric test will naturally be less powerful than the parametric one. However, if it is fairly easy to obtain new observations, the lack of power may be compensated by increase of the sample size, and the non-parametric test appeals just because of its simplicity. In this chapter some of the commoner tests will be described. These tests are particularly interesting to students of the behavioral sciences because so much of the data in psychology, education, etc. is of a kind that can be classified or ranked, but not accurately measured. Good general surveys of non-parametric methods may be found in references [1], [2], [3] and [17].

10.2 **The Chi-Square Test of Hypotheses** Suppose that the members of a population can all be placed in one or other of a set of k categories. These may be nominal like "male" or "female," or may be intervals of the domain of some measured variable like height. Also suppose that according to a certain hypothesis H_0 the probabilities of falling in these classes should be $\pi_1, \pi_2 \ldots \pi_k$. This hypothesis may be tested by observing the actual frequencies $f_1, f_2 \ldots f_k$, in a sample of N items, corresponding to the respective classes. The distribution of the N sample items among the k classes is multinomial (Appendix A.16), and the number of degrees of freedom is $k - 1$, since the k frequencies are connected by the linear relation $\sum f_i = N$. As shown in Appendix A.17, the quantity

$$\chi_s^2 = \sum_{i=1}^{k} \frac{(f_i - N\pi_i)^2}{N\pi_i} \tag{10.2.1}$$

has in the limit the χ^2 distribution with $k - 1$ d.f. It is assumed that the quantities π_i are given by the hypothesis H_0 and are not estimated from the sample. If they have to be estimated, the degrees of freedom are reduced (see § 10.3).

The quantity $N\pi_i$ is the expected frequency in the ith class for a sample of size N, on hypothesis H_0, and may be denoted by ϕ_i. The proof that χ_s^2 is approximately distributed like χ^2 uses Stirling's approximation (Appendix A.2) for $\phi_i!$ and therefore the ϕ_i should not be too small. It has been customary to require that all the ϕ_i should be at least 5, but some studies [4] suggest that values as low as 1 may sometimes be tolerated without causing serious error. In practice, the classes with low expected frequency usually come near the ends of the distribution, and it is common practice to combine or "pool" the end-classes until the ϕ_i reach a satisfactory size. The objection to pooling is that some important differences between f_i and ϕ_i in the end-classes may be hidden by this treatment. Since χ_s^2 depends on the square of $f_i - \phi_i$, the sign of this difference is ignored, and yet it might well be of significance for H_0 if the sign were constant over several classes near the end of the table. For this reason, we recommend that pooling should be done cautiously, only when any expected frequency would otherwise fall close to or below 1. Too much pooling reduces the chance of rejecting H_0 if it really should be rejected.

EXAMPLE 1 The Abbé Mendel, in a now classic experiment on heredity, observed the shape and color of peas from a number of plants in the first-generation progeny of a cross. He found that they could be put into four groups, as follows:

Round and yellow	315
Round and green	108
Wrinkled and yellow	101
Wrinkled and green	32

According to his theory of heredity, these frequencies should be in the ratio 9:3:3:1. The expected frequencies ϕ_i for a total of 556 should therefore be as shown in Table 10.1

TABLE 10.1

f_i	ϕ_i	$f_i - \phi_i$	$(f_i - \phi_i)^2$	$(f_i - \phi_i)^2/\phi_i$
315	312.75	2.25	5.06	0.016
108	104.25	3.75	14.06	0.135
101	104.25	−3.25	10.56	0.101
32	34.75	−2.75	7.56	0.218
556	556	0		0.470

From the data, $\chi_s^2 = 0.470$, with 3 d.f. (since here, $k = 4$). The probability of a value of χ^2 as great as this is about 0.92, so that the agreement of theory and experiment is very good. Considerably larger disagreement might be expected, even when H_0 is true.

Very occasionally, one encounters values of χ_s^2 so low that the corresponding probabilities are as high as 0.99. When these are not due to mistakes in calculation, it may be suspected that the observations are not really random. The hypothesis H_0 should not, of course, be rejected merely because of an agreement that is "too good to be true," but this kind of agreement might well be a ground for critical reappraisal of the data.

10.3 **The Chi-Square Test of Goodness of Fit** An observed frequency distribution in a sample may often, on general theoretical grounds, be supposed to arise from a true binomial, Poisson, normal, or some other known type of distribution in the population. This hypothesis may be tested by comparing the observed frequencies in various classes with those which would be given by the assumed theoretical distribution. Usually, however, the parameters of this distribution will not be known from prior considerations but will have to be estimated from the sample. It may be shown (see for example [5]) that if s parameters are estimated by the method of maximum likelihood, the limiting distribution of χ_s^2 is that of χ^2 with $k - s - 1$ d.f. Each additional parameter estimated from the sample introduces in effect a linear restriction on the variates z_i, namely, $(f_i - \phi_i)/(\phi_i)^{1/2}$, whose squares are added to produce χ_s^2, and so reduces the degrees of freedom by 1. The estimators used for the parameters need not be the maximum likelihood ones, as long as they are asymptotically normal and asymptotically most efficient (see § 5.6).

EXAMPLE 2 Rutherford and Geiger (*Phil. Mag.* **20**, 1910, p. 698) obtained the following distribution of the number (x) of α-particles emitted from a disc in 7.5 sec.

Assuming that the distribution is Poisson with parameter μ, the maximum likelihood estimator of μ is the arithmetic mean $\bar{x}$, which is 3.870. The calculated frequency ϕ for any given x is $Ne^{-\hat{\mu}}\hat{\mu}^x/x!$ with $\hat{\mu} = 3.870$ and $N = 2608$. The last two values of ϕ in Table 10.2 have been pooled to give a total greater than 1. For 13 classes, with one parameter estimated from the sample, the number of degrees of freedom is 11 and the probability P of a value of χ^2 as great as 12.99 is about 0.30. The hypothesis that the distribution in the population is Poisson is therefore not rejected by this test.

TABLE 10.2

x	f	ϕ	$(f - \phi)^2/\phi$
0	57	54.40	0.124
1	203	210.52	0.269
2	383	407.36	1.457
3	525	525.50	0.001
4	532	508.42	1.094
5	408	393.52	0.533
6	273	253.82	1.450
7	139	140.32	0.012
8	45	67.88	7.713
9	27	29.19	0.164
10	10	11.30	0.150
11	4	3.97	0.000
12	2 } 2	1.28 } 1.80	0.022
≥ 13	0 }	0.52 }	
	2608	2608.00	12.99

* 10.4 **The Power of the Chi-Square Test** The power function of the χ^2 test cannot be computed unless a specific alternative hypothesis H_1 is considered. We might for instance suppose in Example 2 that the distribution, instead of being Poisson, is really binomial, and then we could find the probability of rejecting H_0 when it *should* be rejected. Another alternative hypothesis might be that the observations are individually Poisson, but with means that vary in some systematic way.

For very large samples, the chi-square test will usually reject the hypothesis that the distribution follows some simple assumed law, because small variations from this law will tend to show up in so large a sample. With small samples the test is not very sensitive. There is a parametric test of the Poisson distribution which also depends on χ^2, and which is more powerful than the ordinary chi-square test. In a Poisson population the expectation and the variance are equal, and in a sample of size N from such a population the ratio of ns^2/m (where m and s^2 are the sample mean and variance respectively and $n = N - 1$) is distributed as χ^2 with n degrees of freedom. In Example 2 above, the ratio of variance to mean is 0.95, and ns^2/m is 2476. For such a large value, the normal approximation to χ^2 is adequate (§ 4.6), and we find that the probability of a value as low as this with 2607 d.f. is about 0.033. The variance is therefore significantly lower than the mean, and we might be inclined to reject the hypothesis of a Poisson distribution on account of this test, whereas the ordinary chi-square test would lead to acceptance.

10.5 **The Chi-Square Test for a Grouped Distribution** If a measured variate X, which may be supposed to have a continuous distribution in the population, is grouped into classes in a sample, the expected frequencies corresponding to these classes may be calculated if we are prepared to make some assumptions about the population. Generally, we assume that the population distribution follows some relatively simple and plausible law, such as the normal law or one of the other Pearson types, with parameters that are estimated from the sample itself. The agreement between the observed and calculated frequencies may then be tested by the chi-square test. The number of degrees of freedom is $k - s - 1$, where k is the number of classes in the sample (after pooling if necessary) and s is the number of parameters estimated from the sample. (One further degree of freedom is lost because of the forced agreement between the total frequencies, calculated and observed.)

EXAMPLE 3 The data of Table 2.2 may perhaps arise from a normal distribution of weights in the population sampled (eight-year-old Glasgow girls). If they do so, unbiased estimates of the expectation μ and the variance σ^2 for this distribution are provided by the statistics k_1 and k_2 which were calculated in §5.10, namely, 47.71 lb and 33.34 lb^2. In order to find the expected frequencies we need to obtain the standardized z-values corresponding to the class boundaries, these values being given by

$$z = \frac{x_e - \mu}{\sigma} = \frac{x_e - 47.71}{5.774} \tag{10.5.1}$$

For each z a table of the normal law gives the probability of a value not greater than z, i.e.,

$$\Phi(z) = \int_{-\infty}^{z} \phi(u)\, du \tag{10.5.2}$$

TABLE 10.3

Class Boundary (x_e)	Observed Frequency (f_i)	z_i	$\Phi(z_i)$	$\phi_i = N\,\Delta\Phi(z_i)$
$(-\infty)$		$(-\infty)$	0.0000 }	2.5
31.5 lb	1	−2.808	0.0025 }	
35.5	14	−2.115	0.0172	14.7
39.5	56	−1.422	0.0775	60.3
43.5	172	−0.729	0.2330	155.5
47.5	245	−0.0367	0.4854	252.4
51.5	263	0.656	0.7441	258.7
55.5	156	1.349	0.9113	167.2
59.5	67	2.042	0.9774	68.1
63.5	23	2.734	0.9969	17.5
67.5	3	3.427	0.9997 }	3.1
(∞)		(∞)	1.0000 }	
	1000			1000.0

The difference between successive values of $\Phi(z)$, denoted in Table 10.3 by $\Delta\Phi(z)$, are the probabilities, for a random item, of falling in the corresponding classes, so that the expected frequencies are given by

$$\phi_i = N\Delta\Phi(z_i) \tag{10.5.3}$$

These are calculated in Table 10.3.

The first class includes all values of z from $-\infty$ up to -2.808; the last class includes all values from 2.734 up to ∞ (although actually no values were observed beyond 3.427). From the columns for f_i and ϕ_i we obtain

$$\chi_s^2 = 6.78 \text{ with } 10-3 = 7 \text{ d.f.}$$

There is probably no real need to pool, although many writers would advocate pooling the first two and the last two classes. If this is done χ_s^2 becomes 4.82 with only 5 d.f. The value of P with either procedure turns out to be about 0.45. The conclusion is that the normal curve is a good fit to the data; the hypothesis of a normal distribution is certainly not rejected by this test.

Here again, other tests of normality than the chi-square test are possible. One test is based on the observed sample skewness and kurtosis, both of which should, of course, fluctuate around zero (the population values). For this sample we find $g_1 = 0.114$ and $g_2 = 0.104$. For samples of size 1000 from a normal population, the standard errors of g_1 and g_2, found from Eqs. (8.18.4) and (8.18.5), are 0.077 and 0.154 respectively, so that the observed values differ from zero by 1.47 and 0.68 times their standard errors. The probabilities of discrepancies numerically as great as these are 0.14 and 0.50, and therefore, even by these tests, the assumption of normality is justified.

10.6 **The Kolmogorov Test** Like the chi-square test, this test is one of agreement between an empirical distribution and an assumed theoretical one, but it is based on the *cumulative* distribution function rather than on the frequencies in the separate classes. It may also be used to test whether two samples may reasonably be regarded as coming from the same population.

For the one-sample test, suppose that H_0 specifies a distribution function $F(x)$ for the variate x. Also suppose that $S_N(x)$ is the observed cumulative relative frequency in a sample of N corresponding to any given x; that is, if the number of observations $\leq x$ is k, then

$$S_N(x) = k/N \tag{10.6.1}$$

We should expect that if H_0 is true, $S_N(x)$ will be a fairly good approximation to $F(x)$, and will be better as N increases. In fact, according to the strong law of large numbers, $S_N(x)$ tends to $F(x)$ with probability 1. The test function is D_N, the least upper bound (practically, the maximum) of the absolute deviation of $S_N(x)$ from $F(x)$:

$$D_N = \underset{(x)}{\text{l.u.b.}} \left|S_N(x) - F(x)\right| \tag{10.6.2}$$

and the usefulness of the test depends on the fact that the distribution of D_N does not depend on the form of $F(x)$, as long as $F(x)$ is continuous. We can therefore take $F(x) = x, 0 \le x \le 1$.

It was proved by Kolmogorov, and the proof was simplified by Feller [6], that for any given $\lambda > 0$,

$$\lim_{N\to\infty} P(N^{1/2}D_N \ge \lambda) = L(\lambda) \tag{10.6.3}$$

where

$$L(\lambda) = 2\sum_{n=1}^{\infty}(-1)^{n+1}e^{-2n^2\lambda^2}$$

Thus for $\lambda = 1.36$, $L(\lambda) = 0.05$. The asymptotic probability that D_N exceeds $1.36N^{-1/2}$ is therefore 0.05.

Critical values for D_N, for small values of N, were calculated by Massey [7], and a table is given in Appendix B.6. This table gives the values which are *exceeded* with the given probabilities and so corresponds to the upper tail of the distribution of D. For values of N larger than 35 the asymptotic distribution of Eq. (10.6.3) may be used. This is given in the last line of the table.

EXAMPLE 4 (Miller [8]) Can the following five numbers be regarded as a random choice from the interval 0 to 1: 0.52, 0.65, 0.13, 0.71, 0.58?

Here $S_5(x)$ is a step function with steps of equal height at the observed values of x (see Figure 45). The graph of $F(x)$ is a straight line from the origin to (1, 1).

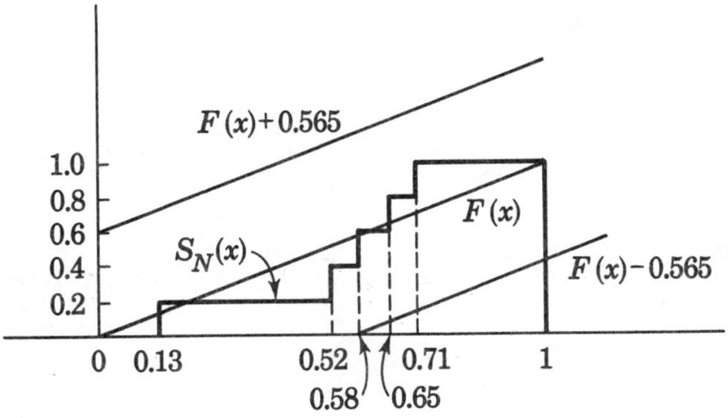

FIG. 45 KOLMOGOROV TEST

The maximum absolute deviation is 0.32, so that the probability is more than 20.0 that this value of D_5 would be exceeded if H_0 were true. There is therefore no reason to reject H_0.

For $N = 5$ there is a probability 0.05 that $D > 0.565$. We should therefore expect that in about 95% of trials a random sample of 5 from a *uniform* distribution on the interval (0, 1) will give a step function lying inside the band bounded by $F(x) = x \pm 0.565$.

Corresponding to an *observed* step function and a given N, we can form a *confidence belt*, by drawing the curves for $S_N(x) \pm D_N$, where D_N is the critical value for the given N and a given level of significance. Any theoretical $F(x)$ which lies wholly within the belt will not be rejected by the data, at this level.

EXAMPLE 5 In Table 10.4, X is the logarithm of soil resistance (in ohms) at a certain depth and k is the cumulative frequency of observations.

TABLE 10.4

x	k	$S_N(x)$
1	0	0
1.699	2	0.056
2	11	0.306
2.699	18	0.500
3	19	0.528
3.699	22	0.611
4	23	0.639
4.477	27	0.750
5	30	0.833
5.699	34	0.944
6	36	1.000

For $N = 36$ we may take $D_N = 0.23$ for a 95% confidence belt. The belt is drawn in Figure 46. Any $F(x)$ which lies wholly within the belt could be accepted with 95% confidence.

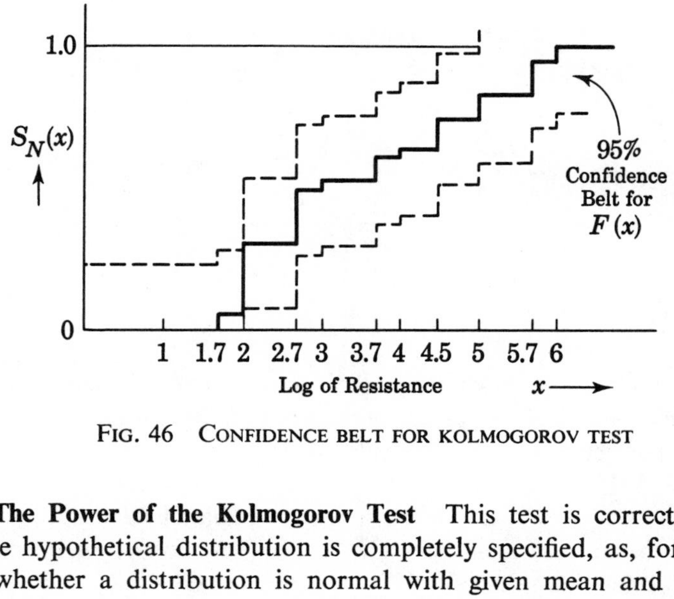

FIG. 46 CONFIDENCE BELT FOR KOLMOGOROV TEST

* 10.7 **The Power of the Kolmogorov Test** This test is correctly used only when the hypothetical distribution is completely specified, as, for instance, in testing whether a distribution is normal with given mean and variance. In

practice, when normality is tested, the mean and variance are generally estimated from the sample itself. The effect of this on the χ^2 test is merely to reduce the degrees of freedom, but the effect on the Kolmogorov test is not precisely known. It may be expected that the general effect will be to reduce the critical level of D, so that the use of the tabulated values in such cases will be conservative.

In order to obtain the power of the Kolmogorov test, we need to have an alternative hypothesis H_1 to the hypothesis H_0 under examination. If H_0 states that the population distribution function is $F_0(x)$ and H_1 states that it is $F_1(x)$ and if the maximum absolute difference between these is δ, then, as Massey [7] has shown, the power of the test is at least

$$1 - \Phi(2\delta N^{1/2} + 2D_N) + \Phi(2\delta N^{1/2} - 2D_N)$$

where D_N is the critical value corresponding to the level of significance α. The actual power is likely to be considerably greater than this.

The power of the χ^2 test in general is not known, but in some cases where comparison with the Kolmogorov test is possible, it appears that the latter is much the more powerful of the two. The least maximum absolute deviation of the true distribution function $F_1(x)$ from an assumed distribution function $F_0(x)$, which will lead to rejection of the latter with probability 0.50, has been calculated for both tests, at the 5% and 1% significance levels, and is smaller for the Kolmogorov test by a factor of nearly 2 (for N between 200 and 2000).

The Kolmogorov test is not applicable to discrete variates, whereas the χ^2 test can be used for these. In other respects the former test seems to have considerable advantages.

10.8 **The Kolmogorov-Smirnov Test for Two Samples** This test is concerned with the agreement between two sets of observed values, and the null hypothesis is that the two samples come from populations with the same distribution function $F(x)$. The test statistic is

$$D_{mn} = \underset{(x)}{\text{l.u.b.}} \left|S_m(x) - S_n(x)\right| \tag{10.8.1}$$

where m and n are the sample sizes and $S_m(x)$ has the same meaning as before.

If it is desired to test the null hypothesis against the alternative hypothesis that the first sample comes from a population in which $F(x)$ is *greater* than it is for the same x in the second population difference, we should use the actual $S_m(x) - S_n(x)$ instead of its absolute value.

The asymptotic distribution of D was worked out by Smirnov, who showed that, as long as $F(x)$ is continuous,

$$\lim_{m,n\to\infty} P(N^{1/2}D_{mn} \geq \lambda) = L(\lambda) \tag{10.8.2}$$

where $L(\lambda)$ is defined as in Eq. (10.6.3), $N = mn/(m + n)$, and it is supposed that m and n both tend to ∞ in such a way that the ratio m/n is finite.

A table of probabilities that $D \leq k/n$, for the case $m = n$ has been compiled by Massey [9] and is the basis for the following table of critical values. The null hypothesis is rejected at significance level $\leq \alpha$ if, for samples of size n, the maximum difference between the cumulative frequencies for any x is k or more.

TABLE 10.5

$\alpha = 0.05$				$\alpha = 0.01$			
n	k	n	k	n	k	n	k
4	4	19	9	5	5	19	10
5	5	20	9	6	6	20	11
6	5	21	9	7	6	21	11
7	6	22	9	8	7	22	11
8	6	23	10	9	7	23	11
9	6	24	10	10	8	24	12
10	7	25	10	11	8	25	12
11	7	26	10	12	8	26	12
12	7	27	10	13	9	27	12
13	7	28	11	14	9	28	13
14	8	29	11	15	9	29	13
15	8	30	11	16	10	30	13
16	8	35	12	17	10	31	13
17	8	40	13	18	10	32	13
18	9						

For large values of m and n, the values of λ in Eq. (2) which would justify rejection of the null hypothesis at level of significance α are given in Table 10.6, calculated by Smirnov [10].

TABLE 10.6

α	0.10	0.05	0.025	0.01	0.005	0.001
λ	1.22	1.36	1.48	1.63	1.73	1.95

EXAMPLE 6 Suppose that in two samples of sizes 55 and 60 respectively we find on drawing the cumulative step functions $S_m(x)$ and $S_n(x)$ that the maximum absolute deviation is 0.25. Then $N^{1/2}D_{mn} = [(55)(60)/115]^{1/2}(0.25) = 1.34$. The probability of a value as great as this is a little more than 0.05, so that the difference is not quite large enough to reject the null hypothesis at the 0.05 level.

10.9 **The Sign Test for Paired Samples** The ordinary t-test for the significance of an observed effect in paired samples (§ 8.9) assumes that all the paired differences can be regarded as independently and normally distributed with a common variance. Sometimes the pairs are observed under widely different conditions and the assumptions of normality and of common variance seem unwarranted. If so, the sign test may be used. This test is very simple to apply and merely assumes that the median of the population of differences is μ, so that the probability that an observed $d_i > \mu$ is the same as the probability that $d_i < \mu$.

The null hypothesis is that $\mu = 0$ and the alternative hypothesis (for a one-tailed test) is that $\mu > 0$, or that $\mu < 0$. For a two-tailed test the alternative is that $|\mu| > 0$.

On the null hypothesis, the expected numbers of positive signs and of negative signs among the differences in a sample of N pairs will be $N/2$. The sampling distribution of the number of positive (or negative) signs will be binomial with $\theta = \frac{1}{2}$. A table of cumulative binomial probabilities for $\theta = \frac{1}{2}$ is included in the Appendix, Table B.7. This gives, for N between 5 and 25 inclusive, the probabilities of occurrence of r or fewer successes, where $r \leq N/2$. For the two-tailed test, the probabilities should be doubled.

In practice, we let r be the number of less frequent signs among the differences d_i $(= x_{1i} - x_{2i})$, so that the condition $r \leq N/2$ is satisfied. If any observed differences happen to be exactly zero, they are not counted, and the sample size is correspondingly reduced.

EXAMPLE 7 The data in Table 10.7 represent yields in bushels for two varieties of apples, A and B, each pair of trees being planted near together under similar conditions of soil, moisture, etc. The separate pairs are, however, scattered over various localities.

TABLE 10.7

$x_1(A)$	$x_2(B)$	$x_1 - x_2$	$x_1 - (x_2 + 1)$
13	16	−3	−4
12	11	1	0
10	8	2	1
6	6	0	−1
13	12	1	0
15	15	0	−1
19	14	5	4
10	9	1	0
11	8	3	2
11	11	0	−1
13	13	0	−1
9	10	−1	−2
14	12	2	1
12	11	1	0
12	9	3	2

In the column of differences, $x_1 - x_2$, there are two minus signs in 11 non-zero items. The probability of two or fewer minus signs is 0.0327, so that the difference is significant, if we assume that A's median yield is certainly not lower than B's. If the difference could be either way, the probability is doubled, and the observed difference would not be significant at the 5% level.

When, as in this case, we are dealing with a type of measurement which has a well-defined unit and a zero, we can employ the sign test to decide whether or not the true difference reaches a certain value. Thus, in Table 10.7, if we add one bushel to each of the x_2 values and re-compute the differences, we get Column 4, which has six minus signs and five plus signs. The value of r is now 5 and the corresponding probability is 0.500. The difference between the yields of A and B is therefore not as great as one bushel, in favour of A.

For values of N larger than 25, the normal approximation to the binomial may be used. The probability of r or fewer successes is approximately $\Phi(z)$, where

$$z = \frac{2r + 1 - N}{N^{1/2}} \tag{10.9.1}$$

The power-efficiency of the sign test is about 95% for $N = 6$, but diminishes as N increases to an asymptotic value of $2/\pi = 63\%$. This means that the sign test has about the same power for a sample of size 100, say, as the most powerful test against the same alternative for a sample of size 63. The most powerful test would in fact be the Student t-test, provided that the assumptions for this test are met. The sign test has the advantage that it can be used in circumstances where the t-test is not applicable. For samples of size 13 the efficiency is still about 75%, so that there is comparatively little loss of power in using the simpler sign test for samples of moderate size, even though a t-test could legitimately be used.

10.10 **The Wilcoxon Signed-Rank Test** This is another test used on matched pairs, [16], more powerful than the sign test because it gives more weight to large numerical differences between the members of a pair than to small differences. The $2N$ subjects are divided into N pairs, each pair as evenly matched as possible. If the effect of a difference of treatments is to be investigated, the choice as to which member of any pair has treatment A and which has treatment B is made at random. The assumptions are that the differences are independent continuous variates from symmetrical populations with common mean μ. The null hypothesis is that $\mu = 0$ and the alternative hypothesis is one of: $\mu > 0$, $\mu < 0$, or $|\mu| > 0$.

The observed differences $d_i = x_{1i} - x_{2i}$ (where x_1 refers to A and x_2 to B) are ranked in increasing order of absolute magnitude and the sum of the ranks is computed for all the differences of like sign. The test statistic T is the smaller of these two rank-sums (one for positive d_i and one for negative d_i). Pairs with $d_i = 0$ are not counted.

Since the variate is supposed continuous, ties should occur only rarely, but will sometimes happen because of the limited accuracy of measurement. If two or more of the d_i have the same magnitude they are given a rank which is the average of the ranks they would have had if they had differed slightly. Thus if the three numerically lowest values of d_i happened to be -1, -1 and 1, they would all be given rank 2, which is the mean of ranks 1, 2 and 3. (The signs are disregarded.)

On the null hypothesis, the expected values of the two rank-sums would be equal. If the positive rank-sum is the smaller, and is equal to or less than the value given for the appropriate N in Table 10.8, the null hypothesis will be rejected at the corresponding level of significance α, in favour of the alternative hypothesis that $\mu > 0$. If the negative rank-sum is the smaller, the alternative will be that $\mu < 0$. If a two-tailed test is required, the alternative being that $|\mu| > 0$, the given levels of significance should be doubled.

TABLE 10.8[a]

N	$\alpha = .025$	$\alpha = .01$	$\alpha = .005$
6	0	–	–
7	2	0	–
8	4	2	0
9	6	3	2
10	8	5	3
11	11	7	5
12	14	10	7
13	17	13	10
14	21	16	13
15	25	20	16
16	30	24	20
17	35	28	23
18	40	33	28
19	46	38	32
20	52	43	38
21	59	49	43
22	66	56	49
23	73	62	55
24	81	69	61
25	89	77	68

[a] Adapted from Table I of reference [16] with the kind permission of the author, F. Wilcoxon, and the publishers, American Cyanamid Co.

For larger values of N, T is approximately normally distributed with mean and variance given by

$$(10.10.1) \qquad \begin{cases} \mu = \dfrac{N(N+1)}{4} \\[2ex] \sigma^2 = N(N+1)\dfrac{2N+1}{24} \end{cases}$$

This means that $[|T - N(N+1)/4| - \frac{1}{2}]/\sigma$ is approximately a standard normal variate. Thus at the 0.025 significance level, for which $z = 1.96$, with $N = 25$, we find $T \approx 162 - 1.96(1381)^{1/2} = 89.1$, in agreement with the last line of Table 10.8.

EXAMPLE 8 For the data of Table 10.7, the ranks of $x_1 - x_2$ (excluding the zero values) are as shown below:

d_i	-3	1	2	1	5	1	3	-1	2	1	3
rank	9	3	$6\frac{1}{2}$	3	11	3	9	3	$6\frac{1}{2}$	3	9

The sum of ranks for the two negative d_i is 12 and for the nine positive d_i is 54. Therefore $T = 12$ and $N = 12$. The hypothesis that the expectation of d_i is *positive* is therefore acceptable at the 2.5% level. The hypothesis that the expectation is *not zero* is acceptable at the 5% level. The latter decision (based on the two-tailed test) is the one that would normally be taken unless there is good reason to believe, before the data are obtained, that if there is any difference it can only be in one direction.

The power-efficiency of the Wilcoxon test is remarkably high. Asymptotically, it is $3/\pi = 95.5\%$, as compared with the t-test, in circumstances where both tests would be applicable.

* 10.11 **The Walsh Test** This is a test with assumptions similar to those for the Wilcoxon signed-rank test, but depending on the averages of pairs of differences. The differences d_i are arranged in increasing order, taking account of sign. The null hypothesis is that the median μ of all these differences is zero. The alternative (two-tailed) hypothesis is that this median is not zero.

The test statistics used are various combinations of the differences. Thus, for $N = 5$, with a two-tailed test, we should reject H_0 at the level $\alpha = 0.125$ if either $\frac{1}{2}(d_4 + d_5) < 0$ or $\frac{1}{2}(d_1 + d_2) > 0$. We should reject at the level $\alpha = 0.062$ if either $d_5 < 0$ or $d_1 > 0$. If we felt *in advance* that μ was bound to be negative if not zero, we could reject H_0 at the level 0.062 if $\frac{1}{2}(d_4 + d_5) < 0$ or at the level 0.031 if $d_5 < 0$. Table B.8 in the Appendix, from Walsh [11], gives for values of N from 5 to 15 the various tests which may be applied at the significance levels indicated, for both one-tailed and two-tailed tests.

Some of these tests are equivalent to the Wilcoxon signed-rank test, but others are not. The efficiency of the tests is, for the most part, from 95 to 99%, and is nowhere below 87.5% (for the first test when $N = 10$, one-tailed).

EXAMPLE 9 Can the following seven observations, arranged in order of size, be considered as coming from populations with the common median 0.5?

$$-2.5, \quad -1.5, \quad -1.3, \quad -0.1, \quad 0.4, \quad 0.7, \quad 0.8$$

If we subtract 0.5 from each of these values, the null hypothesis will be that $\mu = 0$. The values are

d_1	d_2	d_3	d_4	d_5	d_6	d_7
-3.0	-2.0	-1.8	-0.6	-0.1	0.2	0.3

For $N = 7$, there are eight tests altogether, summarized below:

Sig. level (α)	*Tests*	
0.109	$\max[d_5, \frac{1}{2}(d_4 + d_7)] = -0.1,$	$\min[d_3, \frac{1}{2}(d_1 + d_4)] = -1.8$
0.047	$\max[d_6, \frac{1}{2}(d_5 + d_7)] = 0.2,$	$\min[d_2, \frac{1}{2}(d_1 + d_3)] = -2.4$
0.031	$\frac{1}{2}(d_6 + d_7) = 0.25$	$\frac{1}{2}(d_1 + d_2) = -2.5$
0.016	$d_7 = 0.3$	$d_1 = -3.0$

Only the first test, at level $\alpha = 0.109$, leads to rejection of H_0, since the value -0.1 is negative. At the 0.047 level we should accept H_0.

Using the Wilcoxon test on the same data, we have $T = 5$, $N = 7$, which would lead us to accept H_0 at any level up to 0.05. This does not contradict the Walsh test, but the latter is more informative.

10.12 **The Mann-Whitney U-test** This is a test of the null hypothesis that two independent samples A and B come from populations α and β with the same distribution. The alternative (one-tailed) hypothesis is that the variate values in population α are stochastically larger than those in β (or, of course, smaller). This means that if a is any item from α and b any item from β, the probability that $a > b$ (or in the other case that $a < b$) is greater than 0.5. If this is so, the "bulk" of population α has larger (or smaller) variate values than the bulk of population β. As before, a two-tailed test may be used, the alternative hypothesis being that $P(a > b)$ is *not* 0.5.

Let us choose for sample A the one with the smaller size, if the sizes differ. If N_1 and N_2 are the sample sizes and $N = N_1 + N_2$, we rank the *combined* samples in increasing order and then find the sum of ranks, R_1 and R_2, for the two samples separately. The sum $R_1 + R_2$ must be equal to $N(N + 1)/2$, and this fact serves as a check.

The test statistic is given by the smaller of the two quantities:

$$\text{(10.12.1)} \qquad \begin{cases} U = N_1 N_2 + \dfrac{N_1(N_1 + 1)}{2} - R_1 \\ U' = N_1 N_2 - U \end{cases}$$

An equivalent test using R_1 was first proposed by Wilcoxon, but Mann and Whitney gave a more complete treatment, with more extensive tables.

This quantity U is equal to the number of times that an item in A precedes in the ranking an item in B. U' is the number of times that an item in B precedes an item in A. If $P(a > b)$ is large, most items in B will have lower ranks than most items in A and U will be small. The smallness of U determines whether the null hypothesis should be rejected.

EXAMPLE 10 The values of x for two samples are as shown:

Sample A ($N_1 = 8$)		*Sample B* ($N_2 = 9$)	
x_1	*Rank*	x_2	*Rank*
15.2	7	11.5	3
8.6	1	12.6	5
9.3	2	19.4	13
14.4	6	21.3	14
15.6	8	32.5	17
11.8	4	18.6	12
16.3	9	17.0	10
17.8	11	23.4	15
		29.6	16
	$R_1 = 48$		$R_2 = 105$

Here $N = 17$, $\frac{1}{2}N(N + 1) = 153 = R_1 + R_2$.
From Eq. (1), $U = 60$, $U' = 72 - 60 = 12$.
The first item in B (with rank 3) precedes six items in A, the second item precedes five items in A, and the seventh item (with rank 10) precedes one item in A. None of the other items in B precedes anything in A. The total of precedences is 12, agreeing with the calculation of U'. For values of N_1 and N_2 which are moderately large (say 9 or more) the sampling distribution of U (or U') is approximately normal, with mean and variance given by

$$\text{(10.12.2)} \qquad \begin{cases} \mu = \dfrac{N_1 N_2}{2} \\[2ex] \sigma^2 = \dfrac{N_1 N_2 (N + 1)}{12} \end{cases}$$

This implies that the variate

$$\text{(10.12.3)} \qquad z = \frac{|U - \mu| - \frac{1}{2}}{\sigma} = [|N_1(N + 1) - 2R_1| - 1]\left[\frac{N_1 N_2 (N + 1)}{3}\right]^{-1/2}$$

is approximately a standard normal variate.

For small values of N_1 and N_2, special tables due to Mann and Whitney [12] must be used. These give the probabilities of a value of U less than or equal to that observed. For $N_2 \geq 9$, Table B.9 in the appendix may be used. This gives for selected significance levels, and for selected values of N_1 and N_2, critical values of U. Observed values of U (or U') less than or equal to the tabular value are cause for rejection of H_0 at the level quoted. The level should be doubled for a two-tailed test.

In the example above, $N_1 = 8$, $N_2 = 9$, $U' = 12$. The critical values for $\alpha = .05$ and $.01$ are 18 and 11 respectively. The observed U' is therefore significant at the 0.05 level, and almost significant at the 0.01 level, for a one-tailed test.

The normal approximation gives $\mu = 36$, $\sigma^2 = 108$, $z = -2.26$, corresponding to a probability of 0.01. The sample B apparently comes from a population β with on the whole significantly higher x-values than population α, at about the 1% level of significance.

If, before obtaining the samples, we admitted the possibility that either population might have on the whole higher values than the other, we should use a two-tailed test and say that α and β differ at the 2% level of significance.

When ties occur they are treated as in the Wilcoxon test. The effect of ties is to reduce somewhat the value of σ^2 in the normal approximation. If there are t observations tied for a particular rank and if $T = (t^3 - t)/12$, the corrected σ^2 is given by

$$\sigma^2 = \frac{N_1 N_2}{N(N-1)}\left[\frac{N(N^2-1)}{12} - \sum T\right] \tag{10.12.4}$$

the sum being taken over all groups of tied observations. In most cases the correction makes little practical difference.

If applied to data which could be analyzed by the parametric t-test, the Mann-Whitney test has a high power-efficiency, close to 95% and asymptotically equal to $3/\pi$. For some distributions it is even superior to the t-test in its power to reject H_0.

10.13 **Tests of Randomness** It is sometimes desirable to test whether a series of observations can be regarded as random. The residuals in a time series, after removing a trend, may be so tested, for example. According to von Mises, the criterion as to whether a series is random or not is that the relative frequencies in any sub-series of the given series shall be the same as in the original series, providing that the series is very long and that the sub-series is picked out by some pre-assigned rule. The sub-series could consist, for instance, of every third term, or every term corresponding to a prime number, but the rule would obviously have to be independent of the nature of the terms picked. In a series consisting of zeros and ones, the rule could not be to pick all the zeros.

This criterion is clearly not a practical one for testing the randomness of a given finite series. Various tests are actually used in such a case. One is to determine the relative frequencies of different kinds of terms. In the series of digits of the decimal expansion of π (3.14159 . . .), now known to 10,000 places, one can count the numbers of 0's, 1's, 2's, etc. If the distribution were random, one would expect equal numbers of each digit. The actual distribution of 10,001 digits is as shown in the following table:

TABLE 10.9

Digit	0	1	2	3	4	5	6	7	8	9
Frequency	961	1008	1000	1001	1011	1031	1026	1000	953	1010

The χ^2 test for agreement between the observed and calculated values gives $\chi_s^2 = 6.21$ for nine d.f., so that $P = 0.7$. The hypothesis of a random distribution of the digits is certainly not to be rejected, as far as this test goes.

The frequency of *pairs* of digits can also be compared with expectation, or, in sets of four consecutive digits, the frequencies of four or three of a kind, two pairs, etc. (the so-called "poker" test), can be used. The "gap" test uses the average separation between zeros. All these tests are normally made on sets of random numbers produced by some mechanical process, and intended to be used for randomizing in experimental work. It will of course occasionally happen that some sub-group of random numbers will fail to pass a particular test. This group should not be used by itself, but may be quite satisfactory as part of a larger group.

* 10.14 **Runs Up and Down** Several tests of randomness are based on the occurrence of runs in a series. Suppose the observed series consists of numbers $x_1, x_2, x_3 \ldots x_N$, and consider the *signs* of the differences $d_1 = x_2 - x_1$, $d_2 = x_3 - x_2 \ldots d_{N-1} = x_N - x_{N-1}$. The sequence of $N - 1$ signs may look something like this:

$$S: + + - + - - - - + - + + - \ldots .$$

The plus signs correspond to *runs up* in the original series (increasing values of x), the minus signs correspond to *runs down* (decreasing values). In the illustration there are runs up of length 2, 1, 1, and 2, and runs down of length 1, 4, 1 and 1. In this notation three consecutive increasing x terms give a run up of length 2.

Tests of randomness have been suggested based on the total number of runs (r), whether up or down, on the number of plus signs (k) in the sequence S, and on various other properties of the sequence. Moore and Wallis [13] have described such tests, and Levene [14] has discussed their power function.

We may assume that the original observations are all distinct. If two consecutive items happen to be equal, we must suppose that with more accurate measurement they would differ and the difference would be equally likely to be plus or minus. The run lengths are reckoned on both suppositions, and the corresponding probabilities calculated.

The total number of permutations of the N numbers $x_1, x_2 \ldots x_N$ is $N!$ The number producing exactly k positive differences is

$$\phi_N(k) = \sum_{i=0}^{k} (-1)^i \binom{N+1}{i} (k+1-i)^N \tag{10.14.1}$$

The probability of exactly k positive differences is therefore $\phi_N(k)/N!$ The same expression holds for the probability of k negative differences. The expectation of k is given by

$$E(k) = \sum_{k=0}^{N-1} \frac{k\phi_N(k)}{N!} = \frac{N-1}{2} \tag{10.14.2}$$

as is obvious from symmetry. The variance of k is

$$V(k) = \frac{N+1}{12} \tag{10.14.3}$$

The kurtosis of the distribution turns out to be $-\frac{6}{5}\frac{1}{N+1}$ and so tends to zero as $N \to \infty$. For $N > 12$, the distribution of k is approximately normal. In using the normal approximation, the correction for continuity should be applied by diminishing the observed value of $\left|k - \frac{N-1}{2}\right|$ by $\frac{1}{2}$. This is equivalent to taking as a standard normal variate the quantity

$$z = \pm\left(\frac{3}{N+1}\right)^{1/2}(|2k - N + 1| - 1) \tag{10.14.4}$$

Example 11 In a time-series of sweet potato production in the United States over the years 1868–1937, it was found that the 69 differences were positive in 37 cases and negative in 32. With $N = 70$, we find $z = +0.822$. The probability of a value greater numerically than this is 0.41, so that the hypothesis of randomness can be accepted. There is no evidence, as far as this test goes, of a trend in the series. If there is one it is swamped by the extent of the random variations.

A test may also be based on the total number of runs up and down. If r is this number, and if we include the runs at the beginning and end, we find

$$\begin{cases} E(r) = \dfrac{2N-1}{3} \\[2ex] V(r) = \dfrac{16N-29}{90} \end{cases} \tag{10.14.5}$$

Asymptotically, r is normally distributed.

The expected frequencies for runs of given length may also be calculated. If $E(r_p)$ is the expected number of runs of length p exactly, and if $E(r'_p)$ is the expected number of lengths *p or more*,

$$\begin{cases} E(r_1) = \dfrac{5N+1}{12} \\[2ex] E(r_2) = \dfrac{11N-14}{60} \\[2ex] E(r'_3) = \dfrac{4N-11}{60} \end{cases} \tag{10.14.6}$$

The χ^2 test may be used for comparing the observed and expected values, but the distribution of χ_s^2 is not quite that of Pearson's χ^2. For $N > 12$ and the three classes suggested above ($f = 1$, 2 and 3 or more), the observed χ_s^2 may be multiplied by 6/7 and referred to the χ^2 distribution for two degrees of freedom, at least as an approximation.

EXAMPLE 12 In a certain time series of 36 observations, the number of positive differences was 18 and the number of negative ones 17. The total number of runs up and down (r) was 25; and the observed values for r_1, r_2 and r'_3 were 16, 8, and 1 respectively.

Here $E(r) = 23.67$, $V(r) = 6.08$, $\sigma(r) = 2.47$. The observed r is clearly not significantly different from $E(r)$, on the basis of the normal approximation.

The expectations of r_1, r_2 and r'_3 are 15.08, 6.37 and 2.22 respectively. The value of χ_s^2 is 1.143, and 6/7 of this is 0.98. For 2 d.f., this is not significant. The tests agree in permitting us to accept the hypothesis that the given series is random.

Levene [14] has considered the power of tests based on k and r. The null hypothesis H_0 is that the observations are random. The alternative hypothesis H_1 may be that there is a linear trend in the observations, or perhaps that there is a cyclical trend. For detecting a linear trend, the test with k is much more powerful than that with r, but it appears to be less powerful for certain cyclical trends. Since in the limit, k and r have a joint normal distribution, and are uncorrelated under H_0, the statistic

$$\text{(10.14.7)} \qquad \frac{[k - E(k)]^2}{V(k)} + \frac{[r - E(r)]^2}{V(r)} = \chi_s^2$$

is approximately a χ^2 variate with two degrees of freedom.

EXAMPLE 13 An artificial upward linear trend was added to the time series of Example 12. This increased the number of positive differences to 21, and reduced the number of negative ones to 14. The total number of runs (r) was reduced to 21. With the continuity correction, $k - E(k) = 3$, $V(k) = 37/12$, $r - E(r) = -\frac{13}{6}$, $V(r) = 547/90$, so that the expression in Eq. (7) is $108/37 + 845/1094 = 2.92 + 0.77 = 3.69$. For 2 d.f., $P = 0.17$, which indicates that the added trend is not significant even at the 0.1 level. It may be seen that the main contribution to χ_s^2 arises from k rather than from r, thus verifying that the former is more sensitive than the latter to linear trends.

* 10.15 **Jonckheere's Test** This is designed to test the null hypothesis that several samples are randomly drawn from the same population, when the alternative hypothesis specifies a certain rank ordering of the populations. The ordinary F-test may be used to test this null hypothesis, but the alternative does not involve any particular rank ordering of the populations.

If there are only two samples we can, of course, use the Mann-Whitney test, or, if appropriate, a one-tailed t-test, the alternative hypothesis being, say, that $\mu_1 > \mu_2$. Jonckheere's test may be used with k samples. One application is to time-series, when on each occasion the observed event may be one of k possibilities. The null hypothesis would be that each of these occurs at random, and the alternative hypothesis that the different types of event tend to occur in a certain definite order.

For convenience we will suppose that the k samples are all of size r (although it is not necessary to do so). If the ith sample comes from a population with distribution function $F_i(x)$, the null hypothesis is that the $F_i(x)$ are all the same function of x, against the alternative that there is an ordering of the populations such that

$$F_1(x) < F_2(x) < \ldots < F_k(x) \tag{10.15.1}$$

for all x. This condition will be satisfied if there is a real additive treatment effect, different for each sample. We may express the alternative hypothesis as $F_i(x) < F_j(x)$ for all x, where $i = 1, 2 \ldots k - 1$, and $j = 1 + i, \ldots k$.

Let x_{im} be the mth item in the ith sample, and x_{jn} the nth item in the jth sample $(m, n = 1, 2 \ldots r)$. Also let $p_{im,jn} = 1$ if $x_{im} < x_{jn}$ and 0 if $x_{im} > x_{jn}$, and define p_{ij} by the relation

$$p_{ij} = \sum_{m=1}^{r} \sum_{n=1}^{r} p_{im,jn} \tag{10.15.2}$$

The greater the differences between the distribution functions $F_i(x)$ and $F_j(x)$, the larger will tend to be the value of p_{ij}. If we then define S by

$$S = 2\sum_{i<j} p_{ij} - \tfrac{1}{2}k(k-1)r^2 \tag{10.15.3}$$

the statistic S may be used for testing H_0 against H_1. A large value of S will lead to the rejection of H_0.

The following example is given by Jonckheere [15]. There are four measurements on each of four samples:

TABLE 10.10

I	II	III	IV
19	21	40	49
20	61	99	110
60	80	100	151
130	129	149	160
57.25	72.75	97.00	117.50

Considering Samples I and II, we note that the values 19 and 20 are each less than four values in II, 60 is less than three values in II, and 130 is less than none. Therefore,

$$p_{12} = 4 + 4 + 3 = 11$$

In the same way, we can calculate the other five values of p_{ij} and so obtain

$$S = 2(11 + 12 + 13 + 11 + 12 + 12) - 96 = 46.$$

From the tables, Appendix B.10, we find that the probability that $S \geq 46$ is 0.0168, which would suggest rejection of H_0. The usual F-test on the same data gives a probability of 0.346, which would lead to acceptance of H_0. The F-test, however, has a much wider variety of alternative hypotheses than the Jonckheere test.

The quantity S is really a measure of the agreement of the ranking of all the observations with the ranking that they would have if those in each separate sample were tied. The ranks in the example above, along with the tied ranks, are as shown in the Table 10.11, where for each sample the first column gives the actual rank among all 16 observations and the second column the ranks that would be allotted if all the observations in a batch were tied. As usual when dealing with ties, the rank is the arithmetic mean of the ranks that would apply if the ties were broken. Thus $2\frac{1}{2}$ is the mean of 1, 2, 3 and 4 (see § 11.16).

TABLE 10.11

I		II		III		IV	
1	$2\frac{1}{2}$	3	$6\frac{1}{2}$	4	$10\frac{1}{2}$	5	$14\frac{1}{2}$
2	$2\frac{1}{2}$	7	$6\frac{1}{2}$	9	$10\frac{1}{2}$	11	$14\frac{1}{2}$
6	$2\frac{1}{2}$	8	$6\frac{1}{2}$	10	$10\frac{1}{2}$	15	$14\frac{1}{2}$
13	$2\frac{1}{2}$	12	$6\frac{1}{2}$	14	$10\frac{1}{2}$	16	$14\frac{1}{2}$

For each pair of observations we may now allot a score of 1 if they are in the same order on both rankings, -1 if they are in opposite orders, or 0 if they are tied on either ranking. The sum of these scores is the quantity S, and it is easily checked that $S = 46$. The first observation in I and the first in II, for instance, contribute 1, since 1 and 3 are in the same order as $2\frac{1}{2}$ and $6\frac{1}{2}$. This quantity S is used in calculating rank correlation by Kendall's method (see §§ 11.14 and 11.17).

The statistic S has a symmetrical distribution with expectation zero and variance $\sigma^2 = \dfrac{N^2}{18}\{2N + 3 - (2r + 3)/k\}$, where $N = kr$. For large samples, S/σ is approximately normal, especially if a continuity correction is applied by subtracting 1 from S before dividing by σ. A better approximation is the following:

$$S\left[\frac{\nu}{(\nu + 1)\sigma^2 - S^2}\right]^{1/2} \approx t_\nu \tag{10.15.4}$$

i.e., it has the Student-t distribution with ν degrees of freedom. The quantity ν depends on the kurtosis of the distribution of S and is given by

$$\nu + 3 = \frac{1350\sigma^4}{N^3(6N^2 + 15N + 10) - kr^3(6r^2 + 15r + 10)} \tag{10.15.5}$$

In the example, $\sigma = 21.42$ and the approximate z value is $45/\sigma = 2.10$, which corresponds to a probability 0.018. The value of ν turns out to be 36 and

$t_\nu = 2.21$. The probability for a value of t exceeding 2.21 is 0.017 which is very close to the correct value.

When there are only two samples, S reduces to $r^2 - 2U$, or $r^2 - 2U'$, where U is the statistic used in the Mann-Whitney test (§ 10.12).

PROBLEMS

A. (§§ 10.1–10.5)

1. The French naturalist Buffon (1708–1788) once tossed a coin 4040 times and obtained 2048 heads and 1992 tails. Show that this result is not at all surprising with a good coin. *Hint:* Find the probability of a discrepancy from the expected result at least as great as this, using the chi-square test.

2. Over a period of time, the numbers of aircraft accidents that occurred on the different days of the week were noted, with the following result:

Day	M	Tu	W	Th	F	S	Sun
f_0	16	8	12	11	9	14	14

Is there good reason to doubt that an accident is equally likely to occur on any day of the week?

3. In the following table, x is the number of 5's or 6's observed in a throw with five dice. Is this result of 243 throws consistent with the hypothesis that the dice are true?

x	0	1	2	3	4	5
f_0	23	90	81	30	19	0

Hint: Pool the last two classes.

4. A student dealt 26 cards from an ordinary deck and counted the number of honor cards in the hand dealt (A, K, Q, J, and 10 counting as honors). He did this 50 times, with the following result:

x	4	5	6	7	8	9	10	11	12	13	14	15	16	17
f_0	1	0	2	3	7	5	10	8	2	7	2	0	2	1
f_c	0.0	0.2	0.9	2.7	6.0	9.6	11.2	9.6	6.0	2.7	0.9	0.2	0.0	0.0

Would you reject the hypothesis that the cards were well shuffled between each deal? *Hint:* The expected frequency of x honor cards, if the hypothesis is true, is $50\binom{20}{x}\binom{32}{26-x}\Big/\binom{52}{26}$, giving the theoretical frequencies f_c. Pool the first four and the last five classes.

5. In an insecticide test, 20 insects were put into each of 100 jars and subjected to a standard dose of insecticide. The number surviving (x) after three hours was counted for each jar:

x	0	1	2	3	4	5	6	7	8	9
f_0	3	8	11	15	16	14	12	11	9	1

Does this distribution appear to be binomial? *Hint:* If each insect has the same chance θ of surviving, the distribution will be binomial. Estimate θ from the relation $20\theta = \bar{x}$, and calculate the theoretical frequencies. The first two and the last two frequencies should be pooled. Note that the last theoretical frequency is for $x = 9$ *or more*.

6. Samples of 50 balls were taken repeatedly from a mixture of 100 red balls and 1100 white ones, and the number (x) of red balls in each sample was noted. The balls were returned and well mixed after each sampling. In 300 trials the following values of x were observed.

x	0	1	2	3	4	5	6	7	8	9	10 or more
f_0	1	16	36	48	62	51	41	22	18	5	0

Test the agreement of this distribution with a Poisson distribution of parameter $\mu = \frac{50}{12}$.

7. Calculate $\bar{x}$ (the mean of x) for the data of Problem 6, and fit the observations with a Poisson distribution of parameter $\bar{x}$. Test the agreement now. *Hint:* When μ is estimated from the samples, the degrees of freedom are reduced by 1.

8. A classical example (originally given by von Bortkiewicz) of the distribution of rare events is that of the deaths of Prussian cavalrymen from the kicks of horses during the 20 years 1875–1894. The frequency distribution of such deaths in 10 army corps, per corps per annum, was

x	0	1	2	3	4
f_0	109	65	22	3	1

Fit a Poisson distribution and test the goodness of fit.

9. The following table gives a distribution of lengths of time (in seconds) of telephone calls at a certain exchange:

Time	*Number of Calls*
0 – 99	1
100 – 199	28
200 – 299	88
300 – 399	180
400 – 499	247
500 – 599	260
600 – 699	133
700 – 799	42
800 – 899	11
900 – 999	5
	995

The mean and standard deviation are 477.3 sec and 145.7 sec respectively. Fit a normal curve to the data and test the goodness of fit.

10. In a study of plant disease (spotted wilt of tomatoes) the numbers (x) of diseased plants were counted in each of 160 groups of plants. Each group contained nine plants, evenly spaced, so that x could take integral values from 0 to 9 inclusive. The following distribution was found:

x	0	1	2	3	4	5	6	7	8
f_0	36	48	38	23	10	3	1	1	0

Assuming that the probability of being diseased is constant (θ), fit the distribution with a binomial, estimating θ from the sample. Test the agreement by the chi-square method, using three different procedures for pooling—(a) the last three frequencies, (b) the last four, (c) the last five. (Note that, in this example, wider pooling tends to disguise departures from the theoretical distribution).

B. (§§ 10.6–10.8)

1. The cumulative frequencies F_0 corresponding to given values of x (upper class boundaries) in a sampling experiment, are shown in the following table. The theoretical cumulative frequencies F_c from a certain normal population are also given:

Upper Class Boundary (x)	F_0	F_c
20.5	2	0.6
22.5	9	4.5
24.5	24	20.9
26.5	73	68.5
28.5	162	161.7
30.5	300	287.6
32.5	402	401.3
34.5	469	471.5
36.5	505	500.8
38.5	510	509.2
40.5	511	511.0

Use the Kolmogorov test to show that it is reasonable to accept the hypothesis that the population sampled had the normal distribution corresponding to the column F_c.

2. The following table gives cumulative frequencies of correct responses to a psychological test for (a) a group of 24 normal subjects, (b) a group of 24 schizophrenic subjects. The test required the perception of groupings in a design exposed to the subject for a variable time (Kaswan, *British Journ. Psych.*, 1958, p. 131).

Exposure Time	*F(Normal)*	*F(Schizophrenic)*
0.01 sec	4	2
0.04	10	4
0.1	13	5
0.25	17	10
0.75	21	16
5.0	24	21
10.0	24	24

Use the Kolmogorov-Smirnov two-sample test to determine whether there is a significant difference between the two groups.

3. Can the following sample be reasonably regarded as coming from a uniform distribution on the interval (35, 70): 36, 42, 44, 50, 64, 58, 56, 50, 37, 48, 52, 63, 57, 43, 39, 42, 47, 61, 53, 58? Use the Kolmogorov test. *Hint:* Calculate theoretical relative cumulative frequencies, $x/35$, at the given values of x.

4. Are the following observations (on percentage change in systolic pressure in dogs under experimental conditions) consistent with a normal distribution of mean 16.0 and variance 30.0?

20.6, 11.6, 7.5, 10.5, 13.9, 16.2, 14.8
17.8, 26.9, 13.5, 20.1, 22.5, 11.1, 16.7

5. Since $F(x)$ in the Kolmogorov test is assumed to be continuous, there is zero probability of two identical values of x. However, since measurements are made with limited accuracy, ties do occur. What would be in general the effect of this on the value of D_N?

C. (§§ 10.9–10.12)

1. The observed values x_1 and x_2 for two paired samples are given. Is it reasonable to assume that no difference exists between the medians of the two populations from which these samples were taken? Use the sign test.

x_1	15	19	31	36	10	11	19	15	10	16
x_2	19	30	26	8	10	6	17	13	22	8

2. For nine animals, tested under control conditions and experimental conditions, the following values of a measured variable were observed:

Animal No.	1	2	3	4	5	6	7	8	9
Control	21	24	26	32	55	82	46	55	88
Experimental	18	9	23	26	82	199	42	30	62

Test whether a significant difference exists between the medians, using the Wilcoxon signed-ranks test.

3. The following table gives scores of a group of engineering students in (a) mathematics, (b) graphics. Use the signed-ranks test to determine whether the median scores of such students differ significantly in the two subjects.

Student No.	Maths	Graphics	Sudent No.	Maths	Graphics
1	66	51	16	86	72
2	20	51	17	65	73
3	66	45	18	33	69
4	73	77	19	42	51
5	59	68	20	66	66
6	58	51	21	59	82
7	37	50	22	57	74
8	85	81	23	27	44
9	57	66	24	55	65
10	69	65	25	61	65
11	63	54	26	86	71
12	75	53	27	52	65
13	87	73	28	79	62
14	34	40	29	63	62
15	67	73	30	75	64

4. In a comparative study of the effects of oxygen on the peripheral nerve in cats and rabbits, the survival time of a nerve under anoxic conditions was measured. The times (in minutes) are given:

Cat No.	1	2	3	4
Time	45	43	33	25

Rabbit No.	1	2	3	4	5	6	7	8	9	10	11	12	13	14
Time	35	35	30	30	28	28	23	22	22	20	17	16	16	15

Test the hypothesis that these are random samples from populations with the same distribution, as against the alternative hypothesis that the cat times are stochastically larger than the rabbit times.

5. In the following table the variable is the number of trials required by a rat to learn a new pattern of behavior when placed in a new situation. The experimental group of rats had been trained in a certain way; the control group had not. Test whether the previous training significantly affects the ability to learn.

Experimental Group	78	64	75	45	82	54	71			
Control Group	110	70	53	51	62	93	106	88	67	72

D. (§§ 10.13–10.15)

1. The annual marriage rates per 1000 of population in the United States for 1885, 1890, 1895 . . . 1950 are: 9.2, 9.0, 8.9, 9.3, 10.0, 10.3, 10.0, 12.0, 10.3, 9.2, 10.4, 12.1, 12.2, 11.1. Does there appear to be a significant upward trend? (Use the number of positive differences and also the total number of runs.)

2. Apply the approximate chi-square test on runs of a given length to test the randomness of the time series in Problem D-1.

3. A student opened a set of mathematical tables with the entries blocked off in sets of five, and, starting at random, added the five terminal digits in each block of five numbers. Going consecutively through 50 blocks, he obtained the following values:

12, 15, 18, 30, 33, 25, 28, 22, 23, 17
25, 18, 22, 13, 17, 18, 22, 25, 27, 30
28, 32, 24, 27, 20, 22, 15, 18, 20, 23
12, 15, 27, 30, 33, 25, 28, 20, 23, 17
25, 18, 20, 13, 17, 18, 22, 33, 27, 30

Test the sequence for randomness, using runs up and down.

4. Snedecor (*Statistical Methods*, Iowa State College Press, 1956) has given the following table showing the amounts of four different fats (in grams) absorbed by doughnuts, six batches of doughnuts being used for each fat.

Fat			
A	*B*	*C*	*D*
164	178	175	155
172	191	193	166
168	197	178	149
177	182	171	164
195	177	176	168
156	185	163	170

Test for a significant difference between the means by Jonckheere's method. *Hint:* First order the samples in increasing order of their means. Count $\frac{1}{2}$ for ties in computing p_{ij}. Use the normal approximation.

5. The following table is supposed to represent the scores of samples from three different groups of teachers on a certain personality test. Does there seem to be a real ordering of the groups?

Group I	Group II	Group III
96	82	115
128	124	149
83	132	166
61	135	147
101	109	129

REFERENCES

[1] Walker, Helen M., and Lev, J., *Statistical Inference*, Holt, 1953. Includes a chapter by L. E. Moses on non-parametric methods.

[2] Smith, K., "Distribution-Free Statistical Methods and the Concept of Power-Efficiency." This is in a book, *Research Methods in the Behavioral Sciences*, ed. Festinger, L. and Katz, D. Dryden Press, 1953.

[3] Moses, L. E., "Non-Parametric Statistics for Psychological Research," *Psych. Bulletin*, **49,** 1952, pp. 122–143.

[4] Cochran, W. G., "Some Methods for Strengthening the Common Chi-square Tests," *Biometrics*, **10,** 1954, pp. 417–451.

[5] Cramër, H., *Mathematical Methods of Statistics*, Princeton Univ. Press, 1946, pp. 424, 506.

[6] Feller, W., On the Kolmogorov-Smirnov Limit Theorems for Empirical Distributions, *Ann. Math. Stat.*, **19,** 1948, pp. 177–189. See also the table on pp. 279–281 of the same volume.

[7] Massey, F. J., Jr., "The Kolmogorov-Smirnov Test for Goodness of Fit," *J. Amer. Stat. Ass.*, **46,** 1951, pp. 68–78.

[8] Miller, L. H., "Percentage Points of Kolmogorov Statistics," *J. Amer. Stat. Ass.*, **51,** 1956, pp. 111–121.

[9] Massey, F. J., Jr., The Distribution of the Maximum Deviation Between Two Sample Cumulative Step Functions, *Ann. Math. Stat.*, **22,** 1951, pp. 125–128.

[10] Smirnov, N., Table for Estimating the Goodness of Fit of Empirical Distributions, *Ann. Math. Stat.*, **19,** 1948, pp. 279–281.

[11] Walsh, J. E., "Applications of some Significance Tests for the Median Which are Valid Under Very General Conditions," *J. Amer. Stat. Ass.*, **44,** 1949, pp. 342–355.

[12] Mann, H. B., and Whitney, D. R., "On a Test of Whether One of Two Random Variables is Stochastically Larger Than the Other," *Ann. Math. Stat.*, **18,** 1947, pp. 52–54.

[13] Moore, G. H., and Wallis, W. A., "Time Series Significance Tests Based on Signs of Differences," *J. A. Stat. Ass.*, **38,** 1943, pp. 153–164.

[14] Levene, H., "On the Power Function of Tests of Randomness Based on Runs Up and Down," *Ann. Math. Stat.*, **23,** 1952, pp. 34–56.

[15] Jonckheere, A. R., "A Distribution-Free *k*-Sample Test Against Ordered Alternatives," *Biometrika*, **41,** 1954, 133–145.

[16] Wilcoxon, F., *Some Rapid Approximate Statistical Procedures*, American Cyanamid Co., Stamford, Conn., 1949. See also [17].

[17] Siegel, S., *Nonparametric Statistics for the Behavioral Sciences*, McGraw-Hill 1956. This gives a full discussion, with illustrative examples and many tables of all the principal non-parametric tests.

Chapter 11

DISTRIBUTIONS OF PAIRS OF VARIATES

11.1 **The Classical Regression Problem for a Population** We now propose to investigate some measures of the relationship between two random variables (variates), X and Y, both of which are capable of measurement on each member of a given population. For convenience we shall think of them as distributed continuously, but the extension to discrete distributions will usually be obvious —a matter of replacing probability densities by probabilities and integrals by sums.

In general, the joint probability that for a particular member of the population the value of X lies between x and $x + dx$ *and* the value of Y lies between y and $y + dy$ is given by $f(x, y)\, dx\, dy$. The probability density for X alone (regardless of Y) is

(11.1.1) $$g(x) = \int_{-\infty}^{\infty} f(x, y)\, dy$$

and that for Y alone is

(11.1.2) $$h(y) = \int_{-\infty}^{\infty} f(x, y)\, dx$$

The probability density of Y, *for a given value x of X*, is

(11.1.3) $$f(y|x) = f(x, y)/g(x)$$

and similarly for $f(x|y)$. The expectation of Y, for given X, is defined as

(11.1.4) $$\eta_x = E(Y|X) = \int_{-\infty}^{\infty} yf(y|x)\, dy$$

It is, of course, a function of the given value x of X. The graph of η_x as a function of x is called the *true regression curve* of Y on X. If the regression is linear the graph is a straight line, with the equation

(11.1.5) $$\eta_x = \alpha + \beta x$$

The parameter β is the true *regression coefficient* of Y on X, and is the slope of the straight line. The other parameter α represents the intercept on the axis of Y. (See Fig. 47).

The variance of Y, for given X, is similarly defined by

(11.1.6) $$\sigma_{Y|x}{}^2 = V(Y|X) = \int_{-\infty}^{\infty} (y - \eta_x)^2 f(y|x)\, dy$$

This is also in general a function of x, although in some circumstances it turns out to be independent of x. It is the variance of Y for those members of the population whose X-values lie in a thin strip of width dx—these members are said to form an *X-array*. They are represented by dots in Figure 47.

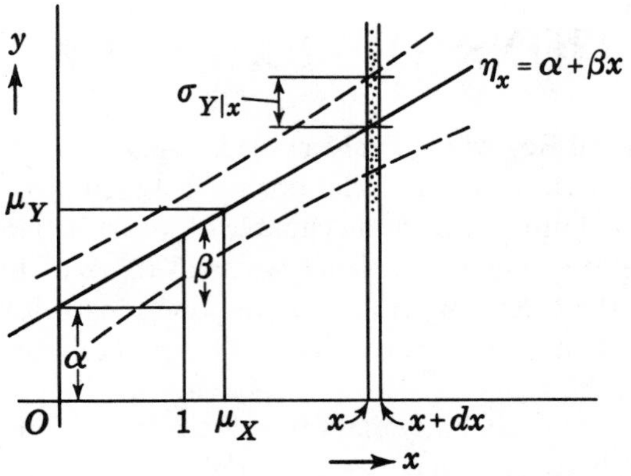

FIG. 47 LINEAR REGRESSION IN A POPULATION

The regression parameters α and β may be expressed in terms of the moments of the distributions of X and Y. Let us define the two means by

$$\text{(11.1.7)} \qquad \begin{cases} \mu_X = \displaystyle\int_{-\infty}^{\infty} g(x)x\,dx = \int_{-\infty}^{\infty}\int_{-\infty}^{\infty} xf(x, y)\,dy\,dx \\ \mu_Y = \displaystyle\int_{-\infty}^{\infty} h(y)y\,dy = \int_{-\infty}^{\infty}\int_{-\infty}^{\infty} yf(x, y)\,dx\,dy \end{cases}$$

and the two variances and the covariance by

$$\text{(11.1.8)} \qquad \begin{cases} \sigma_X^{\,2} = \displaystyle\int_{-\infty}^{\infty} g(x)(x - \mu_X)^2\,dx \\ \sigma_Y^{\,2} = \displaystyle\int_{-\infty}^{\infty} h(y)(y - \mu_Y)^2\,dy \\ \sigma_{XY} = \displaystyle\int_{-\infty}^{\infty}\int_{-\infty}^{\infty} f(x, y)(x - \mu_X)(y - \mu_Y)\,dy\,dx \end{cases}$$

From Eqs. (3) and (4) it follows that

$$\text{(11.1.9)} \qquad g(x)\eta_x = \int_{-\infty}^{\infty} yf(x, y)\,dy$$

so that, on writing $\eta_x = \alpha + \beta x$ and integrating,

$$\text{(11.1.10)} \qquad \int_{-\infty}^{\infty} (\alpha + \beta x)g(x)\,dx = \int_{-\infty}^{\infty}\int_{-\infty}^{\infty} yf(x, y)\,dy\,dx$$

or, using Eq. (7),

(11.1.11) $$\alpha + \beta\mu_X = \mu_Y$$

If we multiply Eq. (9) by x before integrating, we obtain

(11.1.12) $$\int_{-\infty}^{\infty} (\alpha x + \beta x^2) g(x)\, dx = \int_{-\infty}^{\infty}\int_{-\infty}^{\infty} xyf(x, y)\, dy\, dx$$

which may be written

(11.1.13) $$\alpha\mu_X + \beta\mu'_{2X} = \pi_{XY}$$

where μ'_{2X} is the second moment about the origin for X, and π_{XY} is the product moment about the origin for X and Y. Multiplying Eq. (11) by μ_X and subtracting it from Eq. (13), we obtain

$$\beta(\mu'_{2X} - \mu_X{}^2) = \pi_{XY} - \mu_X\mu_Y$$

which is equivalent to

(11.1.14) $$\beta \cdot \sigma_X{}^2 = \sigma_{XY}$$

We have, therefore,

(11.1.15) $$\beta = \sigma_{XY}/\sigma_X{}^2 = \rho_{XY}\sigma_Y/\sigma_X$$

where $\rho_{XY} = \sigma_{XY}/(\sigma_X\sigma_Y)$, the Pearson coefficient of correlation between X and Y.

From Eq. (11),

(11.1.16) $$\alpha = \mu_Y - \beta\mu_X$$

so that the two parameters of the regression line are now expressed in terms of the means, variances and covariance of X and Y. Using these expressions, the equation of the line may be written

(11.1.17) $$\begin{aligned}\eta_x - \mu_Y &= \beta(x - \mu_X)\\ &= \rho_{XY}\sigma_Y \frac{x - \mu_X}{\sigma_X}\end{aligned}$$

which indicates that the line passes through the point with coordinates (μ_X, μ_Y).

A similar equation may be obtained by interchanging X and Y. If ξ_y is the expectation of X, for a given value y of Y,

(11.1.18) $$\xi_y - \mu_X = \rho_{XY}\sigma_X \frac{y - \mu_Y}{\sigma_Y}$$

This line also passes through the point (μ_X, μ_Y) but in general it does not coincide with the first line. There are therefore two regression lines, one of Y on X and one of X on Y, given respectively by Eqs. (17) and (18). The first line represents the expectation of Y for a given X, and the second line the expectation of X for a given Y.

The variance of Y, for a given X, may vary from one value of X to another, but we can define a weighted average of the variances in the different X-arrays by the relation

(11.1.19) $$\sigma_{Ye}^2 = \int_{-\infty}^{\infty} \sigma_{Y|x}^2 g(x)\,dx$$

the variance $\sigma_{Y|x}^2$ for a given x being weighted with the probability density for this value of x. The quantity σ_{Ye}^2 is called the *variance of estimate* of Y. It is a measure of the average variability of Y around the regression line of Y on X.

Using Eqs. (3) and (6) and noting that

$$\begin{aligned}(y - \eta_x)^2 &= \{y - \mu_Y - \beta(x - \mu_X)\}^2 \\ &= (y - \mu_Y)^2 + \beta^2(x - \mu_X)^2 - 2\beta(y - \mu_Y)(x - \mu_X)\end{aligned}$$

we obtain, from Eq. (19),

(11.1.20) $$\begin{aligned}\sigma_{Ye}^2 &= \int_{-\infty}^{\infty}\int_{-\infty}^{\infty} [(y - \mu_Y)^2 + \beta^2(x - \mu_X)^2 \\ &\qquad - 2\beta(y - \mu_Y)(x - \mu_X)] f(x, y)\,dx\,dy \\ &= \sigma_Y^2 + \beta^2\sigma_X^2 - 2\beta\sigma_{XY}.\end{aligned}$$

From Eq. (15), $\beta^2\sigma_X^2 = \sigma_{XY}^2/\sigma_X^2 = \beta\sigma_{XY} = \rho_{XY}^2\sigma_Y^2$, so that

(11.1.21) $$\sigma_{Ye}^2 = \sigma_Y^2(1 - \rho_{XY}^2)$$

It is clear from this relation, since σ_{Ye}^2 and σ_Y^2 are necessarily non-negative, that ρ_{XY}^2 cannot exceed 1. In other words, for all possible distributions of X and Y,

(11.1.22) $$-1 \le \rho_{XY} \le 1$$

As previously indicated (§ 2.13), ρ_{XY} is a measure of the degree of relationship between X and Y. If X and Y are independent, $\rho_{XY} = 0$. If Y is precisely proportional to X, so that all the Y values lie on the straight regression line, then $\sigma_{Ye}^2 = 0$ and $\rho_{XY} = \pm 1$. These are the extreme cases.

11.2 **The Bivariate Normal Surface** An important special case of a two-variate distribution is that with a joint probability density:

(11.2.1) $$f(x, y) = Ke^{-Q}$$

where

(11.2.2) $$K^{-1} = 2\pi\sigma_X\sigma_Y(1 - \rho^2)^{1/2}$$

and

(11.2.3) $$Q = \left[\left(\frac{x - \mu_X}{\sigma_X}\right)^2 + \left(\frac{y - \mu_Y}{\sigma_Y}\right)^2 - 2\rho\left(\frac{x - \mu_X}{\sigma_X}\right)\left(\frac{y - \mu_Y}{\sigma_Y}\right)\right] \div [2(1 - \rho^2)]$$

Here ρ has been written for ρ_{XY}. The quantity Q is a quadratic form in the standardized variates,

(11.2.4)
$$z = \frac{x - \mu_X}{\sigma_X}, \; v = \frac{y - \mu_Y}{\sigma_Y}$$

so that the probability density in terms of the variates z and v is

(11.2.5)
$$g(z, v) = (2\pi)^{-1}(1 - \rho^2)^{-1/2} \exp\left[-\frac{z^2 + v^2 - 2\rho z v}{2(1 - \rho^2)}\right]$$

This represents a surface known as the *bivariate normal surface*, and pictured (in a truncated form) in Figure 48. It is bell-shaped, asymptotic in all directions

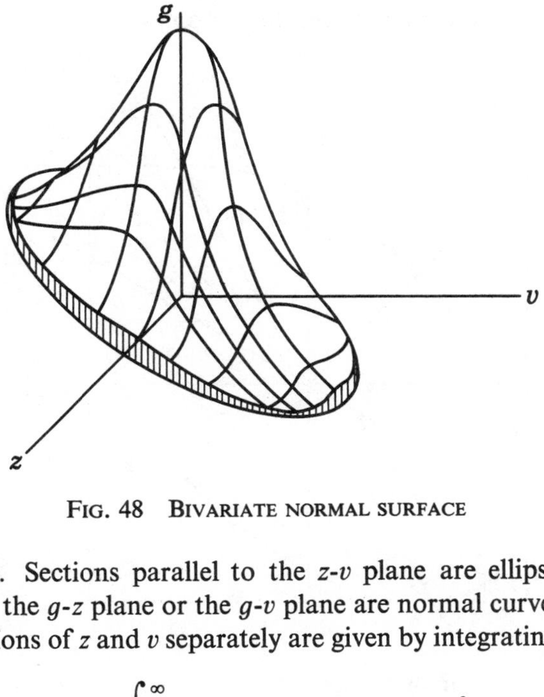

FIG. 48 BIVARIATE NORMAL SURFACE

to the z-v plane. Sections parallel to the z-v plane are ellipses and sections parallel to either the g-z plane or the g-v plane are normal curves.

The distributions of z and v separately are given by integrating Eq. (5). Thus,

(11.2.6)
$$f(z) = \int_{-\infty}^{\infty} g(z, v)\, dv = (2\pi)^{-1/2} e^{-z^2/2}$$

(11.2.7)
$$h(v) = \int_{-\infty}^{\infty} g(z, v)\, dz = (2\pi)^{-1/2} e^{-v^2/2}$$

These are both standard normal distributions. The probability density for v, for a given z, is

(11.2.8)
$$g(v|z) = \frac{g(z, v)}{f(z)} = [2\pi(1 - \rho^2)]^{-1/2} \exp\left[-\frac{(v - \rho z)^2}{2(1 - \rho^2)}\right]$$

The expectation of v for a given z is therefore

$$\eta_z = \int_{-\infty}^{\infty} vg(v|z)\, dv$$
$$= \rho z$$

This represents a straight line through the origin with slope ρ. The regression is therefore linear, and this holds also for the second regression line, with equation

$$\xi_v = \rho v$$

Each array of v's has, for given z, the variance

$$\sigma_{v|z}^{\ 2} = \int_{-\infty}^{\infty} (v - \rho z)^2 g(v|z)\, dv$$

and, on carrying out the integration, this becomes

(11.2.9) $$\sigma_{v|z}^{\ 2} = 1 - \rho^2$$

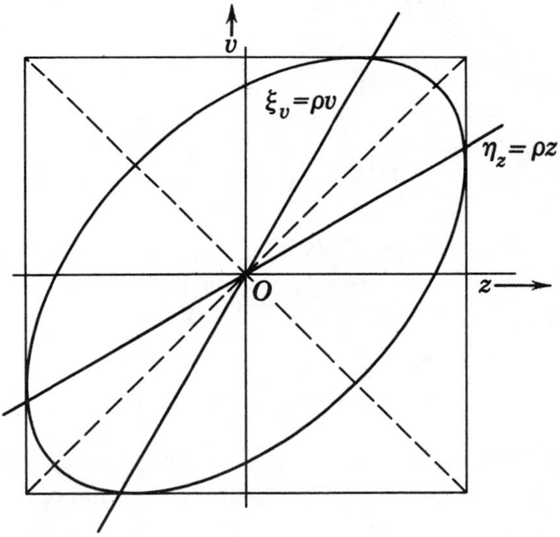

FIG. 49 HORIZONTAL SECTION OF BIVARIATE NORMAL SURFACE

Transforming back to the original variates X and Y, we obtain

(11.2.10) $$\sigma_{Y|x}^{\ 2} = \sigma_Y^{\ 2}(1 - \rho^2)$$

and the variance is therefore independent of x. The weighted average $\sigma_{Ye}^{\ 2}$ is then the same as $\sigma_{Y|x}^{\ 2}$, for all x. A distribution with this property is called "homoscedastic" (from Greek words meaning "equal scattering"). A similar property holds for the y-arrays of X for given values of y.

In the z-v plane the two regression lines have the same slope, one with respect to Oz and the other with respect to Ov. (Figure 49). A section of the bivariate normal surface by a horizontal plane, g = const., is an ellipse with equation

$$z^2 + v^2 - 2\rho zv = c \tag{11.2.11}$$

If this ellipse is drawn on the z-v plane, the tangents at the points where it is cut by the regression lines are parallel to the axes.

The bivariate normal distribution occupies a central position in the theory of two-variate distributions, similar to that of the normal distribution for a single variate. Most of the early classical work of Karl Pearson, Galton and others, on regression and correlation, was based on this distribution.

11.3 **Linear Regression as determined from a Sample** In practice, there is a variety of situations involving two variables. Both X and Y may be random, or only one of them, or neither. One variable, for instance, may be the *time*, as in a time-series of temperatures, stock-market prices, sunspot numbers, etc. Also, in many cases, the values of X are pre-selected instead of being chosen at random, as in a physics experiment where conveniently chosen weights are hung on a wire or spring and the extension produced is measured. Here neither X nor Y is a random variable in the ordinary sense, but both may be subject in different degrees to experimental error. The true relation between X and Y is a *functional* one. Obviously, unless X is a random variable it makes no sense to speak of its distribution, expectation or variance, and of course the same holds for Y.

In the usual regression problem, Y is a random variable and X may be either random or fixed. The classical situation is that in which both are random, a sample being selected randomly from a population such as that considered in § 11.1, and the values of X and Y measured on each of the N selected items. We assume that for a given value x of X, Y is of the form

$$Y = \alpha + \beta x + \varepsilon \tag{11.3.1}$$

where ε is normally distributed with mean 0 and variance σ^2. The relation

$$Y = \alpha + \beta X \tag{11.3.2}$$

which really means

$$E(Y|X = x) = \alpha + \beta x \tag{11.3.3}$$

is sometimes called a *structural relation*. It is the underlying relationship which is disturbed, in an actual sample, by the sampling fluctuation of Y and by the errors of measurement of both X and Y.

We will asume for the present that X and Y are measured without appreciable error, so that we need consider only the sampling fluctuation expressed by the quantity ε in Eq. (1). If a straight line is fitted to the N sample values of X and Y, this line will furnish estimates of the parameters α and β of the true regression line. In fact, as we shall see shortly, when the line is fitted by the usual "least squares" method these estimates are unbiased.

Suppose the observed pairs of sample values are (x_i, y_i), $i = 1, 2 \dots N$. The problem is to find the equation of a straight line which will give the "best" estimate of Y for a given value x of X and to find the standard error of this estimate.

The observed sample values, plotted in the x-y plane, form a *scatter diagram* (Fig. 50). The least-squares method of fitting a straight line chooses the constants a and b of the equation

(11.3.4) $$y_c = a + bx$$

in such a way that the sum of squares of the deviations of the sample points from this line, measured parallel to the y-axis, will be a minimum. In Figure 50 the

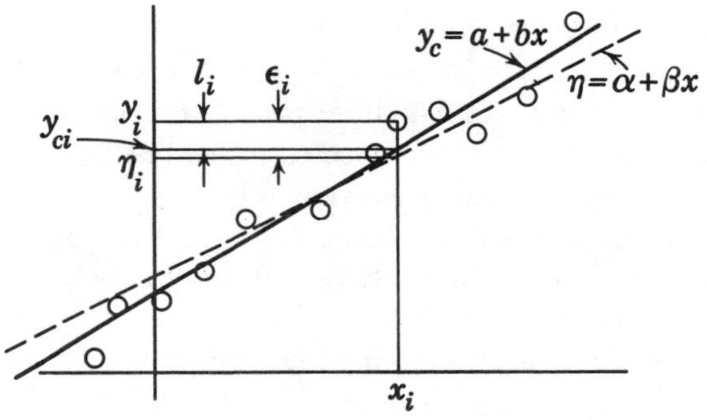

FIG. 50 SAMPLE LINEAR REGRESSION

deviation of y_i from the least-squares line is denoted by e_i, and the deviation from the true regression line $\eta = \alpha + \beta x$ by ε_i.

If $S = \sum_i e_i^2 = \sum_i (y_i - a - bx_i)^2$, the minimum value of S will be given by solving the simultaneous equations $\dfrac{\partial S}{\partial a} = 0$, $\dfrac{\partial S}{\partial b} = 0$. These may be written, after cancelling a factor -2,

(11.3.5) $$\begin{cases} \sum\limits_i (y_i - a - bx_i) = 0 \\ \sum\limits_i x_i(y - a - bx_i) = 0 \end{cases}$$

which are called the *normal equations* of the problem. They can be rearranged as

(11.3.6) $$\begin{cases} Na + \sum x_i b = \sum y_i \\ \sum x_i a + \sum x_i^2 b = \sum x_i y_i \end{cases}$$

Solving for a and b, we obtain

(11.3.7) $$b = \frac{s_{XY}}{s_X^{\,2}}$$

(11.3.8) $$a = \bar{y} - b\bar{x}$$

where $\bar{x} = \sum x_i/N$, $\bar{y} = \sum y_i/N$, $(N-1)s_X{}^2 = \sum x_i{}^2 - N\bar{x}^2$, and $(N-1)s_{XY} = \sum x_i y_i - N\bar{x}\bar{y}$. Here $s_X{}^2$ is the sample variance of X and s_{XY} is the sample covariance of X and Y.

The Pearson coefficient of correlation for the sample is defined by

(11.3.9) $$r_{XY} = \frac{s_{XY}}{s_X s_Y} = b\frac{s_X}{s_Y}$$

so that the equation of the least-squares line may be written, using Eqs. (7) and (8), as

(11.3.10) $$y_c - \bar{y} = b(x - \bar{x}) = \frac{rs_Y}{s_X}(x - \bar{x})$$

where r stands for r_{XY}.
For any given value x of X, y_c provides an estimate of the corresponding value of Y.

All the above argument can be carried through with X and Y interchanged. If we wish to find the best straight line for estimating X for a given value y of Y, the least squares criterion for the deviations of the sample points from this line, *measured parallel to the x-axis*, will give

(11.3.11) $$x_c - \bar{x} = b'(y - \bar{y}) = \frac{rs_X}{s_Y}(y - \bar{y})$$

where

(11.3.12) $$b' = \frac{s_{XY}}{s_Y{}^2}$$

It may be noted that $bb' = r^2$, so that r is the geometric mean of the two slopes (one measured from the x-axis, one from the y-axis). The two regression lines intersect at the point whose coordinates are $(\bar{x}, \bar{y})$.

In most practical situations one of the two variates will, for non-statistical reasons, be the one we would like to estimate, and this is the one we label Y. It is therefore hardly necessary to treat the second regression line in detail. What is said in subsequent paragraphs about the regression of Y on X can be applied, with minor changes, to the regression of X on Y.

11.4 **Computation of the Regression and Correlation Coefficients** Calculation of b and r requires, in effect, the determination of two variances and a covariance, since $b = s_{XY}/s_X{}^2$ and $r = s_{XY}/(s_X s_Y)$. The most convenient formulas for ungrouped variates are the following:

(11.4.1) $$b = \frac{N\sum xy - \sum x \sum y}{N\sum x^2 - (\sum x)^2}$$

(11.4.2) $$r^2 = \frac{(N\sum xy - \sum x \cdot \sum y)^2}{[N\sum x^2 - (\sum x)^2][N\sum y^2 - (\sum y)^2]}$$

If, for the sake of simplifying the calculations, we put $u = (x - x_0)/h$ and $v = (y - y_0)/k$, where x_0, y_0, h and k are chosen arbitrarily. it will make no difference to the value of r if we replace x and y by u and v throughout Eq. (2). The value of b, however, obtained by using u and v in Eq. (1) must be multiplied by k/h to give the value in terms of x and y.

EXAMPLE 1 Specimens of steels containing various percentages of nickel were tested for toughness with the following results (X is toughness in arbitrary units, Y is percentage of nickel):

X	47	50	52	52	54	56	58	59	60	60	62	64	65	66
Y	2.5	2.7	2.8	2.8	2.9	3.2	3.2	3.3	3.4	3.5	3.5	3.6	3.7	3.8

Suppose it is desired to estimate percentage of nickel from measured toughness in further specimens and to estimate the correlation between these variables. It is assumed that a random sample of nickel-steel alloys was selected for testing, and both X and Y were measured on each specimen.

If we let $u = X - 50$ and $v = 10(Y - 30)$, we get Table 11.1. Then,

TABLE 11.1

u	v	u^2	v^2	uv
−3	−5	9	25	15
0	−3	0	9	0
2	−2	4	4	−4
2	−2	4	4	−4
4	−1	16	1	−4
6	2	36	4	12
8	2	64	4	16
9	3	81	9	27
10	4	100	16	40
10	5	100	25	50
12	5	144	25	60
14	6	196	36	84
15	7	225	49	105
16	8	256	64	128
105	29	1235	275	525

$$N \sum uv - \sum u \sum v = 14(525) - 105(29) = 4305$$

$$N \sum u^2 - \left(\sum u\right)^2 = 14(1235) - (105)^2 = 6265$$

$$N \sum v^2 - \left(\sum v\right)^2 = 14(275) - (29)^2 = 3009$$

Since $h = 1$ and $k = 0.1$, we have

$$b = 0.1\,\frac{4305}{6265} = 0.0687$$

$$r^2 = \frac{(4305)^2}{6265 \times 3009} = 0.9831$$

$$r = 0.9915$$

$$\bar{x} = 50 + \frac{105}{14} = 57.5$$

$$\bar{y} = 3.0 + 0.1\,\frac{29}{14} = 3.207$$

so that for any given x

$$y_c = 3.207 + 0.0687(x - 57.5)$$

This is the relation required.

EXAMPLE 2 When the sample to be studied is large, it is usually convenient to replace the scatter diagram by a two-way frequency table, the frequency in any cell of the table being the number of individuals in the sample falling within the corresponding class intervals for both X and Y. In Table 11.2, X represents the grade achieved on a mental test by applicants for a certain type of industrial job, and Y the productive ability of these applicants after hiring (measured as a percentage of a certain standard of production). The auxiliary variables u and v are here defined as $u = (x - 42.5)/5$, $v = (y - 85)/10$. The values of x and y shown in the table headings are the centers of the class-intervals. The marginal column totals are denoted by f_u (the frequency for a given value of u), and the marginal row totals are similarly denoted by f_v. Clearly,

$$\sum_u f_u = \sum_v f_v = N = 260$$

which is also the sum of all the frequencies in all the cells in the main body of the table (the part surrounded by a double line). The values of f_v are added vertically and the values of f_u horizontally.

The rows headed uf_u and u^2f_u are obtained by multiplying each f_u by the corresponding u and then multiplying the product again by u. Similarly for the columns headed vf_v and v^2f_v. These rows and columns give the means and variances of u and v.

$$\bar{u} = \sum \frac{uf_u}{N} = \frac{-17}{260} = -0.06538$$

$$\bar{v} = \sum \frac{vf_v}{N} = \frac{60}{260} = 0.23077$$

$$(N-1)s_u^{\,2} = \sum u^2 f_u - \frac{(\sum u f_u)^2}{N} = 735 - \frac{289}{260}$$
$$= 733.89$$
$$(N-1)s_v^{\,2} = \sum v^2 f_v - \frac{(\sum v f_v)^2}{N} = 802 - \frac{3600}{260}$$
$$= 788.15$$

The means and variances of X and Y are

$$\bar{x} = 42.5 + 5\bar{u} = 42.2$$
$$\bar{y} = 85 + 10\bar{v} = 87.3$$
$$s_X^{\,2} = 25 s_u^{\,2} = 70.84$$
$$s_Y^{\,2} = 100 s_v^{\,2} = 304.3$$

There are various methods in use for obtaining the covariance s_{uv}. One method, which has the advantage of providing convenient checks on the calculations, is

TABLE 11.2

	x	22.5	27.5	32.5	37.5	42.5	47.5	52.5	57.5					
y	v \\ u	−4	−3	−2	−1	0	1	2	3	f_v	vf_v	v^2f_v	U	vU
125	4					2	3	2		7	28	112	7	28
115	3			1	3	1	4	4	4	17	51	153	19	57
105	2			5	7	8	11	8	7	46	92	184	31	62
95	1		2	1	10	12	9	8	2	44	44	44	13	13
85	0	1	3	12	11	7	12	7	1	54	0	0	−19	0
75	−1	2	1	5	6	16	8	5		43	−43	43	−9	9
65	−2	2	5	5	8	8	6	1		35	−70	140	−33	66
55	−3	2	3	3	4	1	1			14	−42	126	−26	78
	f_u	7	14	32	49	55	54	35	14	260	60	802	−17	313
	uf_u	−28	−42	−64	−49	0	54	70	42	−17	Checks			
	u^2f_u	112	126	128	49	0	54	140	126	735				
	V	−12	−18	−10	−1	4	32	37	28	60				
	uV	48	54	20	1	0	32	74	84	313				

indicated in the table. The row headed V is obtained by multiplying each cell-frequency in a given column by the value of v corresponding to that cell and adding along the column. Thus, for the first column, $V = 1(0) + 2(-1) + 2(-2) + 2(-3) = -12$. Each value of V is then multiplied by the corresponding u to give the row headed uV.

A similar procedure is used for the columns U and vU. For the first row $U = 2(0) + 3(1) + 2(2) = 7$, and for this row $v = 4$, so that $vU = 28$. The checks are

$$\sum V = \sum vf_v = 60$$

$$\sum U = \sum uf_u = -17$$

$$\sum uV = \sum vU = 313$$

From the method of calculation it is clear that $\sum uV$ or $\sum vU$ gives the same result as we might have obtained by multiplying each individual cell-frequency by its own u and v, and adding over all the cells of the table. That is, it gives the quantity $\sum fuv$ which we need in calculating the covariance. In fact,

$$\begin{aligned}(N-1)s_{uv} &= \sum uV - \left(\sum uf_u\right)\frac{\sum vf_v}{N} \\ &= 313 - (-17)\frac{60}{260} \\ &= 316.92\end{aligned}$$

so that $s_{XY} = 50s_{uv} = 61.18$.

The regression line of Y on X is given by $y_c - \bar{y} = b(x - \bar{x})$, where $b = 61.18/70.84 = 0.864$.

The coefficient of correlation between X and Y is given by

$$\begin{aligned}r^2 &= \frac{61.18^{\ 2}}{(70.84)(304.3)} \\ &= 0.1737\end{aligned}$$

so that

$$r = 0.417$$

11.5 **Variance about the Regression Line** When the values of a and b are chosen according to the equations of (11.3.5), the minimum value of S is given by

$$\begin{aligned}S_{\min} &= \sum_i e_i^2 = \sum_i [y_i - \bar{y} - b(x_i - \bar{x})]^2 \\ &= \sum_i (y_i - \bar{y})^2 + b^2 \sum_i (x_i - \bar{x})^2 \\ &\qquad - 2b \sum_i (x_i - \bar{x})(y_i - \bar{y}) \\ &= (N-1)(s_Y^2 + b^2 s_X^2 - 2bs_{XY})\end{aligned} \tag{11.5.1}$$

where s_X^2, s_Y^2, and s_{XY} are the sample variances of X and Y and the sample covariance, respectively. Using Eq. (11.3.9), we may write this

$$S_{\min} = (N-1)s_Y^2(1-r^2) \tag{11.5.2}$$

which may be compared with (11.1.21) and suggests that $S_{\min}/(N-1)$ is an estimator of σ_{Ye}^2. It turns out, however, that a better (unbiased) estimator is $S_{\min}/(N-2)$.

Corresponding to the sample value x_i, we have three values of Y, namely, the actual sample value y_i, the estimated value y_{ci} and the true expected value η_i, these being connected by the relations

$$\begin{aligned} y_i &= \eta_i + \varepsilon_i = \alpha + \beta x_i + \varepsilon_i \\ &= y_{ci} + e_i = a + bx_i + e_i \end{aligned} \tag{11.5.3}$$

Now

$$\begin{aligned} (N-1)s_{XY} &= \sum_i (y_i - \bar{y})(x_i - \bar{x}) \\ &= \sum_i y_i(x_i - \bar{x}) \end{aligned}$$

since $\sum_i (x_i - \bar{x}) = 0$. Writing $y_i = \alpha + \beta x_i + \varepsilon_i = \alpha + \beta(x_i - \bar{x}) + \beta\bar{x} + \varepsilon_i$, we obtain $(N-1)s_{XY} = \beta \sum_i (x_i - \bar{x})^2 + \sum \varepsilon_i(x_i - \bar{x})$ since the other terms vanish. Therefore $(N-1)s_{XY} = \beta(N-1)s_X^2 + \sum \varepsilon_i(x_i - \bar{x})$ so that, on dividing by $(N-1)s_X^2$, $\frac{s_{XY}}{s_X^2} = \beta + \sum \varepsilon_i(x_i - \bar{x})/[(N-1)s_X^2]$. We have then

$$b = \beta + e_b \tag{11.5.4}$$

where

$$e_b = \sum_i \frac{\varepsilon_i(x_i - \bar{x})}{(N-1)s_X^2} \tag{11.5.5}$$

which may be regarded as the "error" or sampling fluctuation of b. Since we assumed that $E(\varepsilon_i) = 0$ for all i, it follows that $E(e_b) = 0$ and therefore b is an unbiased estimator of β. We have also assumed that the variance of ε_i is the same (σ^2) for all i—in other words, that Y is *homoscedastic*. If so, we can write

$$\begin{aligned} V(b) = V(e_b) &= \sigma^2 \sum_i \frac{(x_i - \bar{x})^2}{(N-1)^2 s_X^4} \\ &= \frac{\sigma^2}{(N-1)s_X^2} \end{aligned} \tag{11.5.6}$$

Moreover, if ε_i is normally distributed for each i, so is b, with expectation β and variance $\sigma^2/[(N-1)s_X^2]$.

From Eq. (3),

$$\varepsilon_i - e_i = (a - \alpha) + (b - \beta)x_i \tag{11.5.7}$$

Also, from the first equation of (11.3.5), $\sum e_i = 0$, so that, on summing Eq. (7) over i and dividing by N, we obtain

(11.5.8) $$a - \alpha + (b - \beta)\bar{x} = \bar{\varepsilon}$$

From Eqs. (7) and (4),

(11.5.9) $$\varepsilon_i - e_i = \bar{\varepsilon} + e_b(x_i - \bar{x})$$

Now, for any fixed value x,

$$\begin{aligned} y_c - \eta &= a - \alpha + (b - \beta)x \\ &= \bar{\varepsilon} + e_b(x - \bar{x}) \\ &= \sum_i \varepsilon_i \left[\frac{1}{N} + \frac{(x - \bar{x})(x_i - \bar{x})}{(N-1)s_X^2}\right]. \end{aligned}$$

Since the ε_i are independent, with mean 0 and common variance σ^2, it follows that

(11.5.10) $$E(y_c) = \eta$$

and

(11.5.11) $$\begin{aligned} V(y_c) &= \sigma^2 \sum_i \left[\frac{1}{N} + \frac{(x - \bar{x})(x_i - \bar{x})}{(N-1)s_X^2}\right]^2 \\ &= \sigma^2\left[\frac{1}{N} + \frac{(x - \bar{x})^2}{(N-1)s_X^2}\right] \end{aligned}$$

since $\sum_i (x_i - \bar{x})^2 = (N - 1)s_X^2$.

This expression contains the unknown variance σ^2, which we need to estimate. The minimum sum of squares $S_{\min}$, by Eq. (9), may be written

(11.5.12) $$\begin{aligned} S_{\min} &= \sum_i e_i^2 = \sum_i [\varepsilon_i - \bar{\varepsilon} - e_b(x_i - \bar{x})]^2 \\ &= \sum_i \varepsilon_i^2 - N\bar{\varepsilon}^2 + (N-1)e_b^2 s_X^2 - 2e_b \sum_i \varepsilon_i(x_i - \bar{x}) \\ &= \sum_i \varepsilon_i^2 - N\bar{\varepsilon}^2 - (N-1)s_X^2 e_b^2 \end{aligned}$$

Now ε_i/σ is by hypothesis a standard normal variate, and the sum of N squares of such variates is distributed as χ^2 with N degrees of freedom. Also, $N^{1/2}\bar{\varepsilon}/\sigma$ is a standard normal variate, and its square is distributed as χ^2 with 1 d.f. Finally, $(N - 1)^{1/2}s_X e_b/\sigma$ is also a standard normal variate and its square is a χ^2 variate with 1 d.f. These last two normal variates are both linear functions of the ε_i/σ and it is easily verified that they are orthogonal (see Appendix A.10). It follows from Fisher's Theorem (§ 4.7) that $S_{\min}/\sigma^2$ is distributed as χ^2 with $N - 2$ degrees of freedom and is independent of the last two terms on the right-hand side of (12).

From the known properties of the chi-square distribution,

$$E\,\frac{S_{\min}}{\sigma^2} = N - 2$$

or

$$E\,\frac{S_{\min}}{N-2} = \sigma^2 \tag{11.5.13}$$

An unbiased estimator of σ^2 is therefore

$$\hat{\sigma}^2 = \frac{S_{\min}}{N-2} = \frac{N-1}{N-2}\,s_Y^{\,2}(1 - r^2) \tag{11.5.14}$$

and on substituting this for σ^2 in Eq. (11) we obtain for the estimated variance of y_c,

$$\hat{V}(y_c) = \hat{\sigma}^2\left[\frac{1}{N} + \frac{(x-\bar{x})^2}{(N-1)s_X^{\,2}}\right] \tag{11.5.15}$$

This is a function of x, having its least value when $x = \bar{x}$.

We may be interested in the variance of Y *about the estimated regression line*, that is, in the variance of $Y - y_c$ for some new assumed value x of X. Since the variance of y_c depends only on the N observations already made and the new observation is independent of these, we may write

$$\begin{aligned} V(Y - y_c \mid x) &= V(Y \mid x) + V(y_c) \\ &= \sigma^2\left[1 + \frac{1}{N} + \frac{(x-\bar{x})^2}{(N-1)s_X^{\,2}}\right] \end{aligned} \tag{11.5.16}$$

and an estimate of this variance is given by putting $\hat{\sigma}^2$, from Eq. (14), in place of σ^2. The square root of $\hat{V}(y - y_c)$ is called the *standard error of estimate.*

11.6 Confidence Limits for the Parameters of Regression and for Estimated Y

Since the slope b of the regression line (on the hypotheses stated above) is a normal variate with expectation β and variance $\sigma^2/[(N-1)s_X^{\,2}]$, and since an independent unbiased estimate of σ^2 is furnished by $\hat{\sigma}^2$, with $N - 2$ degrees of freedom, the quantity $(b - \beta)\left[\frac{(N-1)s_X^{\,2}}{\hat{\sigma}^2}\right]^{1/2}$ has the Student-t distribution with $N - 2$ d.f.

Using Eq. (11.5.14) we can write the $100(1 - \alpha)\%$ confidence limits for β as

$$\beta = b \pm t_\alpha\left(\frac{s_Y^{\,2}(1 - r^2)}{(N-2)s_X^{\,2}}\right)^{1/2} \tag{11.6.1}$$

where t_α is the value of t exceeded numerically with probability α. This may also be written

$$\beta = b \pm t_\alpha\,\frac{b}{r}\left(\frac{1 - r^2}{N-2}\right)^{1/2} \tag{11.6.2}$$

The variance of a is $\sigma^2\left[\frac{1}{N} + \frac{\bar{x}^2}{(N-1)s_X^2}\right]$ and confidence limits for the true regression parameter α may be found in a similar manner. However, it is seldom important to know α. In most regression problems it is the slope that matters.

EXAMPLE 3 For a sample of size 27, we find that $b = 0.163$, $r = 0.582$. What are the 95% confidence limits for β?

For 25 d.f., $t_{0.05} = 2.060$. Therefore the limits are

$$\begin{aligned}\beta &= 0.163 \pm 2.060(0.0456)\\ &= 0.069 \text{ and } 0.257\end{aligned}$$

If $\beta = 0$ (and therefore $\rho = 0$), the quantity $r\left(\frac{N-2}{1-r^2}\right)^{1/2}$ has the Student-t distribution with $N - 2$ d.f. This is sometimes useful in deciding whether an observed value of r differs significantly from 0.

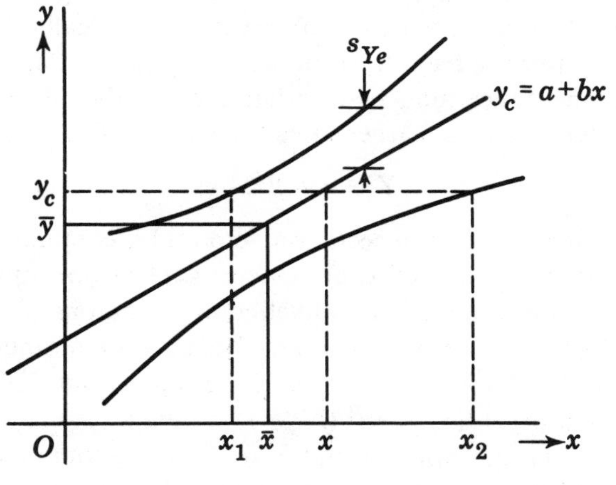

FIG. 51 CONFIDENCE BELT FOR ESTIMATE FROM LINEAR REGRESSION

EXAMPLE 4 For a sample of size 27, suppose that $r = 0.348$. Is this significantly different from zero?

Here $t = 1.856$, with 25 d.f. The probability of a value *numerically* as great as this is between 0.05 and 0.1, so that at the 5% level of significance the answer to the question is "No." It takes a fairly large value of r to be significant with a sample size as small as 27.

Since the expectation of $Y - y_c$ is zero and its variance is given by Eq. (11.5.16), it follows that

$$\frac{(Y - y_c)}{\left\{\hat{\sigma}^2\left[1 + \frac{1}{N} + \frac{(x-\bar{x})^2}{(N-1)s_X^2}\right]\right\}^{1/2}}$$

where $\hat{\sigma}^2$ is substituted for σ^2, has the t-distribution with $N - 2$ d.f. The $100(1 - \alpha)\%$ confidence limits for Y for a given x are therefore

$$Y = y_c \pm t_\alpha s_{Ye} \tag{11.6.3}$$

where $y_c = a + bx$ and

$$\begin{aligned} s_{Ye}^2 &= \hat{\sigma}^2\left[1 + \frac{1}{N} + \frac{(x - \bar{x})^2}{(N - 1)s_X^2}\right] \\ &= \frac{N - 1}{N - 2} s_Y^2(1 - r^2)\left[1 + \frac{1}{N} + \frac{(x - \bar{x})^2}{(N - 1)s_X^2}\right] \end{aligned}$$

The curves bounding the confidence belt are hyperbolas (Figure 51). For large N, $s_{Ye}^2 \approx s_Y^2(1 - r^2)$, and the belt is almost of uniform width.

11.7 **Regression when the Variable X Is Not Random** As mentioned in § 11.3, it often happens in practice that the values of X in an experiment are pre-selected, so as to be convenient numbers instead of being chosen at random. Sometimes, also, the circumstances of observation practically dictate the values of X, so that the observer has very little choice in the matter. The problem is then not really one concerning two variates, but rather it concerns a single variate Y which depends on a mathematical variable x. The assumption is that

$$Y = \alpha + \beta x + \varepsilon_x \tag{11.7.1}$$

where the ε_x are normal, with expectation zero and a common variance σ^2 for all x, and, for different values of x, are independent of one another. With this assumption (see § 11.8) the maximum likelihood estimators of α and β turn out to be the a and b of § 11.3. Since, however, X is now not a random variable, we must understand by s_X^2 not the estimator of the true variance of X but merely a symbol for the quantity $\sum (x_i - \bar{x})^2/(N - 1)$, where $\bar{x} = \sum x_i/N$.

As we have seen earlier, when X and Y are both random there are in general two distinct regression lines, one for estimating Y for given values of X, and the other for estimating X for given values of Y. The former is still good when X is pre-selected and not random (the least squares argument of § 11.3 is still valid) but the latter has no meaning in this case.

It sometimes happens that we would like to estimate X for a given value of Y even though X is not random. In testing an insecticide, for example, we may wish to estimate the median lethal dose (the dose that will kill 50% of the time) from observations made on the proportions of insects killed with various known doses. The method is to invert the regression equation $y_c = a + bx$ and write

$$x = \frac{y_c - a}{b} \tag{11.7.2}$$

where y_c is the given value of Y (see Figure 51). The $100(1 - \alpha)\%$ confidence limits for x are x_1 and x_2, these being the abscissae of the points where the line

$Y = y_c$ cuts the confidence band. The values of x_1 and x_2 can be calculated from (11.6.3) by putting $y_c = a + bx \pm t_\alpha s_{Ye}$ and treating this equation as a quadratic in x. The two roots are the required limits, provided that b is sufficiently large. For values of b too small to be significantly different from zero at the level α, it may happen that no finite confidence interval for x exists, but this case is not of much practical importance.

* 11.8 **Maximum Likelihood Estimation of the Regression Coefficients** We consider first the case when X is "fixed" (pre-selected), Y being given by Eq. (11.7.1), with the assumptions there stated. If ε_i denotes the value of ε_x when $x = x_i$, $i = 1, 2 \ldots N$, the joint distribution of the N values ε_i has the density function

$$g(\varepsilon_1, \ldots \varepsilon_N) = (2\pi\sigma^2)^{-N/2} \exp\left(-\sum_i \frac{\varepsilon_i^2}{2\sigma^2}\right) \tag{11.8.1}$$

Let the observed values of Y corresponding to the fixed values x_i be denoted by y_i. The joint density function for the y_i is $f(y_1, y_2 \ldots y_N)$, where

$$f(y_1, y_2 \ldots y_N)\, dy_1\, dy_2 \ldots dy_N = g(\varepsilon_1, \varepsilon_2 \ldots \varepsilon_N)\, d\varepsilon_1 \ldots d\varepsilon_N \tag{11.8.2}$$

The y_i are related to the ε_i by equations

$$y_i = \alpha + \beta x_i + \varepsilon_i, \qquad i = 1, 2 \ldots N \tag{11.8.3}$$

and the Jacobian of the transformation from the ε_i to the y_i, for fixed x_i, is equal to 1. It follows that

$$f(y_1, y_2 \ldots y_N) = (2\pi\sigma^2)^{-N/2} \exp\left[-\sum_i \frac{(y_i - \alpha - \beta x_i)^2}{2\sigma^2}\right] \tag{11.8.4}$$

The principle of maximum likelihood suggests that we choose α and β to maximize this function $f(y_1 \ldots y_N)$. This is equivalent to minimizing $\sum_i (y_i - \alpha - \beta x_i)^2$, and on differentiating partially with respect to α and β and equating the derivatives to zero, we arrive at the following equations for the maximum likelihood estimators $\hat{\alpha}$ and $\hat{\beta}$:

$$\begin{cases} \sum_i (y_i - \hat{\alpha} - \hat{\beta} x_i) = 0 \\ \sum_i x_i(y_i - \hat{\alpha} - \hat{\beta} x_i) = 0 \end{cases} \tag{11.8.5}$$

These are identical with the equations of (11.3.5) so that $\hat{\alpha} = a$, $\hat{\beta} = b$. We get the same result as by the method of least squares.

The above argument does not hold if X is a random variable. If the joint probability density for X and Y is $f(x, y)$, the likelihood for the observed sampl[e] is

$$L = f(x_1, y_1) f(x_2, y_2) \ldots f(x_N, y_N) \tag{11.8.6}$$

If we transform to new variates U and ε by the relations

$$(11.8.7)\qquad \begin{cases} X = U \\ Y = \alpha + \beta U + \varepsilon \end{cases}$$

the Jacobian of the transformation is 1, and therefore Eq. (6) becomes

$$(11.8.8)\qquad L = f(u_1, \alpha + \beta u_1 + \varepsilon_1) f(u_2, \alpha + \beta u_2 + \varepsilon_2) \ldots f(u_N, \alpha + \beta u_N + \varepsilon_N)$$

If this splits into two factors, one of which depends on the ε_i alone and the other on the u_i alone, we must have

$$f(u, \alpha + \beta u + \varepsilon) = g(u) \cdot h(\varepsilon)$$

or, equivalently,

$$(11.8.9)\qquad f(x, y) = g(x) \cdot h(y - \alpha - \beta x)$$

If and only if this condition is satisfied, we can feel justified in estimating α and β from the distribution $h(\varepsilon)$ of ε alone. When this distribution is normal we get back to the equations (5). The condition (9) is satisfied, for example, if the population is bivariate normal. Using the standardized variates z and v, we see from Eqs. (11.2.5), (11.2.6) and (11.2.8) that

$$(11.8.10)\qquad \begin{aligned} g(z, v) &= f(z) \cdot g(v|z) \\ &= [(2\pi)^{-1/2} e^{-z^2/2}](2\pi)^{-1/2}(1 - \rho^2)^{-1/2} \exp\left[-\frac{(v - \rho z)^2}{2(1 - \rho^2)}\right] \end{aligned}$$

The first factor is a function of z alone and the second of $v - \rho z$ alone. In these variates $v - \rho z$ is equivalent to $y - \alpha - \beta x$.

It should be observed that when condition (9) is satisfied the regression is *necessarily* linear. For

$$(11.8.11)\qquad \begin{aligned} E(Y|X = x) &= \int y \cdot h(y - \alpha - \beta x)\, dy \\ &= \int (y - \alpha - \beta x) h(y - \alpha - \beta x)\, dy \\ &\qquad + (\alpha + \beta x) \int h(y - \alpha - \beta x)\, dy \\ &= E(\varepsilon) + (\alpha + \beta x) \end{aligned}$$

since $h(\varepsilon)$ is a density function.

The first term can be taken as zero by suitably adjusting α, and we thus have the ordinary equation for linear regression.

11.9 Functional Relation Between Variables Subject to Error It often happens that the variables X and Y, whether "fixed" or random, are subject to experimental error. With random variables this error is mixed up with the fluctuation due to the underlying probability distribution. We will therefore suppose to begin with that, apart from the error, X and Y are fixed and that Y is a linear function of X. That is,

$$\begin{cases} X = \xi + u \\ Y = \eta + v \end{cases} \tag{11.9.1}$$

where

$$\eta = \alpha + \beta\xi \tag{11.9.2}$$

The quantities ξ and η are regarded as the true values of the variables, and u and v as the errors. We suppose that u and v are uncorrelated with each other, and, for any value of ξ, are distributed normally with means zero and variances σ_u^2 and σ_v^2 respectively.

The joint density function for X and Y is then

$$f(x, y) = (2\pi\sigma_u\sigma_v)^{-1} \exp\left[-\frac{(x-\xi)^2}{2\sigma_u^2} - \frac{(y-\eta)^2}{2\sigma_v^2}\right] \tag{11.9.3}$$

The likelihood function for a set of N pairs of observed values (x_i, y_i) is

$$L = (2\pi\sigma_u\sigma_v)^{-N} \exp\left[-\frac{\sum_i (x_i-\xi_i)^2}{2\sigma_u^2} - \frac{\sum_i (y_i-\alpha-\beta\xi_i)^2}{2\sigma_v^2}\right] \tag{11.9.4}$$

so that

$$\log L = C - N\log\sigma_u - N\log\sigma_v - \frac{1}{2\sigma_u^2}\sum_i (x_i-\xi_i)^2 - \frac{1}{2\sigma_v^2}\sum_i (y_i-\alpha-\beta\xi_i)^2 \tag{11.9.5}$$

The right-hand side contains $N + 4$ unknown parameters, namely α, β, σ_u, σ_v and the N values ξ_i. The maximum likelihood equations, found by differentiating partially with respect to each of these parameters and setting the derivatives equal to zero, are

$$\sum (y_i - \alpha - \beta\xi_i) = 0 \tag{11.9.6}$$

$$\sum \xi_i(y_i - \alpha - \beta\xi_i) = 0 \tag{11.9.7}$$

$$\sum (x_i - \xi_i)^2 = N\sigma_u^2 \tag{11.9.8}$$

$$\sum (y_i - \alpha - \beta\xi_i)^2 = N\sigma_v^2 \tag{11.9.9}$$

$$(x_i - \xi_i)/\sigma_u^2 + \beta(y_i - \alpha - \beta\xi_i)/\sigma_v^2 = 0, \qquad i = 1, 2 \ldots N \tag{11.9.10}$$

From Eqs. (8) and (10) we find

$$N\sigma_u^2 = \sum \beta^2\sigma_u^4 \frac{(y_i - \alpha - \beta\xi_i)^2}{\sigma_v^4}$$

and, on substituting from Eq. (9),

$$\beta^2 = \frac{\sigma_v^{\ 2}}{\sigma_u^{\ 2}} \tag{11.9.11}$$

Since this relation cannot be supposed to hold in general, the maximum likelihood method is not satisfactory unless some further assumption is made. The most convenient one to make is that the *ratio* of $\sigma_v^{\ 2}$ to $\sigma_u^{\ 2}$ is definitely known. If this ratio is denoted by λ, and if we let $\sigma_u^{\ 2} = \sigma^2$, $\sigma_v^{\ 2} = \lambda\sigma^2$, we obtain, instead of Eqs. (8) and (9), the one relation

$$2N\lambda\sigma^2 = \lambda \sum (x_i - \xi_i)^2 + \sum (y_i - \alpha - \beta\xi_i)^2 \tag{11.9.12}$$

and instead of Eq. (10)

$$\lambda(x_i - \xi_i) + \beta(y_i - \alpha - \beta\xi_i) = 0 \tag{11.9.13}$$

Substituting from Eq. (13) in Eqs. (6) and (7), we find, after some rearrangement of terms, that

$$N\alpha + \sum x_i\beta = \sum y_i \tag{11.9.14}$$

and

$$\sum x_i\alpha + \sum x_i^2\beta = \sum x_iy_i + \frac{\beta}{\lambda}\left(\sum y_i^2 - \alpha \sum y_i - \beta \sum x_iy_i\right) \tag{11.9.15}$$

By eliminating α from these two equations we obtain a quadratic equation in β, which reduces to

$$s_{XY}\beta^2 + (\lambda s_X^{\ 2} - s_Y^{\ 2})\beta - \lambda s_{XY} = 0 \tag{11.9.16}$$

where $s_X^{\ 2}$, $s_Y^{\ 2}$ and s_{XY} are the variances and covariance of the observed sample values x_i and y_i. The estimator of β is

$$\hat{\beta} = \{s_Y^{\ 2} - \lambda s_X^{\ 2} + [(s_Y^{\ 2} - \lambda s_X^{\ 2})^2 + 4\lambda s_{XY}^{\ \ 2}]^{1/2}\}/(2s_{XY}) \tag{11.9.17}$$

The estimator of α is found from Eq. (14) and that of σ^2 from Eq. (12). It turns out that

$$\hat{\sigma}^2 = \frac{N-1}{4N\lambda}\{s_Y^{\ 2} + \lambda s_X^{\ 2} - [(s_Y^{\ 2} - \lambda s_X^{\ 2})^2 + 4\lambda s_{XY}^{\ \ 2}]^{1/2}\} \tag{11.9.18}$$

Unfortunately, as Lindley [1] has shown, this is not a consistent estimator of σ^2. It tends in probability, as N increases, to the value $\sigma^2/2$.

If we use the *least squares method*, it is no longer correct, as in the classical regression problem, to minimize the sum of squares of deviations parallel to the y-axis. Since the values of X observed are no longer the true values, the variance of X must be taken into consideration. One way is to minimize the sum of squares of the X deviations, weighted inversely as σ_u^2, plus the sum of squares of the Y deviations, weighted inversely as σ_v^2. That is, we minimize

$$\lambda \sum (x_i - \xi_i)^2 + \sum (y_i - \alpha - \beta\xi_i)^2$$

with respect to α, β and the ξ_i. Another method is to weight the squared deviation of y_i from $\alpha + \beta x_i$ with a weight inversely proportional to the variance of $y_i - \alpha - \beta x_i$. Since this variance is $\sigma_v^2 + \beta^2\sigma_u^2 = \sigma_u^2(\lambda + \beta^2)$, the expression to be minimized is

$$\sum \frac{(y_i - \alpha - \beta x_i)^2}{\lambda + \beta^2}$$

Differentiating partially with respect to α and β (λ is supposed known, as in the maximum likelihood method), we obtain

(11.9.19) $$\bar{y} = \hat{\alpha} + \hat{\beta}\bar{x}$$

and

(11.9.20) $$\hat{\beta} \sum (y_i - \hat{\alpha} - \hat{\beta}x_i)^2 + (\lambda + \hat{\beta}^2) \sum x_i(y_i - \hat{\alpha} - \hat{\beta}x_i) = 0$$

If we imagine the origin shifted to the point $(\bar{x}, \bar{y})$, which will not affect the slope of the line, we can put $\hat{\alpha} = 0$—from Eq. (19)—and then

(11.9.21) $$\hat{\beta} \sum (y_i - \hat{\beta}x_i)^2 + (\lambda + \hat{\beta}^2) \sum x_i(y_i - \hat{\beta}x) = 0$$

which can be written

$$(\lambda - \hat{\beta}^2) \sum x_i y_i + \hat{\beta} \sum (y_i^2 - \lambda x_i^2) = 0$$

With the new origin, $\sum x_i^2$, $\sum y_i^2$, and $\sum x_i y_i$ are proportional to s_X^2, s_Y^2 and s_{XY} respectively, so that

(11.9.22) $$(\lambda - \hat{\beta}^2)s_{XY} + \hat{\beta}(s_Y^2 - \lambda s_X^2) = 0$$

which is the same as Eq. (16). The estimator $\hat{\beta}$ is therefore the same as that furnished by the method of maximum likelihood, but now a consistent estimator of σ^2 is obtainable by dividing the minimum sum of squares by the number of degrees of freedom, $N - 2$. That is,

$$\hat{\sigma}^2 = \frac{1}{N-2} \sum \frac{(y_i - \hat{\beta}x_i)^2}{\lambda + \hat{\beta}^2}$$
$$= \frac{1}{N-2} \frac{1}{\hat{\beta}} \sum x_i(\hat{\beta}x_i - y_i)$$

by Eq. (21). Therefore

(11.9.23) $$\hat{\sigma}^2 = \frac{N-1}{N-2}\left(s_X^2 - \frac{1}{\hat{\beta}} s_{XY}\right)$$
$$= \frac{N-1}{N-2}\frac{1}{\lambda}(s_Y^2 - \hat{\beta}s_{XY}), \text{ by Eq. (22)}$$

A good discussion of the functional relation between variables subject to error may be found in reference [2].

*** 11.10 The Regression Relation Between Variables Subject to Error** If Y is a random variable, subject also to error, and if X is "fixed," that is, pre-selected, the relations corresponding to Eqs. (11.9.1) and (11.9.2) are

$$(11.10.1) \qquad \begin{cases} X = \xi + u \\ Y = \eta + \varepsilon + v \\ \eta = \alpha + \beta\xi \end{cases}$$

If (x_i, y_i) are the observed pairs of values,

$$(11.10.2) \qquad \begin{cases} x_i = \xi_i + u_i \\ y_i = \eta_i + \varepsilon_i + v_i \\ \eta_i = \alpha + \beta\xi_i \end{cases}$$

where successive observations are independent, the u_i and v_i are uncorrelated with each other and with the ε_i, and where u_i, v_i and ε_i have zero expectations and variances σ_u^2, σ_v^2, σ_ε^2, for all i. Then

$$(11.10.3) \qquad \sigma_X^2 = \sigma_u^2, \qquad \sigma_Y^2 = \sigma_\varepsilon^2 + \sigma_v^2$$

The analysis of § 11.9 then holds with σ_Y^2 in place of σ_v^2. The only difference is that the variance of Y is now partly due to its inherent nature as a random variable and partly due to experimental error.

If we are interested in estimating the regression of Y on the *observed* X rather than on the *true* ξ, we simply treat the X as being without error. That is, we use the technique of § 11.7, and our estimator of β is the ordinary one obtained in § 11.3, namely,

$$(11.10.4) \qquad \hat{\beta} = b = \frac{s_{XY}}{s_X^2}$$

This is obtained as a special case of (11.9.16) when $\sigma_u^2 = 0$ and therefore $\lambda \to \infty$.

If both X and Y are random variables (the structural situation mentioned in § 11.3), the equations (1) still hold, but ξ is now a random variable with expectation μ and variance σ_ξ^2. Then η is also a random variable with expectation $\alpha + \beta\mu$ and variance $\beta^2\sigma_\xi^2$. On the assumption that u, v and ε are uncorrelated with each other and with ξ, we have

$$(11.10.5) \qquad \sigma_X^2 = \sigma_\xi^2 + \sigma_u^2, \qquad \sigma_Y^2 = \beta^2\sigma_\xi^2 + \sigma_\varepsilon^2 + \sigma_v^2, \qquad \sigma_{XY} = \beta\sigma_\xi^2$$

The sum $\sigma_\varepsilon^2 + \sigma_v^2$ may be denoted by $\sigma_{v'}^2$, as the two components cannot be distinguished. The maximum likelihood treatment of this case is discussed in reference [3]. Estimators of μ, α and β are connected by the relations

$$\hat{\mu} = \bar{x}, \qquad \hat{\alpha} + \hat{\beta}\bar{x} = \bar{y} \tag{11.10.6}$$

but in order to find $\hat{\beta}$ it is necessary to assume that either σ_u^2, $\sigma_{v'}^2$ or the ratio $\lambda = \sigma_{v'}^2/\sigma_u^2$ is known. If λ is known, $\hat{\beta}$ is given by

$$\hat{\beta} = \tau + (\tau^2 + \lambda)^{1/2}, \qquad \tau = \frac{s_Y^2 - \lambda s_X^2}{2s_{XY}} \tag{11.10.7}$$

This is the same as Eq. (11.9.17). An estimator of σ_ξ^2 is given by

$$\hat{\beta}\hat{\sigma}_\xi^2 = \hat{\sigma}_{XY} = (N-1)\frac{s_{XY}}{N} \tag{11.10.8}$$

11.11 **The Method of Grouping** A very simple method of fitting a straight line, when X and Y are functionally related but subject to error, has been suggested by several writers, notably A. Wald and M. S. Bartlett (see references [4] and [5]).

The observed N pairs are ordered, usually by reference to the X values, and divided into three groups. A number p is chosen ($p \leq \frac{1}{2}$ and such that Np is integral) and then the first Np observations are put in group G_1, the last Np in group G_3, and the remainder in group G_2. Wald suggested taking p as near as possible to $\frac{1}{2}$, so that G_2 was either empty or contained one observation. Bartlett suggested taking p approximately $\frac{1}{3}$, which in general gives greater accuracy. The exact value is not very important, but studies [5] indicate that the numbers in the three groups G_1, G_2 and G_3 should be nearly in the ratio 1:2:1 for maximum efficiency.

The method consists in plotting the points A and B, which are the centroids of the two groups of points G_1 and G_3, and joining AB. The coordinates of A and B are the group means $(\bar{x}_1, \bar{y}_1)$ and $(\bar{x}_3, \bar{y}_3)$. The slope of AB is an estimator of β. The line parallel to AB through the over-all mean $(\bar{x}, \bar{y})$, at C, is the line required, with equation

$$y - \bar{y} = \hat{\beta}(x - \bar{x}) \tag{11.11.1}$$

where

$$\hat{\beta} = \frac{\bar{y}_3 - \bar{y}_1}{\bar{x}_3 - \bar{x}_1} \tag{11.11.2}$$

(see Figure 52).

This quantity $\hat{\beta}$ is a consistent estimator of β if two conditions are satisfied: (1) the grouping should be independent of the errors of observation, (2) the quantity $\bar{x}_3 - \bar{x}_1$ should not approach zero as $N \to \infty$. The second condition is obviously satisfied if the observations are ordered according to their increasing true values ξ. Unfortunately we do not know the true values and the order of the *observed* values x may not be independent of the errors.

The precise conditions for the consistency of β are difficult to satisfy, particularly if we assume that the error in X is *normally* distributed. Theoretically an error of any magnitude whatever is possible, since the range of a normal variate is infinite. Practically the probability of an error numerically greater than $\delta = 4\sigma_u$ is negligible. If the values of ξ corresponding to cumulative

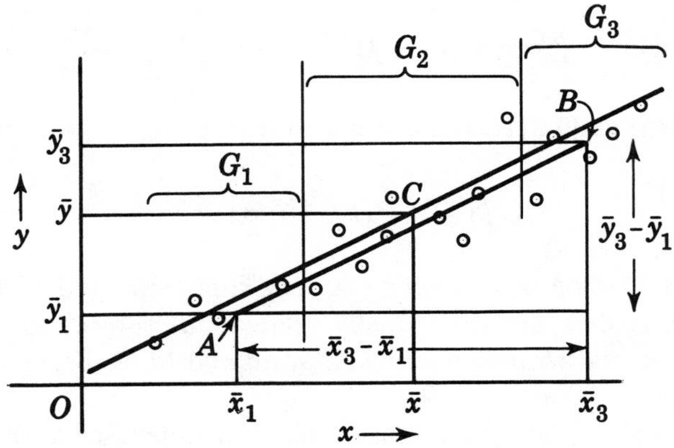

FIG. 52 FITTING A STRAIGHT LINE BY THE METHOD OF GROUPING

probabilities p and $1 - p$ are denoted by ξ_p and ξ_{1-p}, the grouping by *observed* values will be practically the same as grouping by *true* values, provided that scarcely any observed values fall in the intervals

$$[\xi_p - \delta, \xi_p + \delta] \quad \text{and} \quad [\xi_{1-p} - \delta, \xi_{1-p} + \delta].$$

As in § 11.9, we assume that the errors $u = X - \xi$ are independent and normal with common variance σ_u^2, the errors $v = Y - \eta$ are independent and normal with common variance σ_v^2, and u and v are uncorrelated with each other. Then it is possible to determine confidence limits for β.

From Eq. (2), we have

$$(11.11.3) \qquad \begin{aligned}(\bar{x}_3 - \bar{x}_1)(\hat{\beta} - \beta) &= \bar{y}_3 - \bar{y}_1 - \beta(\bar{x}_3 - \bar{x}_1) \\ &= (\bar{v}_3 - \beta\bar{u}_3) - (\bar{v}_1 - \beta\bar{u}_1)\end{aligned}$$

since $y_i = \eta_i + v_i = \alpha + \beta\xi_i + v_i$ and $x_i = \xi_i + u_i$. The variance of $\bar{v}_3$ or $\bar{v}_1$ is σ_v^2/k, where $k = Np$, and that of $\bar{u}_3$ or $\bar{u}_1$ is σ_u^2/k, so that the variance of the right-hand side of Eq. (3) is $(\sigma_v^2 + \beta^2\sigma_u^2)(2/k)$.

For the points in group G_1, $y_i - \bar{y}_1 - \beta(x_i - \bar{x}_1) = v_i - \bar{v}_1 - \beta(u_i - \bar{u}_1)$, so that

$$(11.11.4) \qquad \begin{aligned}\sum_{G_1} [y_i - \bar{y}_1 - \beta(x_i - \bar{x}_1)]^2 &= \sum (v_i - \bar{v}_1)^2 \\ &+ \beta^2 \sum (u_i - \bar{u}_1)^2 - 2\beta \sum (v_i - \bar{v}_1)(u_i - \bar{u}_1)\end{aligned}$$

and the left-hand side is therefore an estimator of $(k - 1)(\sigma_v^2 + \beta^2\sigma_u^2)$. The corresponding sums over G_2 and G_3 are estimators of $(N - 2k - 1)(\sigma_v^2 + \beta^2\sigma_u^2)$

and $(k - 1)(\sigma_v^2 + \beta^2\sigma_u^2)$ respectively. The three sums combined give an estimator of $(N - 3)(\sigma_v^2 + \beta^2\sigma_u^2)$.

If we write

$$(11.11.5)\qquad (N-3)s_x^2 = \sum_{G_1}(x_i - \bar{x}_1)^2 + \sum_{G_2}(x_i - \bar{x}_2)^2 + \sum_{G_3}(x_i - \bar{x}_3)^2$$

and similar expressions for $(N - 3)s_y^2$ and $(N - 3)s_{xy}$, the estimator of $\sigma_v^2 + \beta^2\sigma_u^2$ is $s_y^2 - 2\beta s_{xy} + \beta^2 s_x^2 = S(\beta)$, say.

Since on our assumptions—see Eq. (3) above—the statistic $(\bar{x}_3 - \bar{x}_1)(\hat{\beta} - \beta)$ is normally distributed about zero, and since an independent estimator of its variance with $N - 3$ d.f. is $2S(\beta)/k$, the quantity

$$(11.11.6)\qquad t = \frac{(\bar{x}_3 - \bar{x}_1)(\hat{\beta} - \beta)}{[2S(\beta)/k]^{1/2}}$$

has the Student-t distribution with $N - 3$ degrees of freedom. If t_α is the value of t exceeded in absolute value with probability α, the quadratic equation

$$(11.11.7)\qquad \frac{2t_\alpha^2}{k}[s_y^2 - 2\beta s_{xy} + \beta^2 s_x^2] = (\bar{x}_3 - \bar{x}_1)^2(\hat{\beta} - \beta)^2$$

where $\hat{\beta}$ is given by Eq. (2), provides $100(1 - \alpha)\%$ confidence limits for β. If β_u is the upper 95% limit, a rough estimate of the standard deviation of $\hat{\beta}$ is given by $(\beta_u - \hat{\beta})/t_{0.05}$.

EXAMPLE 5 The friction (ounces) in a simple machine is y when the load is x ounces.

x	23.4	44.7	65.4	86.8	107.5	128.8	149.6	171.0
y	3.4	4.7	5.4	6.8	7.5	8.8	9.6	11.0

Taking the first two and the last two measurements, we have $\bar{x}_1 = 34.05$, $\bar{x}_3 = 160.3$, $\bar{y}_1 = 4.05$, $\bar{y}_3 = 10.3$; also $\bar{x} = 97.15$, $\bar{y} = 7.15$. Therefore $\hat{\beta} = 0.0495$, and the line fitted is $y = 0.0495x + 2.34$. We also find from Eq. (5) that $5s_x^2 = 226.8 + 2224.0 + 229.0 = 2679.8$, $5s_y^2 = 7.8525$, $5s_{xy} = 143.85$, so that $S(\hat{\beta}) = 1.5705 - 57.54\hat{\beta} + 535.96\hat{\beta}^2 = 0.0354$.

The 95% confidence limits for β are given by Eq. (7) with $k = 2$, $t_\alpha = 2.571$. This equation reduces to $\beta^2 - 0.09661\beta + 0.002313 = 0$, the roots of which are $\beta = 0.0437$ and 0.0529.

If we use Wald's method with two groups of four observations each, the central group G_2 is empty and $N - 3$ in Eq. (5) must be replaced by $N - 2$. We find $\bar{x}_1 = 55.075$, $\bar{x}_3 = 139.225$, $\bar{y}_1 = 5.075$, $\bar{y}_3 = 9.225$, with $\bar{x}$ and $\bar{y}$ as before. The line fitted is $y = 0.0493x + 2.36$. Also $6s_x^2 = 4456.5$, $6s_y^2 = 12.48$, $6s_{xy} = 234.48$, so that $S(\hat{\beta}) = 2.08 - 78.16\hat{\beta} + 742.75\hat{\beta}^2 = 0.0321$ with $\hat{\beta} = 0.0493$.

The 95% limits given by Eq. (7), with $k = 4$, are 0.0430 and 0.0526; this method, therefore, in spite of using all the observations, gives a slightly less reliable value of the slope than the method which uses only the first two and last two observations.

11.12 **The Distribution of the Pearson Correlation Coefficient** We have seen that if the parent population is uncorrelated (i.e., $\rho = 0$) and if X and Y are normally distributed, the quantity

$$t = r(N-2)^{1/2}(1-r^2)^{-1/2} \tag{11.12.1}$$

has the Student-t distribution with $N - 2$ d.f. If $u = r^2/(1-r^2) = t^2/(N-2)$, it follows from the result of § 8.5 that u is distributed like the ratio of two independent χ^2 variates with 1 and $N - 2$ d.f. respectively. This in turn means (see § 4.5) that u is a beta-prime variate with parameters $\frac{1}{2}$ and $(N-2)/2$. Its density function is therefore

$$f(u) = u^{-1/2}(1+u)^{-(N-1)/2}/B\left(\frac{1}{2}, \frac{N-2}{2}\right) \tag{11.12.2}$$

The density function for r is obtained by putting $2g(r)\,dr = f(u)\,du$. The factor 2 arises because, as r goes from -1 to 1, u goes from $+\infty$ to 0 and back to $+\infty$. Since $du/dr = 2r(1-r^2)^{-2}$, we have

$$g(r) = (1-r^2)^{(N-4)/2}/B\left(\frac{1}{2}, \frac{N-2}{2}\right) \tag{11.12.3}$$

The graph of $g(r)$ is a symmetrical bell-shaped curve for $N \geq 5$. Because of the symmetry, $E(r) = 0$, and it is easily proved that

$$E(r^2) = (N-1)^{-1} \tag{11.12.4}$$

The standard deviation of r is therefore $(N-1)^{-1/2}$.

The kurtosis is $-6/(N-1)$, which tends to zero as N increases. For very large N the distribution is approximately normal.

It is not necessary to assume a bivariate normal distribution. Provided that X and Y are independent and that at least one is a random sample from a univariate normal distribution, the distribution of Eq. (3) holds.

When ρ is not zero, the exact distribution of r is quite complicated. It was first found by Fisher, using an essentially geometrical argument. An analytical treatment may be found in [7], and a full discussion by Hotelling [8] is probably the last word for some time on this subject. The density function for r is

$$f(r, \rho) = \frac{N-2}{\pi}(1-\rho^2)^{(N-1)/2}(1-r^2)^{(N-4)/2}I(\rho r) \tag{11.12.5}$$

where $I(\rho r)$ is given by

$$I(\rho r) = \int_0^\infty \frac{du}{(\cosh u - \rho r)^{N-1}} \tag{11.12.6}$$

The integral can be expressed as a rapidly convergent series, and we obtain

$$f(r, \rho) = \frac{(N-2)\Gamma(N-1)(1-\rho^2)^{(N-1)/2}(1-r^2)^{(N-4)/2}}{(2\pi)^{1/2}\Gamma(N-\frac{1}{2})(1-\rho r)^{(2N-3)/2}}S(\rho r) \tag{11.12.7}$$

where

$$S(\rho r) = 1 + \frac{\rho r + 1}{4(2N-1)} + \frac{9}{32}\cdot\frac{(\rho r + 1)^2}{(2N-1)(2N+1)} + O\left(\frac{1}{N^3}\right) \tag{11.12.8}$$

Tables of the function $f(r, \rho)$ and of its integral have been prepared by Miss F. N. David [9], who included also several charts from which approximate confidence limits for ρ may readily be obtained.

The distribution of r is far from normal, particularly when ρ is near $+1$ or -1. Series expressions may be found for the cumulants of this distribution, as follows (with n written for $N - 1$):

$$\begin{cases} \kappa_1 = E(r) = \rho - \dfrac{\rho(1-\rho^2)}{2n}\left[1 - \dfrac{1-9\rho^2}{4n} + \dots\right] \\ \kappa_2 = V(r) = \dfrac{(1-\rho^2)^2}{n}\left[1 + \dfrac{11\rho^2}{2n} + \dots\right] \\ \gamma_1 = \dfrac{\kappa_3}{\kappa_2{}^{3/2}} = \dfrac{-6\rho}{n^{1/2}}\left[1 + \dfrac{77\rho^2 - 30}{12n} + \dots\right] \\ \gamma_2 = \dfrac{\kappa_4}{\kappa_2{}^2} = \dfrac{6}{n}(12\rho^2 - 1) + \dots \end{cases} \tag{11.12.9}$$

Thus if $\rho = 0.8$ and $N = 50$ we find that $\gamma_1 = -0.71$ and $\gamma_2 = 0.82$.

11.13 **Fisher's Transformation** Fisher showed that if we transform to a new variable z' by the relation

$$z' = \tanh^{-1} r = \tfrac{1}{2}\log_e \frac{1+r}{1-r} \tag{11.13.1}$$

then z' is approximately normally distributed with variance $1/(N - 3)$, whatever the value of ρ. This remark enables us to assess readily the significance of an observed value of r, without having to use David's tables.

If $\zeta = \tanh^{-1} \rho$, it may be proved that:

$$E(z') = \xi + \frac{\rho}{2n} + O\left(\frac{1}{n^2}\right) \tag{11.13.2}$$

where $n = N - 1$. Also

$$V(z') = \frac{1}{n} + \frac{4-\rho^2}{2n^2} + O\left(\frac{1}{n^3}\right) \tag{11.13.3}$$

$$\gamma_1(z') = \frac{\rho^3}{n^{3/2}} + O\left(\frac{1}{n^{5/2}}\right) \tag{11.13.4}$$

$$\gamma_2(z') = \frac{2}{n} + O\left(\frac{1}{n^2}\right) \tag{11.13.5}$$

For $\rho = 0.8$ and $N = 50$, $\gamma_1(z') = 0.0015$ and $\gamma_2(z') = 0.042$. This shows clearly, when compared with the values given at the end of § 11.12, how much more nearly normal z' is than r.

From Eq. (3), the variance depends to some extent on ρ. For $\rho = 0$ it is approximately $1/(n - 2)$ and for $\rho = \pm 1$ approximately $1/(n - 3/2)$. The Fisher value $1/(n - 2)$ is therefore reasonably close for any ρ.

EXAMPLE 6 For 20 students the correlation coefficient between scores on two tests was 0.65. What are the 95% confidence limits for ρ?

Assuming a normal distribution for z', the 95% limits will be $z' \pm 1.96/\sqrt{17}$. Since the expectation of z' is $\zeta + \rho/38$, the limits for ζ will be $z' - \rho/38 \pm 1.96/\sqrt{17}$. Now $z' = \tanh^{-1} 0.65 = 0.775$, but ρ is unknown. However, since $\rho/38$ is small, we can substitute for ρ the sample value 0.65, and this gives $\zeta \approx 0.775 - 0.017 \pm 0.475 = 0.283$ to 1.233. The corresponding limits for ρ $(= \tanh \zeta)$ are 0.276 to 0.844. Direct reading from the chart in David's tables gives 0.28 and 0.84. Appendix B.11 gives a table of $r = \tanh z'$.

The Fisher transformation is another example of the transformations of variables considered in Chapter 4 (§ 4.3). It achieves at the same time approximate constancy of variance and approximate normality. It is convenient in taking the average of the correlations obtained from several samples, supposedly from the same population. The values of r are transformed to values of z' and each is given the weight $N - 3$, inversely proportional to its variance. The weighted mean of the z' is then transformed back to r.

EXAMPLE 7 Separate samples of sizes 50, 70 and 100 give correlation coefficients 0.72, 0.68 and 0.77. What is the best value to use as an average?

The z' are 0.908, 0.829, and 1.020. The weighted mean is

$$\bar{z}' = \frac{1}{211}[47(0.908) + 67(0.829) + 97(1.020)] = 0.934$$

The corresponding r is 0.733.

11.14 **Rank Correlation** It is sometimes possible to place a group of individuals in order with respect to some characteristic without having to measure this characteristic numerically for each one. For instance, a judge in a contest may have to rank a group of young ladies for beauty or a sales-manager may rank a group of salesmen for keenness or efficiency. One of the classic methods of estimating hardness for minerals was by a rank order—mineral A was said to be harder than B if a piece of A would scratch a piece of B when the two were rubbed together. If A scratched C but C scratched B, then C was intermediate in hardness between A and B. A simple scale could be established by taking some standard minerals and labelling them 1, 2, 3, and so on, in the proper order, but this does not give any true measurement of relative hardness.

If the *same* individuals are ranked in two ways, say by different judges or according to different criteria, the degree of concordance between the two rankings may be of interest. The coefficient of rank correlation is intended to measure this concordance. If the two judges agree perfectly in their rankings, so that the ith individual in one ranking is also i^{th} in the other ranking ($i = 1, 2 \ldots N$), we should expect the coefficient of correlation to be 1; if the judges are diametrically opposite, so that the ith individual in one ranking is the $(N - i + 1)^{\text{th}}$ in the other, the coefficient of correlation should be -1. If the ranks are assigned by pure chance we should expect a value of the correlation coefficient near zero. The two principal methods of calculating rank correlation agree in these extreme cases, but differ in the values assigned to intermediate degrees of concordance.

Spearman's coefficient r_S depends on the *differences* $d_i = x_i - y_i$ between the ranks x_i and y_i of the same individual on the two rankings. In fact,

(11.14.1) $$r_S = 1 - 6 \sum d_i^2/[N(N^2 - 1)]$$

EXAMPLE 8 Suppose that seven bathing beauties, labelled A to G, were ranked by two judges as in the following table:

TABLE 11.3

Contestant	A	B	C	D	E	F	G
(x) Judge 1	2	1	4	5	3	7	6
(y) Judge 2	3	4	2	5	1	6	7
$d^2 = (x - y)^2$	1	9	4	0	4	1	1

The differences of the ranks are squared in the last row. Here $N = 7$ and $\sum d_i^2 = 20$, so that

$$r_S = 1 - \frac{120}{7(48)} = 0.64$$

This suggests that the judges agree reasonably well, but the question of significance will be taken up later.

Kendall's method [6] depends on giving each possible pair of individuals in the sample a score of $+1$ or -1 according as their ranks are in the same order or in the opposite order on the two rankings. Thus, in Example 8 above, A ranks above D according to both judges, and so the pair AD gets a score $+1$. On the other hand A is below B on one ranking but above on the other, so the pair AB gets a score -1. The total number of pairs is $\binom{N}{2} = N(N - 1)/2$. If S is the total score,

(11.14.2) $$r_K = S \bigg/ \binom{N}{2}$$

It is not necessary to consider every one of the pairs in this way. The same result is obtained by writing the x_i in their natural order and then for each y_i

counting how many numbers there are to the right of this y_i and greater than it. Let this number be n_i and let $P = \sum_{i=1}^{N} n_i$. Then

$$r_K = \frac{4P}{N(N-1)} - 1 \tag{11.14.3}$$

Rewriting the above table, we have

TABLE 11.4

Contestant	B	A	E	C	D	G	F
x_i	1	2	3	4	5	6	7
y_i	4	3	1	2	5	7	6
n_i	3	3	4	3	2	0	0

so that $P = 15$ and $r_K = 60/42 - 1 = 0.43$. (The number 4 for y_i in the column headed B has three numbers to the right which are greater than 4. This gives $n_i = 3$.)

The reason that Eq. (3) gives the same result as Eq. (2) is that only pairs with their y_i increasing from left to right will give a positive contribution to the score (since the x_i always increase from left to right). Then P is the total positive contribution to S. If the total negative contribution is $-Q$, $P + Q = N(N-1)/2$, so that S, which is defined as $P - Q$, is $2P - N(N-1)/2$. On substituting this value of S in Eq. (2) we arrive at Eq. (3). The modification of this procedure due to ties in the ranking will be considered in § 11.16.

* 11.15 **Relation Between Rank Correlation and Pearson Correlation** Suppose we have N individuals, or sample items, which are numbered for identification and are given ratings on two attributes X and Y. These ratings may be ranks or numerical values, and for the ith item will be denoted by x_i, y_i. For any *pair* of items, numbered i and j, we suppose that a score a_{ij} is allotted on attribute X and a score b_{ij} on Y. These scores will naturally depend on the values x_i and x_j or y_i and y_j, but we merely require that $a_{ij} = -a_{ji}$ and therefore that $a_{ij} = 0$ when $i = j$, with a similar condition on the b_{ij}. We can then define a generalized correlation coefficient r_G by the equation

$$r_G = \frac{\sum (a_{ij}b_{ij})}{(\sum a_{ij}^2 \cdot \sum b_{ij}^2)^{1/2}} \tag{11.15.1}$$

the sums being over all values of i and j from 1 to N ($i \neq j$).

If we define a_{ij} as $+1$ when the X-rank of the ith item exceeds that of the jth item and -1 in the contrary case, then a_{ij}^2 will always be 1 for $i \neq j$, and $\sum a_{ij}^2$ will be $N(N-1)$, the total number of ordered pairs. The same holds for $\sum b_{ij}^2$. But $\sum (a_{ij}b_{ij})$ is twice S, since each pair is counted twice, once in the order ij and once in the order ji. (Each term in the sum is $+1$ if a_{ij} and b_{ij} are both $+1$ or both -1, and is -1 if they have different signs.) It follows that

r_G is equal to r_K as defined by Eq. (11.14.2). If x_i is the *X-rank* of item number i and y_i its *Y-rank*, and if in Eq. (1) we let

$$a_{ij} = x_i - x_j, \quad b_{ij} = y_i - y_j \tag{11.15.2}$$

we arrive at the Spearman coefficient r_S. To show this, we note first that x_i and y_i both run through the set of integers from 1 to N, so that

$$\sum x_i = \sum y_i = \tfrac{1}{2}N(N + 1) \tag{11.15.3}$$

Now

$$\begin{aligned}\sum (a_{ij}b_{ij}) &= \sum_{ij} (x_i - x_j)(y_i - y_j)\\ &= \sum_{ij} x_i y_i + \sum_{ij} x_j y_j - \sum_{ij} (x_i y_j + x_j y_i)\end{aligned}$$

The first two terms on the right-hand side are each equal to $N \sum x_i y_i$. The third term is the same as $-2(\sum_i x_i)(\sum_j y_j)$, since i and j can be interchanged. Therefore, by Eq. (3),

$$\sum (a_{ij}b_{ij}) = 2N \sum x_i y_i - 2\left[\frac{N(N + 1)}{2}\right]^2 \tag{11.15.4}$$

Putting $x_i = y_i$ we obtain

$$\sum a_{ij}^2 = \sum b_{ij}^2 = 2N \sum x_i^2 - \frac{N^2(N + 1)^2}{2} \tag{11.15.5}$$

Since $\sum x_i^2$ is the sum of the squares of the integers from 1 to N, which is $N(N + 1)(2N + 1)/6$, the denominator of r_G is

$$\begin{aligned}(\textstyle\sum a_{ij}^2 \sum b_{ij}^2)^{1/2} &= \frac{N^2(N + 1)(2N + 1)}{3} - \frac{N^2(N + 1)^2}{2}\\ &= \frac{N^2(N^2 - 1)}{6}\end{aligned} \tag{11.15.6}$$

Now, if we write $d_i = x_i - y_i$,

$$\begin{aligned}\sum_{i=1}^{N} d_i^2 &= \sum (x_i^2 + y_i^2 - 2x_i y_i)\\ &= 2\sum x_i^2 - 2\sum x_i y_i\\ &= \frac{N(N + 1)(2N + 1)}{3} - 2\sum x_i y_i\end{aligned} \tag{11.15.7}$$

Substituting this expression in Eq. (4), we obtain

$$\begin{aligned}\sum a_{ij}b_{ij} &= \frac{N^2(N + 1)(2N + 1)}{3} - N\sum d_i^2 - \frac{N^2(N + 1)^2}{2}\\ &= \frac{N^2(N^2 - 1)}{6} - N\sum d_i^2\end{aligned} \tag{11.15.8}$$

Dividing by Eq. (6), we find that $r_G = r_S$.

Lastly, if the scores a_{ij} and b_{ij} are based on *measured* values x_i, y_i, and if $a_{ij} = x_i - x_j$, $b_{ij} = y_i - y_j$, the numerator of r_G is

$$(11.15.9) \qquad \sum_{ij} (x_i - x_j)(y_i - y_j) = 2N \sum_i x_i y_i - 2 \left(\sum_i x_i\right)\left(\sum_j y_j\right) = 2N(N-1)s_{XY}$$

where s_{XY} is the sample covariance of X and Y. In the same way, the denominator of r_G is equal to $2N(N - 1)s_X s_Y$, and therefore r_G reduces to the ordinary Pearson coefficient. Thus the Spearman coefficient is simply the Pearson coefficient calculated as if the ranks were the actual variates.

11.16 **Tied Ranks** In practice it is often difficult to distinguish between two or more individuals with regard to the attribute considered, and they are reckoned as "tied." In such cases the tied individuals are given a rank which is the mean of the ranks they would have had if they had been distinguishable. If, for example, the 3rd and 4th items are tied, they are both given the rank $3\frac{1}{2}$. If the 3rd, 4th and 5th are tied they are all given rank 4. This preserves the sum of ranks but reduces somewhat the sum of squares, as compared with a ranking in which there are no ties.

If in the X ranking there are t items tied and in the Y ranking u items tied, the denominator of Kendall's coefficient in Eq. (11.14.2) is replaced by

$$(11.16.1) \qquad [\tfrac{1}{2}N(N-1) - \tfrac{1}{2}t(t-1)]^{1/2}[\tfrac{1}{2}N(N-1) - \tfrac{1}{2}u(u-1)]^{1/2}$$

and if there are several sets of ties the appropriate amount is subtracted for each such set. The numerator is calculated as before, all tied pairs contributing zero to the total score S. Thus, suppose the ranks are as given in the following table:

EXAMPLE 9

TABLE 11.5

x_i	1	$2\frac{1}{2}$	$2\frac{1}{2}$	4	5	7	7	7	9	10
y_i	2	1	4	4	$7\frac{1}{2}$	9	10	4	6	$7\frac{1}{2}$
n_i	8	$7\frac{1}{2}$	6	$5\frac{1}{2}$	$2\frac{1}{2}$	1	$\frac{1}{2}$	2	1	0

There are two sets of ties, with $t = 2$ and $t = 3$, in the first ranking and two sets, with $u = 2$ and $u = 3$, in the second ranking. The expression (1) is $(45 - 1 - 3)^{1/2}(45 - 1 - 3)^{1/2} = 41$. In the shorter method of calculation of S, any number to the right of y_i and *equal* to it counts $\frac{1}{2}$ towards n_i and any number (greater or smaller) counts $\frac{1}{2}$ if its x rank is the same. Then $P = 34$ and $S = 23$, giving $r_K = 23/41 = 0.56$.

In Spearman's method, the formula as corrected for one set of t ties in X and one set of u ties in Y is

$$(11.16.2) \qquad r_S = \frac{N^3 - N - 6\sum d^2 - \frac{1}{2}(t^3 - t) - \frac{1}{2}(u^3 - u)}{[N^3 - N - (t^3 - t)]^{1/2}[N^3 - N - (u^3 - u)]^{1/2}}$$

and each set of ties gives a separate correction.

In Example 9, $\sum d^2 = 49$, and $N^3 - N = 990$. Therefore, with $t = 2$ and 3, and $u = 2$ and 3,

$$r_S = \frac{990 - 294 - 3 - 12 - 3 - 12}{990 - 6 - 24}$$

$$= \frac{666}{960} = 0.69$$

The uncorrected value is 0.70.

11.17 **Significance of the Rank Correlation Coefficient** The rank correlation coefficient can be used as a test of the association between X and Y, and has the advantage over the Pearson coefficient of not requiring the assumption that one or both variables are normally distributed. If we suppose that in the parent population X and Y are independent of each other, then if the N items in a sample are placed in their natural order of ranking for X (i.e., 1, 2, 3 . . . N), the order of ranking for Y is equally likely to be any one of the $N!$ permutations of the numbers 1 to N. (For the present, we are assuming that there are no ties.) For small values of N it is possible to calculate for each of these permutations the corresponding value of r_K and so form a probability distribution. Thus if $N = 5$, there are 120 possible rankings for Y, and these give the following possible values of S and of r_K (different rankings may correspond to the same value of S in (11.14.2)).

TABLE 11.6

S	10	8	6	4	2	0	−2	−4	−6	−8	−10
r_K	1.0	0.8	0.6	0.4	0.2	0	−0.2	−0.4	−0.6	−0.8	−1.0
f	1	4	9	15	20	22	20	15	9	4	1

A frequency polygon drawn for the distribution of S lies fairly close to a normal curve, and in fact it may be proved that, as N increases, this distribution tends to normality, with variance $N(N - 1)(2N + 5)/18$. For $N > 10$ the approximation is quite good.

From the above table it is evident that the probability, when $N = 5$, of a value of r_K numerically as high as 0.8, when X and Y are independent, is $10/120 = 0.083$, so that even a value as high as this is not significant at the 5% level. For a sample of 10, the standard deviation of S is $(125)^{1/2} = 11.2$, so that a significant value at the 5% level, obtained from the normal approximation would be $1.96 \times 11.2 = 22.0$. This corresponds to a value of $r_K = \pm 0.49$. The exact probability for $|S| \geq 23$ is 0.046, a little under 5%.

In estimating the significance of an observed S we should make a correction for continuity similar to that made in replacing the binomial distribution by a normal distribution. Since the distribution of S is discrete, successive values differing by 2 units, the observed S should be replaced by $S - 1$ if S is positive or by $S + 1$ if it is negative.

Spearman's coefficient r_S also tends to normality as N increases, but more slowly than r_K. Whether there are ties or not, the variance of $\sum d^2$ is given by

$$V(\textstyle\sum d^2) = \left(\frac{N^3 - N}{6}\right)^2 \cdot \frac{1}{N-1} \tag{11.17.1}$$

and that of r_S by

$$V(r_S) = \frac{1}{N-1} \tag{11.17.2}$$

When $N > 20$ the distribution may be taken as approximately normal.

Down to somewhat lower values of N (say 10) the distribution of $r_S[(N-2)/(1-r_S{}^2)]^{1/2}$ is approximately that of Student's t with $N-2$ degrees of freedom. This is the same as the distribution which was shown earlier to hold exactly (on certain assumptions) for Pearson's coefficient in a sample from an uncorrelated parent population.

11.18 **Contingency Tables** Often in medical, biological, psychological or econometric research we encounter characteristics or attributes which we cannot measure accurately and according to which we may not even be able to rank the individuals of a sample, but which do permit us to divide the sample into classes and count the numbers in each class. We might, for example, classify a sample of women students by the color of their hair, as "fair-haired," "red-haired," "brown-haired" or "black-haired," or a sample of housewives by their place of residence as "rural" or "urban."

A frequency table in which a sample is classified according to two different attributes (whether quantitative or not) is called a *contingency table*. It looks rather like a correlation table, except that the columns and rows do not necessarily correspond to any numerical values of the attributes X and Y. If a sample of N is divided into s X-classes (denoted by $X_1, X_2 \ldots X_s$) and into t Y-classes (denoted by $Y_1, Y_2 \ldots Y_t$), the frequency f_{ij} of individuals falling into class X_i and also into class Y_j is entered in the ith row and j^{th} column of the table. Thus, for $s = 4$ and $t = 3$,

TABLE 11.7

		Y_1	Y_2	Y_3	
X	X_1	f_{11}	f_{12}	f_{13}	r_1
	X_2	f_{21}	f_{22}	f_{23}	r_2
	X_3	f_{31}	f_{32}	f_{33}	r_3
	X_4	f_{41}	f_{42}	f_{43}	r_4
		c_1	c_2	c_3	N

The marginal total for the ith row is $r_i = \sum_j f_{ij}$ and that for the jth column is $c_j = \sum_i f_{ij}$. The grand total is $N = \sum r_i = \sum c_j$.

We may assume that in the population there is a *probability* π_{ij} that an individual selected at random will fall in classes X_i and Y_j. The relative frequency f_{ij}/N will be an approximation to π_{ij}. If X and Y are *independent* we shall have

(11.18.1)
$$\pi_{ij} = \pi_i \cdot \pi_j$$

where π_i $(= \sum_j \pi_{ij})$ is the probability of X_i regardless of the Y classes, and π_j $(= \sum_i \pi_{ij})$ is the probability of Y_j regardless of the X classes. We can define the *mean square contingency*, which is a measure of the degree of association between X and Y in the population, by

(11.18.2)
$$\phi^2 = \sum_{ij} \frac{\pi_{ij}^2}{\pi_i \pi_j} - 1$$

This is 0 if and only if X and Y are independent. Its greatest possible value is $q - 1$ where q is the smaller of the numbers s and t (or their common value if they are equal). The quantity $\phi^2/(q - 1)$ may therefore be used as a measure of the degree of association, and, like r^2, it varies between 0 and 1.

The *expected* frequency in the i, j^{th} cell of the contingency table is $N\pi_{ij}$, and the deviation of the table from expectation can be measured by calculating the quantity

(11.18.3)
$$\chi_s^2 = \sum_{i,j} \frac{(f_{ij} - N\pi_{ij})^2}{N\pi_{ij}}$$

the sum being extended over all the cells of the table. Since we usually wish to test the hypothesis that X and Y are independent, we can replace π_{ij} by $\pi_i\pi_j$, but these marginal probabilities are unknown, and must be estimated from the sample. It is natural to estimate π_i and π_j by the relative marginal frequencies r_i/N and c_j/N respectively, and in fact these are the estimators given by the method of maximum likelihood.

The likelihood of the observed sample of N picked from the assumed population is given by

(11.18.4)
$$L = \prod_{i,j} (\pi_i \pi_j)^{f_{ij}}$$

where the π_i and π_j are subject to the restrictions $\sum \pi_i = 1, \sum \pi_j = 1$. Using the method of Lagrange multipliers (Appendix A.15) and maximizing $\log L - \lambda \sum \pi_i - \mu \sum \pi_j$, we obtain the relations

(11.18.5)
$$\frac{r_i}{\hat{\pi}_i} - \lambda = 0, \quad \frac{c_j}{\hat{\pi}_j} - \mu = 0$$

It follows, since $\sum r_i = \sum c_j = N$, that $r_i = N\hat{\pi}_i$ and $c_j = N\hat{\pi}_j$.

Using these estimators, we can take the expected frequency in the i, j^{th} cell as

$$\phi_{ij} = N \cdot \frac{r_i}{N} \cdot \frac{c_j}{N} = \frac{r_i c_j}{N} \tag{11.18.6}$$

Therefore,

$$\begin{aligned} \chi_s^2 &= \sum_{ij} \frac{(f_{ij} - \phi_{ij})^2}{\phi_{ij}} \\ &= \sum_{ij} \frac{f_{ij}^2}{\phi_{ij}} - 2 \sum f_{ij} + \sum \phi_{ij} \\ &= \sum_{ij} \frac{f_{ij}^2}{\phi_{ij}} - N \\ &= N \left[\sum_{ij} \frac{f_{ij}^2}{r_i c_j} - 1 \right] \end{aligned} \tag{11.18.7}$$

It is shown in Appendix A.17 that this quantity has approximately the χ^2 distribution (discussed in § 4.6) with $(s - 1)(t - 1)$ degrees of freedom. Since the population probabilities π_i and π_j are estimated from the marginal frequencies r_i and c_j, we must assume that in all samples these marginal frequencies remain constant. The observed frequencies f_{ij} in the contingency table can therefore be varied only in $(s - 1)(t - 1)$ of the cells, and when these are filled the frequencies in the remaining cells are automatically determined. The distribution of the N observations among the cells is then multinomial (Appendix A.16), subject to the restriction just mentioned.

Since the χ^2 distribution applies strictly only in the limiting case, as the expected frequencies increase indefinitely, the approximation should not be used if some of these frequencies are very small. Some investigations (see [10]) suggest that a minimum value of ϕ_{ij} as low as 1 may be tolerated if values below 5 do not occur in more than about 20% of the cells.

If a larger proportion of the cells have expected frequencies below 5, it is wise to use an exact method [11].

EXAMPLE 10 Table 11.8 gives some results obtained by Woo (*Biometrika*, 1928) on the association between "left-handedness" and "left-eyedness." The X-categories are left-handed, ambidextrous and right-handed, the Y-categories left-eyed, ambiocular, and right-eyed.

The number of degrees of freedom is $2 \times 2 = 4$, so that this value of χ^2 is certainly not significant. The hypothesis that there is no association between the attributes A and B is not rejected.

The sample statistic which corresponds to the mean square contingency ϕ^2 defined in Eq. (11.18.2) is

$$f^2 = \sum_{ij} \left(\frac{f_{ij}^2}{r_i c_j} \right) - 1 = \frac{\chi_s^2}{N} \tag{11.18.8}$$

This, divided by $q - 1$ (q being the lesser of s and t), is a measure of the degree of association indicated by the sample. The upper limit 1 is attained if (for $s \leq t$) each column contains just one non-zero frequency, or (for $s \geq t$) if each row contains just one non-zero frequency. The quantity $C = [f^2/(q-1)]^{1/2}$ is a coefficient of contingency.

TABLE 11.8

	Y_1	Y_2	Y_3	
X_1	34	62	28	124
X_2	27	28	20	75
X_3	57	105	52	214
	118	195	100	413

Here $\chi_s^2 = 413[(34)^2/(124 \times 118) + (62)^2/(124 \times 195) + \ldots + (52)^2/(214 \times 100) - 1] = 4.02.$

In Example 10 above, $q = 3$ and $f^2/(q-1) = 4.02/826 = 0.0049$, so that $C = 0.07$. There is very little association between X and Y.

11.19 **The Contingency Table with Two Rows or Two Columns** For a $2 \times n$ table the calculation of χ_s^2 may be somewhat simplified. As shown by Brandt and Snedecor, the value for a sample, with frequencies as given in the following table, is:

(11.19.1)
$$\chi_s^2 = \frac{N^2}{r_1 r_2}\left(\sum_j \frac{a_j^2}{c_j} - \frac{r_1^2}{N}\right)$$

where $c_j = a_j + b_j$.

TABLE 11.9

	Y_1	Y_2	Y_3	...	Y_n	
X_1	a_1	a_2	a_3	...	a_n	r_1
X_2	b_1	b_2	b_3	...	b_n	r_2
	c_1	c_2	c_3	...	c_n	N

Either row in the table can, of course be chosen as the a_j.

EXAMPLE 11 (Lindstrom) The variable X is the presence (or absence) of a sugar-producing gene in ears of corn, Y is the number of rows of kernels in the ear.

TABLE 11.10

No. of Rows of Kernels

		8	10	12	14	
X	Present	18	37	27	0	82
	Absent	15	26	43	4	88
		33	63	70	4	170

Since the numbers in the last column ($Y = 14$) are so small, it is better to group the last two columns together. We then have a 2×3 table, and Eq. (1) gives

$$\begin{aligned}\chi_s^2 &= \frac{(170)^2}{(82)(88)}\left[\frac{18^2}{33} + \frac{37^2}{63} + \frac{27^2}{74} - \frac{82^2}{170}\right] \\ &= 7.39\end{aligned}$$

With 2 d.f., the probability of a value as large as this is about 0.025, so thas association between presence of the sugar gene and few rows of kernels it definitely indicated, although not strongly so.

The value of the contingency coefficient C is 0.21, which is fairly high for this coefficient.

With a 2×2 table the calculation of χ_s^2 is still simpler. If the frequencies in the four cells are denoted by a, b, c, d, the value of χ_s^2 is given by

$$\frac{\chi_s^2}{N} = \frac{a^2/r_1 + c^2/r_2}{c_1} + \frac{b^2/r_1 + d^2/r_2}{c_2} - 1$$

	Y_1	Y_2	
X_1	a	b	r_1
X_2	c	d	r_2
	c_1	c_2	N

On substituting $a + b$ for r_1, etc., and carrying out some algebraic manipulations, this becomes

$$\chi_s^2 = \frac{N(ad - bc)^2}{r_1 r_2 c_1 c_2} \qquad (11.19.2)$$

EXAMPLE 12 A drug said to prevent sea-sickness was tested as follows: 25 men were given the drug and 25 others were not. Both groups were tested in a rocking machine and the numbers who became sick were noted. The results were as shown:

TABLE 11.11

	Sick	Not Sick	
With drug	10	15	25
Without drug	19	6	25
	29	21	50

$$\chi_s^2 = \frac{50\cdot(60 - 285)^2}{25\cdot 25\cdot 29\cdot 21} = 6.65$$

The probability of a value as high as this with 1 d.f. is a little less than 0.01, so that an association between the use of the drug and immunity from sickness is pretty definitely indicated. The coefficient of contingency is 0.37.

Note that with one degree of freedom, the distribution of χ_s (the square root of χ_s^2) is *normal*. The probability can therefore be found from a table of the normal law, using both tails.

11.20 **The Yates Correction for Continuity** The cell-frequencies in a contingency table are necessarily integers, so that χ_s^2 is a discrete variate, whereas χ^2 varies continuously. The situation is something like that in approximating a binomial distribution by a continuous normal distribution, where the sum of terms from $x = a$ to $x = b$ (inclusive) is approximated by an integral from $a - \frac{1}{2}$ to $b + \frac{1}{2}$ (see § 3.11).

In the 2×2 table, as pointed out by Yates, the approximation to χ^2 is much improved by replacing d by $d - \frac{1}{2}$ or $d + \frac{1}{2}$, according as $ad > bc$ or $ad < bc$, and adjusting the other frequencies so as to keep the marginal totals constant. The effect of this is to replace $(ad - bc)^2$ in Eq. (11.19.2) by $(|ad - bc| - N/2)^2$, and thereby to reduce somewhat the apparent significance of the result.

In Example 12, above, the rearranged table would be as shown. Since $[(10\frac{1}{2})(6\frac{1}{2}) - (14\frac{1}{2})(18\frac{1}{2})]^2 = (200)^2 = (225 - 25)^2$, the value of χ^2, with the correction, is reduced to 5.25, the probability for which is more than 0.02.

$10\frac{1}{2}$	$14\frac{1}{2}$	25
$18\frac{1}{2}$	$6\frac{1}{2}$	25
29	21	50

If the total frequency $N \leq 20$, or if $20 < N < 40$ and the smallest expected frequency is less than 5, it is better to use Fisher's exact method, given in the next section.

* 11.21 **Fisher's Exact Method for 2 × 2 Tables** Fisher has pointed out that exact probabilities can be calculated for all the possible 2×2 tables which have

the same set of marginal frequencies. Let the observed cell-frequencies be a, b, c, d, and suppose the table so arranged that d is the smallest of these. The distribution of the N items in the sample among the four cells (on the hypothesis that there is no association between X and Y) is a hypergeometric one (see § 3.5). It has the following mathematical model: Given N balls in an urn, of which r_1 are black (corresponding to X_1) and r_2 are white (corresponding to X_2), and given N boxes of which c_1 are red (for Y_1) and c_2 are green (for Y_2), withdraw the balls one at a time and place them at random in the boxes, one ball to a box. The number of black balls in red boxes will be a, and similarly for the other frequencies.

The probability that there are just b black balls and d white ones in the c_2 green boxes $= \binom{r_1}{b}\binom{r_2}{d}\Big/\binom{N}{c_2}$ since the numerator is the number of ways of choosing b black balls out of r_1 and d white balls out of r_2, while the denominator is the total number of ways of picking c_2 balls out of the urn. But once the green boxes have been filled, the numbers a and c for the red boxes are fixed, since $a = r_1 - b$ and $c = r_2 - d$. The probability of the whole observed set of frequencies a, b, c, d, is, therefore,

$$p(d) = \frac{\binom{r_1}{b}\binom{r_2}{d}}{\binom{N}{c_2}} = \frac{r_1!\,r_2!\,c_1!\,c_2!}{a!\,b!\,c!\,d!\,N!} \tag{11.21.1}$$

Now the theoretical frequency δ, corresponding to d, is r_2c_2/N. If $d < \delta$, all smaller values of d down to zero will be even less likely than d itself. We can therefore calculate the probability P *of the observed distribution and of all less likely ones in one direction*, given by

$$P = \sum_{d=0}^{d_1} p(d) \tag{11.21.2}$$

where d_1 is the observed d. (The value of d determines the whole table, the marginal frequencies being fixed.) If $d_1 > \delta$, the sum will go from $d = d_1$ up to $d = c_2$, since we are now concerned with values of d *larger* than the expected one. The P so calculated corresponds to one tail of the distribution, whereas the χ^2 method takes account of both tails. It is to be expected, therefore, that the probability calculated from χ_s^2 will be an approximation to $2P$ and not to P.

In Example 12 above, $d = 6$ and $\delta = 10.5$, so that

$$P = \sum_{d=0}^{6} p(d)$$

The values of $p(d)$ are given in the following table:

d	0	1	2	3	4,5,6
$p(d)$	0.00860	0.00161	0.00020	0.00002	0.00000

and we find that $P = 0.0104$. The χ^2 value, with Yates's correction, gives $P = 0.022$, which is quite close to $2P$.

The chief objection to Fisher's method is the considerable amount of computation usually involved. Tables recently published by Mainland and others [12] enable the significance (at 5% and 1% levels) to be estimated very quickly, without calculation.

An alternative procedure is to use a normal approximation with a continuity correction. The variance of d, as given by Eq. (3.5.7), is

$$V(d) = \frac{N - c_2}{N - 1} \cdot c_2 \cdot \frac{r_2}{N}\left(1 - \frac{r_2}{N}\right) = \frac{c_1 c_2 r_1 r_2}{N^2(N - 1)} \tag{11.21.3}$$

and if

$$z = \frac{|d - \delta| - \frac{1}{2}}{[V(d)]^{1/2}} \tag{11.21.4}$$

is treated as a normal variate, the probability of a value at least as great as this can be found.

Thus, in Example 12, $|d - \delta| - \frac{1}{2} = 4$, $V(d) = 3.11$, so that $z = 4/1.76 = 2.27$, giving $P = .0116$. This is fairly close to the exact value.

11.22 **The Chi-Square Test as a Test of Homogeneity** It sometimes happens that a table which looks like a contingency table really reflects a different situation. The rows of the table represent each a different set of observations, r_i in number, the individuals in each set being classified according to the attribute Y. The numbers r_i are selected arbitrarily and do not depend at all on the population. The hypothesis to be tested is that each sample (represented by a row of the table) comes from the *same* population in which the probability of attribute Y_j is π_j (with $\sum \pi_j = 1$).

The value of π_j is estimated as before by c_j/N. It may be shown that the limiting distribution of χ_s^2, calculated in the ordinary way, is still the χ^2 distribution with $(s - 1)(t - 1)$ degrees of freedom.

EXAMPLE 13 In order to see if the age-distribution of whitefish in Lake Wabamun, Alberta, had changed significantly between 1957 and 1958, samples of the catches in these two years were classified in age-groups as follows:

TABLE 11.12

Age (Years)

Year	3–4	5	6	7	8	≥9	
1957	6	15	10	38	62	26	157
1958	16	12	9	22	36	5	100
	22	27	19	60	98	31	257

Here

$$\chi_s^2 = \frac{(257)^2}{(100)(157)}\left(\frac{16^2}{22} + \frac{12^2}{27} + \cdots + \frac{5^2}{31} - \frac{100^2}{257}\right)$$
$$= 18.6$$

The number of degrees of freedom is 5, so that this value of χ^2 is highly significant. The value of f^2 is 0.072, giving $C = 0.27$.

PROBLEMS

A (§§ 11.1–11.2)

1. If the joint probability density for X and Y is $f(x, y) = 2/a^2$, $0 \leq x \leq y$, $0 \leq y \leq a$, find (a) the marginal probability densities $g(x)$ and $h(y)$ (b) the regression equations of Y on X and of X on Y (c) the means and variances of X and Y, the covariance, and the coefficient of correlation between X and Y. *Hint:* $f(x, y)$ is constant over a triangular area in the xy plane.

2. Is it true that a necessary and sufficient condition for two variates X and Y to have a bivariate normal distribution is that the two regression equations are linear? *Hint:* See Problem 1 above.

3. For the bivariate normal distribution, Eq. (11.2.5), show that the variance of η_z is equal to ρ^2 and that the correlation coefficient between η_z and v is equal to ρ.

4. Show that the coefficient of correlation for two variates X and Y is the geometric mean of the slopes of the two regression lines, one reckoned from the X-axis and one from the Y-axis. (The geometric mean of a and b is $\sqrt{ab}$.)

5. Show that if X and Y are independent variates they are necessarily uncorrelated. (The condition for independence implies that $E(XY) = E(X)\cdot E(Y)$.)

6. If X is uniformly distributed on $(-1,1)$ and if $Y = X^2$, show that X and Y are uncorrelated. (Note that X and Y are certainly not independent.)

7. Prove that the acute angle between the two lines of regression is given by $\tan\theta = \dfrac{1-\rho^2}{\rho}\cdot\dfrac{\sigma_X\sigma_Y}{\sigma_X^2+\sigma_Y^2}$.

8. Let the variate X have the marginal distribution $g(x) = 1$, $-\frac{1}{2} < x < \frac{1}{2}$, and let the conditional density of Y, given $X = x$, be $f(y|x) = 1$, $x < y < x + 1$, $-\frac{1}{2} < x < 0$, $f(y|x) = 1$, $-x < y < 1 - x$, $0 < x < \frac{1}{2}$, and $f(y|x) = 0$ otherwise. Prove that X and Y are uncorrelated.

9. If X_1, X_2, X_3 are uncorrelated variates, each with the same standard deviation σ, find the coefficient of correlation between $X_1 + X_2$ and $X_2 + X_3$.

10. If X and Y are uncorrelated, with means zero and variances σ_X^2, σ_Y^2, show that the variates $U = X\cos\alpha + Y\sin\alpha$ and $V = X\sin\alpha - Y\cos\alpha$ have a correlation coefficient

$$\rho_{UV} = \frac{\sigma_X^2 - \sigma_Y^2}{[(\sigma_X^2 - \sigma_Y^2)^2 + 4\sigma_X^2\sigma_Y^2\operatorname{cosec}^2 2\alpha]^{1/2}}$$

B (§§ 11.3–11.6)

1. The following data represent the ages of husband (X) and wife (Y) for 20 couples selected at random from a certain population.

X	22	24	26	26	27	27	28	28	29	30	30	30	31	32	33	34	35	35	36	37
Y	18	20	20	24	22	24	27	24	21	25	29	32	27	27	30	27	30	31	30	32

Find the equations of the two regression lines. Make a scatter diagram for the data and draw the two lines on it.

2. Calculate the coefficient of correlation of X and Y from the data of Problem B.1. On the assumption that the population is bivariate normal, find 95% confidence limits for the two regression coefficients, β (for the first regression line) and β' (for the second).

3. In studying a set of pairs of values of related variates X and Y, a statistician has computed the following quantities: $N = 100$, $\sum x = 12{,}500$, $\sum y = 8{,}000$, $\sum x^2 = 1{,}585{,}000$, $\sum y^2 = 648{,}100$, $\sum xy = 1{,}007{,}425$. Calculate $\bar{x}$, $\bar{y}$, s_X, s_Y, s_{XY} and r for these data.

4. In the following table, X is the weight (to nearest half pound) and Y the height (to nearest tenth of an inch) for 200 freshmen at a university.

Y \ X	90 –99.5	100 –109.5	110 –	120 –	130 –	140 –	150 –	160 –	170 –	180 –	190 –	200 –209.5
76–77.9				1								
74–75.9							1	1	1	1		
72–				1	1	4		1				
70–			1	2	6	7	6	2	1	2	1	1
68–			2	8	17	8	9	2	1	1	1	
66–			8	16	14	13	6	2	1			1
64–		3	8	7	7	3	3	1	1			
62–	1	4	1	7	1							
60–												
58–59.9		1										

Find the regression equation of height on weight, and give 95% confidence limits for the regression coefficient β. Calculate the correlation coefficient between height and weight.

5. A coefficient of correlation calculated from a sample of size 25 is found to be 0.37. Is this value significantly different (at the 5% level) from zero?

6. A sample correlation coefficient of 0.561 is said to be highly significant. Assuming that this means that the probability of getting a value numerically as great is less than 0.01, what is the smallest sample size that would warrant the statement? *Hint:* $r[(N-2)/(1-r^2)]^{1/2}$ is to be the same as $t_{0.01}$ for $N-2$ d.f. Find $N-2$ by trial, using the table of t.

7. Is it true that a correlation coefficient of $r = 0.6$ indicates a relationship twice as close as that indicated by $r = 0.3$? *Hint:* Consider the relative accuracy of estimation of Y from a given X in the two cases, as measured by the reciprocal of the standard error of estimate.

8. The marks of a class of 12 students on a mid-term test (x) and on the final examination (y) were:

x	41	45	50	68	47	77	90	100	80	100	40	43
y	60	63	60	48	85	56	53	91	74	98	65	43

What is the regression estimate of the final mark of a student who obtained 60 on the test but was ill at the time of the final examination? What is the standard error of this estimate?

9. The two regression lines for variates X and Y have been computed as $4x - 5y + 33 = 0$ and $20x - 9y - 107 = 0$. Given that the variance of X is 9, calculate the variance of Y, the means for X and Y and the coefficient of correlation between X and Y.

C (§§ 11.7–11.11)

1. The following table gives death-rates per 100,000 in the United States from typhoid fever for the years from 1900–1920:

Year	*Rate*	*Year*	*Rate*	*Year*	*Rate*
1900	31.3	1907	20.5	1914	10.8
1901	27.5	1908	19.6	1915	9.2
1902	26.3	1909	17.2	1916	8.8
1903	24.6	1910	18.0	1917	8.1
1904	23.9	1911	15.3	1918	7.0
1905	22.4	1912	13.2	1919	4.8
1906	22.0	1913	12.6	1920	5.0

Find the best-fitting straight line for these data. If the linear trend had continued, estimate the date at which typhoid fever would have been wiped out in the United States. *Hint:* Take the origin of X (the date) at 1910, so that the values of x are -10, -9, etc.

2. In the following table, x is the amount of irrigation water (inches) applied to an experimental farm in India and y is the yield of rice in tons/acre [13].

x	12	18	24	30	36	42	48
y	5.27	5.68	6.25	7.21	8.02	8.71	8.42

The values of x are fixed and those of y are random. Estimate how much water would be necessary for a yield of 7.5 tons/acre and obtain 95% confidence limits for this estimate. (*Note:* An approximation for moderately large N to the standard error of estimate for x is given by $s_{Xe} = s_{Ye}/b$. This is much simpler than solving the quadratic for x as suggested in § 11.7.)

3. The following values were obtained in an experiment intended to show a linear functional relationship between X and Y. Both variables are subject to error and it is considered that the variance of the error in Y is 16 times that of the error in X. Obtain values for the estimators of α and β in the linear relation $\eta = \alpha + \beta\xi$, and for a consistent estimator of σ^2, the variance of X.

x	1	2	3	4	5	6	7	8	9	10
y	9.9	13.2	16.4	19.7	22.5	26.1	29.2	32.5	35.7	38.8

4. Find the best-fitting line for the data of Problem C.3 by the method of grouping, using (a) two groups of 5, (b) three groups of 3, 4, and 3 respectively. Find 90% confidence intervals for β in both cases.

In method (b), obtain an estimator for σ_u^2, assuming that $\sigma_v^2 = 16\,\sigma_u^2$.

D (§§ 11.12–11.13)

1. [14] Over a period of 20 years the mean wheat yield of eastern England was found to be correlated with the autumn rainfall, with $r = -0.629$. Is this significantly different from zero at the 1% level? Is it significantly different from -0.3?

2. In a sample of 25 pairs of individuals (parent and child), the correlation in a certain character was found to be 0.60. Obtain 90% confidence limits for the population correlation coefficient. Could one conclude, with 90% confidence, that the true value (in the population sampled) was at least 0.40?

3. One random sample of 28 from a certain bivariate population gave $r = 0.60$; another independent random sample of 23 gave $r = 0.40$. Is the difference significant at the 5% level? *Hint:* Use a two-tailed test for the difference, after making the Fisher transformation.

4. For a sample of size 30 from a bivariate normal population, r was found to be 0.684. An independent sample of size 40 gave $r = 0.719$. What estimate would you suggest for the true value of ρ?

5. Obtain an estimator of ρ from the sampling distribution of r (Eqs. 11.12.7 and 11.12.8) by finding that value of ρ for which $f(r, \rho)$ is a maximum, with a given r. Is this estimator unbiased? *Hint:* Put $d(\log f)/d\rho = 0$ and solve the quadratic equation for ρ as far as terms of order $1/N$. Neglect all terms in $S(\rho r)$ except the first.

E (§§ 11.14–11.17)

1. (*Garrett*) Twelve salesmen were ranked in order of merit for efficiency (X) by their manager. The ranking (Y) in accordance with length of service is also given in the following table: What correlation is there between length of service and efficiency?

Salesmen	*A*	*B*	*C*	*D*	*E*	*F*	*G*	*H*	*J*	*K*	*L*	*M*
X	6	12	1	9	8	5	2	10	3	7	4	11
Y	7.5	11.5	2	4	6	9	1	11.5	5	7.5	3	10

Calculate both Spearman's and Kendall's coefficient, correcting for the ties in the Y ranking.

2. The scores of 10 students on two tests are given in the following table. Calculate the Pearson coefficient of correlation for the actual scores, and the Spearman coefficient for the ranks.

Student	*A*	*B*	*C*	*D*	*E*	*F*	*G*	*H*	*J*	*K*
X	92	89	87	86	83	77	71	62	53	40
Y	88	85	93	79	70	87	52	84	41	64

3. If a sample of seven pairs is drawn from a population of values of independent variates X and Y, it is known that the computed Spearman coefficient will exceed 0.714 in not more than 5% of cases and will exceed 0.893 in not more than 1% of cases. What conclusion may be drawn regarding the judges in Example 8 of § 11.14? Apply the Student-t approximation of § 11.17 to the same problem.

4. In a drama competition, ten plays were ranked independently by two adjudicators, as follows:

Play	*A*	*B*	*C*	*D*	*E*	*F*	*G*	*H*	*J*	*K*
Rank (X)	5	2	6	8	1	7	4	9	3	10
Rank (Y)	1	7	6	10	4	5	3	8	2	9

Calculate the coefficient of rank correlation by both the Spearman and Kendall formulas. Would you say that there is a significant measure of agreement between the two adjudicators? *Hint:* Use the normal approximation to the variance of S, and make the correction for continuity.

F (§§ 11.18–11.22)

1. In the accompanying contingency table, X represents a rating given to each of a group of university freshmen on the basis of high school reports and Y represents the final standing in degree examinations for the same group. Discuss the association between these two attributes, and calculate the coefficient of contingency C defined in § 11.18.

Y \ X	Fair	Good	Excellent
3rd class	73	67	10
2nd class	64	84	15
1st class	5	24	28

2. In a public opinion survey the following questions were asked: (1) Do you drink beer? (2) Are you in favour of local option on the sale of liquor? In one district the results (excluding those who had no opinions) were as indicated.

	For Local Option	Against
Drinkers	18	39
Non-drinkers	45	37

Does this provide good evidence of an association between drinking habits and opinion on the subject of local option?

3. Two batches of 12 experimental animals, one batch inoculated and the other not inoculated, were exposed to infection under comparable conditions. Of the inoculated group, 2 died and 10 survived; of the other group 8 died and 4 survived. Does this observation provide evidence (at the 5% level of significance) of the value of the inoculation in increasing the chances of survival when exposed to infection? *Hint:* Calculate the probability of a result at least as extreme as that actually observed (a) by the χ^2 test, using Yates's correction, and (b) by Fisher's exact method.

4. In a certain community a random sample of 50 men and 50 women over 21 years of age were asked about their educational background, classified as junior high, senior high or college. The results were:

	Junior High	Senior High	College
Male	13	25	12
Female	23	20	7

Is there a significant association between sex and educational level? *Hint:* Use the Brandt-Snedecor formula.

5. Two groups of freshmen applying to enter a university took the same college aptitude test. The groups (A and B) differed in the type of high school education they had experienced. The frequency distributions of scores for the two groups were as follows:

Score	0–9	10–19	20–29	30–39	40–49	50–59	60–69	70–79	80–89	90–99
Group A	71	68	66	47	51	39	43	39	33	18
Group B	22	8	14	12	3	13	3	14	12	10

Calculate the value of χ^2 and determine whether there is a significant difference in college aptitude between the groups.

6. Prove that if $ad < bc$, then χ^2 for the table

$a + \frac{1}{2}$	$b - \frac{1}{2}$
$c - \frac{1}{2}$	$d + \frac{1}{2}$

is given by $N(|ad - bc| - N/2)^2/(r_1 r_2 c_1 c_2)$ where $r_1 = a + b$, $r_2 = c + d$, $c_1 = a + c$, $c_2 = b + d$.

7. It has been suggested by V. M. Dandekar (see [13], p. 388) that a better approximation than that given by Yates, to the true probability P for a 2×2 table, may be obtained by subtracting from the uncorrected value χ_0^2 the term $(\chi_{-1}^2 - \chi_0^2)(\chi_0^2 - \chi_1)^2/(\chi_{-1}^2 - \chi_1^2)$ where χ_1^2 and χ_{-1}^2 are the values obtained by respectively increasing and decreasing the smallest frequency in the table by unity. Test this suggestion on the data of Problem F-3.

8. Prove the Brandt-Snedecor formula for χ_s^2, Eq. (11.19.1).

REFERENCES

[1] Lindley, D. V., "Regression Lines and the Linear Functional Relationship," *Journ. R. Stat. Soc. Suppl.*, **9,** 1947, pp. 218–244.

[2] Kendall, M. G., "Regression, Structure, and Functional Relationships," Part I, *Biometrika*, **38,** 1951, pp. 11–25, Part II, *Biometrika*, **39,** 1952, pp. 96–108.

[3] Madansky, A., "The Fitting of Straight Lines When Both Variables Are Subject to Error," *J. Amer. Stat. Assoc.*, **54,** 1959, pp. 173–205.

[4] Bartlett, M. S., "Fitting a Straight Line When Both Variables Are Subject to Error," *Biometrics*, **5,** 1949, pp. 207–212.

[5] Gibson, W. M., and Jowett, G. H., "Three-Group Regression Analysis," Part I, *Applied Statistics*, **6,** 1957, pp. 114–122.

[6] Kendall, M. G., *Rank Correlation Methods*, Chas. Griffin, 1948.

[7] Kenney, J. F., and Keeping, E. S., *Mathematics of Statistics, Part II*, Van Nostrand, 1951, p. 217. See also [9].

[8] Hotelling, Harold, "New Light on the Correlation Coefficient and Its Transforms," *J. Roy. Stat. Soc.*, **B15,** 1953, pp. 193–225 (also discussion following this paper, pp. 225–232).

[9] David, F. N., *Tables of the Correlation Coefficient*, Biometrika Office, University College, London, 1938.

[10] Cochran, W. G., "Some Methods for Strengthening the Common χ^2 Tests," *Biometrics*, **10,** 1954, pp. 417–451.

[11] Freeman, G. H., and Halton, J. H., "Note on an Exact Treatment of Contingency, Goodness of Fit, and Other Problems of Significance," *Biometrika*, **38,** 1951, pp. 141–149.

[12] Mainland, D., Herrera, L., and Sutcliffe, M. I., *Statistical Tables for Use with Binomial Samples*, Department of Medical Statistics, N.Y. Univ. College of Medicine, 1956.

[13] Kapur, J. N., and Saxena, H. C., *Mathematical Statistics*, S. Chand and Co., Delhi, 1960.

[14] Fisher, Sir R. A., *Statistical Methods for Research Workers*, Oliver and Boyd, 13th ed., 1958.

Chapter 12

REGRESSION ANALYSIS AND CURVE FITTING

12.1 **The Equations of Multiple Regression** In the last chapter we considered the relations between two variates X and Y. We now suppose that Y depends on p other variates which will be denoted by $X_1, X_2 \ldots X_p$. These need not be independent, and in fact may all be powers of a single variate X. We shall call $X_1 \ldots X_p$ the *predictors* and Y the *predicted* (or *dependent*) variate. The usual problem is to find the best linear predicting equation for Y (in the least squares sense) of the form:

$$y_c = \sum_i b_i x_i, \qquad i = 0, 1, 2 \ldots p \tag{12.1.1}$$

To avoid introducing a separate constant term, the first variate X_0 is a dummy which always takes the value 1. The coefficient b_0 is then the constant, denoted in Chapter 11 by a, and b_1 is the previous b. The results of this chapter reduce to those of Chapter 11 when $p = 1$.

The coefficients b_i are called *partial regression coefficients*. They are estimators of the true regression coefficients β_i which are supposed to characterize the population, and they are calculated from a set of observations of each of the $p + 1$ variates made on N individuals from the population. We shall denote the observed value of X_i for the individual numbered α by $x_{i\alpha}$, ($i = 1, 2 \ldots p$, $\alpha = 1, 2 \ldots N$). The set of all $N(p + 1)$ observations may be written as a *matrix*—

$$\begin{bmatrix} x_{11} x_{12} & \cdots & x_{1N} \\ x_{21} x_{22} & \cdots & x_{2N} \\ \cdot \quad \cdot & & \cdot \\ \cdot \quad \cdot & & \cdot \\ \cdot \quad \cdot & & \cdot \\ x_{p1} x_{p2} & \cdots & x_{pN} \\ y_1 \; y_2 & \cdots & y_N \end{bmatrix}$$

with $p + 1$ rows and N columns and much of the material in this chapter is most conveniently expressed in the notation of matrix algebra. For those who are unfamiliar with this subject, a brief discussion of the principal ideas will be found in the Appendix, §§ A.18–23.

The true regression equation in the population is supposed to be

$$\eta = \sum_i \beta_i x_i, \qquad i = 0, 1, 2 \dots p \tag{12.1.2}$$

The x_i are fixed numbers, or at least the errors in x_i are small compared with the error in y. The b_i will therefore be chosen to minimize the sum of squares of the differences between the observed y_α and the theoretical η_α. We shall use the symbol $\sum$ to denote summation over variates with respect to i (sometimes j or k) and S to denote summation over individuals with respect to α (sometimes β or γ). The least-squares condition becomes

$$\underset{\alpha}{S}\,(y_\alpha - \sum \beta_i x_{i\alpha})^2 = \text{minimum} \tag{12.1.3}$$

On differentiating with respect to the β_i and equating the derivatives to zero, we have for the estimators $\hat{\beta}_i$ the equations

$$\underset{\alpha}{S}\, x_{i\alpha}\left(y_\alpha - \sum_j \hat{\beta}_j x_{j\alpha}\right) = 0, \qquad i, j = 0, 1, 2 \dots p$$

or, equivalently,

$$\underset{\alpha}{S} \sum_{j=0}^{p} x_{i\alpha} x_{j\alpha} \hat{\beta}_j = \underset{\alpha}{S}\, x_{i\alpha} y_\alpha, \qquad i = 0, 1, 2 \dots p \tag{12.1.4}$$

This is a system of $p + 1$ linear equations in the $p + 1$ unknowns $\hat{\beta}_j$. Written out in full, with $\hat{\beta}_j = b_j$, they are

$$\begin{cases} b_0 a_{00} + b_1 a_{01} + \dots + b_p a_{0p} = g_0 \\ b_0 a_{10} + b_1 a_{11} + \dots + b_p a_{1p} = g_1 \\ b_0 a_{20} + b_1 a_{21} + \dots + b_p a_{2p} = g_2 \\ \vdots \qquad\quad \vdots \qquad\qquad\quad \vdots \qquad \vdots \\ b_0 a_{p0} + b_1 a_{p1} + \dots + b_p a_{pp} = g_p \end{cases} \tag{12.1.5}$$

where

$$a_{ij} = \underset{\alpha}{S}\, x_{i\alpha} x_{j\alpha}, \qquad g_i = \underset{\alpha}{S}\, x_{i\alpha} y_\alpha \tag{12.1.6}$$

This system is called the set of *normal equations* of the regression problem. It is clear from the definition of a_{ij} in (6) that $a_{ij} = a_{ji}$. The set of coefficients a_{ij} in the normal equations therefore forms a symmetric matrix. If we denote this square symmetric matrix by A and let b and g denote the one-column matrices (usually called column vectors)

$$b = \begin{bmatrix} b_0 \\ \vdots \\ b_p \end{bmatrix}, \qquad g = \begin{bmatrix} g_0 \\ \vdots \\ g_p \end{bmatrix}$$

the equations of (5) may be written in the compact matrix form

$$Ab = g \tag{12.1.7}$$

The matrix solution of these equations is

$$b = A^{-1}g \tag{12.1.8}$$

where A^{-1} is the *inverse* of A, that is, the matrix which when multiplied by A becomes the unit matrix. One method of inverting a matrix is given in Appendix A.23. There are other, and perhaps speedier, methods, but this one is straightforward and systematic.

12.2 **The Regression Equations and Maximum Likelihood** If $y_\alpha = \eta_\alpha + \varepsilon_\alpha$, and if the ε_α may be assumed to be independently and normally distributed about zero with a common variance σ^2, the joint probability density for the set of ε's is $L = \sigma^{-N}(2\pi)^{-N/2}\exp[-S(\varepsilon_\alpha{}^2)/2\sigma^2]$. Therefore,

$$\log L = -N \log \sigma - \frac{N}{2}\log(2\pi) - \frac{S(\varepsilon_\alpha{}^2)}{2\sigma^2} \tag{12.2.1}$$

The condition for maximum L is clearly the same as for minimum $S(\varepsilon_\alpha{}^2)$, which is equivalent to Eq. (12.1.3). The method of maximum likelihood leads therefore to the same normal equations as the method of least squares.

We may also consider the b's as linear functions of the y_α, chosen so as to be the best unbiased estimators of the β's. If by "best" we mean having minimum variance (and therefore maximum precision), it may be shown [1] that by using this criterion we again arrive at the same set of normal equations.

12.3 **The Solution of the Normal Equations** The normal equations are

$$\sum_j a_{ij}b_j = g_i, \qquad i = 0, 1, 2 \ldots p \tag{12.3.1}$$

Besides the matrix solution of Eq. (12.1.8), there is an elegant theoretical solution provided by *Cramer's rule*, namely,

$$b_j = \frac{d_j(A)}{d(A)} \tag{12.3.2}$$

where $d(A)$ is the determinant of the matrix A (assumed to be non-singular) and $d_j(A)$ is the determinant of the matrix derived from A by replacing its jth column by the column of g's. However, for $p \geq 3$, it is generally safer to use a process of systematic elimination of the unknowns, one at a time. One such process is the Square Root Method, often called Choleski's method ([2], [3]). Details of the process, with an illustrative example, are given in Appendix A.24. There are several other methods available, but this is one of the more compact schemes.

12.4 **The Variances and Covariances of the Regression Coefficients** From the assumptions mentioned in § 12.2, it follows that the expectation of b_i is equal to β_i and that the covariance of b_i and b_j is $\sigma^2\, a^{ij}$, where a^{ij} is an element of the inverse matrix A^{-1}.

Since $g_i = \underset{\alpha}{S} x_{i\alpha} y_\alpha$ and $b_i = \sum_j a^{ij} g_j$, we have

$$b_i = \sum_j a^{ij} \underset{\alpha}{S} x_{j\alpha} y_\alpha \tag{12.4.1}$$

Now the expectation of y_α is $\eta_\alpha = \sum_k \beta_k x_{k\alpha}$, so that

$$\begin{aligned} E(b_i) &= \sum_j a^{ij} \underset{\alpha}{S} x_{j\alpha} \sum_k \beta_k x_{k\alpha} \\ &= \sum_k \beta_k \sum_j a^{ij} a_{jk} \\ &= \sum_k \beta_k \delta_{ik} \end{aligned} \tag{12.4.2}$$

where $\delta_{ik} = 1$ when $i = k$ and 0 when $i \neq k$. Therefore,

$$E(b_i) = \beta_i \tag{12.4.3}$$

Since ε_α and ε_β are supposed to be independent, the covariance of y_α and y_β is given by

$$C(y_\alpha, y_\beta) = \delta_{\alpha\beta}\sigma^2 \tag{12.4.4}$$

The x's being fixed,

$$\begin{aligned} C(g_i, g_j) &= \underset{\alpha}{S}\,\underset{\beta}{S}\, x_{i\alpha} x_{j\beta} C(y_\alpha, y_\beta) \\ &= \sigma^2 \underset{\alpha}{S} x_{i\alpha} x_{j\alpha} \\ &= \sigma^2 a_{ij} \end{aligned} \tag{12.4.5}$$

since the only non-zero term in the sum over β arises when $\beta = \alpha$. Then

$$\begin{aligned} C(b_i, b_j) &= \sum_k a^{ik} \sum_l a^{jl} C(g_k, g_l) \\ &= \sum_{k,l} a^{ik} a^{jl} \sigma^2 a_{kl} \\ &= \sigma^2 \sum_k a^{ik} \delta_{jk} = \sigma^2 a^{ij} \end{aligned} \tag{12.4.6}$$

The elements of the matrix A^{-1}, multiplied by σ^2, give therefore the variances and covariances of the regression coefficients. The diagonal terms in particular give the variances (and hence the estimated standard errors) of the regression coefficients.

The variance of the predicted value y_c, given by Eq. (12.1.1) for some new set of values $x_1, x_2 \ldots x_p$ of the predictors, is obtained from Eq. (6). In fact,

$$\begin{aligned} V(y_c) &= E\Big[\sum_i (b_i - \beta_i) x_i\Big]^2 = \sum_{i,j} x_i x_j E[(b_i - \beta_i)(b_j - \beta_j)] \\ &= \sum_{i,j} x_i x_j C(b_i, b_j) \\ &= \sigma^2 \sum_{i,j} a^{ij} x_i x_j \end{aligned} \tag{12.4.7}$$

The variance of the *observed* y which would correspond to the new observed set $x_1 \ldots x_p$ is found by adding σ^2, which is the variance about the regression plane (or hyperplane). It becomes

$$V(y) = \sigma^2\left[1 + \sum_{i,j} a^{ij}x_ix_j\right] \tag{12.4.8}$$

This is a generalization to $p + 1$ dimensions of the two-dimensional relation of Eq. (11.5.16). For if $x_0 = 1$ and $x_1 = x$, the 2 by 2 matrix A is

$$A = \begin{bmatrix} N & N\bar{x} \\ N\bar{x} & Sx_\alpha^2 \end{bmatrix}$$

and its determinant is $d(A) = N(N - 1)s_x^2$. The inverse matrix is

$$A^{-1} = \frac{1}{(N-1)s_x^2}\begin{bmatrix} \frac{N-1}{N}s_x^2 + \bar{x}^2 & -\bar{x} \\ -\bar{x} & 1 \end{bmatrix}$$

Therefore,

$$\sum_{i,j} a^{ij}x_ix_j = a^{00} + 2a^{01}x + a^{11}x^2$$

$$= \frac{\frac{N-1}{N}s_x^2 + \bar{x}^2 - 2x\bar{x} + x^2}{(N-1)s_x^2}$$

$$= \frac{1}{N} + \frac{(x-\bar{x})^2}{(N-1)\,s_x^2}$$

as in § 11.5.

12.5 **Residuals** The difference between the observed value y_α and the computed value y_c, for a given individual in the sample is called a *residual*, and is usually denoted by v_α.

$$v_\alpha = y_\alpha - y_c = y_\alpha - \sum_i b_ix_{i\alpha} \tag{12.5.1}$$

This is not the same as the *true error* δ_α which may be defined by

$$\delta_\alpha = y_\alpha - \eta_\alpha = y_\alpha - \sum_i \beta_ix_{i\alpha} \tag{12.5.2}$$

If v is the column vector of the v_α ($\alpha = 1, 2 \ldots N$) and X is the $(p + 1)$-by-N matrix $(x_{i\alpha})$, then

$$Xv = X(y - X'b) = g - Ab$$

where X' is the transpose of X (with rows and columns interchanged). This follows from the definitions of a_{ij} and g_i in (12.1.6), which in matrix notation become $XX' = A$ and $Xy = g$. But by (12.1.7), $g = Ab$, and therefore

$$Xv = 0 \tag{12.5.3}$$

This is equivalent to the set of equations

$$\underset{\alpha}{S} x_{i\alpha}\, v_\alpha = 0, \quad i = 0, 1, 2 \ldots p \tag{12.5.4}$$

The residuals are therefore said to be *orthogonal* to each of the predictors $x_1, x_2 \ldots x_p$.

The sum of squares of the residuals, which is the minimum sum of squares in (12.1.3), may be written

$$\underset{\alpha}{S} v_\alpha^2 = v'v = (y - X'b)'v$$
$$= (y' - b'X)v = y'v$$

since $Xv = 0$. Therefore,

$$\underset{\alpha}{S} v_\alpha^2 = y'(y - X'b) = y'y - g'b \tag{12.5.5}$$

since $g' = y'X'$. In scalar notation this is

$$\underset{\alpha}{S} v_\alpha^2 = \underset{\alpha}{S}\, y_\alpha^2 - \sum_i b_i g_i \tag{12.5.6}$$

This equation gives another method of computing the sum of squares of residuals. It should be noted that since the two terms on the right-hand side are often nearly equal in magnitude, the values of b_i used should be correct to several more significant figures than are required in the final sum of squares.

12.6 **Distribution of the Sum of Squares of Residuals** Since $y_\alpha = \eta_\alpha + \delta_\alpha$, and since the δ_α are supposed to be distributed with expectation zero and variance σ^2, we have

$$E(y_\alpha^2) = E(\eta_\alpha^2 + 2\eta_\alpha\delta_\alpha + \delta_\alpha^2)$$
$$= \eta_\alpha^2 + \sigma^2$$

Now

$$\eta_\alpha = \sum \beta_i x_{i\alpha},$$

so that

$$\eta_\alpha^2 = \sum_{i,j} \beta_i\beta_j x_{i\alpha}x_{j\alpha}$$

Therefore,

$$E\left(\underset{\alpha}{S}\, y_\alpha^2\right) = \sum_{i,j} \beta_i\beta_j a_{ij} + N\sigma^2 \tag{12.6.1}$$

Also, $\sum_i b_i g_i = \sum_{i,j} b_i a_{ij} b_j$, so that

$$E\left(\sum_i b_i g_i = \sum_{i,j} a_{ij}E(b_i b_j)\right) \tag{12.6.2}$$

By Eqs. (12.4.3) and (12.4.6), $E(b_i b_j) = \beta_i\beta_j + \sigma^2 a^{ij}$ and therefore,

$$E\left(\sum_i b_i g_i\right) = \sum_{i,j} (\beta_i\beta_j + \sigma^2 a^{ij})a_{ij} \tag{12.6.3}$$
$$= \sum_{ij} \beta_i\beta_j a_{ij} + \sigma^2(p + 1)$$

since $\sum_{i,j} a^{ij}a_{ij} = \sum_j \delta_{jj} = p + 1$, δ_{jj} being 1 for each of its $p + 1$ values.

Substituting Eqs. (1) and (3) in Eq. (12.5.6), we obtain

$$E\left(\underset{\alpha}{S}\, v_\alpha^2\right) = \sigma^2(N - p - 1) \tag{12.6.4}$$

which means that an unbiased estimator of σ^2 is furnished by

$$\hat{\sigma}^2 = \frac{1}{N - p - 1}\left(\underset{\alpha}{S}\, v_\alpha^2\right) \tag{12.6.5}$$

It may be proved (e.g., [4]) that if the δ_α are assumed also to be normal, then $(N - p - 1)\hat{\sigma}^2/\sigma^2$ has the χ^2 distribution with $N - p - 1$ degrees of freedom. Moreover, on this assumption the b_i are normally distributed and are independent of $\hat{\sigma}^2$. This means that the Student-t distribution can be used to fix confidence intervals for the b_i. In fact,

$$t = \frac{b_i - \beta_i}{(a^{ii}\hat{\sigma}^2)^{1/2}} \tag{12.6.6}$$

with $N - p - 1$ degrees of freedom, so that if t_α corresponds to a confidence coefficient of $100(1 - \alpha)\%$,

$$b_i - \hat{\sigma}(a^{ii})^{1/2}t_\alpha < \beta_i < b_i + \hat{\sigma}(a^{ii})^{1/2}t_\alpha \tag{12.6.7}$$

The variance of the *difference* of two coefficients b_i and b_j is given by

$$\begin{aligned} V(b_i - b_j) &= V(b_i) + V(b_j) - 2C(b_i b_j) \\ &= \sigma^2(a^{ii} + a^{jj} - 2a^{ij}) \end{aligned} \tag{12.6.8}$$

and this may be used to test whether two coefficients differ significantly.

Example 1 The following artificial data are supposed to represent the yield y of a chemical reaction under different conditions of (a) time of reaction, (b) temperature, (c) amount of an added ingredient. Each variate x_i ($i = 1, 2, 3$) takes only two values, which we may code as -1 and 1. The variate x_0 is a dummy which always has the value 1. The matrix X therefore has the form

$$X = \begin{bmatrix} 1 & 1 & 1 & 1 & 1 & 1 & 1 & 1 \\ -1 & 1 & -1 & 1 & -1 & 1 & -1 & 1 \\ -1 & -1 & 1 & 1 & -1 & -1 & 1 & 1 \\ -1 & -1 & -1 & -1 & 1 & 1 & 1 & 1 \end{bmatrix}$$

and y' (the row vector of observations) is $y' = [61, 83, 51, 70, 66, 92, 56, 83]$. Then

$$A = XX' = \begin{bmatrix} 8 & 0 & 0 & 0 \\ 0 & 8 & 0 & 0 \\ 0 & 0 & 8 & 0 \\ 0 & 0 & 0 & 8 \end{bmatrix}$$

and

$$g = Xy = \begin{bmatrix} 562 \\ 94 \\ -42 \\ 32 \end{bmatrix}$$

Therefore.

$$A^{-1} = \begin{bmatrix} \frac{1}{8} & 0 & 0 & 0 \\ 0 & \frac{1}{8} & 0 & 0 \\ 0 & 0 & \frac{1}{8} & 0 \\ 0 & 0 & 0 & \frac{1}{8} \end{bmatrix}$$

and

$$b = A^{-1}g = \begin{bmatrix} 70.25 \\ 11.75 \\ -5.25 \\ 4.00 \end{bmatrix}$$

The fitted equation is

$$y = 70.25 + 11.75x_1 - 5.25x_2 + 4.00x_3$$

The residuals are given by

$$v' = [1.25, -0.25, 1.75, -2.75, -1.75, 0.75, -1.25, 2.75]$$

and $\underset{\alpha}{S} v_\alpha{}^2 = 400/16 = 25$.

This sum of squares of residuals represents both experimental error and the inadequacy of the linear model. Unless the experimental error can be independently estimated (for example, by replicated observations) there is no good way of telling whether the linear model is satisfactory.

The estimator of σ^2 in this example is

$$\hat{\sigma}^2 = \frac{25}{8-4} = 6.25$$

For four degrees of freedom, t_α corresponding to a confidence coefficient of 95% is 2.78, and for each value of i, $a^{ii} = \frac{1}{8}$. Therefore the confidence interval for each b_i is $b_i \pm 2.78\,(0.78)^{1/2} = b_i \pm 2.45$.

12.7 **Fitting a Polynomial of Second or Higher Degree** Since the predictors $X_1 \ldots, X_p$ of § 12.1 are not assumed to be independent, they may be taken as powers of a single variate X, say $X, X^2, \ldots X^p$, and the method of least squares may then be used to fit a polynomial of degree p to a set of N observations of pairs of values (x_α, y_α). The values of x_α may be chosen arbitrarily and the computations will be simplified if they can be taken as equally spaced along the x-axis. Instead of Eq. (12.1.6) we now have

$$(12.7.1)\qquad \begin{cases} a_{ij} = \underset{\alpha}{S}\, x_\alpha^{\,i+j}, & i, j = 0, 1 \dots p \\ g_i = \underset{\alpha}{S}\, x_\alpha^{\,i} y_\alpha \end{cases}$$

Thus if we wish to fit the quadratic

$$(12.7.2)\qquad y_c = b_0 + b_1 x + b_2 x^2$$

to a set of N pairs (x_α, y_α), the equation giving the b_i is

$$Ab = g$$

or, written out,

$$(12.7.3)\qquad \begin{bmatrix} N & Sx_\alpha & Sx_\alpha^2 \\ Sx_\alpha & Sx_\alpha^2 & Sx_\alpha^3 \\ Sx_\alpha^2 & Sx_\alpha^3 & Sx_\alpha^4 \end{bmatrix} \begin{bmatrix} b_0 \\ b_1 \\ b_2 \end{bmatrix} = \begin{bmatrix} Sy_\alpha \\ Sx_\alpha y_\alpha \\ Sx_\alpha^2 y_\alpha \end{bmatrix}$$

If we choose the unit of x so that the values (assumed equally spaced) increase by 1 from one observation to the next, and if we take the origin of x midway between the first and last observations, we shall have $Sx_\alpha = Sx_\alpha^3 = 0$, and this will considerably shorten the calculations. The equations of (3) then become:

$$(12.7.4)\qquad \begin{cases} Nb_0 + Sx_\alpha^2 \cdot b_2 = Sy_\alpha \\ Sx_\alpha^2 \cdot b_1 = Sx_\alpha y_\alpha \\ Sx_\alpha^2 \cdot b_0 + Sx_\alpha^4 \cdot b_2 = Sx_\alpha^2 y_\alpha \end{cases}$$

EXAMPLE 2 Suppose corresponding values of x_α and y_α are as given in the following table:

x_α	5	15	25	35	45	55	65	75	85	95
y_α	10.0	8.1	9.3	12.1	13.6	17.5	20.0	24.0	30.0	42.5
u_α	−4.5	−3.5	−2.5	−1.5	−0.5	0.5	1.5	2.5	3.5	4.5

If we replace x by $u = (x - 50)/10$, the conditions $Su_\alpha = Su_\alpha^3 = 0$ will be satisfied. Also $Su_\alpha^2 = \frac{330}{4} = 82.5$, $Su_\alpha^4 = \frac{19{,}338}{16} = 1208.6$, $Sy_\alpha = 187.1$, $Su_\alpha y_\alpha = 273.45$, and $Su_\alpha^2 y_\alpha = 1817.98$. The equations for b_0, b_1, b_2 (in $y_c = b_0 + b_1 u + b_2 u^2$) are, therefore,

$$\begin{aligned} 10b_0 + 82.5b_2 &= 187.1 \\ 82.5b_1 &= 273.45 \\ 82.5b_0 + 1208.6b_2 &= 1817.98 \end{aligned}$$

from which $b_0 = 14.4225$, $b_1 = 3.3145$, and $b_2 = 0.5197$. In terms of x, the best-fitting quadratic (or parabola) is

$$\begin{aligned} y_c &= 14.4225 + 0.33145(x - 50) + 0.005197(x - 50)^2 \\ &= 10.842 - 0.1882x + 0.005197x^2 \end{aligned}$$

The goodness of fit may be estimated from the sum of squares of residuals, which in this case amounts to 22.36. If a straight line were fitted by the same least squares process the equation would be

$$\begin{aligned} y_c &= 18.71 + 3.3145u \\ &= 2.138 + 0.33145x \end{aligned}$$

The sum of squares of residuals is 164.95, so that the fit of the parabola is apparently considerably better than that of the straight line. The relation between quadratic and linear regression may be brought out by an analysis of variance, as in Table 12.1.

TABLE 12.1

Variation	S.S.	D.F.	M.S.
Total $[S(y_\alpha - \bar{y})^2]$	1071.33	9	
Linear regression $[S(y_c - \bar{y})^2]$	906.38	1	906.38
About linear regression $[Sv_\alpha^2]$	164.95	8	20.62
Quadratic regression $[S(y_c - \bar{y})^2]$	1048.97	2	524.48
About quadratic regression $[Sv_\alpha^2]$	22.36	7	3.19

The variations "about regression" are the sums of squares of the residuals for the two fitted lines. The variations "due to regression" are calculated by difference from the total S.S., which is $Sy_\alpha^2 - \frac{1}{N}(Sy_\alpha)^2$. Since there are two constants for the straight line and three for the parabola, calculated from the data, the degrees of freedom for variation about regression are $N - 2$ and $N - 3$ respectively.

The reduction in S.S., due to replacing the straight line by the parabola, is $164.95 - 22.36 = 142.59$, with 1 d.f. This reduction may be compared with the S.S. about the parabola (22.36 with 7 d.f.). The F-value is clearly highly significant ($F = 44.7$ with 1 and 7 d.f.).

* 12.8 **Orthogonal Polynomials** The method of § 12.7 has the disadvantage that if we want to improve the fit by using a higher degree polynomial than one already fitted (a cubic instead of a quadratic, for example) the coefficients for the new polynomial have to be calculated afresh from the beginning. The

method of *orthogonal polynomials*, suggested by R. A. Fisher, allows us to add new terms independently of those already calculated. Incidentally, tests of significance of the coefficients are simplified.

Two polynomials $P_1(x)$ and $P_2(x)$ are said to be *orthogonal* for the set of values x_α ($\alpha = 1, 2 \ldots N$) if

$$\underset{\alpha}{S}\,[P_1(x_\alpha)\cdot P_2(x_\alpha)] = 0 \tag{12.8.1}$$

For example, the polynomials $P_1 = x - 4$, $P_2 = x^2 - 8x + 12$, $P_3 = x^3 - 12x^2 + 41x - 36$, are orthogonal to each other and to $P_0 = 1$ for $x = 1, 2, 3 \ldots 7$, as is evident from the following table of values:

TABLE 12.2

x	P_0P_1	P_0P_2	P_0P_3	P_1P_2	P_1P_3	P_2P_3
1	−3	5	−6	−15	18	−30
2	−2	0	6	0	−12	0
3	−1	−3	6	3	−6	−18
4	0	−4	0	0	0	0
5	1	−3	−6	−3	−6	18
6	2	0	−6	0	−12	0
7	3	5	6	15	18	30
	—	—	—	—	—	—
	0	0	0	0	0	0

It can be proved [5] that any polynomial of degree p can be expressed as a linear function of $p + 1$ polynomials

$$P(x) = A_0\xi_0 + A_1\xi_1 + \ldots + A_p\xi_p \tag{12.8.2}$$

where ξ_i is a polynomial in x of degree i. The equality holds for N distinct values of x, denoted by x_α, and the polynomials ξ_i are all orthogonal to each other. If x takes the values $1, 2 \ldots N$, the first few orthogonal polynomials are

$$\begin{cases} \xi_0 = 1 \\ \xi_1 = \lambda_1(x - \bar{x}) \\ \xi_2 = \lambda_2[(x - \bar{x})^2 - (N^2 - 1)/12] \\ \xi_3 = \lambda_3[(x - \bar{x})^3 - (x - \bar{x})(3N^2 - 7)/20] \\ \xi_4 = \lambda_4[(x - \bar{x})^4 - (x - \bar{x})^2(3N^2 - 13)/14 + 3(N^2 - 1)(N^2 - 9)/560] \end{cases} \tag{12.8.3}$$

where $\bar{x} = (N + 1)/2$ and the λ's are constants chosen so as to make the ξ_i integers (as small as possible) for all the N values of x. Thus if $N = 7$ we have $\bar{x} = 4$ and the constants are: $\lambda_1 = 1$, $\lambda_2 = 1$, $\lambda_3 = 1/6$, $\lambda_4 = 7/12$.

The sets of values of these polynomials for $N = 7$ are given in the following table:

TABLE 12.3

x	ξ_1	ξ_2	ξ_3	ξ_4
1	−3	5	−1	3
2	−2	0	1	−7
3	−1	−3	1	1
4	0	−4	0	6
5	1	−3	−1	1
6	2	0	−1	−7
7	3	5	1	3

On comparing Tables 12.2 and 12.3 it is clear that ξ_1, ξ_2 and ξ_3 are the same as the polynomials previously called P_1, P_2 and P_3, except that they are now multiplied by the corresponding λ's. All the polynomials with even subscripts (like ξ_2 and ξ_4) have a set of values which is symmetric about the middle, while all those with odd subscripts (like ξ_1 and ξ_3) are skew-symmetric. It is therefore unnecessary in a table to record *all* the values, and usually the lower half of the table (with the middle line when N is odd) is all that is actually printed. For Table 12.3, this would be the values from $x = 4$ to $x = 7$.

If a polynomial of degree p is to be fitted to a set of N observations (x_α, y_α), the method of least squares applied to the equation $y_c = P(x)$, where $P(x)$ is given by Eq. (2), leads to the set of normal equations:

$$(12.8.4)\qquad \begin{cases} NA_0 + S(\xi_{1\alpha})A_1 + \dots + S(\xi_{p\alpha})A_p = S(y_\alpha) \\ S(\xi_{1\alpha})A_0 + S(\xi_{1\alpha}^2)A_1 + \dots + S(\xi_{1\alpha}\xi_{p\alpha})A_p = S(y_\alpha\xi_{1\alpha}) \\ \cdot\ \cdot\ \cdot\ \cdot\ \cdot\ \cdot\ \cdot\ \cdot\ \cdot\ \cdot\ \cdot\ \cdot\ \cdot\ \cdot \\ S(\xi_{p\alpha})A_0 + \dots + S(\xi_{p\alpha}^2)A_p = S(y_\alpha\xi_{p\alpha}) \end{cases}$$

However, because of the orthogonal property of these polynomials (including $\xi_0 = 1$), all the terms but one on the left-hand side vanish in each of these equations. The set therefore reduces to

$$(12.8.5)\qquad A_iS(\xi_{i\alpha}^2) = S(y_\alpha\xi_{i\alpha}), \qquad i = 0, 1 \dots p$$

from which the A_i are immediately obtainable.

The sum of squares of the residuals is given by

$$(12.8.6)\qquad S(v_\alpha^2) = S(y_\alpha^2) - A_0S(y_\alpha) - A_1S(y_\alpha\xi_{1\alpha}) \dots - A_pS(y_\alpha\xi_{p\alpha})$$

Since $A_0 = S(y_\alpha)/N = \bar{y}$, the first two terms of $S(v_\alpha^2)$ give the total S.S. about the mean. The third term gives the reduction due to linear regression, the fourth term the additional reduction due to quadratic regression, and so on.

The numerical work of calculating the A_i is greatly facilitated by tables giving the values of ξ_1 to ξ_5 for different values of N. Such tables up to $N = 75$

may be found in Fisher and Yates' *Statistical Tables* (Oliver and Boyd). More extensive tables up to $N = 104$ have been given by Anderson and Houseman [6].

EXAMPLE 3 Suppose it is required to fit polynomials up to the fourth degree to the data of Example 2, § 12.7. We replace x by $u = (x + 5)/10$, so that u takes the values 1, 2 . . . 10. The values of ξ_1, ξ_2, ξ_3, ξ_4 for $N = 10$ are read from the tables.

TABLE 12.4

x	u	y	ξ_1	ξ_2	ξ_3	ξ_4
5	1	10.0	−9	6	−42	18
15	2	8.1	−7	2	14	−22
25	3	9.3	−5	−1	35	−17
35	4	12.1	−3	−3	31	3
45	5	13.6	−1	−4	12	18
55	6	17.5	1	−4	−12	18
65	7	20.0	3	−3	−31	3
75	8	24.0	5	−1	−35	−17
85	9	30.0	7	2	−14	−22
95	10	42.5	9	6	42	18

We calculate $S(y_\alpha) = 187.1$, $S(y_\alpha^2) = 4571.97$, $S(y_\alpha\xi_{1\alpha}) = 546.9$, $S(y_\alpha\xi_{2\alpha}) = 137.2$, $S(y_\alpha\xi_{3\alpha}) = 252.2$, $S(y_\alpha\xi_{4\alpha}) = 196.8$. The values of $S(\xi_1^2) = 330$, $S(\xi_2^2) = 132$, $S(\xi_3^2) = 8580$, $S(\xi_4^2) = 2860$ are read from the tables, as are also the values $\lambda_1 = 2$, $\lambda_2 = 1/2$, $\lambda_3 = 5/3$, $\lambda_4 = 5/12$. Then

$$\begin{aligned} A_0 &= 187.1/10 = 18.71, & A_0S(y_\alpha) &= 3500.64 \\ A_1 &= 546.9/330 = 1.6573, & A_1S(y_\alpha\xi_{1\alpha}) &= 906.38 \\ A_2 &= 137.2/132 = 1.0394, & A_2S(y_\alpha\xi_{2\alpha}) &= 142.61 \\ A_3 &= 252.2/8580 = 0.029394, & A_3S(y_\alpha\xi_{3\alpha}) &= 7.41 \\ A_4 &= 196.8/2860 = 0.068811, & A_4S(y_\alpha\xi_{4\alpha}) &= 13.54 \end{aligned}$$

The polynomial is

$$y_c = 18.71 + 1.6573\xi_1 + 1.0394\xi_2 + 0.029394\xi_3 + 0.068811\xi_4 \tag{12.8.7}$$

where, by Eq. (3) with $\bar{u} = 5.5$ and $N = 10$,

$$\begin{aligned} \xi_1 &= 2(u - 5.5) \\ \xi_2 &= \tfrac{1}{2}[(u - 5.5)^2 - 8.25] \\ \xi_3 &= \tfrac{5}{3}[(u - 5.5)^3 - 14.65(u - 5.5)] \\ \xi_4 &= \tfrac{5}{12}[(u - 5.5)^4 - 20.5(u - 5.5)^2 + 48.2625] \end{aligned} \tag{12.8.8}$$

The best-fitting straight line is given by the first two terms only of Eq. (7), the best-fitting parabola by the first three terms, and so on. On replacing u by $(x + 5)/10$ we recover the results of § 12.7.

The analysis of variance is set out in Table 12.5.

TABLE 12.5

Variation	S.S.	D.F.	M.S.
Total	1071.33	9	
Linear regression	906.38	1	906.38
About linear regression	164.95	8	20.62
Additional for quadratic regression	142.61	1	142.61
About quadratic regression	22.34	7	3.19
Additional for cubic regression	7.41	1	7.41
About cubic regression	14.93	6	2.49
Additional for quartic regression	13.54	1	13.54
About quartic regression	1.39	5	0.28

Compared with the deviation about the cubic regression line, the additional sum of squares for cubic regression is not significant ($F = 3.0$ with 1 and 6 d.f.). However, compared with the deviation about quartic regression, the additional S.S. for quartic regression is highly significant. ($F = 48$, with 1 and 5 d.f.). This indicates that the cubic curve is not appreciably better than the parabola, but the quartic curve is a much better fit than both.

Of course, by using a ninth degree curve we could fit the given 10 points exactly, but such a complicated curve is obviously not desirable. We have to compromise between the desire for simplicity and the desire to get a good fit. The second-degree curve (the parabola) is probably as satisfactory as any polynomial in this example.

The matrix which corresponds to A in Eq. (12.1.7) is now diagonal, so that it is very easy to invert. In fact

$$A = \begin{bmatrix} N & 0 & \cdots & 0 \\ 0 & S(\xi_{1\alpha}^{2}) & \cdots & 0 \\ \cdot & \cdot & \cdots & \cdot \\ 0 & 0 & \cdots & S(\xi_{p\alpha}^{2}) \end{bmatrix}$$

Therefore $a^{ii} = [S(\xi_{i\alpha}^{2})]^{-1}$, and the estimated variance of the coefficient A_i in Eq. (2), $i = 0, 1, 2 \ldots p$, is given by

$$V(A_i) = a^{ii} \frac{Sv_\alpha^{2}}{N - p - 1} \qquad (12.8.9)$$

Thus in Eq. (7), the estimated standard error of A_0 is $[1.39/50]^{1/2} = 0.17$ and that of A_1 is $[1.39/(5 \times 330)]^{1/2} = 0.029$. These values apply, of course, only if a fourth-degree curve is fitted, since the residuals v_α relate to such a curve. The sum Sv_α^2 is given by Eq. (6), with $p = 4$.

* 12.9 **A Test for Linearity of Regression with Grouped Variates** When the number of observations is sufficiently large to warrant grouping, we may be presented with a two-way table of data like that in § 11.4, Example 2. Each column (or x-array) includes all the observations with x-values lying in one particular class-interval, and these are all assumed to have the same value, namely that at the centre of the interval.

Within one column the y-values are also grouped in classes, and in each class (i.e., in each row of the table) all the observations are assumed to have the central value of y. We suppose that there are p columns in the table and that the total frequency in the ith column is f_i where $\sum f_i = N$.

We can then define for each column the arithmetic mean $\bar{y}_i$ of the y-values in that column. If these column means are plotted against the central x-values for the columns, the result of joining them is a sort of empirical trend line. In fact, if a straight line is fitted by least squares to this set of column means, each one being weighted with the corresponding column frequency, the result is precisely the ordinary regression line of y on x.

The sum of squares of deviations from the column mean within one column is $\underset{\alpha}{S}(y_{i\alpha} - \bar{y}_i)^2$, where the $y_{i\alpha}$ are the y values in the ith column, $(\alpha = 1, 2 \ldots f_i)$. The ratio of this sum, added up for all columns, to the total sum of squares for y about the over-all mean $\bar{y}$, defines a quantity called the *correlation ratio* of y on x (E_{yx}) by the relation:

$$1 - E_{yx}^2 = \frac{\sum_i \underset{\alpha}{S} (y_{i\alpha} - \bar{y}_i)^2}{\sum_i \underset{\alpha}{S} (y_{i\alpha} - \bar{y})^2} \tag{12.9.1}$$

The denominator is simply $(N - 1)\, s_Y^2 = S_y$ (say), since the sum is over all y values in the table. The above expression may be compared with one obtained in Chapter 11—see Eq. (11.5.2)—namely,

$$1 - r^2 = \frac{\sum_i \underset{\alpha}{S} (y_{i\alpha} - y_{ci})^2}{(N - 1)s_Y^2} \tag{12.9.2}$$

where y_{ci} is the calculated value of y for the center of the ith column, according to the linear regression equation of y on x. This indicates that E_{yx} is similar in nature to the Pearson coefficient r. If the regression is in fact nearly linear, the two agree quite closely, but the more the regression (as indicated by the line of column means) departs from a straight line the more do E_{yx} and r differ. The difference may be used to estimate the significance of an apparent departure from linearity.

The quantity E_{yx}^2 may also be written

$$E_{yx}^2 = \frac{S_{\bar{y}}}{S_y} \tag{12.9.3}$$

where $S_{\bar{y}}$ is the weighted sum of squares of the column means about the over-all mean and S_y is the total sum of squares for y about the over-all mean. That is,

$$\begin{cases} S_{\bar{y}} = \sum_i f_i(\bar{y}_i - \bar{y})^2 \\ S_y = \sum_i S_\alpha(y_{i\alpha} - \bar{y})^2 = (N-1)s_Y^2 \end{cases} \tag{12.9.4}$$

In the notation of § 11.4, with the auxiliary variables u and v,

$$\begin{cases} S_{\bar{y}} = k^2 \sum_u f_u\left(\frac{V}{f_u} - \bar{v}\right)^2 \\ \quad\;\; = k^2\left(\sum_u \frac{V^2}{f_u} - N\bar{v}^2\right) \\ S_y = (N-1)k^2 s_v^2 \end{cases} \tag{12.9.5}$$

This gives the most convenient formula in practice for calculating E_{yx}^2. Analogously to Eq. (3) we can rewrite Eq. (2) in the form

$$r^2 = \frac{S_{y_c}}{S_y} \tag{12.9.6}$$

where $S_{y_c} = \sum_i f_i(y_{ci} - \bar{y})^2$ which is the weighted sum of squares for the calculated linear regression values y_{ci}. Therefore,

$$\begin{aligned} (E_{yx}^2 - r^2)S_y &= S_{\bar{y}} - S_{y_c} \\ &= \sum_i f_i[(\bar{y}_i - \bar{y})^2 - (y_{ci} - \bar{y})^2] \end{aligned} \tag{12.9.7}$$

and so represents that part of the sum of squares for column means which is not accounted for by linear regression. If this part is large compared with the sum of squares within columns about the column means, $\sum_i S_\alpha(y_{i\alpha} - \bar{y}_i)^2 = (1 - E_{yx}^2)S_y$, we may reasonably reject the hypothesis that the true regression is linear. The test ratio is therefore $(E_{yx}^2 - r^2)/(1 - E_{yx}^2)$.

If the values of y within each column are normally distributed with a variance σ^2 common to all the columns, then $S_\alpha(y_{i\alpha} - \bar{y}_i)^2$ for any column is distributed as $\chi^2\sigma^2$ with $f_i - 1$ degrees of freedom. Therefore $(1 - E_{yx}^2)S_y$ is distributed as $\chi^2\sigma^2$ with $\sum (f_i - 1) = N - p$ degrees of freedom.

Furthermore, the p values of $\bar{y}_i$ are each normal with variance σ^2/f_i, so that $\sum_i f_i(\bar{y}_i - \bar{y})^2$ is distributed as $\chi^2\sigma^2$ with $p - 1$ d.f. Also

$$\begin{aligned} \sum_i f_i(y_{ci} - \bar{y})^2 &= b^2 \sum_i f_i(x_{ci} - \bar{x})^2 \\ &= b^2(N-1)s_x^2 \end{aligned}$$

Now, as shown in § 11.5, b is normal with varianee $\sigma^2/[(N-1)s_x^2]$, so that $b^2(N-1)s_x^2$ is distributed as $\chi^2\sigma^2$ with 1 d.f. It follows from Eq. (7) and Theorem 4.3 that $(E_{yx}^2 - r^2)S_y$ is distributed as $\chi^2\sigma^2$ with $p-2$ d.f., and is independent of $S_{\bar{y}}$. The ratio

$$F = \frac{N-p}{p-2}\cdot\frac{E_{yx}^2 - r^2}{1 - E_{yx}^2} \tag{12.9.8}$$

is therefore distributed as Snedecor's F with $p-2$ and $N-p$ degrees of freedom. A significant value of F indicates a significant departure from linearity. The test is one-tailed.

EXAMPLE 4 The following table represents some results on the relation between the percentage protein in wheat (y) and the yield in bushels per acre (x)

TABLE 12.6

$v \backslash u$	−3	−2	−1	0	1	2	3	4	f_v	vf_v	v^2f_v
5		1							1	5	25
4									0	0	0
3	3	1							4	12	36
2	2	2							4	8	16
1	3	4	1	1					9	9	9
0	2	4	15	2	1				24	0	0
−1			4	3	1	2	2		12	−12	12
−2				5	3	3	1		12	−24	48
−3			2	1	1	7	7	2	20	−60	180
−4				1	1	2	1		5	−20	80
f_u	10	12	22	13	7	14	11	2	91	−82	406
uf_u	−30	−24	−22	0	7	28	33	8	0		
u^2f_u	90	48	22	0	7	56	99	32	354		
V	16	16	−9	−19	−14	−37	−29	−6	−82		
Vu	−48	−32	9	0	−14	−74	−87	−24	−270		
$\bar{v}_u = V/f_u$	1.60	1.33	−0.41	−1.46	−2.00	−2.64	−2.64	−3.00			
V^2/f_u	25.60	21.33	3.68	27.77	28.00	97.79	76.45	18.00	298.62		

for a set of 91 experimental plots. We suppose that it is desired to predict y for a given x, so that the x values may be regarded as fixed.

The coded u and v values are given by

$$u = \frac{x - 22}{5}, \qquad v = y - 13.45$$

From the table and Eq. (5) we obtain

$$S_{\bar{y}} = 298.62 - 91\left(\frac{-82}{91}\right)^2 = 224.73$$

$$S_y = 406 - 91\left(\frac{-82}{91}\right)^2 = 332.11$$

whence

$$E_{yx}^{\ 2} = 0.677$$

Also the sum of squares for x and the sum of products for x and y are given by

$$S_x = 25(354 - 0) = 8850$$

$$S_{xy} = 5(-270 - 0) = -1350$$

so that

$$r^2 = \frac{(-1350)^2}{(8850)(332.11)} = 0.620$$

Therefore

$$F = \frac{N - p}{p - 2} \cdot \frac{E_{yx}^{\ 2} - r^2}{1 - E_{yx}^{\ 2}} = \frac{83}{6} \cdot \frac{0.057}{0.323}$$
$$= 2.44$$

with 6 and 83 d.f. The 5% point is about 2.21 so that there is a significant departure from linearity.

The original data (ungrouped) are given in Snedecor's *Statistical Methods* (4th ed., p. 380), where a parabola is fitted by the method of § 12.7. The difference between the S.S. about the parabola and the S.S. about the best straight line is significant as compared with the former S.S. itself. This confirms the departure from linearity. The curve of column means (in units of v) is plotted in Fig. 53, the necessary data being obtained from the last row but one of Table 12.6. For comparison the best-fitting parabola is also shown.

* 12.10 **The Distribution of the Correlation Ratio** The correlation ratio E_{yx} may be used as a measure of the degree to which the observations (as grouped) tend to cluster around the curve of column means, just as Pearson's r measures the degree of clustering around the straight regression line of Y on X. A similar, but usually different, ratio, E_{xy}, measures the clustering around the curve of row means. It may be calculated in the same way as E_{yx}, with x and y interchanged throughout.

On the hypotheses mentioned in the previous section, and on the assumption that there is really no association between the variates in the parent population,

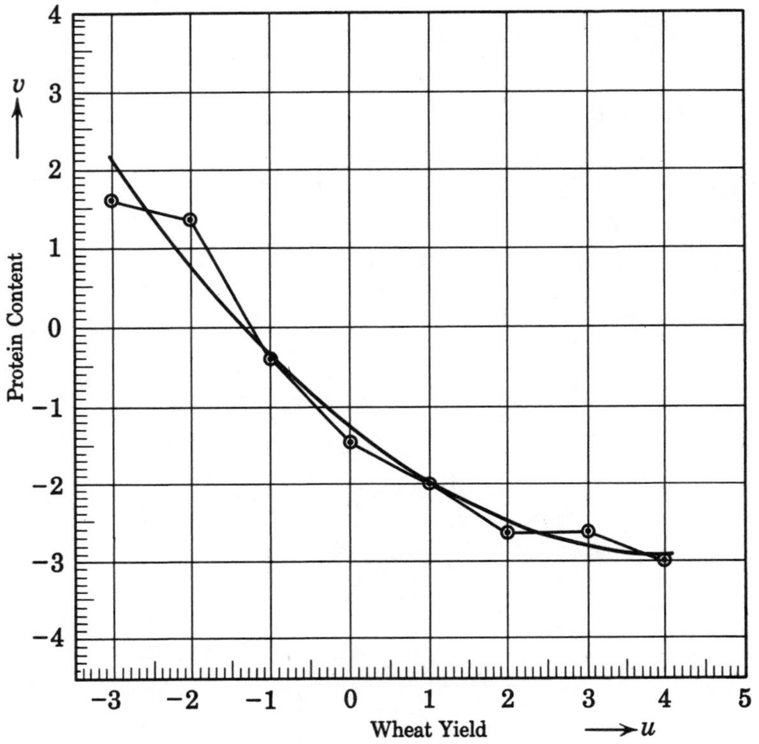

FIG. 53 CURVE OF COLUMN MEANS AND FITTED PARABOLA, FOR DATA ON WHEAT YIELD AND PROTEIN CONTENT

the distribution of E_{yx}^2 was worked out by Hotelling [7]. He showed that E_{yx}^2 is a beta-variate with parameters $n_1 = p - 1$ and $n_2 = N - p$. It follows therefore that $n_2 E_{yx}^2/[n_1(1 - E_{yx}^2)]$ has the F-distribution with n_1 and n_2 d.f.

The significance of an observed E_{yx} may be tested by means of Pearson's Tables of the Incomplete Beta Function or ordinary tables of F. A special table for large values of N, 50(1)1000, was prepared by Woo [8].

If the population correlation ratio η_{yx} is not zero, but if we assume that in all samples there are the same set of frequencies f_i, the density function for E^2 is

$$f(E^2) = e^{-\lambda}(E^2)^{a-1}(1 - E^2)^{b-1}\frac{H(\lambda E^2)}{B(a, b)} \tag{12.10.1}$$

where E^2 is written for E_{yx}^2, η^2 for η_{yx}^2, $a = n_1/2$, $b = n_2/2$, and $\lambda = N\eta^2/[2(1 - \eta^2)]$. The function H, with argument λE^2, is called the *confluent hypergeometric function*, defined by the series:

(12.10.2) $$H(x) = 1 + \frac{a+b}{1!a}x + \frac{(a+b)(a+b+1)}{2!a(a+1)}x^2 + \ldots$$

Tang [9] has tabulated the distribution function for E^2, namely, the probability that $E^2 \leq E_\alpha{}^2$ for certain values of λ, $E_\alpha{}^2$ being fixed by the condition:

(12.10.3) $$\int_{E_\alpha{}^2}^{1} f(E^2|\lambda = 0)\, dE^2 = \alpha$$

for $\alpha = 0.01$ or 0.05. The tabulated probability is therefore that of an error of the second kind (§ 6.6), the chance of an error of the first kind, namely, a wrong rejection of the null hypothesis that $\lambda = 0$, being fixed at the value α.

It may be noted that E^2 has the same distribution as x in § 9.12, where x is the ratio of the S.S. between treatments to the total S.S. in a one-way analysis-of-variance problem. The difference is that the number of treatments is replaced by the number of columns and the S.S. between treatments by the S.S. between column means. The null hypothesis of no treatment effects becomes the hypothesis that in the population the column means are all equal. Under this hypothesis, and if the variance of Y is the same within each column, $S_{\bar{y}}/\sigma^2$ has the χ^2 distribution with $n_1(= p - 1)$ d.f. Under the alternative hypothesis (that η, and therefore λ, is not zero) it has the *non-central chi-square* distribution (Appendix A.13) with non-centrality parameter

(12.10.3) $$\lambda = \frac{N\eta^2}{2(1-\eta^2)} = \frac{N(\sigma_Y{}^2 - \sigma^2)}{2\sigma^2}$$

where $\sigma_Y{}^2$ is the variance of y in the population and σ^2 is the variance in each column about the column mean. When $\lambda = 0$ this distribution becomes the ordinary (central) chi-square distribution.

* 12.11 **Exponential Regression** It is not uncommon for a variable Y to increase or decrease with time at an approximately uniform percentage rate. This holds, for example, for money accumulating at a fixed rate of compound interest or for a bacterial population growing in an ample supply of culture medium. In fact this type of increase is often referred to as "the law of growth." It may be expressed mathematically by the relation:

(12.11.1) $$\frac{dY}{Y} = k\, dX$$

or, in integral form,

(12.11.2) $$Y = Ae^{kX}$$

If we wish to fit such an exponential curve by least squares to a set of N pairs of observed values (x_α, y_α), $\alpha = 1, 2 \ldots N$, we have to calculate A and k from the relation:

(12.11.3) $$\underset{\alpha}{S}\,(y_\alpha - Ae^{kx_\alpha})^2 = \text{minimum}$$

from which we get, by differentiating,

$$(12.11.4)\qquad \begin{cases} \underset{\alpha}{S}\, e^{kx_\alpha}(Ae^{kx_\alpha} - y_\alpha) = 0 \\ \underset{\alpha}{S}\, x_\alpha e^{kx_\alpha}(Ae^{kx_\alpha} - y_\alpha) = 0 \end{cases}$$

The exact solution of these equations for the unknowns A and k is tedious and time-consuming. It is customary instead to write Eq. (2) in the form

$$(12.11.5)\qquad \log Y = \log A + kX$$

and to fit a straight line by the method of Chapter 11 to the observed values of $\log y_\alpha$ and x_α. This, of course, means that the sum of squares of deviations for $\log Y$ is minimized instead of the corresponding S.S. for Y. If the standard deviation for Y is proportional to Y itself, as seems to be nearly true for many types of data in economics, this procedure is quite reasonable, since $\delta(\log Y) \approx \delta Y/Y$ and the standard deviation of $\log Y$ is therefore approximately constant. If, however, there is reason to believe that the standard deviation of Y itself is constant, the effect of the customary procedure is to give undue weight to the smaller values of Y.

A method of allowing (at least approximately) for this effect is to fit a straight line to the observed $\log y_\alpha$ but to weight each observation proportionately to y_α. The weighted least-squares condition is

$$(12.11.6)\qquad S(\log y_\alpha - a - kx_\alpha)^2 \cdot y_\alpha = \min$$

where $a = \log A$ in Eq. (5). This furnishes the normal equations:

$$(12.11.7)\qquad \begin{cases} aSy_\alpha + kS(x_\alpha y_\alpha) = S(y_\alpha \log y_\alpha) \\ aS(x_\alpha y_\alpha) + kS(x_\alpha{}^2 y_\alpha) = S(x_\alpha y_\alpha \log y_\alpha) \end{cases}$$

from which a and k can be found. If common logarithms instead of natural logarithms are used, the equation for Y will be of the form $Y = A\cdot 10^{kx}$ instead of that in Eq. (2).

TABLE 12.7

x	y	$\log_{10} y$
6	0.029	−1.538
7	.052	−1.284
8	.079	−1.102
9	.125	−0.903
10	.181	−0.742
11	.261	−0.583
12	.425	−0.372
13	.738	−0.132
14	1.130	0.053
15	1.882	0.275
16	2.812	0.449

EXAMPLE 5 (Snedecor [11]) In Table 12.7, x represents the age in days of chick embryos and y the dry weight in grams.

The calculated y_c for a given x, obtained by fitting a straight line to the weighted values of $\log y$, is $y_c = 0.001875(10)^{0.1989x} = 0.001875e^{0.4581x}$.

Without weighting, the result is $y_c = 0.002046e^{0.4511x}$, while the exact least-squares solution is $y_c = 0.001895e^{0.4573x}$. The method of weighting the observed values of $\log y$ gives (at least in this example) a very good approximation.

In some problems the data follow more or less a *modified exponential* curve, expressed by

$$Y = C + Ae^{kx} \tag{12.11.8}$$

The exact least-squares solution of Eq. (8) for a set of sample values x_α, y_α, is even more difficult than for Eq. (2), and when plotted on semilog graph paper the points (x_α, y_α) do not lie nearly on a straight line. However it is possible to use a graphical method due to Cowden [10] to obtain approximate values of C, A and k, and these may be improved, if necessary, by using Seidel's process (§ 12.12).

The data are plotted on ordinary or semilog graph paper and a tentative trend line is drawn in by hand. Three equidistant ordinates Y_0, Y_1 and Y_2 of the curve at convenient values of X ($X - h$, X and $X + h$, say) are measured, and C is estimated from the relation:

$$C = \frac{Y_0 Y_2 - Y_1{}^2}{Y_0 + Y_2 - 2Y_1} \tag{12.11.9}$$

Values of $y_\alpha - C$ are now plotted on semilog graph paper. If C is correct, these points should lie close to a straight line; if there still appears to be some curvature the value of C may be readjusted slightly by trial. From the resulting straight line, A and k may be estimated, A being the ordinate at $X = 0$ and e^{kx_1} the ratio of the ordinates at $X = x_1$ and $X = 0$.

* 12.12 **Seidel's Method of Successive Approximations** Sometimes it is convenient to obtain approximate values of the regression coefficients from a graph. With these as a start, Seidel's method permits better values to be obtained by a least-squares procedure.

Suppose the regression curve is of the form

$$Y = f(x, \beta_0, \beta_1)$$

where β_0 and β_1 are the true parameters. (The method can easily be extended to more than two parameters.) If the preliminary approximations are b_0 and b_1, let $\delta b_0 = \beta_0 - b_0$ and $\delta b_1 = \beta_1 - b_1$. Then if the approximations are reasonably good we should be able to neglect squares, products and higher powers of δb_0 and δb_1, so that

$$\begin{aligned} Y &= f(x, b_0 + \delta b_0, b_1 + \delta b_1) \\ &\approx f(x, b_0, b_1) + \frac{\partial f}{\partial b_0}\,\delta b_0 + \frac{\partial f}{\partial b_1}\,\delta b_1 \end{aligned} \tag{12.12.1}$$

where $\partial f/\partial b_0$, $\partial f/\partial b_1$ mean the partial derivatives of $f(x, \beta_0, \beta_1)$ with respect to β_0 and β_1, evaluated at $\beta_0 = b_0$ and $\beta_1 = b_1$. Then δb_0 and δb_1 are selected so as to minimize the sum of squares of residuals

$$\underset{\alpha}{S}\left[y_\alpha - f(x_\alpha, b_0, b_1) - \frac{\partial f}{\partial b_0}\,\delta b_0 - \frac{\partial f}{\partial b_1}\,\delta b_1\right]^2$$

The normal equations for δb_0 and δb_1 may therefore be written

$$\text{(12.12.2)}\qquad \begin{cases} \underset{\alpha}{S}\dfrac{\partial f}{\partial b_0}\left(y_\alpha - f - \dfrac{\partial f}{\partial b_0}\,\delta b_0 - \dfrac{\partial f}{\partial b_1}\,\delta b_1\right) = 0 \\[2ex] \underset{\alpha}{S}\dfrac{\partial f}{\partial b_1}\left(y_\alpha - f - \dfrac{\partial f}{\partial b_0}\,\delta b_0 - \dfrac{\partial f}{\partial b_1}\,\delta b_1\right) = 0 \end{cases}$$

where, in f and its partial derivatives, x is replaced by x_α. On solving these equations and adding δb_0 to b_0 and δb_1 to b_1, the preliminary approximations are improved. The process can be repeated if necessary and usually converges quite rapidly.

EXAMPLE 6 Suppose that we have plotted on semilog paper the data in Example 5 above and have drawn by eye an approximately best-fitting straight line. From the line we estimate $k_1 = 0.20$ and $a_1 = \log A_1 = -2.70$, these being first approximations to k and $a(= \log A)$ respectively. Then

$$\log Y = a + kx$$

so that $\partial f/\partial a = 1$, $\partial f/\partial k = x$. The weighted normal equations, weighted according to the values of y_α, are

$$S[y_\alpha(\log y_\alpha - a_1 - k_1x_\alpha - \delta a_1 - x_\alpha\delta k_1)] = 0$$

$$S[y_\alpha x_\alpha(\log y_\alpha - a_1 - k_1x_\alpha - \delta a_1 - x_\alpha\delta k_1)] = 0$$

On substituting the values of x_α and y_α, these become

$$7.714\delta a_1 + 110.712\delta k_1 = -0.32786$$

$$110.712\delta a_1 + 1619.18\delta k_1 = -4.73783$$

from which

$$\delta a_1 = -0.0271, \qquad \delta k_1 = -0.0011$$

Therefore the improved values of a and k are

$$a_2 = -2.70 - 0.0271 = -2.7271$$

$$k_2 = 0.20 - 0.0011 = 0.1989$$

The fitted curve is

$$\log Y = -2.7271 + 0.1989x$$

or

$$Y = 0.001875(10)^{0.1989x}$$

which agrees with the result quoted in § 12.11.

PROBLEMS

A. (§§ 12.1–12.3)

1. Write out the normal equations (12.1.5) for the case of two predictors ($X_0 = 1$, $X_1 = X$, $X_2 = Z$) and show that these equations can be put into the form:

$$\bar{y} = b_0 + b_1\bar{x} + b_2\bar{z}$$
$$s_{YX} = b_1 s_X^2 + b_2 s_{XZ}$$
$$s_{YZ} = b_1 s_{XZ} + b_2 s_Z^2$$

where $\bar{x}$, $\bar{y}$, $\bar{z}$, are the means, and s_X^2, s_Y^2, s_Z^2, s_{YX}, s_{YZ}, s_{XZ} are the variances and covariances, for the variates X, Y and Z.

2. Show that the equation of the regression plane of Y on X and Z may be written $(y_c - \bar{y})(d_1/s_Y) + (x - \bar{x})(d_2/s_X) + (z - \bar{z})(d_3/s_Z) = 0$ where d_1, d_2, d_3 are the cofactors of the elements of the first row in the determinant

$$d = \begin{vmatrix} 1 & r_{XY} & r_{ZY} \\ r_{XY} & 1 & r_{XY} \\ r_{ZY} & r_{XZ} & 1 \end{vmatrix}$$

Hint: See Problem 1. The quantities r_{XY}, r_{XZ}, r_{ZY} are the respective coefficients of correlation.

3. (*Hooker*) For a certain district in England, records were kept over 20 years of the following variates:

Y = seed-hay crop (cwt/acre)
X = spring rainfall (inches)
Z = accumulated temperature above 42°F in spring.

From the data the following statistics were calculated:
$\bar{x} = 4.91$, $\bar{y} = 28.02$, $\bar{z} = 594$, $s_X = 1.10$, $s_Y = 4.42$, $s_Z = 85$, $r_{XY} = 0.80$, $r_{XZ} = -0.56$, $r_{ZY} = -0.40$.

Use the result of Problem 2 to find the regression of hay crop on spring rainfall and accumulated temperature.

4. At an experimental farm in Alberta, records were kept over 35 years of the evaporation (Y) from an open tank. It was thought that this might be related to the date of observation (X) and to the annual rainfall (Z). With the date coded as an integer from 1 to 35, the following observations were recorded: $Sx_\alpha = 630$, $Sx_\alpha^2 = 14{,}910$, $Sz_\alpha = 286.90$, $Sz_\alpha^2 = 2563.47$, $Sy_\alpha = 452.53$, $Sy_\alpha^2 = 5980.79$, $Sx_\alpha y_\alpha = 7814.63$, $Sz_\alpha y_\alpha = 3626.96$, $Sx_\alpha z_\alpha = 5287.95$.

Write out the matrices A and g and find, by Cramer's rule, the predicting equation for Y in terms of X and Z.

5. (*P. O. Johnson*). As part of a study dealing with the prediction of freshman achievement in college, the following scores were noted for each of a random sample of 50 students:

Y = honor-point ratio at end of freshman year,
X_1 = score on an English test,
X_2 = score on an algebra test,
X_3 = percentile ranking at high school graduation, transformed to probits (this provides in effect a normal variate with mean 5).

From the data obtained,

$Sy_\alpha = 36.19$	$Sy_\alpha^2 = 40.6393$	$N = 50$,
$Sx_{1\alpha} = 4802$	$Sx_{2\alpha} = 1560$	$Sx_{3\alpha} = 248.22$
$Sx_{1\alpha}^2 = 487{,}798$	$Sx_{2\alpha}^2 = 66{,}942$	$Sx_{3\alpha}^2 = 1260.7630$
$Sx_{1\alpha}y_\alpha = 3533.58$	$Sx_{2\alpha}y_\alpha = 1213.74$	$Sx_{3\alpha}y_\alpha = 189.1539$
$Sx_{1\alpha}x_{2\alpha} = 157{,}863$	$Sx_{1\alpha}x_{3\alpha} = 23{,}926.22$	$Sx_{2\alpha}x_{3\alpha} = 7804.57$

Write out and solve the normal equations and so obtain the equation for predicting Y from X_1, X_2 and X_3.

B. (§§ 12.4–12.6)

1. In Problem A-4, invert the matrix A. Use Eq. (12.5.6) to find the sum of squares of residuals, and so obtain an estimate of the variance σ^2 of Y about the true regression plane.

2. In Problem A-4, obtain 90% confidence intervals for the regression coefficients β_0 β_1 and β_2. (For 32 degrees of freedom, $t_\alpha = 1.694$.)

3. Write down the matrix A for the data of Problem A-5. Calculate the diagonal terms in the inverse matrix and hence obtain the variances of the three regression coefficients b_1, b_2 and b_3 in the predicting equation for Y. Show that only the regression on X_3 is significant.

4. From the data of Problem A-3, calculate approximately the values of the matrix elements a_{ij}, and so obtain an estimate of the variance σ^2 about the true regression plane. Estimate also the variance of the predicted value of Y for a new pair of observed values of X and Z.

C. (§§ 12.7–12.8)

1. Fit a second-degree parabola to the following data:

x	1.0	1.5	2.0	2.5	3.0	3.5	4.0
y	1.1	1.3	1.6	2.3	2.7	3.4	4.1

2. Construct an analysis of variance table for the data in Problem C-1, and determine whether there is a significant reduction in the sum of squares about the regression line when the straight-line regression is replaced by parabolic regression.

3. If a *correlation index* r_c is defined by $r_c^2 = 1 - (Sv_\alpha^2)/[(N-1)s_Y^2]$, where v_α is a residual for parabolic regression, find the value of r_c for the data of Problem C-1. Compare with the Pearson coefficient of correlation for the same data.

4. (*Holzinger*) In the following table, X represents mean age in years for a group of men, and Y their mean vital capacity. Use the method of orthogonal polynomials to find the equation of the best-fitting cubic curve.

X	Y	X	Y	X	Y
19.5	227	37.5	223	55.5	201
22.5	230	40.5	218	58.5	185
25.5	230	43.5	216	61.5	200
28.5	237	46.5	210	64.5	169
31.5	227	49.5	205	67.5	160
34.5	229	52.5	193	70.5	163

Note: The following extract from Anderson and Houseman's tables is relevant:

$N = 18$

ξ_1	1	3	5	7	9	11	13	15	17
ξ_2	−40	−37	−31	−22	−10	5	23	44	68
ξ_3	−8	−23	−35	−42	−42	−33	−13	20	68

$$S(\xi_1^2) = 1938, \quad S(\xi_2^2) = 23{,}256, \quad S(\xi_3^2) = 23{,}256$$
$$\lambda_1 = 2, \quad \lambda_2 = 3/2, \quad \lambda_3 = 1/3$$

5. Draw up an analysis of variance table for the regression of Problem C-4. Find estimates of the standard error for each of the coefficients of the orthogonal polynomials obtained in this problem.

D. (§§ 12.9–12.10)

1. In the following table X is the amount of irrigation water (inches) applied to a crop, and Y is the crop yield in bushels per acre. The numbers in the headings are the class-marks (central values) of the respective classes. Test the regression of Y on X for linearity.

Y \ X	12	15	18	21	24	27	30	
90						1	2	3
85					2	3		5
80			2	5	4	1		12
75		2	4	6	1			13
70			4	3	1			8
65		2		3				5
60	2			2				4
	2	4	10	19	8	5	2	50

2. In Problem B-4 of Chapter 11, calculate the two correlation ratios E_{yx} and E_{xy}. Does either regression (of height on weight or of weight on height) depart significantly from linearity?

3. Prove the statement in § 12.9 that if a straight line is fitted by least squares to the weighted column means in a grouped two-way table, the result is the ordinary regression line of Y on X.

4. Show that when $\lambda = 0$, the variate E^2 in Eq. (12.10.1) becomes a beta-variate. Show also that in this case $n_2E^2/[n_1(1 - E^2)]$ has the F distribution with n_1 and n_2 degrees of freedom.

E. (§§ 12.11–12.12)

1. The uniform horizontal scale on a sheet of semilog paper ranges from 0 to 10. The vertical logarithmic scale (on the left side) ranges from 100 to 1000. A straight line is drawn from the upper end of the vertical scale to the midpoint of the horizontal scale. What is the equation of the line (a) in the coordinates x and y', where $y' = \log_{10} y$, (b) in the coordinates x and y?

2. A straight line is drawn on semilog paper through the points (2, 1) and (4, 100). What is the equation of this line?

3. Fit an exponential curve to the following data (a) without weighting, and (b) with weighting:

x	1	2	3	4	5
y	2.1	4.3	14.5	42.2	123.1

Hint: For part (b), use Eq. (12.11.7).

4. Prove that the exact least-squares solution of the problem of fitting the modified exponential curve $y = C + Ae^{kx}$ requires the determination of A, C and k from the equations:

$$\begin{aligned} Sy_\alpha &= NC + ASe^{kx_\alpha} \\ S(y_\alpha e^{kx_\alpha}) &= CSe^{kx_\alpha} + ASe^{2kx_\alpha} \\ S(y_\alpha x_\alpha e^{kx_\alpha}) &= CS(x_\alpha e^{kx_\alpha}) + AS(x_\alpha e^{2kx_\alpha}) \end{aligned}$$

5. Use the method of least squares to fit the curve $y = ax^2 + b/x$ to the following data:

x	1	2	3	4
y	-1.51	0.99	3.88	7.66

6. The logistic curve $y_c = a(1 + bq^x)^{-1}$ has been used to represent population growth. Fit such a curve, by Cowden's method, to the following data on the population (in millions) of the United States, 1790 to 1950:

x	1790	1800	1810	1820	1830	1840	1850	1860	
y	3.93	5.31	7.24	9.64	12.87	17.07	23.19	31.44	
x	1870	1880	1890	1900	1910	1920	1930	1940	1950
y	39.82	50.16	62.95	76.00	91.97	105.71	122.78	131.67	150.70

Hint: Write the equation $1/y_c = A + Bq^x$, where $A = 1/a$, $B = b/a$, $q = e^p$, and plot values of $1/y$ instead of y. Use coded x values.

7. Show that the Gompertz curve, $y_c = ab^{q^x}$, may be approximately fitted by Cowden's method if $\log y$ is plotted instead of y. This curve is used in actuarial work.

8. The following data were obtained in a physical experiment, where E represents the energy radiated from a carbon filament lamp per cm^2 per sec, and T the absolute temperature of the filament in thousands of degrees K.

T	1.309	1.471	1.490	1.565	1.611	1.680
E	2.138	3.421	3.597	4.340	4.882	5.660

By plotting on log-log graph paper it is seen that the data follow apparently a law of the type $E = aT^b$, with $a = 0.725$ and $b = 4.0$ approximately. Use the Seidel method to improve these values of a and b.

REFERENCES

[1] Hotelling, Harold, "Problems of Prediction," *Amer. J. Sociology*, **48,** 1942, pp. 61–76.

[2] Dwyer, P. S., The Solution of Simultaneous Equations, *Psychometrika*, **6,** 1941, pp. 101–129.

[3] Householder, A. S., "Some Numerical Methods for Solving Systems of Linear Equations," *American Math. Monthly*, **57,** 1950, pp. 453–459.

[4] Kenney, J. F., and Keeping, E. S., *Mathematics of Statistics, Part II*, Van Nostrand, 1951, pp. 313–315.

[5] Allan, F. E., "The General Form of the Orthogonal Polynomials for Simple Series with Proofs of Their Simple Properties," *Proc. Roy. Soc. Edinburgh*, **50,** 1935, pp. 310–320.

[6] Anderson, R. L., and Houseman, E. E., *Tables of Orthogonal Polynomials Extended to N* = 104, Research Bulletin 297, Iowa State College, 1942.

[7] Hotelling, Harold, "The Distribution of Correlation Ratios calculated from Random Data," *Proc. Nat. Acad. Sci.*, **11,** 1925, pp. 657–662.

[8] Woo, L. L., "Tables for Ascertaining the Significance or Non-Significance of Association Measured by the Correlation Ratio, *Biometrika*, **21,** 1929, pp. 1–66. The tables are reprinted in *Tables for Statisticians and Biometricians, Part II*, University College, London, 1931.

[9] Tang, P. C., "The Power Function of the Analysis of Variance Tests, with Tables and Illustrations of Their Use," *Statistical Research Memoirs*, **2,** 1938 pp. 126–157, University College, London.

[10] Cowden, D. J., "Simplified Methods of Fitting Certain Types of Growth Curves," *Journ. Amer. Stat. Ass.*, **42,** 1947, pp. 585–590.

[11] Snedecor, G. W., *Statistical Methods*, 5th ed., Iowa State University Press, 1956.

[12] Plackett, R. L., *Regression Analysis*, Clarendon Press, Oxford, 1960. This gives a theoretical treatment, using matrix methods, of least squares and polynomial regression.

Chapter 13

SOME REMARKS ON MULTIVARIATE PROBLEMS AND STOCHASTIC PROCESSES

13.1 **Multiple Regression in Terms of Correlation** In Chapter 12 we considered the *multiple regression* of one variate on a number of others, and the present treatment is closely connected with that in §§ 12.1 to 12.6.

For simplicity we will first assume that the predicted variate Y depends on just two predictors X_1 and X_2. (The generalization to any larger number is easily made.) The linear predicting equation for Y is of the form

(13.1.1) $$y_c = b_0 + b_1x_1 + b_2x_2$$

Note that we are not now using the dummy variate (which is always equal to 1) as in § 12.1. The new notation will be more convenient in the present context. Geometrically, Eq. (1) represents a plane in the three-dimensional sample space with coordinates x_1, x_2 and y. This is called the *regression plane* of Y on X_1 and X_2.

Suppose that N sets of observations $(x_{1\alpha}, x_{2\alpha}, y_\alpha)$ are made on the three variates. For the sake of uniformity we will let the variate Y be called X_0, with values denoted by $x_{0\alpha}$. The normal equations corresponding to Eq. (12.1.5) will now be

(13.1.2) $$\begin{cases} Nb_0 + b_1Sx_{1\alpha} + b_2Sx_{2\alpha} = Sx_{0\alpha} \\ b_0Sx_{1\alpha} + b_1S(x_{1\alpha}^2) + b_2S(x_{1\alpha}x_{2\alpha}) = S(x_{0\alpha}x_{1\alpha}) \\ b_0Sx_{2\alpha} + b_1S(x_{1\alpha}x_{2\alpha}) + b_2S(x_{2\alpha}^2) = S(x_{0\alpha}x_{2\alpha}) \end{cases}$$

To simplify these equations we can suppose that the origin is chosen at the arithmetic mean of each of the variates X_0, X_1 and X_2. Then $Sx_{1\alpha} = Sx_{2\alpha} = Sx_{0\alpha} = 0$. Also, if s_1^2, s_2^2 and s_0^2 denote the sample variances of X_1, X_2 and X_0 respectively, and if r_{ij} is the sample Pearson coefficient of correlation between X_i and X_j $(i, j = 0, 1, 2)$, we have

(13.1.3) $$\begin{cases} S(x_{1\alpha}^2) = (N-1)s_1^2, \qquad S(x_{2\alpha}^2) = (N-1)s_2^2 \\ S(x_{1\alpha}x_{2\alpha}) = (N-1)r_{12}s_1s_2, \qquad \text{etc.} \end{cases}$$

so that the equations of (2) become

(13.1.4) $$\begin{cases} b_0 = 0 \\ b_1s_1 + b_2r_{12}s_2 = r_{01}s_0 \\ b_1r_{12}s_1 + b_2s_2 = r_{02}s_0 \end{cases}$$

From these we get

(13.1.5) $$b_1 = \frac{s_0(r_{01} - r_{02}r_{12})}{s_1(1 - r_{12}^2)}, \qquad b_2 = \frac{s_0(r_{02} - r_{01}r_{12})}{s_2(1 - r_{12}^2)}$$

If we let R_{ij} be the cofactor of r_{ij} in the determinant of the *correlation matrix*

(13.1.6) $$R = \begin{bmatrix} r_{00} & r_{01} & r_{02} \\ r_{01} & r_{11} & r_{12} \\ r_{02} & r_{12} & r_{22} \end{bmatrix}$$

where, of course, $r_{ij} = 1$ when $i = j$, we may express b_1 and b_2 in the equivalent forms:

(13.1.7) $$\begin{cases} b_1 = -\dfrac{s_0 R_{01}}{s_1 R_{00}} \\ b_2 = -\dfrac{s_0 R_{02}}{s_2 R_{00}} \end{cases}$$

The equation of the regression plane may therefore be written in the symmetrical notation:

(13.1.8) $$\frac{x_0 R_{00}}{s_0} + \frac{x_1 R_{01}}{s_1} + \frac{x_2 R_{02}}{s_2} = 0$$

If we have p variates on which X_0 may depend, the equation of the regression hyperplane of X_0 on $X_1, X_2 \ldots X_p$ is

(13.1.9) $$\sum_{i=0}^{p} \frac{x_i R_{0i}}{s_i} = 0$$

where R_{0i} is the cofactor of r_{0i} in the determinant of R, the matrix with typical element r_{ij}. The predicted x_0 can be written, when $R_{00} \neq 0$,

(13.1.10) $$x_{0c} = -\frac{s_0}{R_{00}} \sum_{i=1}^{p} \frac{R_{0i} x_i}{s_i}$$

so that the relative contribution of the variate X_i to the prediction of X_0 is measured by the coefficient $(R_{0i}s_0)/(R_{00}s_i)$. Equation (10) is, of course, precisely equivalent to Eqs. (12.1.1) and (12.3.2) but in a different notation. The purpose of giving this alternative form is to show the relation of the regression coefficients to the coefficients of correlation between the different variates.

13.2 **Multiple Correlation** If v_α is the difference between the observed $x_{0\alpha}$ and the computed value given by Eq. (13.1.10) for the observed $x_{i\alpha}$ $(i = 1, 2 \ldots p)$,

(13.2.1) $$Sv_\alpha^2 = S\left(x_{0\alpha} + \sum_{i=1}^{p} \frac{s_0 R_{0i} x_{i\alpha}}{R_{00} s_i}\right)^2$$

This is the sum of squares of residuals. On using Eq. (13.1.3) it becomes

$$(13.2.2)\quad Sv_\alpha^{\ 2} = \frac{(N-1)s_0^{\ 2}}{R_{00}^{\ \ 2}}\left(R_{00}^{\ 2} + \sum_i R_{0i}^{\ 2} + R_{00}\sum_i R_{0i}r_{0i} + \sum_{i\neq j} R_{0i}R_{0j}r_{ij}\right)$$

$$= \frac{(N-1)s_0^{\ 2}}{R_{00}^{\ \ 2}} \sum_{i,j=0}^{p} R_{0i}R_{0j}r_{ij}$$

But we know from the properties of cofactors that $\sum_j R_{0j}r_{ij} = 0$ if $i \neq 0$ and $= d(R)$ if $i = 0$, where $d(R)$ is the determinant of R, so that

$$(13.2.2)\qquad Sv_\alpha^{\ 2} = (N-1)s_0^{\ 2}\frac{d(R)}{R_{00}}$$

The quantity $S(v_\alpha^{\ 2})/(N-1)$ is the approximate *variance of estimate* of x_0, denoted by $s_{0,12\cdot\cdot p}^2$. Therefore

$$(13.2.3)\qquad s_{0,12\cdot\cdot p}^2 = s_0^{\ 2}\frac{d(R)}{R_{00}}$$

As in § 12.6, it may be proved that $(N-1)s_{0,12\cdot\cdot p}^2/(N-p-1)$ is an unbiased estimate of the corresponding population parameter $\sigma_{0,12\cdot\cdot p}^2$, usually denoted simply by σ^2.

The variance *due to regression* may be defined by

$$(13.2.4)\qquad s_{012\cdot\cdot p}^2 = s_0^{\ 2} - s_{0,12\cdot\cdot p}^2$$

$$= s_0^{\ 2}\left(1 - \frac{d(R)}{R_{00}}\right)$$

The ratio of the variance due to regression to the total variance of X_0 (namely, $s_0^{\ 2}$) is the square of the *multiple correlation coefficient*

$$(13.2.5)\qquad r_{0,12\cdot\cdot p}^2 = 1 - \frac{d(R)}{R_{00}}$$

For the case $p = 1$, $d(R) = 1 - r_{01}^{\ 2}$ and $R_{00} = 1$, so that $r_{0,12\cdot\cdot p}$ reduces to the ordinary correlation coefficient between X_0 and X_1.

EXAMPLE 1. The variates X_0, X_1 and X_2 have pairwise correlation coefficients $r_{01} = 0.8$, $r_{02} = -0.7$, $r_{12} = -0.9$. The matrix R is

$$R = \begin{bmatrix} 1.0 & 0.8 & -0.7 \\ 0.8 & 1.0 & -0.9 \\ -0.7 & -0.9 & 1.0 \end{bmatrix}$$

so that $d(R) = 0.068$, $R_{00} = 0.19$. Therefore $r_{0,12}^2 = 1 - 0.36 = 0.64$, so that $r_{0,12} = 0.80$.

EXAMPLE 2 If $r_{01} = 0.6$ and $r_{02} = 0.4$, find r_{12} so that $r_{0,12} = 1$.

If we write $r_{12} = r$, and substitute the known values in R, we find that $d(R) = r^2 - 0.48r - 0.48 = 0$. The solution of this quadratic equation gives $r = 0.97$. In this example there is perfect multiple correlation of X_0 with X_1

and X_2, in the sense that all the observed points lie in one regression plane, even though the individual correlations of X_0 with X_1 and X_2 separately are not large.

The variance of some future observed value x_0 corresponding to assigned values $x_1, x_2 \ldots x_p$ of the predictors (this set not being any of the N sets of values already used in computing the correlations) is given by Eq. (12.4.8). In the notation of the present chapter, and with the origin placed at the sample mean, the matrix A of § 12.1 is

$$A = (N-1)\begin{bmatrix} \frac{N}{N-1} & 0 & \cdots & 0 \\ 0 & s_1^2 & \cdots & r_{1p}s_1s_p \\ 0 & r_{12}s_1s_2 & \cdots & r_{2p}s_2s_p \\ \vdots & \vdots & & \vdots \\ 0 & r_{1p}s_1s_p & \cdots & s_p^2 \end{bmatrix} \tag{13.2.6}$$

For the special case $p = 2$, $d(A) = N(N-1)^2 s_1^2 s_2^2 (1 - r_{12}^2)$ and

(13.2.7)

$$A^{-1} = (N-1)^{-1}\begin{bmatrix} \frac{N-1}{N} & 0 & 0 \\ 0 & s_1^{-2}(1-r_{12}^2)^{-1}, & -s_1^{-1}s_2^{-1}r_{12}(1-r_{12}^2)^{-1} \\ 0 & -s_1^{-1}s_2^{-1}r_{12}(1-r_{12}^2)^{-1}, & s_2^{-2}(1-r_{12}^2)^{-1} \end{bmatrix}$$

so that

$$V(x_0) = \sigma^2\left[1 + \frac{1}{N} + \frac{x_1^2/s_1^2 + x_2^2/s_2^2 - 2r_{12}x_1x_2/(s_1s_2)}{(N-1)(1-r_{12}^2)}\right] \tag{13.2.8}$$

The multiple correlation coefficient may be regarded as the ordinary correlation coefficient between the observed and the computed values of X_0, the latter being given by Eq. (13.1.10). By using Eq. (4) for the variance of the N computed values, we obtain for this correlation coefficient:

$$r = -\left[\frac{s_0}{(N-1)R_{00}} \underset{\alpha}{S} \sum_{i=1}^{p} R_{0i}x_{i\alpha}x_{0\alpha}s_i^{-1}\right] \Big/ \left[s_0 s_{012\cdots p}\right] \tag{13.2.9}$$

Since $\underset{\alpha}{S} x_{i\alpha}x_{0\alpha} = (N-1)r_{0i}s_0s_i$ and $\sum_{i=1}^{p} R_{0i}r_{0i} = d(R) - R_{00}$, this reduces to

$$r = \frac{s_0}{s_{012\cdots p}}\left[1 - \frac{d(R)}{R_{00}}\right] \tag{13.2.10}$$

$$= \left[1 - \frac{d(R)}{R_{00}}\right]^{1/2}$$

which is the same as $r_{0,12\cdots p}$.

* 13.3 **The Distribution of the Multiple Correlation Coefficient** From Eqs. (13.2.2) and (13.2.5),

$$\text{(13.3.1)} \qquad \frac{S(v_\alpha^2)}{S(x_{0\alpha}^2)} = \frac{d(R)}{R_{00}} = 1 - r_{0,12\cdots p}^2$$

whence

$$\text{(13.3.2)} \qquad \frac{r_{0,12\cdots p}^2}{1 - r_{0,12\cdots p}^2} = \frac{S(x_{0\alpha}^2) - S(v_\alpha^2)}{S(v_\alpha^2)} = \frac{S(x_{0c\alpha}^2)}{S(v_\alpha^2)}$$

The numerator is the sum of squares of the calculated values of the variate X_0 (the sum of squares *due to* regression), while the denominator is the sum of squares *about* the regression plane. If the variates $X_1, X_2 \ldots X_p$ all have fixed sets of values and if X_0 is a random normal variate independent of $X_1, X_2 \ldots X_p$ (which means that the true multiple correlation coefficient is zero), the numerator and denominator are independently distributed as $\chi^2\sigma^2$ with p and $N - p - 1$ d.f. respectively. Here σ^2 stands for $\sigma_{0,12\cdots p}^2$, the population variance about the true regression plane. It follows that

$$\text{(13.3.3)} \qquad F = \frac{(N - p - 1)S(x_{0c\alpha}^2)}{pS(v_\alpha^2)}$$

has the F distribution with p and $N - p - 1$ d.f. This means that $r_{0,12\cdots p}^2$ is a beta-variate with parameters $p/2$ and $(N - p - 1)/2$ and so its distribution is identical with that of the squared correlation ratio E_{yx}^2 (see § 12.10) with $p + 1$ instead of p. The slight change arises from the fact that we are now dealing with $p + 1$ variates altogether, namely, X_0 (or Y) and $X_1, X_2 \ldots X_p$.

The distribution of the multiple correlation coefficient, when the corresponding coefficient $\rho_{0,12\cdots p}$ in the population is not zero, was worked out by Fisher [1]. The density function is

$$\text{(13.3.4)} \qquad f(r^2) = (1 - \rho^2)^{(N-1)/2}(1 - r^2)^{(N-p-3)/2}(r^2)^{(p-2)/2}g(r^2)$$

where r^2 and ρ^2 are written for the squares of the multiple correlation coefficient in the sample and in the population, and where

$$g(r^2) = \frac{F[(N - 1)/2, (N - 1)/2, p/2; \rho^2 r^2]}{B[p/2, (N - p - 1)/2]}$$

the numerator being a hypergeometric function and the denominator a beta function.

The definition of the hypergeometric function as a series is

$$\text{(13.3.5)} \qquad F(a, b, c; z) = 1 + \frac{ab}{c}\cdot\frac{z}{1!} + \frac{a(a + 1)b(b + 1)}{c(c + 1)}\frac{z^2}{2!} + \ldots$$

$$= \sum_{n=0}^{\infty} \frac{[a]_n[b]_n}{[c]_n}\frac{z^n}{n!}$$

where $[a]_n = a(a + 1)(a + 2) \ldots (a + n - 1)$. For the properties of this function see references [2] and [3].

The expected value of r^2 is given by

$$E(r^2) = 1 - \frac{N - p - 1}{N - 1}(1 - \rho^2)F\left(1, 1, \frac{N + 1}{2}; \rho_2\right) \tag{13.3.6}$$

$$= 1 - \frac{N - p - 1}{N - 1}(1 - \rho^2)\left(1 + \frac{2\rho^2}{N + 1} + \ldots\right)$$

When $\rho = 0$ this reduces to

$$E(r^2) = \frac{p}{N - 1} \tag{13.3.7}$$

It may be noted that Fisher's z'-transformation (§ 11.13) can be applied to the multiple correlation coefficient and brings about approximate normality, for moderately large N. The variance of z' is $p/(N - p - 2)$, approximately.

13.4 **Partial Correlation** Sometimes we would like to know what the correlation would be between say X_0 (or Y) and X_1 if the influence of all other variates such as $X_2, X_3 \ldots X_p$ were eliminated. This is called the partial correlation of X_0 and X_1, and the coefficient is written $r_{01,2 \cdots p}$. It is in general different from the ordinary correlation coefficient r_{01} for X_0 and X_1. For simplicity we consider the case of three variates, X_0, X_1 and X_2, for which the three pairwise Pearson coefficients of correlation are r_{01}, r_{02}, r_{12}, and we suppose that X_2 is the variate to be eliminated. The effect of X_2 on X_0 is estimated from the ordinary regression of X_0 on X_2, ignoring X_1, and is given by $r_{02}s_0x_2/s_2$ for a measured value x_2. (Each variate is supposed measured from its own mean as origin.) The residual part of the observed x_0, after subtracting the part due to X_2, is

$$x_{0\cdot 2} = x_0 - r_{02}s_0\frac{x_2}{s_2} \tag{13.4.1}$$

In the same way, the residual part of the observed x_1 is

$$x_{1\cdot 2} = x_1 - r_{12}s_1\frac{x_2}{s_2} \tag{13.4.2}$$

The partial correlation coefficient $r_{01,2}$ is defined as the ordinary correlation coefficient for $x_{0\cdot 2}$ and $x_{1\cdot 2}$. That is

$$r_{01,2} = \frac{\underset{\alpha}{S}(x_{0\cdot 2})_\alpha(x_{1\cdot 2})_\alpha}{(N - 1)s_{0\cdot 2}s_{1\cdot 2}} \tag{13.4.3}$$

the numerator being summed over all the N sets of observations. In the denominator, $s_{0\cdot 2}^2$ is the residual variance of x_0 after eliminating the regression on x_2, so that

$$s_{0\cdot 2}^2 = s_0^2(1 - r_{02}^2) \tag{13.4.4}$$

Similarly,

(13.4.5) $$s_{1\cdot 2}^2 = s_1^{\,2}(1 - r_{12}^{\,2})$$

Using Eqs. (1) and (2), we find

(13.4.6) $$\begin{aligned} S_\alpha(x_{0\cdot 2})_\alpha(x_{1\cdot 2})_\alpha &= Sx_{0\alpha}x_{1\alpha} - r_{02}\frac{s_0}{s_2}Sx_{1\alpha}x_{2\alpha} \\ &\quad - r_{12}\frac{s_1}{s_2}Sx_{0\alpha}x_{2\alpha} + r_{02}r_{12}\frac{s_0s_1}{s_2^{\,2}}Sx_{2\alpha}^{\,2} \\ &= (N-1)\left(r_{01}s_0s_1 - r_{02}\frac{s_0}{s_2}r_{12}s_1s_2\right. \\ &\quad \left. - r_{12}\frac{s_1}{s_2}r_{02}s_0s_2 + r_{02}r_{12}s_0s_1\right) \\ &= (N-1)s_0s_1(r_{01} - r_{02}r_{12}) \end{aligned}$$

Therefore,

(13.4.7) $$r_{01,2} = \frac{r_{01} - r_{02}r_{12}}{[(1 - r_{02}^{\,2})(1 - r_{12}^{\,2})]^{1/2}}$$

This expresses the partial correlation coefficient in terms of the three ordinary correlation coefficients. In terms of the correlation matrix R, as defined in Eq. (13.1.6),

(13.4.8) $$r_{01,2} = -\frac{R_{01}}{(R_{00}R_{11})^{1/2}}$$

and this form can be generalized for p variates $X_1, X_2 \ldots X_p$. Thus

(13.4.9) $$r_{01,2\cdots p} = -\frac{R_{01}}{(R_{00}R_{11})^{1/2}}$$

where the correlation matrix now has $p + 1$ rows and columns.

In certain circumstances the partial correlation coefficient $r_{01,2}$ is the same as the ordinary coefficient r_{01} when the third variate X_2 is held constant. That is, if we select out of the set of all observations a subset in which $X_2 = x_2$ (very nearly), and calculate r_{01} for this subset, we get in effect $r_{01,2}$. In general this is not true, and the result depends on the chosen value x_2; the calculated r_{01} is equal to $r_{01,2}$ if and only if (a) the bivariate regression of X_0 on X_2 (ignoring X_1) is linear, with the variance of X_0 constant for all X_2; (b) the trivariate regression of X_0 on X_1 and X_2 is linear with the variance of X_0 constant for all X_1 and X_2. These conditions are not likely to be satisfied very precisely in practical applications, and the calculated value of $r_{01,2}$ will be a sort of average of the correlations r_{01} that would be obtained for different assigned values of x_2 (see Problem A-6).

EXAMPLE 3 The following results were obtained at Syracuse University in an investigation by M. A. May of the factors influencing "academic success." The sample consisted of 450 students and the variates were X_0 (honor points), X_1 (general intelligence) and X_2 (hours of study per week). One object of the investigation was to find to what extent honor points were related to general intelligence, when the effect of varying study periods was eliminated. From the data,

$$\bar{X}_0 = 18.5, \qquad \bar{X}_1 = 100.6, \qquad \bar{X}_2 = 24$$

$$s_0 = 11.2, \qquad s_1 = 15.8, \qquad s_2 = 6.0$$

$$r_{01} = 0.60, \qquad r_{02} = 0.32, \qquad r_{12} = -0.35$$

We find from Eq. (7) that $r_{01,2} = 0.80$. The multiple correlation of X_0 on X_1 and X_2, as given by Eq. (13.2.5), is $r_{0,12} = 0.82$. The regression coefficients, from Eq. (13.1.7), are $b_1 = 0.58$, $b_2 = 1.13$.

The sampling distribution of the partial correlation coefficient $r_{01,2}$ is the same as that of r_{01} (see §11.12) but with $N - 1$ instead of N and with $\rho_{01,2}$ instead of ρ. With $p + 1$ variates, $X_0, X_1 \ldots X_p$, the density function for $r_{01;2\cdots p}$ has $N - p + 1$ instead of N.

13.5 **The Multivariate Normal Distribution** The univariate normal distribution has the density function (when the variable is standardized so as to have mean zero and variance unity)

$$f(x) = (2\pi)^{-1/2} \exp(-x^2/2) \tag{13.5.1}$$

The importance and central position of this distribution in statistical theory have already been emphasized in earlier chapters. A corresponding position in multivariate theory is taken by the standardized multivariate normal distribution, with joint density function

$$f(x_0, x_1 \ldots x_p) = (2\pi)^{-(p+1)/2}|d(A)|^{1/2} \exp(-Q/2) \tag{13.5.2}$$

where

$$Q = x'Ax \tag{13.5.3}$$

Here x' is the row vector $(x_0\, x_1 \ldots x_p)$ and x is its transpose (a column vector), while A is a symmetric positive definite matrix of $p + 1$ rows and columns. It is, in fact, the inverse of the correlation matrix P.

$$A^{-1} = P = \begin{bmatrix} 1 & \rho_{01} & \cdots & \rho_{0p} \\ \rho_{01} & 1 & \cdots & \rho_{1p} \\ \vdots & \vdots & & \vdots \\ \rho_{0p} & \rho_{1p} & \cdots & \rho_{pp} \end{bmatrix} \tag{13.5.4}$$

For the bivariate case, in agreement with the customary notation, we will write $x_1 = x$, $x_0 = y$, $\rho_{01} = \rho$. Then

$$P = \begin{bmatrix} 1 & \rho \\ \rho & 1 \end{bmatrix}, \qquad A = (1 - \rho^2)^{-1} \begin{bmatrix} 1 & -\rho \\ -\rho & 1 \end{bmatrix}$$

$$d(A) = (1 - \rho^2)^{-1}$$

$$(1 - \rho^2)Q = [yx] \begin{bmatrix} 1 & -\rho \\ -\rho & 1 \end{bmatrix} \begin{bmatrix} y \\ x \end{bmatrix} = x^2 + y^2 - 2\rho xy$$

so that

$$f(x, y) = (2\pi)^{-1}(1 - \rho^2)^{-1/2} \exp\left[-\frac{x^2 + y^2 - 2\rho xy}{2(1 - \rho^2)} \right] \tag{13.5.5}$$

This is the bivariate normal distribution in standardized form. Tables of this function for selected values of ρ may be found in references [4] and [5]. A method has been given [6] for reducing the integral of the multivariate function to a bivariate integral and thereby obtaining numerical values.

* 13.6 **The Relation of the Multivariate and Multinomial Distributions** We saw in Chapter 3 that with increasing sample size the binomial distribution with parameter θ tends to a normal distribution with mean $N\theta$ and variance $N\theta(1 - \theta)$. A similar result holds for the multinomial distribution (Appendix A.16 and 17). If the probability that a random item from the population belongs to the ith class is π_i ($i = 1, 2 \ldots k$), the observed frequencies f_i in the various classes will tend, as the total sample size $N(= \sum f_i)$ increases, to a multivariate normal distribution with means $N\pi_i$ and variance-covariance matrix V, where

$$V = N \begin{bmatrix} \pi_1(1 - \pi_1) & -\pi_1\pi_2 & \cdots & -\pi_1\pi_k \\ -\pi_2\pi_1 & \pi_2(1 - \pi_2) & \cdots & -\pi_2\pi_k \\ \vdots & \vdots & & \vdots \\ -\pi_k\pi_1 & -\pi_k\pi_2 & \cdots & \pi_k(1 - \pi_k) \end{bmatrix} \tag{13.6.1}$$

Because of the fact that $\sum \pi_i = 1$, we have $d(V) = 0$, which means that the matrix V is singular and cannot be inverted. However, we may omit one of the variates f_i (say f_k) and express it in terms of the remaining $k - 1$ variates by the relation $f_k = N - f_1 - f_2 - \ldots - f_{k-1}$. These $k - 1$ frequencies are multinormally distributed with means $N\pi_i$ and variance-covariance matrix V^*, which is just V with the last row and the last column omitted. The inverse of V^* is

$$(V^*)^{-1} = \frac{1}{N} \begin{bmatrix} (\pi_1{}^{-1} + \pi_k{}^{-1}) & \pi_k{}^{-1} & \cdots & \pi_k{}^{-1} \\ \pi_k{}^{-1} & (\pi_2{}^{-1} + \pi_k{}^{-1}) & \cdots & \pi_k{}^{-1} \\ \vdots & & & \\ \pi_k{}^{-1} & \pi_k{}^{-1} & \cdots & (\pi_{k-1}{}^{-1} + \pi_k{}^{-1}) \end{bmatrix} \tag{13.6.2}$$

as may be checked by multiplying it with V^*, bearing in mind that $\pi_k = 1 - \sum_1^{k-1}\pi_i$. The quantity Q of Eq. (13.5.3) which appears in the exponent of the multivariate normal distribution is, in this case,

$$Q = (f_i - N\pi_i)'(V^*)^{-1}(f_i - N\pi_i) \tag{13.6.3}$$

where $(f_i - N\pi_i)'$ is a row vector of $k - 1$ elements and $(f_i - N\pi_i)$ is the corresponding column vector. On carrying out the multiplication, we find

$$\begin{aligned} Q &= \sum_{i=1}^{k-1} \frac{(f_i - N\pi_i)^2}{N\pi_i} + \sum_{i,j=1}^{k-1} \frac{(f_i - N\pi_i)(f_j - N\pi_j)}{N\pi_k} \\ &= \sum_{i=1}^{k-1} \frac{(f_i - N\pi_i)^2}{N\pi_i} + \frac{\left[\sum_{i=1}^{k-1}(f_i - N\pi_i)\right]^2}{N\pi_k} \\ &= \sum_{i=1}^{k} \frac{(f_i - N\pi_i)^2}{N\pi_i} \end{aligned} \tag{13.6.4}$$

As shown in Appendix A.17, Q is approximately distributed as χ^2 with $k - 1$ degrees of freedom. This fact is the basis of the ordinary χ^2 test for goodness of fit, already discussed in § 10.3.

* 13.7 **Hotelling's Generalization of Student's *t*** Hotelling [7] investigated the properties of a statistic T which generalizes Student's t for p variates. We recall that Student's t is the ratio of the difference between the sample mean and the population mean to the estimated standard deviation of the sample mean, so that

$$t^2 = \frac{N(\bar{x} - \mu)^2}{s_x^2} \tag{13.7.1}$$

Hotelling's T is a standardized measure of the departure of all the p sample means from their population values. Let

$$S_{ij} = \underset{\alpha}{S}\,(x_{i\alpha} - \bar{x}_i)(x_{j\alpha} - \bar{x}_j) \tag{13.7.2}$$

and let (S^{ij}) be the inverse of the matrix (S_{ij}). Then T is defined by

$$T^2 = N(N - 1)\sum_{i,j} S^{ij}(\bar{x}_i - \mu_i)(\bar{x}_j - \mu_j) \tag{13.7.3}$$

where N is the sample size. It is easily seen that when $p = 1$, T^2 reduces to t^2. For in this case, if $x_1 = x$,

$$S_{11} = \underset{\alpha}{S}\,(x_\alpha - \bar{x})^2 = (N - 1)s_x^2$$

so that

$$S^{11} = [(N - 1)s_x^2]^{-1}$$

and

$$T^2 = \frac{N(\bar{x} - \mu)^2}{s_x^{\ 2}} = t^2$$

This statistic T, given by Eq. (3), may be used, just as t is used, to test the null hypothesis that $\mu_i = \mu_{i0} (i = 1, 2 \ldots p)$ in the population, μ_i being the true value of the mean of X_i, and μ_{i0} an assumed value. On the assumption that the population is multivariate normal, with covariance matrix (σ_{ij}), the null hypothesis H_0 is rejected when $T^2 \geq T_0^{\ 2}$, where

(13.7.4) $$T^2 = N(N-1) \sum_{i,j} S^{ij}(\bar{x}_i - \mu_{i0})(\bar{x}_j - \mu_{j0})$$

and where $T_0^{\ 2}$ is chosen so that the probability that $T^2 \geq T_0^{\ 2}$, when H_0 is true, is equal to some assigned value α.

Hotelling showed that under H_0 the quantity $u = T^2/(N-1)$ is a beta-prime variate with density function

(13.7.5) $$f(u) = \frac{1}{B\left(\frac{p}{2}, \frac{N-p}{2}\right)} \cdot \frac{u^{(p-2)/2}}{(1+u)^{N/2}}$$

This is equivalent to the statement that $[T^2(N-p)]/[(N-1)p]$ has the ordinary F distribution with p and $N - p$ degrees of freedom. When H_0 is not true, and therefore $\mu_i - \mu_{i0}$ is not zero, the distribution is non-central F, with the same degrees of freedom and with non-centrality parameter $\lambda = \frac{N}{2} \sum_{i,j} \sigma^{ij}$ $(\mu_i - \mu_{i0})(\mu_j - \mu_{j0})$. The non-central F distribution has the density function

(13.7.6) $$f(F) = \frac{p}{N-p} \frac{e^{-\lambda}}{\Gamma[(N-p)/2]} \sum_{\beta=0}^{\infty} \frac{\lambda^\beta \Gamma\left(\frac{N}{2} + \beta\right)}{\beta! \Gamma\left(\frac{p}{2} + \beta\right)} \frac{\left(\frac{pF}{N-p}\right)^{p/2+\beta-1}}{\left(1 + \frac{pF}{N-p}\right)^{N/2+\beta}}$$

This reduces when $\lambda = 0$ to the ordinary central form as in Eq. (8.15.5). Tables calculated by Tang (see §§ 9.12 and 12.10) give the probability of accepting the null hypothesis when it is not true, for various values of λ and for significance levels 0.05 and 0.01. His number of degrees of freedom f_1 is our p, his f_2 is our $N - p$, and his non-centrality parameter ϕ is related to our λ by

(13.7.7) $$(p+1)\phi^2 = 2\lambda$$

Also the variate which Tang denotes by E^2 is our $T^2/(T^2 + N - 1)$. This has the same distribution as the correlation ratio, which is the reason for Tang's notation. Further details of the T test and its optimum properties may be found in [8].

13.8 **Discriminant Functions** Suppose we wish to assign several individuals, on the basis of a measured variate X, to one or other of two populations A

and B which differ in their means but whose distributions may overlap (Fig. 54). If the mean of A (μ_1) is greater than the mean of B (μ_2), we would naturally assign an individual with a high value of X to population A and one with a low X to population B. If the curves representing the density functions for the two populations intersect at $X = \alpha$, we might well take α as the dividing point. There will, of course, be a certain risk of mis-classification. The probability of classifying an individual who is really an A as belonging to B is $F_A(\alpha)$, where $F_A(x)$ is the distribution function for population A. Similarly the probability of classifying a B as an A is $1 - F_B(\alpha)$.

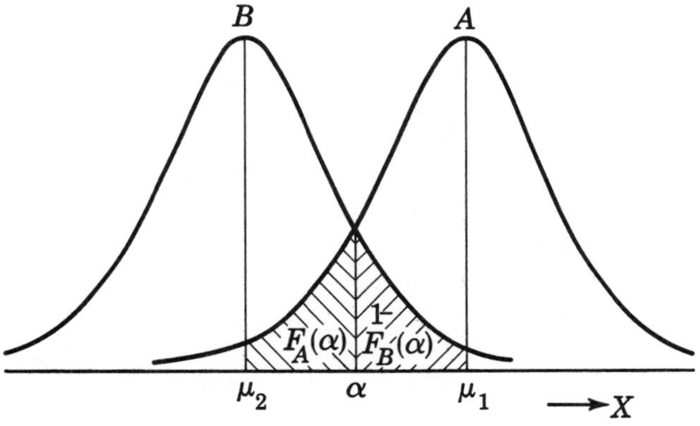

FIG. 54 DISCRIMINATION BETWEEN TWO POPULATIONS

In practice we often have several variates $X_1, X_2 \ldots X_p$ which may be used for discrimination, and the problem then arises of choosing the best function of these variates for discriminating with the least error between populations A and B. An example is the use of intelligence, aptitude and achievement tests of various kinds, along with high school records, for attempting to assess whether a student planning to enter a university is, or is not, likely to graduate in, say, engineering. A student adviser would be glad to have available a suitable function of the various test scores to assist him in coming to a decision. A function of this kind is called a *discriminant function*. Individuals with a value of the function greater than some fixed value α will be classed as A, those with a smaller value as B.

Let us assume that we want a *linear* function of the measured x_i, say

$$L = \sum b_i x_i, \qquad i = 1, 2 \ldots p \tag{13.8.1}$$

and would like to choose the b_i so that L will be as efficient as possible in discriminating between the two populations. Suppose that of N sample items available (on each of which p measurements are made and for each of which the proper classification is known) there are N_1 from population A and N_2 from population B.

The measurements on variate X_i from population A will be denoted by $x_{1i\alpha}$, $\alpha = 1, 2 \ldots N_1$, and those from population B by $x_{2i\beta}$, $\beta = 1, 2 \ldots N_2$, and it is assumed that the two sets x_{1i} and x_{2i} have each a p-variate normal distribution, independent of each other, with means μ_{1i}, μ_{2i} respectively and a common covariance matrix (σ_{ij}). A single later sample item gives the set of observations x_i ($i = 1, 2 \ldots p$) and it is known that this item belongs to either A or B but it is not known to which. The discriminant function helps us to make the decision.

The null hypothesis H_1 is that the new item belongs to A; the alternative hypothesis H_2 is that it belongs to B. By the Neyman-Pearson theorem (see § 6.9) the most powerful critical region for testing H_1 against H_2 is given by

$$\frac{p_1(x_1, x_2 \ldots x_p)}{p_2(x_1, x_2 \ldots x_p)} < k \tag{13.8.2}$$

where p_1 and p_2 denote the joint probability density functions for $(x_1, x_2 \ldots x_p)$ under H_1 and H_2 respectively, and where k is a constant determined by the size of the critical region.

With the assumptions we have made,

$$\begin{cases} p_1 = (2\pi)^{-p/2}[d(\sigma_{ij})]^{-1/2} \exp\left[-\frac{1}{2}\sum_{i,j} \sigma^{ij}(x_i - \mu_{1i})(x_j - \mu_{1j})\right] \\ p_2 = (2\pi)^{-p/2}[d(\sigma_{ij})]^{-1/2} \exp\left[-\frac{1}{2}\sum_{i,j} \sigma^{ij}(x_i - \mu_{2i})(x_j - \mu_{2j})\right] \end{cases} \tag{13.8.3}$$

where $d(\sigma_{ij})$ is the determinant of the matrix (σ_{ij}). Then Eq. (2) is equivalent (on taking logs of both sides) to

$$\sum_{i,j} \sigma^{ij}[(x_i - \mu_{2i})(x_j - \mu_{2j}) - (x_i - \mu_{1i})(x_j - \mu_{1j})] < 2 \log k \tag{13.8.4}$$

However, the population parameters μ_{1i}, μ_{2i}, σ_{ij} are unknown, and we must replace them by estimators. The optimum estimators of μ_{1i} and μ_{2i} are

$$\bar{x}_{1i} = N_1{}^{-1} \underset{\alpha}{S}\, x_{1i\alpha}, \qquad \bar{x}_{2i} = N_2{}^{-1} \underset{\beta}{S}\, x_{2i\beta} \tag{13.8.5}$$

respectively, while that of σ_{ij} is

$$s_{ij} = \frac{S_{ij}}{N_1 + N_2 - 2} \tag{13.8.6}$$

where

$$S_{ij} = \underset{\alpha}{S}\,(x_{1i\alpha} - \bar{x}_{1i})(x_{2j\alpha} - \bar{x}_{2j}) + \underset{\beta}{S}\,(x_{2i\beta} - \bar{x}_{2i})(x_{2j\beta} - \bar{x}_{2j}) \tag{13.8.7}$$

On substituting these estimates in Eq. (4) we obtain the test statistic

$$R = \sum_{i,j} s^{ij}[(x_i - \bar{x}_{2i})(x_j - \bar{x}_{2j}) - (x_i - \bar{x}_{1i})(x_j - \bar{x}_{1j})] \tag{13.8.8}$$

where (s^{ij}) is the inverse of (s_{ij}). Now R can be written

$$R = \sum_{i,j} s^{ij}[x_i(\bar{x}_{1j} - \bar{x}_{2j}) + x_j(\bar{x}_{1i} - \bar{x}_{2i}) - \bar{x}_{1i}\bar{x}_{1j} + \bar{x}_{2i}\bar{x}_{2j}]$$

and because $s^{ij} = s^{ji}$,

$$\sum_{i,j} s^{ij}x_i(\bar{x}_{1j} - \bar{x}_{2j}) = \sum_{i,j} s^{ij}x_j(\bar{x}_{1i} - \bar{x}_{2i})$$

Also the last two terms in the square bracket do not depend on x_i. The test $R < c$ is therefore equivalent to

(13.8.9) $$L = \sum_{i,j} s^{ij}x_i(\bar{x}_{1j} - \bar{x}_{2j}) < k'$$

where k' is a new constant, suitably chosen. Let us denote the difference between the two sample means for the variate X_i by d_i, so that

(13.8.10) $$d_i = \bar{x}_{1i} - \bar{x}_{2i}$$

Then

$$L = \sum_i b_i x_i$$

with

(13.8.11) $$b_i = \sum_j s^{ij}\, d_j$$

This function L is a discriminant function for assigning the new item to either A or B. If $L < k'$, we shall assign it to B and if $L > k'$ to A. It is convenient to take $k' = \frac{1}{2}\sum_i b_i(\bar{x}_{1i} + \bar{x}_{2i})$.

The same discriminant function is obtained by using a different approach (due to Fisher). The constants b_i in the function L are chosen so as to make the sum of squares *between* populations for the given sample items as great as possible, relative to the sum of squares and products *within* populations. Of course, it is only the *ratios* of the b_i that matter, so a constant multiplier makes no essential difference.

It may be noted that the discriminant function is related to Hotelling's statistic T. We can define a generalized statistic T_2 for the two-population case by

(13.8.12) $$T_2^2 = \sum_{i,j} s^{ij}[N_1(\bar{x}_{1i} - \bar{x}_i)(\bar{x}_{1j} - \bar{x}_j) + N_2(\bar{x}_{2i} - \bar{x}_i)(\bar{x}_{2j} - \bar{x}_j)]$$

where

(13.8.13) $$\bar{x}_i = \frac{N_1\bar{x}_{1i} + N_2\bar{x}_{2i}}{N_1 + N_2}$$

which is the combined sample mean for X_i, and s_{ij} is given by Eqs. (6) and (7) above. The matrix (s^{ij}) is, as usual, the inverse of (s_{ij}). The quantity S_{ij} is the sum of squares and products within populations. By substituting from Eq. 13 in Eq. (12) we obtain, after a little reduction,

$$T_2^2 = \nu \sum_{i,j} s^{ij} \, d_i \, d_j \qquad (13.8.15)$$

where $\nu = (N_1 N_2)/(N_1 + N_2)$ and d_i is given by Eq. (10). It is seen that, apart from the constant ν, T_2^2 is the same as L with d_i in place of x_i.

EXAMPLE 4 In a certain experiment (the details have been modified for convenience of presentation) 18 rabbits each received a high dose of insulin and 18 received a low dose. The blood-sugar was measured at 1, 2 and 3 hours after each dose. The three readings are denoted by x_1, x_2 and x_3. The aim is to find what linear combination of these readings would be expected to discriminate most effectively between a high dose and a low dose of insulin in another rabbit.

The S.S. and S.P. within populations were as shown in the following table:

S.S.			S.P.		
x_1^2	x_2^2	x_3^2	x_1x_2	x_1x_3	x_2x_3
2677	2358	3223	1278	1814	1966

and the values of d_i (mean low-insulin value − mean high-insulin value) were 7.594, 19.73 and 25.04, respectively. The matrix (S_{ij}) is

$$\begin{bmatrix} 2677 & 1278 & 1814 \\ 1278 & 2358 & 1966 \\ 1814 & 1966 & 3223 \end{bmatrix}$$

and its inverse (S^{ij}) is proportional to

$$\begin{bmatrix} 3.735 & -0.553 & -1.765 \\ -0.553 & 5.337 & -2.947 \\ -1.765 & -2.945 & 4.679 \end{bmatrix}$$

The values of b_1, b_2, b_3 are therefore proportional to $d_1S^{11} + d_2S^{21} + d_3S^{31}$, $d_1S^{12} + d_2S^{22} + d_3S^{32}$, and $d_1S^{13} + d_2S^{23} + d_3S^{33}$ respectively, that is, to −26.7, 27.4 and 45.6. A good approximation to the best discriminant would accordingly be

$$L = -3x_1 + 3x_2 + 5x_3$$

since 26.7, 27.4 and 45.6 are approximately in the ratio 3:3:5.

EXAMPLE 5 (*Nature*, 168, 1951, p. 794). In order to discriminate between fossil skulls of men and chimpanzees, a discriminant function was calculated using four distinct measurements on the lower milk canine teeth. On the basis of 44 chimpanzee and 40 human teeth, the function obtained was

$L = x_1 - 7.49x_2 + 2.34x_3 + 4.70x_4$. The average value of L turned out to be $+17.6$ for the chimpanzee teeth and -5.0 for the human, with a standard deviation of 2.45. It was therefore concluded that if a tooth of unknown origin gave an L between 11.5 and 23.7 it might very reasonably be classed as a chimpanzee's, and if L lay between 1.1 and -11.1 as human. For the famous Taungs skull, of great antiquity, L turned out to be -7.9 and for the Kromdraai skull -2.6, so that both these are probably human.

* 13.9 **The Distance Between Two Populations** The discriminant function and Hotelling's generalized T_2 are both closely related to a measure of "generalized distance" between two populations, proposed by Mahalanobis.

Suppose p variates X_i are measured on each sample item from each population. Let the population means of X_i be μ_{1i} and μ_{2i}, and let

(13.9.1)
$$\begin{cases} X_{1i} = \mu_{1i} + \varepsilon_i \\ X_{2i} = \mu_{2i} + \varepsilon'_i \end{cases}$$

where the ε_i have a multivariate normal distribution (the same for both populations) with means 0 and covariance matrix (σ_{ij}). If

(13.9.2)
$$\delta_i = \mu_{1i} - \mu_{2i}$$

the generalized distance is given by

(13.9.3)
$$\Delta^2 = \sum_{i,j} \sigma^{ij} \delta_i \delta_j$$

where (σ^{ij}) is the inverse of (σ_{ij}). A factor p^{-1} is sometimes included on the right-hand side of Eq. (3) but this makes no essential difference.

Since in practice we have to estimate the population means from the sample means, the formula for the observed distance becomes

(13.9.4)
$$D^2 = \sum_{i,j} \sigma^{ij} d_i d_j$$

where

(13.9.5)
$$d_i = \bar{x}_{1i} - \bar{x}_{2i}$$

The first mean is calculated from a sample of size N_1, say, and the second from a sample of size N_2. If σ_{ij} is not known, the estimator s_{ij}, defined as in Eqs. (13.8.6) and (13.8.7) must be used. In this case, we speak of the *studentized distance*, D_s, which is given by

(13.9.6)
$$D_s^2 = \sum_{i,j} s^{ij} d_i d_j$$

and so is identical, apart from the constant ν, with Hotelling's generalized T_2^2, in Eq. (13.8.15).

If the true Mahalanobis distance Δ is zero (so that the two populations are really identical), the quantity νD^2 has an ordinary χ^2 distribution with p d.f.

If Δ is not zero, νD^2 is distributed like non-central χ^2 with parameter of non-centrality $\nu\Delta^2/2$. (see Appendix A.13).

For the studentized distance, it was shown by Bose and Roy that if $\Delta = 0$, the quantity

$$F = \frac{N_1 + N_2 - p - 1}{p} \cdot \frac{D_s^2}{N_1 + N_2 - 2} \tag{13.9.7}$$

has the ordinary F distribution with p and $N_1 + N_2 - p - 1$ d.f. If Δ is not zero, it has the non-central F distribution (see (13.7.6)) with non-centrality parameter $\nu\Delta^2/2$.

For the bivariate case, if we denote x_1 by x and x_2 by y, and if the correlation between x and y in the population is measured by ρ, we have

$$(\sigma_{ij}) = \begin{bmatrix} \sigma_x^2, & \rho\sigma_x\sigma_y \\ \rho\sigma_x\sigma_y, & \sigma_y^2 \end{bmatrix}$$

$$(\sigma^{ij}) = [\sigma_x^2\sigma_y^2(1 - \rho^2)]^{-1}\begin{bmatrix} \sigma_y^2 & -\rho\sigma_x\sigma_y \\ -\rho\sigma_x\sigma_y & \sigma_x^2 \end{bmatrix}$$

so that D^2 becomes

$$D^2 = (1 - \rho^2)^{-1}\left[\frac{(\bar{x}_1 - \bar{x}_2)^2}{\sigma_x^2} + \frac{(\bar{y}_1 - \bar{y}_2)^2}{\sigma_y^2} - \frac{2\rho(\bar{x}_1 - \bar{x}_2)(\bar{y}_1 - \bar{y}_2)}{\sigma_x\sigma_y}\right] \tag{13.9.8}$$

which, when $\rho = 0$, reduces to the square of the geometrical distance between the two sample means in the x-y plane (if standardized units for x and y are used).

13.10 **Stochastic Processes** An important part of modern statistical theory is concerned with events which change in a random way as time goes on. An example is the position of a microscopic inert particle suspended in a fluid, and exhibiting the so-called Brownian motion. Another is the size of a population affected by births, deaths, immigration, emigration, etc. These and many other such linked chains of events, proceeding in time and subject to random fluctuations, are called *stochastic processes*.

The mathematical model of a stochastic process is a variate $X(t)$ depending on a parameter t which is usually (in physical applications) the time. Since t is continuous, the set of possible values of $X(t)$ is non-countable, but in practice observations are taken at a finite set of values t_r $(r = 1, 2 \ldots n)$. The corresponding values of X, say $x_1, x_2 \ldots x_n$, are the components of an n-dimensional vector, having a distribution function $F(x_1 \ldots x_n;\ t_1 \ldots t_n)$, so that X is a multivariate random, or stochastic, variable. One of the simplest examples of a stochastic process is the "random walk" mentioned in § 7.6. The random step X_r taken at time t_r in either the positive or negative direction of the x-axis is independent of all the previous X_i at times $t_1, t_2 \ldots t_{r-1}$. The cumulative sum

$$S_r = X_1 + X_2 + \ldots + X_r \tag{13.10.1}$$

is a stochastic process, representing the position of the individual taking the "walk" at time t_r. Thus if A plays a set of gambling games with B, receiving one dollar from B each time he wins and paying one dollar to B each time he loses, and if the result of any game is independent of the preceding games, the total sum held by A at the end of r games, supposing he started off with a fixed sum m dollars, is a random walk process. Here X_r is always either $+1$ or -1, and sooner or later either A or his opponent (if similarly situated) will be ruined.

Sequential binomial sampling is another, and more respectable, example, in which each step represents the result of sampling one more item from the population (usually referred to as a "lot"). The step is in one direction or another according as the result of the inspection is a "success" or a "failure." The total number of steps taken is a stochastic variable representing the total number of sample items required to arrive at one or other of two possible decisions regarding the population sampled (for instance, to accept the whole lot or to reject the whole lot).

13.11 **Markov Processes** A stochastic process is said to be of the Markov type if the value of X at any time t_r depends at most on the value at the immediately preceding available time t_{r-1}. No earlier history of the process adds anything further to the probable future history of X. The joint probability for the observed set of values $x_1, x_2 \ldots x_n$ is therefore given by

(13.11.1) $$P(x_1, x_2 \ldots x_n) = P(x_1)\cdot P(x_2|x_1)\cdot P(x_3|x_2) \ldots P(x_n|x_{n-1}) = P(x_1) \prod_{r=2}^{n} P(x_r|x_{r-1})$$

A Markov process is defined by the initial probability distribution $P(x_1)$ and the conditional distribution $P(x_r|x_{r-1})$ for any arbitrary choice of the times t_r. It follows that

(13.11.2) $$P(x_r|x_{r-2}) = \int_{(x_{r-1})} P(x_r|x_{r-1})\cdot P(x_{r-1}|x_{r-2})$$

where the integral stands for a sum over the possible values of x_{r-1} (if X is discrete) or the ordinary integral over the whole domain of x_{r-1} (if X is continuous). This relation is known as the Chapman-Kolmogorov equation. By a repeated application of the equation we may obtain $P(x_r|x_1)$ and then

(13.11.3) $$P(x_r) = \int_{(x_1)} P(x_r|x_1)P(x_1)$$

When the variate X is discrete (as in many applications), the process is called a *Markov chain*, and a matrix notation is convenient. We suppose that at each time t_r, X can take one of a finite number n of possible values $x_1, x_2 \ldots x_n$, representing n possible states of the system. The probability that X changes from x_i to x_j between t_{r-1} and t_r will be denoted by p_{ij} (called a *transition probability*). If we know the matrix P of transition probabilities, and the initial value of X, we can calculate the probability for any possible value at any future time.

Note that in this matrix

$$P = \begin{bmatrix} p_{11} & \cdots & p_{1n} \\ \vdots & & \vdots \\ p_{n1} & & p_{nn} \end{bmatrix}$$

the probabilities in each row (but not necessarily in each column) add up to 1. This is because in whatever state the system may be at time t_{r-1} it must be in one and only one of the n possible states at time t_r. However, since the transition probabilities refer only to transitions *from* x_i *to* x_j, a similar argument does not apply to the columns.

For a Markov chain, Eqs. (2) and (3) become

(13.11.4)
$$P(t_r|t_{r-2}) = P^2$$
$$P(t_r) = p'P^{r-1}$$

where $P(t_r)$ denotes the probability of the various states at time t_r and p' is a row vector of the probabilities at t_1 for the same n states.

EXAMPLE 6 A system can be in one of two possible states. Initially the chance is the same for each, and at each transition the probability matrix is

$i \backslash j$	1	2
1	$\frac{1}{3}$	$\frac{2}{3}$
2	$\frac{1}{2}$	$\frac{1}{2}$

What are the probabilities for the two states after three steps?

We have $p' = [\frac{1}{2}, \frac{1}{2}]$

$$P = \begin{bmatrix} \frac{1}{3} & \frac{2}{3} \\ \frac{1}{2} & \frac{1}{2} \end{bmatrix}, \qquad P^2 = \begin{bmatrix} \frac{4}{9} & \frac{5}{9} \\ \frac{5}{12} & \frac{7}{12} \end{bmatrix}, \qquad P^3 = \begin{bmatrix} \frac{23}{54} & \frac{31}{54} \\ \frac{31}{72} & \frac{41}{72} \end{bmatrix}$$

$$p'P^3 = [\tfrac{185}{432}, \tfrac{247}{432}]$$

The required probabilities are therefore $\frac{185}{432}$ and $\frac{247}{432}$.

If written as decimals, the rows of P^3 are [0.426, 0.574] and [0.431, 0.570], which are nearly the same. A basic theorem on Markov chains states that if the matrix P is regular (that is, if some power of P has no zero elements) then, as n increases, P^n tends to a matrix Q, each row of which consists of the *same* probability vector q', with no zero element. Furthermore, whatever the initial probability vector p', $p'P^n \to q'$ and q' is a unique fixed probability vector satisfying the relation

(13.11.5)
$$q'P = q'$$

In the example above, the vector q' is $[\frac{3}{7}, \frac{4}{7}]$. The interpretation of this theorem is that after a great many steps, the probability that the system is in a state x_j is very nearly equal to the jth element of q' regardless of the initial probabilities of the various states.

Another theorem states that in a regular Markov chain, with transitions at unit time intervals, the average time it takes to return to a given state, having once been there, is the reciprocal of the limiting probability of being in that state.

13.12 **Stationary Processes** A stationary process is one whose distribution function $F_n(x_1, x_2 \ldots x_n; t_1, t_2 \ldots t_n)$, for any set of times $t_1, t_2 \ldots t_n$, depends only on the $n - 1$ *intervals* $t_r - t_1$ $(r = 2, 3 \ldots n)$ and so is independent of the absolute times. Translation along the time axis makes no difference to the probabilities in a stationary process. An example is a quality control chart where the variate concerned is satisfactorily "in control." Also a Markov chain, after a considerable number of steps, has become practically stationary. Stationary processes represent a kind of stochastic equilibrium and are therefore of importance in many practical situations, such as arise in problems of communication engineering.

A process in which the mean is constant but the variance may change is said to be *stationary to the first order.*

If the mean and variance are constant, as well as the covariance between values at a fixed interval apart, the process is *stationary to the second order*, and so on.

Let us consider the case when all the observations are taken at regularly spaced times, so that $t_r - t_{r-1} = T$ for all r. For a second-order stationary process

(13.12.1)
$$E(X_r) = \mu$$
$$E[(X_r - \mu)(X_{r-1} - \mu)] = \sigma^2\rho(T)$$

where σ^2 is the variance of any X_r and $\rho(T)$ is a function of the fixed interval T. This function is called the *autocorrelation coefficient*, and is the correlation coefficient for pairs of values of X separated by the interval T. Since a multivariate normal distribution depends only on the first and second moments, it follows that a *normal process* which is stationary to the second order is also completely stationary.

The simplest case of a stationary process is a sequence of independent observations such as heads or tails in repeated tosses of a coin. Here $\rho(T) = 0$ for every non-zero interval T. Another example is a *linear Markov process*, defined by

(13.12.2)
$$X_{r+s} = \lambda_s X_r + Y_{r+s}$$

where Y_{r+s} is a sequence of uncorrelated variates independent of X_r, for all $s > 0$.

Since $E(X_r) = \mu$ for all r, we have from Eq. (2),

$$\mu = \lambda_s \mu + \mu_{y,s}$$

where $\mu_{y,s}$ is the expectation of Y_{r+s}. Then

$$\mu_{y,s} = \mu(1 - \lambda_s) \tag{13.12.3}$$

If we multiply Eq. (2) by X_r and then take expectations, we get, on putting σ^2 for the variance of X_r, and ρ_s for the correlation coefficient between X_r and X_{r+s},

$$\sigma^2\rho_s + \mu^2 = \lambda_s(\sigma^2 + \mu^2) + \mu\cdot\mu_{y,s}$$

This gives, after substituting from Eq. (3),

$$\sigma^2\rho_s = \lambda_s\sigma^2$$

or

$$\lambda_s = \rho_s \tag{13.12.4}$$

Since the process is Markov, the *partial* correlation between, say, X_1 and X_3 (eliminating X_2) is zero. By Eq. (13.4.7) this implies that $\rho_{13} - \rho_{12}\rho_{23} = 0$, where ρ_{13} is the ordinary correlation coefficient for X_1 and X_3. Since ρ_{12} and ρ_{23} are both equal to ρ_1 (s being 1 in both cases), it follows that $\rho_{13} = \rho_1^2$, and, in general,

$$\rho_s = \rho_1^s \tag{13.12.5}$$

This relationship among the correlation coefficients is characteristic of the stationary linear Markov process.

We cannot enter further into the topic of stochastic processes, with its many applications to economic time-series, population and genetic problems, communication theory and traffic engineering, cosmic ray showers and thermodynamics, to mention only a few of the directions in which the theory has been applied. Those who wish some further insight may consult reference [9] which contains a fairly full bibliography. See also [10].

PROBLEMS

A. (§§ 13.1–13.4)

1. Given that, for a group of children between the ages of 8 and 14, the ordinary coefficients of correlation between intelligence and school achievement, between intelligence and age, and between school achievement and age, are 0.80, 0.70 and 0.60, respectively, calculate the correlation coefficient between intelligence and school achievement, eliminating the effect of age.

2. In Problem A-3 of Chapter 12, calculate the three partial correlation coefficients and also the multiple correlation coefficient of Y on X and Z. Is this multiple correlation significantly different from zero?

3. In calculating correlation coefficients between three variables, a student obtained the values $r_{01} = 0.6$, $r_{02} = -0.4$, $r_{12} = 0.7$. Is there good reason to suspect these values? Why? *Hint:* Calculate $r_{0,12}$ from the data.

4. In Example 3 of § 13.4, calculate the partial coefficients of correlation between X_0 and X_2 and between X_1 and X_2.

5. From the data of Problem A-5 of Chapter 12, calculate all the ordinary coefficients of correlation, the partial coefficients of correlation between Y and each of X_1, X_2, and X_3, and the multiple correlation of Y on X_1, X_2 and X_3.

6. (*Pearl and Surface*) In a biometric study of egg production in the domestic fowl, measurements of length, breadth and weight (X_0, X_1, X_2, respectively) were made on 453 eggs. From all these the value of $r_{01.2}$ was -0.8955. If the 42 eggs weighing from 53 to 53.9 gm are considered alone, the ordinary coefficient of correlation r_{01} between length and breadth is -0.9117; similarly for the 46 eggs between 56 and 56.9 gm, $r_{01} = -0.8911$, and for the 13 eggs between 62 and 62.9 gm, $r_{01} = -0.8739$. Show that the weighted mean of these values of r_{01} is nearly equal to $r_{01.2}$ (compare § 13.4, immediately before Example 3).

7. If all the ordinary correlation coefficients in a set of $p + 1$ variates $X_0, X_1 \ldots X_p$ are equal to r, show that the partial correlation coefficients $r_{01.2\ldots p}$, $r_{02.1\ldots p}$, etc. are each equal to $r/[1 + (p - 1)r]$ and that the multiple correlation of X_0 on $X_1, X_2 \ldots X_p$ is given by

$$1 - r^2_{0,12\ldots p} = (1 - r)\frac{1 + pr}{1 + (p - 1)r}$$

Hint: Show that the determinant $d(R)$ is equal to $(1 - r)^p(1 + pr)$ and that $R_{00} = (1 - r)^{p-1}[1 + (p - 1)r]$, $R_{01} = -r(1 - r)^{p-1}$, etc.

8. With three variates X_0, X_1 and X_2, show that the correlation coefficient between the residuals $x_{0.12}$ and $x_{1.20}$ is equal and opposite to that between $x_{0.2}$ and $x_{1.2}$. *Hint:* $x_{0.12}$ is the same as the v_α of § 13.2, and $x_{1.20}$ is the residual for the multiple regression of X_1 on X_2 and X_0.

B. (§§ 13.5–13.7)

1. Write out the joint density function for the trivariate normal distribution, taking $x_0 = x$, $x_1 = y$, $x_2 = z$ and putting $\rho_{01} = \rho_{xy}$, etc. The variables may be supposed all standardized.

2. Show that, if X and Y are independent normal variates with zero means and variances σ_X^2, σ_Y^2 respectively, the bivariate normal surface is cut by a plane through the z-axis in a curve for which the points of inflexion lie on the elliptic cylinder $x^2/\sigma_X^2 + y^2/\sigma_Y^2 = 1$. *Hint:* The equation of the plane is $y = mx$. The points of inflexion are given by $d^2z/dx^2 = 0$, where $z = f(x, y)$.

3. If the variates $X_1, X_2 \ldots X_N$ $(-\infty < X_i < \infty)$ are independent and have a joint density function which is a function of $x_1^2 + x_2^2 + \ldots + x_N^2$ only, show that the X_i must be normal with mean zero and common variance. *Hint:* The functional equation $f(x)f(y) = f(x + y)$ has the solution $f(x) = e^{cx}$.

4. Let X_1, X_2 have a joint bivariate normal distribution with means zero and covariance matrix $C = \begin{bmatrix} C_{11} & C_{12} \\ C_{21} & C_{22} \end{bmatrix}$. Write out the density function for X_2, given X_1. *Hint:* $f(x_2|x_1) = f(x_1, x_2)/f(x_1)$. When the variates are not standardized, the matrix A^{-1} of Eq. (13.5.4) is replaced by the covariance matrix.

5. If X_1, X_2, X_3 have a joint trivariate normal distribution, with covariance matrix

$$K = \begin{bmatrix} C_{11} & C_{12} & C_{13} \\ C_{21} & C_{22} & C_{23} \\ C_{31} & C_{32} & C_{33} \end{bmatrix}$$

calculate the expectation of X_1, given X_2 and X_3.

6. In the following table from Student's 1908 paper [11], x_1 and x_2 represent additional hours of sleep obtained by the use of soporific drugs A and B respectively on certain patients.

Patient	x_1	x_2
1	1.9	0.7
2	0.8	−1.6
3	1.1	−0.2
4	0.1	−1.2
5	−0.1	−0.1
6	4.4	3.4
7	5.5	3.7
8	1.6	0.8
9	4.6	0.0
10	3.4	2.0

Assuming that each pair of observations of x_1 and x_2 for a given patient is from a bivariate normal distribution, use the Hotelling T test to test the hypothesis at significance level 0.01 that neither drug really produces any soporific effect. (The following results of computation may be used:

$$\bar{x}_1 = 2.33, \quad \bar{x}_2 = 0.75, \quad S_{11} = 36.08, \quad S_{22} = 28.80, \quad S_{12} = 25.63$$

The null hypothesis is that $\mu_{10} = \mu_{20} = 0$).

C. (§§ 13.8–13.9)

1. Two treatments were applied to experimental forage plots, in 15 randomized blocks, each consisting of two plots, so that both treatments were used once in each block. The variable was the amount of Dutch clover in the forage stand, and this was estimated by two methods—(1) using a mechanical counter and (2) by eye. The means for the differences of the two treatments by the two methods were $d_1 = 1.34$, $d_2 = 1.06$, and the sums of squares and products within the two treatments were $S_{11} = 20.44$, $S_{22} = 6.41$, $S_{12} = 4.89$. Calculate the best discriminant function of the form $b_1x_1 + b_2x_2$ for distinguishing between the two treatments.

2. (Johnson [12]) Two populations of students taking university physics are distinguished as (1) those taking the standard elementary course and (2) those taking a somewhat more advanced course intended for better-prepared students. Discrimination is made on the basis of three measurements, X_1 (a mathematics test score), X_2 (the A.C.E. test score) and X_3 (the student's honor-point ratio). For 111 students in the first course and 257 students in the second, the following results were recorded:

	Course (1)	Course (2)
N	111	257
$\bar{x}_1$	87.640	92.397
$\bar{x}_2$	31.081	56.074
$\bar{x}_3$	1.1586	1.2689
$S(x_1 - \bar{x}_1)^2$	53136	194356
$S(x_2 - \bar{x}_2)^2$	11616	15864
$S(x_3 - \bar{x}_3)^2$	51.85	120.39
$S(x_1 - \bar{x}_1)(x_2 - \bar{x}_2)$	4863	17878
$S(x_1 - \bar{x}_1)(x_3 - \bar{x}_3)$	485.5	1844.0
$S(x_2 - \bar{x}_2)(x_3 - \bar{x}_3)$	243.8	836.6

Calculate the best discriminant function for distinguishing between the courses. If a new student comes along with the scores $x_1 = 80$, $x_2 = 40$, $x_3 = 1.5$, to which course should he be assigned? *Hint:* Calculate $L_1 = \sum_i b_i \bar{x}_i$ for course (1), and $L_2 = \sum_i b_i \bar{x}_i$ for course (2). If the observed $L < \frac{1}{2}(L_1 + L_2)$, the student should be assigned to course (1).

3. R. A. Fisher [13] has discussed the separation of two species of iris, namely, (1) versicolor and (2) setosa. The criteria are X_1 (sepal length), X_2 (sepal width), X_3 (petal length) and X_4 (petal width), all in centimeters. The data on 50 specimens of (1) and 50 specimens of (2) are summarized as follows:

$$(\bar{x}_i^{(1)}) = \begin{bmatrix} 5.936 \\ 2.770 \\ 4.260 \\ 1.326 \end{bmatrix} \qquad (\bar{x}_i^{(2)}) = \begin{bmatrix} 5.006 \\ 3.428 \\ 1.462 \\ 0.246 \end{bmatrix}$$

$$98(s_{ij}) = \begin{bmatrix} 19.1434 & 9.0356 & 9.7634 & 3.2394 \\ 9.0356 & 11.8658 & 4.6232 & 2.4746 \\ 9.7634 & 4.6232 & 12.2978 & 3.8794 \\ 3.2394 & 2.4746 & 3.8794 & 2.4604 \end{bmatrix}$$

The inverse matrix is

$$(s^{ij}) = 98 \begin{bmatrix} 0.11872 & -0.06687 & -0.08162 & 0.03964 \\ -0.06687 & 0.14527 & 0.03341 & -0.11075 \\ -0.08162 & 0.03341 & 0.21936 & -0.27202 \\ 0.03964 & -0.11075 & -0.27202 & 0.89455 \end{bmatrix}$$

Calculate the discriminant function, and state the criterion for allotting another specimen of iris to the one species or the other.

4. Calculate Hotelling's T_2^2 for the data of Problem C-3, and test its significance by the F test. *Hint:* On the null hypothesis that the two populations have the same vectors of means, $(\mu_i^{(1)})$ and $(\mu_i^{(2)})$, the quantity $T_2^2\,(N_1 + N_2 - p - 1)/[p(N_1 + N_2 - 2)]$ has the F distribution with p and $N_1 + N_2 - p - 1$ d.f.

D. (§§ 13.10–13.12)

1. In Example 6 of § 13.11, calculate P^4, and hence obtain the probabilities for the two possible states of the system after four transitions.

2. If the transition probabilities for a Markov chain are given by

$$\begin{bmatrix} 0 & 1 & 0 & 0 \\ 1 & 0 & 0 & 0 \\ 0 & 0 & \frac{1}{2} & \frac{1}{2} \\ 0 & 0 & \frac{1}{2} & \frac{1}{2} \end{bmatrix}$$

and the initial state has probabilities $(\frac{1}{2}, \frac{1}{3}, \frac{1}{12}, \frac{1}{12})$ what are the probabilities after one, two and three transitions? Why is there no limiting probability distribution in this example?

3. Compute the fixed probability vector for the following matrices:

$$\begin{bmatrix} 0.1 & 0.9 \\ 0.6 & 0.4 \end{bmatrix}, \qquad \begin{bmatrix} \frac{3}{4} & \frac{1}{4} & 0 \\ 0 & \frac{2}{3} & \frac{1}{3} \\ \frac{1}{4} & \frac{1}{4} & \frac{1}{2} \end{bmatrix}$$

4. Two urns, labelled (1) and (2), each contain n balls, either white or black. There are n black balls and n white balls altogether, but the compositions of the urns may differ. A transition consists in choosing a ball at random from each urn and interchanging them, putting the ball from (1) into (2) and the ball from (2) into (1). Each state is completely specified by the number of black balls in urn number (1). If at any state of the process there are j black balls in urn (1), what are the transition probabilities?

If initially $j = 0$ and $n = 4$, what are the probable compositions of the urns after three transitions?

Show that the vector of probabilities $q' = (p_0, p_1, p_2 \ldots p_n)$, where $p_j = \binom{n}{j}^2 \Big/ \binom{2n}{n}$, is the fixed vector for this transition matrix.

5. A Markov chain is said to be *ergodic* if it is possible to go from every state to every other state. A regular chain is necessarily ergodic, since if the nth power of P contains no zeros, there is a non-zero probability of every possible transition in n steps. Show that the chain represented by the transition matrix

$$\begin{bmatrix} 0 & \frac{1}{2} & 0 & \frac{1}{2} \\ \frac{1}{2} & 0 & \frac{1}{2} & 0 \\ 0 & \frac{1}{2} & 0 & \frac{1}{2} \\ \frac{1}{2} & 0 & \frac{1}{2} & 0 \end{bmatrix}$$

is ergodic but not regular.

6. Suppose that the following is an extract from a stationary process, the values being taken at successive times separated by the fixed interval T: -5, -6, -2, 4, 7, 3, 1, -5, -1, 2. Estimate the auto-correlation coefficient by calculating the sample variance of the observations and the sample covariance between successive pairs. *Hint:* Take the variance as $[V(X_r)V(X_{r-1})]^{1/2}$. To estimate $V(X_r)$ use the last nine observations, and to estimate $V(X_{r-1})$ use the first nine. There are nine pairs in the expression for the covariance.

REFERENCES

[1] Fisher, R. A., "The General Sampling Distribution of the Multiple Correlation Coefficient," *Proc. Roy. Soc. A*, **121,** 1928, pp. 654–673.

[2] Erdélyi, A., et al, *Higher Transcendental Functions, Vol. I*, McGraw-Hill, 1953, Chap. 2.

[3] Rainville, E. D., *Special Functions*, Macmillan, 1960.

[4] Pearson, K., *Tables for Statisticians and Biometricians, Part II.* University College, London, 1931.

[5] *Tables of the Bivariate Normal Distribution Function and Related Functions*, National Bureau of Standards, Washington, D.C., 1959.

[6] Plackett, R. L., "A Reduction Formula for Normal Multivariate Integrals," *Biometrika*, **41,** 1954, pp. 351–360.

[7] Hotelling, Harold, "The Generalization of Student's Ratio," *Ann, Math. Stat.*, **2,** 1931, pp. 360–378.

[8] Anderson, T. W., *An Introduction to Multivariate Statistical Analysis*, Wiley, 1958.

[9] Bartlett, M. S., *An Introduction to Stochastic Processes*, Cambridge U.P., 1955.

[10] Doob, J. L., "What is a Stochastic Process?" *Amer. Math. Monthly*, **49,** 1942, pp. 648–653.

[11] Student, "On The Probable Error of the Mean," *Biometrika*, **6,** 1908, pp. 1–25.

[12] Johnson, P. O., *Statistical Methods in Research*, Prentice-Hall, 1949.

[13] Fisher, R. A., "The Use of Multiple Measurements in Taxonomic Problems," *Ann. Eugenics*, **7,** 1936, pp. 179–188.

Appendix A

MATHEMATICAL APPENDIX

A.1 The Limit of $(1 + x/n)^n$ for Fixed x, as $n \to \infty$

By the binomial theorem,

$$(1 + 1/n)^n = 1 + n\left(\frac{1}{n}\right) + \frac{n(n-1)}{2!}\left(\frac{1}{n}\right)^2 + \ldots + \frac{n(n-1)\ldots 1}{n!}\left(\frac{1}{n}\right)^n \tag{A.1.1}$$

$$= 1 + 1 + \frac{1}{2!}\left(1 - \frac{1}{n}\right) + \frac{1}{3!}\left(1 - \frac{1}{n}\right)\left(1 - \frac{2}{n}\right)$$

$$+ \ldots + \frac{1}{n!}\left(1 - \frac{1}{n}\right)\left(1 - \frac{2}{n}\right)\ldots\left(1 - \frac{n-1}{n}\right)$$

The $(p + 1)^{\text{th}}$ term in this expansion (the term involving $1/p!$) is

$$\frac{1}{p!}\left(1 - \frac{1}{n}\right)\left(1 - \frac{2}{n}\right)\ldots\left(1 - \frac{p-1}{n}\right), \qquad p \leq n \tag{A.1.2}$$

This is always positive, and increases for fixed p as n increases (since the subtracted terms get smaller). Also the number of such terms in $(1 + 1/n)^n$ increases with n. For both reasons $(1 + 1/n)^n$ increases with n, so that it must either have a limit or tend to $+\infty$. But we can show that it has a limit, as follows:

All the factors $1 - \frac{1}{n}, 1 - \frac{2}{n} \ldots 1 - \frac{n-1}{n}$ are less than 1, so that

$$\left(1 + \frac{1}{n}\right)^n < 1 + 1 + \frac{1}{2!} + \frac{1}{3!} + \ldots + \frac{1}{n!}$$

$$< 1 + 1 + \frac{1}{2} + \frac{1}{2^2} + \ldots + \frac{1}{2^{n-1}}$$

$$< 1 + \left(1 + \frac{1}{2} + \frac{1}{2^2} + \ldots\right)$$

But the sum of this geometric progression (in parentheses) to infinity is $1/(1 - \frac{1}{2}) = 2$, so that $(1 + 1/n)^n < 3$, however large n may be. Also, the sum is obviously greater than 2 (which is the sum of the first two terms alone in Eq. (1)). The limit is therefore a number between 2 and 3 which may be denoted by e. It is actually an irrational number, 2.71828 . . . .

The expression (A.1.2), with p fixed, tends to the value $1/p!$ as $n \to \infty$. The number e is therefore the sum $1 + 1 + 1/2! + 1/3! + \ldots$, and this sum may be calculated to any required degree of accuracy by taking enough terms. On an electronic digital computer it has been obtained to about 60,000 decimal places.

If we define $\log x$ by the equation

(A.1.3) $$\log x = \int_1^x \frac{dt}{t}, \qquad x > 0$$

and define the exponential function $\exp x$ as the inverse of the logarithmic function, so that

(A.1.4) $$x = \exp y \quad \text{if} \quad y = \log x,$$

then the following argument indicates that $\exp x$ is the same as the above number e raised to the power x, written e^x.

By Eq. (3), $\log(1 + xt) = \int_1^{1+xt} \frac{du}{u}$, so that, on differentiating the integral with respect to t, (see § A.9)

$$\frac{d}{dt}\log(1 + xt) = \frac{1}{1 + xt}\cdot\frac{d}{dt}(1 + xt) = \frac{x}{1 + xt}$$

for $t > 0$.

But by the definition of the derivative this relation merely states that

$$\lim_{h\to 0}\frac{\log(1 + xt + xh) - \log(1 + xt)}{h} = \frac{x}{1 + xt}$$

and if we now let $t \to 0$ from above,

$$\lim_{h\to 0} h^{-1}\log(1 + xh) = x$$

On writing $h = 1/k$, this becomes

$$\lim_{k\to\infty} k\log(1 + x/k) = x$$

that is,

$$\lim_{k\to\infty}\log(1 + x/k)^k = x$$

Hence, by Eq. (4),

(A.1.5) $$\lim_{k\to\infty}(1 + x/k)^k = \exp x$$

If we suppose that $k \to \infty$ through integral values only, we have the result

(A.1.6) $$\lim_{n\to\infty}(1 + x/n)^n = \exp x$$

It follows from Eq. (6) and the earlier part of this section that $\exp 1 = e$, so that $\log e = 1$. The quantity e^x for any real x is defined by $\log e^x = x\log e = x$, in accordance with the usual convention for indices, so that $e^x = \exp x$ and we obtain the required result, namely,

(A.1.7) $$\lim_{n\to\infty}(1 + x/n)^n = e^x$$

A.2 Stirling's Approximation to $n!$

Factorials are not very convenient for mathematical manipulation, and it is often useful to replace $n!$ by an approximation. The most common approximation is Stirling's, namely,

(A.2.1) $$n! \approx (2\pi)^{1/2}n^{n+1/2}e^{-n}$$

or, equivalently,

(A.2.2) $$\log n! \approx \tfrac{1}{2}\log(2\pi) + (n + \tfrac{1}{2})\log n - n$$

The meaning of the approximate equality here is that the ratio of the two sides tends to 1 as $n \to \infty$. The accuracy of the approximation may be gauged by the following examples:

$$n = 5,\quad 5! = 120,\ (2\pi)^{1/2}5^{5\frac{1}{2}}e^{-5} = 118.02$$

$$n = 10,\ 10! = 3{,}628{,}800,\ (2\pi)^{1/2}10^{10\frac{1}{2}}e^{-10} = 3{,}598{,}700$$

The *relative* error is roughly $1/(12n)$, and therefore diminishes as n increases.

We will establish Stirling's result in the following form:

(A.2.3) $$\log n! = (n + \tfrac{1}{2})\log n - n + C_n$$

where $\frac{3}{4} + (4n)^{-1} < C_n < 1$ and then show that $\lim_{n\to\infty} C_n = \frac{1}{2}\log(2\pi)$.

Consider the curve $y = \log x$ between $x = 1$ and $x = n$ (Figure 55). The area under the curve is given by

(A.2.4) $$A = \int_1^n \log x\, dx = n\log n - n + 1$$

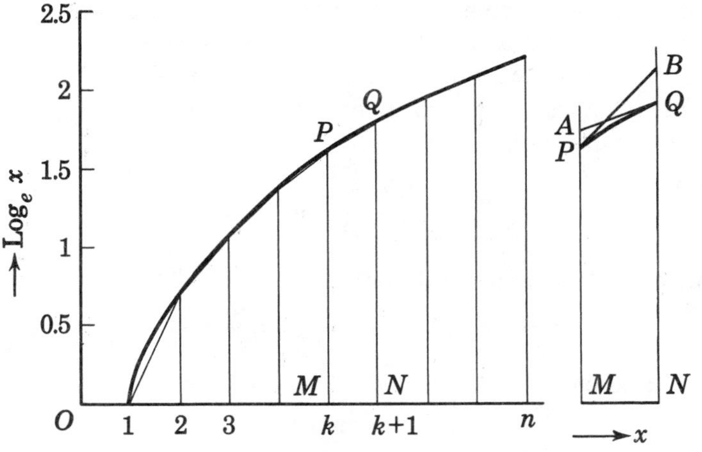

FIG. 55 STIRLING APPROXIMATION TO N!

If the tops of the ordinates at $x = 1, 2, 3, \ldots$ are joined by chords, the area under the chords will be less than that under the curve, since the curve is everywhere concave to the x-axis. This area is a sum of trapeziums, the area of the trapezium between k and $k + 1$ being $\frac{1}{2}[\log k + \log(k + 1)]$. The total area under the chords is

$$\begin{aligned}&\tfrac{1}{2}[(\log 1 + \log 2) + (\log 2 + \log 3) + \ldots + (\log(n-1) + \log n)]\\ &\quad= \log 1 + \log 2 + \log 3 + \cdots + \log n - \tfrac{1}{2}(\log 1 + \log n)\\ &\quad= \log n! - \tfrac{1}{2}\log n.\end{aligned}$$

Since this is less than A, given by Eq. (4), we have the inequality

$$\text{(A.2.5)} \qquad \log n! < (n + \tfrac{1}{2})\log n - n + 1$$

This establishes the upper bound on C_n in Eq. (3).

If we draw tangents to the curve at $P(x = k)$ and $Q(x = k + 1)$ and if the tangent at P meets the ordinate at Q in B, the slope of PB is $1/k$ $\left(\text{since } \frac{d}{dx}\log x = 1/x\right)$ and therefore $NB = MP + 1/k = \log k + 1/k$. Similarly $MA = \log(k + 1) - 1/(k + 1)$. The areas $MAQN$ and $MPBN$ are both greater than the area under the curve PQ between MP and NQ, and the mean of the two will also be greater. Since $MAQN = \frac{1}{2}[2\log(k + 1) - 1/(k + 1)]$ and $MPBN = \frac{1}{2}[2\log k + 1/k]$, the mean of these two is $\frac{1}{2}[\log k + \log(k + 1)] + \frac{1}{4}[1/k - 1/(k + 1)]$. Summing for all k from 1 to $n - 1$, we see that the area under the curve is less than $\log n! - \frac{1}{2}\log n + \frac{1}{4}(1 - 1/n)$. We have therefore from Eq. (4) the inequality

$$\text{(A.2.6)} \qquad \log n! > (n + \tfrac{1}{2})\log n - n + \tfrac{3}{4} + \frac{1}{4n}$$

which establishes the lower bound on C_n. Clearly C_n lies between $\frac{7}{8}(n = 2)$ and 1. Also as n increases C_n decreases (since the difference between the area under the curve and that under the set of chords is $1 - C_n$ and this difference increases with n). Hence C_n must approach a limit C as $n \to \infty$. By using Wallis's formula (see Appendix A.8), namely,

$$\text{(A.2.7)} \qquad \lim_{n\to\infty} \frac{2^{2n}(n!)^2}{n^{1/2}(2n)!} = \pi^{1/2}$$

we can evaluate C as $\frac{1}{2}\log(2\pi) = 0.9189$. For, by Eq. (3),

$$n! = n^{n+1/2}e^{-n}e^{C_n}, \qquad (2n)! = (2n)^{2n+1/2}e^{-2n}e^{C_{2n}}$$

so that Eq. (7) becomes

$$\lim_{n\to\infty} \frac{2^{2n}n^{2n+1}e^{-2n}e^{2C_n}}{n^{1/2}(2n)^{2n+1/2}e^{-2n}e^{C_{2n}}} = \pi^{1/2}$$

Since $\lim C_n = \lim C_{2n} = C$ this becomes

$$e^C = (2\pi)^{1/2}$$

or

$$C = \tfrac{1}{2}\log(2\pi)$$

A.3 Improper Integrals

If $f(x)$ is a continuous function of x for $x \geq a$ and if

$$\lim_{b\to\infty}\int_a^b f(x)\,dx\,(=l) \tag{A.3.1}$$

exists, then the improper integral $\int_a^\infty f(x)\,dx$ is said to converge and has the value l. If it does not converge, the integral either diverges to $+\infty$ or $-\infty$, or oscillates. Similarly, if $f(x)$ is continuous for $x \leq b$ and if

$$\lim_{a\to-\infty}\int_a^b f(x)\,dx\,(=k) \tag{A.3.2}$$

exists, the integral $\int_{-\infty}^b f(x)\,dx$ converges and is equal to k.

If both integrals $\int_{-\infty}^a$ and $\int_a^\infty$ converge and have the values k and l respectively, then $\int_{-\infty}^\infty f(x)\,dx$ converges and has the value $k + l$.

The *Cauchy principal value* of the integral $\int_{-\infty}^\infty f(x)\,dx$ is given by

$$\lim_{c\to\infty}\int_{-c}^c f(x)\,dx \tag{A.3.3}$$

This limit may exist even though the two separate integrals do not. Thus $\displaystyle\int_{-c}^c \frac{x\,dx}{x^2+1} = 0$ for all real values of c, but both integrals $\displaystyle\int_{-\infty}^a \frac{x\,dx}{x^2+1}$ and $\displaystyle\int_a^\infty \frac{x\,dx}{x^2+1}$ diverge, since the indefinite integral of $f(x)$ is here $\frac{1}{2}\log(x^2 + 1)$. The improper integral $\displaystyle\int_{-\infty}^\infty \frac{x\,dx}{x^2+1}$ therefore diverges, but its Cauchy principal value is zero.

Another type of improper integral is that in which $f(x)$ becomes infinite at some point or points of the range of integration; by splitting up the range into sub-intervals marked off by these points we need consider only the cases when $f(x)$ becomes infinite at either the lower bound or the upper bound of integration.

If $f(x) \to \infty$ as $x \to a$ from above, but is otherwise continuous in the interval from a to A, and if

$$\lim_{\varepsilon\to+0}\int_{a+\varepsilon}^A f(x)\,dx = l \tag{A.3.4}$$

then the integral $\int_a^A f(x)\,dx$ converges and has the value l. Similarly, if $f(x) \to \infty$ as $x \to A$ from below, but is otherwise continuous from a to A, and if

$$\text{(A.3.5)} \qquad \lim_{\varepsilon \to +0} \int_a^{A-\varepsilon} f(x)\,dx = k$$

then $\int_a^A f(x)\,dx$ converges and has the value k.

If $f(x)$ becomes infinite at a value $x = c$ between $x = a$ and $x = b$, we can define the *Cauchy principal value* of the integral, if it exists, by

$$\text{(A.3.6)} \qquad \int_a^b f(x)\,dx = \lim_{\varepsilon \to +0} \left(\int_a^{c-\varepsilon} f(x)\,dx + \int_{c+\varepsilon}^b f(x)\,dx \right)$$

A.4 Change of Variables in Integration

It is often convenient, in order to perform an integration, to change the variables from one set to another. With a single variable the process is probably familiar to students who have had a course of calculus. If the variable is changed from x to u, where $x = g(u)$, if $g(u)$ is a differentiable function of u, and if $f(x)$ is integrable from a to b, then

$$\text{(A.4.1)} \qquad \int_a^b f(x)\,dx = \int_\alpha^\beta f[g(u)]g'(u)\,du$$

where $g'(u) = d\,g(u)/du$, $a = g(\alpha)$ and $b = g(\beta)$. Thus,

$$\int_0^{1/2} \frac{dx}{\sqrt{1-x^2}} = \int_0^{\pi/6} \frac{\cos u\,du}{\sqrt{1-\sin^2 u}} = \int_0^{\pi/6} du = \frac{\pi}{6}$$

where the change of variable is expressed by $x = \sin u$.

Care is necessary in determining the new bounds of integration α and β to make sure that the interval from α to β for u does correspond to the interval from a to b for x, particularly when x is not a single-valued function of u. Thus, for example, if $u = x^2$, either interval of x, from -2 to -1 or from 1 to 2, would correspond to the same interval of u from 1 to 4. In the first case, $g'(u)$ is negative, and in the second, positive, but the bounds of integration for u are interchanged in the two cases.

If there are two variables x and y, and the integral is a double one, we may need to change to a new pair of variables u and v, where $x = g(u, v)$ and $y = h(u, v)$. A double integral over a region R of the (x, y) plane is calculated by means of a repeated integral

$$\text{(A.4.2)} \qquad \int_{(R)} f(x, y)\,dA = \int_a^b \int_{g_1(x)}^{g_2(x)} f(x, y)\,dy\,dx$$

The region of integration is considered as bounded by the curves $y = g_1(x)$ and $y = g_2(x)$ between $x = a$ and $x = b$. The corresponding region R in the (u, v) plane is bounded by the curves $v = \gamma_1(u)$, $v = \gamma_2(u)$, between $u = \alpha$ and $u = \beta$

(Figure 56). The element of area $dy\,dx$ in the (x, y) plane becomes in the (u, v) plane

$$dA_1 = \left|J\left(\frac{g, h}{u, v}\right)\right| du\,dv \tag{A.4.3}$$

where

$$J\left(\frac{g, h}{u, v}\right) = \begin{vmatrix} \dfrac{\partial g}{\partial u} & \dfrac{\partial g}{\partial v} \\ \dfrac{\partial h}{\partial u} & \dfrac{\partial h}{\partial v} \end{vmatrix} = \frac{\partial g}{\partial u}\cdot\frac{\partial h}{\partial v} - \frac{\partial g}{\partial v}\cdot\frac{\partial h}{\partial u} \tag{A.4.4}$$

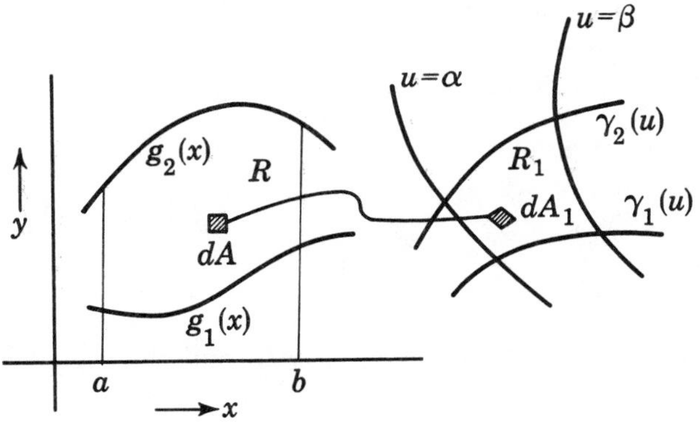

FIG. 56 CHANGE OF VARIABLES IN INTEGRATION

The functions g and h are supposed to possess continuous first partial derivatives throughout the region R_1, and it is also supposed that $J\left(\frac{g, h}{u, v}\right)$ does not vanish anywhere in R_1. This function J is called the *Jacobian* of g and h with respect to u and v. The double integral (A.4.2) can then be written as the repeated integral

$$\int_{(R)} f(x, y)\,dA = \int_\alpha^\beta \int_{\gamma_1(u)}^{\gamma_2(u)} f(g, h)\left|J\left(\frac{g, h}{u, v}\right)\right| dv\,du \tag{A.4.5}$$

An example is the change from Cartesian coordinates x, y to polar coordinates r, θ, where $x = g(r, \theta) = r\cos\theta$, $y = h(r, \theta) = r\sin\theta$. Here

$$J\left(\frac{g, h}{r, \theta}\right) = \frac{\partial g}{\partial r}\cdot\frac{\partial h}{\partial \theta} - \frac{\partial g}{\partial \theta}\cdot\frac{\partial h}{\partial r} = \cos\theta(r\cos\theta) - (-r\sin\theta)\sin\theta = r$$

If $f(x, y)$ is integrable over the whole first quadrant in the (x, y) plane, we have

(A.4.6) $$\int_0^\infty \int_0^\infty f(x, y)\, dy\, dx = \int_0^\infty \int_0^{\pi/2} f(r \cos \theta, r \sin \theta)\, r\, d\theta\, dr$$

since in the first quadrant θ ranges from 0 to $\pi/2$ and r from 0 to ∞.

The equations $x = g(u, v)$, $y = h(u, v)$ can be solved to express u and v in terms of x and y. If the Jacobian does not vanish, the resulting functions, $u = \phi(x, y)$ and $v = \psi(x, y)$, will themselves be differentiable. We can then calculate $J\left(\frac{\phi, \psi}{x, y}\right)$. It is sometimes convenient to note that

(A.4.7) $$J\left(\frac{g, h}{u, v}\right) = \left[J\left(\frac{\phi, \psi}{x, y}\right)\right]^{-1}$$

since one of these Jacobians may be easier to calculate than the other.

The above considerations may be extended to triple or n-dimensional integrals.

A.5 The Gamma Function

The improper convergent integral

(A.5.1) $$\Gamma(n) = \int_0^\infty x^{n-1} e^{-x}\, dx, \quad n > 0$$

is called the gamma function of n. Using the formula for integration by parts we easily obtain

(A.5.2) $$\begin{aligned} \Gamma(n + 1) &= \int_0^\infty x^n e^{-x}\, dx \\ &= \left[-x^n e^{-x}\right]_0^\infty + n \int_0^\infty x^{n-1} e^{-x}\, dx \\ &= n\Gamma(n) \end{aligned}$$

since $\lim_{x \to \infty} x^n e^{-x} = 0$ for all $n > 0$.

If we put $n = 1$ in Eq. (1), we obtain $\Gamma(1) = \int_0^\infty e^{-x}\, dx = 1$ so that $\Gamma(2) = 1 \cdot \Gamma(1) = 1$, $\Gamma(3) = 2 \cdot \Gamma(2) = 2$, and in general for any positive integral value of n,

(A.5.3) $$\Gamma(n + 1) = n(n-1) \ldots 1 = n!$$

The gamma function may therefore be regarded as a generalized factorial, and indeed the notation $n!$ is often used for the integral denoted here by $\Gamma(n + 1)$ for any $n > -1$, whether integral or not.

For $n = 0$, the integral of Eq. (1) diverges.

For negative values of n, except negative integers, $\Gamma(n)$ may be defined by means of Eq. (2),

$$\Gamma(n) = \frac{1}{n}\,\Gamma(n+1)$$

Thus

$$\Gamma(-\tfrac{1}{2}) = -2\Gamma(\tfrac{1}{2})$$

The graph of $\Gamma(n)$ is shown in Figure 57. The function is discontinuous at all negative integral values and at $n = 0$.

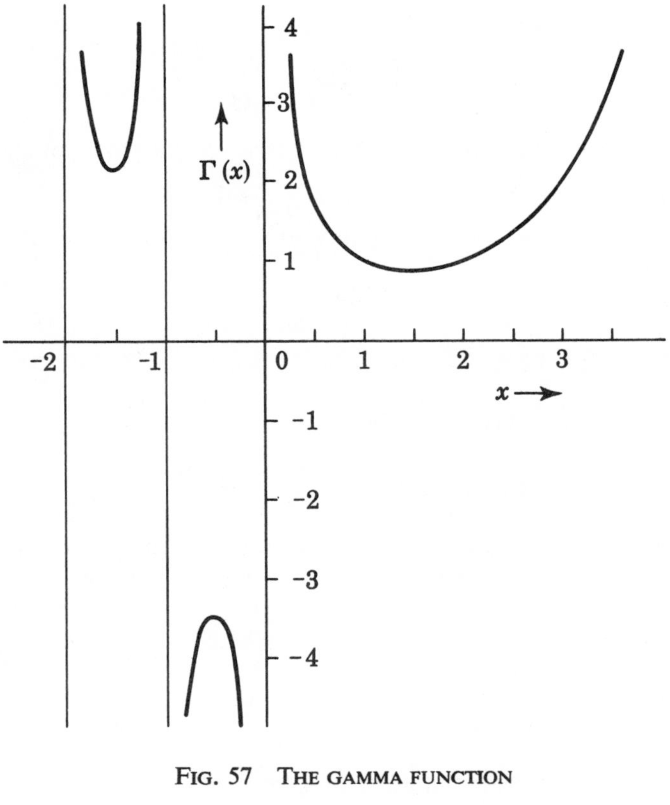

FIG. 57 THE GAMMA FUNCTION

An alternative form for the gamma function is obtained by a change of variable. If $x = u^2$, we have by Eq. (A.4.1)

(A.5.4) $$\Gamma(n) = 2\int_0^\infty u^{2n-1}e^{-u^2}\,du$$

A.6 **The Beta Function**

The definite integral

(A.6.1) $$B(m, n) = \int_0^1 x^{m-1}(1-x)^{n-1}\,dx, \quad m > 0,\ n > 0$$

is called the *beta function* of m and n. Clearly, $B(1, 1) = 1$. If we write $1 - y$

for x, we obtain

(A.6.2) $$B(m, n) = -\int_1^0 (1-y)^{m-1}y^{n-1}\,dy = \int_0^1 y^{n-1}(1-y)^{m-1}\,dy$$
$$= B(n, m)$$

so that, in the beta function, m and n may be interchanged at will. Alternative forms for the beta function may be obtained by change of variable. Thus, if $x = \sin^2\theta$,

(A.6.3) $$B(m, n) = 2\int_0^{\pi/2} \sin^{2m-1}\theta\cos^{2n-1}\theta\,d\theta$$

and if $x = (1+y)^{-1}$

(A.6.4) $$B(m, n) = \int_0^\infty y^{n-1}(1+y)^{-m-n}\,dy$$

An important relation between the beta function and the gamma function is the following:

(A.6.5) $$B(m, n) = \frac{\Gamma(m)\Gamma(n)}{\Gamma(m+n)}$$

To prove this, we note from Eq. (A.5.4) that

$$\Gamma(m)\Gamma(n) = 4\int_0^\infty x^{2m-1}e^{-x^2}\,dx\int_0^\infty y^{2n-1}e^{-y^2}\,dy$$

This repeated integral may be written as a double integral:

$$4\int_0^\infty\int_0^\infty x^{2m-1}y^{2n-1}e^{-(x^2+y^2)}\,dy\,dx$$

and interpreted as an integral over the first quadrant of the (x, y) plane. Changing to polar coordinates (r, θ) and using Eq. (A.4.6), we find

$$\Gamma(m)\Gamma(n) = 4\int_0^\infty\int_0^{\pi/2}(r\cos\theta)^{2m-1}(r\sin\theta)^{2n-1}e^{-r^2}r\,d\theta\,dr$$
$$= 4\int_0^\infty r^{2m+2n-1}e^{-r^2}\,dr\int_0^{\pi/2}\cos^{2m-1}\theta\sin^{2n-1}\theta\,d\theta$$

where the double integral is now written as a repeated integral. Using Eqs. (A.5.4) and (A.6.3), we obtain

$$\Gamma(m)\cdot\Gamma(n) = \Gamma(m+n)\cdot B(n, m)$$

whence, by Eq. (2),

$$B(m, n) = \frac{\Gamma(m)\Gamma(n)}{\Gamma(m+n)}$$

A.7 The Integral $\int_0^\infty e^{-u^2}\,du$ and Related Integrals

From Eq. (A.6.3),

(A.7.1) $$B(\tfrac{1}{2}, \tfrac{1}{2}) = 2\int_0^{\pi/2} d\theta = \pi$$

But, by Eq. (A.6.5),

$$B(\tfrac{1}{2}, \tfrac{1}{2}) = \frac{[\Gamma(\frac{1}{2})]^2}{\Gamma(1)} = [\Gamma(\tfrac{1}{2})]^2$$

so that

(A.7.2) $$\Gamma(\tfrac{1}{2}) = \pi^{1/2}$$

Therefore, from Eq. (A.5.4), with $n = \frac{1}{2}$,

(A.7.2) $$2\int_0^\infty e^{-u^2}\,du = \pi^{1/2}$$

or, since e^{-u^2} is an even function of u,

(A.7.3) $$\int_{-\infty}^\infty e^{-u^2}\,du = \pi^{1/2}$$

Writing $\sqrt{2}u = v$, we obtain

(A.7.4) $$\int_{-\infty}^\infty e^{-v^2/2}\,dv = (2\pi)^{1/2}$$

or

(A.7.5) $$\int_{-\infty}^\infty \phi(v)\,dv = 1, \qquad \phi(v) = (2\pi)^{-1/2}e^{-v^2/2}$$

This expresses the fact that the total area under the standard normal curve is 1, as it must be if $\phi(v)$ is to be a probability density function.

A useful related integral is

(A.7.6) $$I_k = \int_0^\infty v^k e^{-v^2/2}\,dv, \qquad k = 1, 2, 3 \ldots$$

Putting $v = \sqrt{2}u$, we get

(A.7.7) $$\begin{aligned} I_k &= \int_0^\infty 2^{k/2}u^k e^{-u^2}2^{1/2}\,du \\ &= 2^{(k+1)/2}\cdot\tfrac{1}{2}\Gamma\left(\frac{k+1}{2}\right) \text{ by Eq. (A.5.4)} \\ &= 2^{(k-1)/2}\Gamma\left(\frac{k+1}{2}\right) \end{aligned}$$

Thus,

(A.7.8) $$\begin{cases} I_1 = \Gamma(1) = 1, \\ I_2 = 2^{1/2}\Gamma(\frac{3}{2}) = 2^{1/2}\frac{1}{2}\Gamma(\frac{1}{2}) = \left(\dfrac{\pi}{2}\right)^{1/2} \\ I_3 = 2\Gamma(2) = 2 \\ I_4 = 2^{3/2}\Gamma(\frac{5}{2}) = 2^{3/2}\cdot\frac{3}{2}\cdot\frac{1}{2}\cdot\Gamma(\frac{1}{2}) \\ \qquad = 3\left(\dfrac{\pi}{2}\right)^{1/2} \end{cases}$$

and so on.

A.8 Wallis's Formula

By Eq. (A.1.5), $\lim_{n\to\infty}\left(1 - \frac{t}{n}\right)^n = e^{-t}$, so that, for $x > 0$,

(A.8.1) $$\lim_{n\to\infty}\int_0^n\left(1 - \frac{t}{n}\right)^n t^{x-1}\,dt = \int_0^\infty e^{-t}t^{x-1}\,dt = \Gamma(x)$$

Putting $t/n = u$, we get

$$\lim_{n\to\infty}\int_0^1 (1-u)^n(un)^{x-1}n\,du = \Gamma(x)$$

and therefore,

(A.8.2) $$\lim_{n\to\infty} n^x \int_0^1 u^{x-1}(1-u)^n\,du = \Gamma(x)$$

The integral in Eq. (2) is $B(x, n+1) = \Gamma(x)\Gamma(n+1)/\Gamma(n+x+1)$, so that

(A.8.3) $$\lim_{n\to\infty}\frac{n^x\Gamma(n+1)}{\Gamma(n+x+1)} = 1$$

Putting $x = \frac{1}{2}$, and noting that

$$\Gamma(n+1) = n!$$

and

$$\begin{aligned}\Gamma(n+\tfrac{3}{2}) &= (n+\tfrac{1}{2})(n-\tfrac{1}{2})(n-\tfrac{3}{2})\ldots(\tfrac{1}{2})\Gamma(\tfrac{1}{2})\\ &= \frac{2n+1}{2}\,\frac{2n-1}{2}\,\frac{2n-3}{2}\ldots\tfrac{1}{2}\pi^{1/2}\\ &= \frac{(2n+1)!\pi^{1/2}}{2n(2n-2)\ldots(2)2^{n+1}}\\ &= \frac{(2n+1)!\pi^{1/2}}{n!2^{2n+1}}\end{aligned}$$

we obtain from Eq. (3)

(A.8.4) $$\lim_{n\to\infty}\frac{n^{1/2}(n!)^2 2^{2n+1}}{(2n+1)!\pi^{1/2}} = 1$$

or

(A.8.5) $$\lim_{n\to\infty}\frac{2^{2n}.(n!)^2}{(2n)!n^{1/2}} = \lim_{n\to\infty}\frac{2n+1}{2n}\pi^{1/2} = \pi^{1/2}$$

This is Wallis's formula.

A.9 Differentiation Under the Sign of Integration

Let $f(x, \theta)$ be a function of x depending on a parameter θ and continuous over an interval of x between a and b, where a and b are themselves differentiable functions of θ. Suppose also that the partial derivative $\partial f/\partial\theta$ exists and is continuous over the same interval of x, for all admissible values of θ. Then if

(A.9.1) $$I(\theta) = \int_{a(\theta)}^{b(\theta)} f(x, \theta)\,dx$$

the derivative of $I(\theta)$ with respect to θ is given by

$$\frac{dI(\theta)}{d\theta} = \int_a^b \frac{\partial f}{\partial\theta}\,dx + f(b,\theta)\frac{db}{d\theta} - f(a,\theta)\frac{da}{d\theta} \tag{A.9.2}$$

This is known as *Leibniz's formula.* A proof may be found in textbooks of advanced calculus, such as Sokolnikoff's (McGraw-Hill, 1939), page 121.

If a and b are independent of θ, the last two terms vanish and Eq. (2) becomes

$$\frac{dI(\theta)}{d\theta} = \int_a^b \frac{\partial f}{\partial\theta}\,dx \tag{A.9.3}$$

EXAMPLE Let $f(x, \theta) = e^{-x^2/2\theta}$, and let $a = -(2\theta)^{1/2}$, $b = (2\theta)^{1/2}$. Then $da/d\theta = -(2\theta)^{-1/2}$, $db/d\theta = (2\theta)^{-1/2}$, and $\partial f/\partial\theta = (x^2/2\theta^2)e^{-x^2/2\theta}$. Also $f(a, \theta) = f(b, \theta) = e^{-1}$, so that if

$$I(\theta) = \int_{-(2\theta)^{1/2}}^{(2\theta)^{1/2}} e^{-x^2/2\theta}\,dx$$

we have

$$\frac{dI}{d\theta} = \frac{1}{2\theta^2}\int_{-(2\theta)^{1/2}}^{(2\theta)^{1/2}} x^2 e^{-x^2/2\theta}\,dx + 2e^{-1}(2\theta)^{-1/2}$$

A.10 Orthogonal Linear Transformations

The linear transformation

$$Y_i = \sum_{j=1}^{n} c_{ij}X_j, \qquad i = 1, 2 \ldots n \tag{A.10.1}$$

is called *orthogonal* if the constants c_{ij} satisfy the conditions

$$\sum_k c_{ik}c_{jk} = \begin{cases} 1, \text{ when } i = j \\ 0, \text{ when } i \neq j \end{cases} \tag{A.10.2}$$

If the determinant of the coefficients c_{ij} is multiplied by itself, with rows and columns transposed, (transposing does not alter its value), the result, using Eq. (2), is

$$\begin{vmatrix} 1 & 0..0 \\ 0 & 1..0 \\ \cdot & \cdot \\ \cdot & \cdot \\ 0 & 0..1 \end{vmatrix} = 1$$

The value of the determinant is therefore ± 1, and it can be taken as 1 by changing, if necessary, the sign of one of the Y_i. Since

$$\frac{\partial Y_i}{\partial X_j} = c_{ij} \tag{A.10.3}$$

this determinant is also the Jacobian of the Y's with respect to the X's (see Appendix A.4). Therefore,

(A.10.4) $$dY_1 \ldots dY_n = dX_1 \ldots dX_n$$

It can be proved by using some matrix algebra (see § A.22) that Eq. (2) implies

(A.10.5) $$\sum_k c_{ki}c_{kj} = \begin{cases} 1, \text{ when } i = j \\ 0, \text{ when } i \neq j \end{cases}$$

From Eq. (1), by squaring,

(A.10.6) $$Y_k^2 = \sum_j c_{kj}^2 X_j^2 + \sum_{i \neq j} c_{ki}c_{kj}X_iX_j$$

and therefore, by Eq. (5),

(A.10.7) $$\sum_k Y_k^2 = \sum_j X_j^2$$

The orthogonal transformation with determinant 1 is equivalent geometrically to a rotation of the coordinate axes about the origin. Such a rotation, of course, leaves the distance of any point from the origin unchanged, and this is the meaning of Eq. (7).

A.11 Angle Brackets and k-Statistics

As in § 5.8 we define

$$S_r = \sum_{i=1}^{N} X_i^r = N\langle r \rangle$$

$$N(N-1)\langle pq \rangle = \sum_{ij}{}' X_i^p X_j^q$$

$$N(N-1)(N-2)\langle pqr \rangle = \sum_{ijk}{}' X_i^p X_j^q X_k^r$$

and so on, where the symbol $\sum'$ indicates that in every term of the sum the subscripts $i, j \ldots$ are to be *different*.

Then

$$\begin{aligned} S_1^2 &= \left(\sum_i X_i\right)\left(\sum_j X_j\right) \\ &= \sum_i X_i^2 + \sum_{ij}{}' X_iX_j \end{aligned}$$

where we have separated the terms of the product in which $i = j$ from the terms in which i and j are different. Since $\sum_i X_i^2 = S_2$, and $\sum_{ij}' X_iX_j = N(N-1)\langle 11 \rangle$, we have

(A.11.1) $$S_1^2 = S_2 + N(N-1)\langle 11 \rangle$$

Again,

$$\begin{aligned} S_1S_2 &= \left(\sum_i X_i\right)\left(\sum_j X_j^2\right) \\ &= \sum_i X_i^3 + \sum_{ij}{}' X_iX_j^2 \end{aligned}$$

from which we get

(A.11.2) $$S_1S_2 = S_3 + N(N-1)\langle 12\rangle$$

Other results given in Eq. (5.8.[illegible]) may be found similarly. Also,

$$N(N-1)S_1\langle 11\rangle = \left(\sum_i X_i\right)\left(\sum_{jk}{}' X_jX_k\right)$$

In the terms of this product there will be some in which $i = j$, some in which $i = k$, and some in which i, j, k are all different. Therefore,

$$\begin{aligned} N(N-1)S_1\langle 11\rangle &= \sum{}' X_j{}^2X_k + \sum{}' X_jX_k{}^2 + \sum{}' X_iX_jX_k \\ &= 2N(N-1)\langle 12\rangle + N(N-1)(N-2)\langle 111\rangle \end{aligned}$$

Therefore,

(A.11.3) $$(N-2)\langle 111\rangle = S_1\langle 11\rangle - 2\langle 12\rangle$$

Similar arguments will give the other results of Eq. (5.8.3). The checks Eq. (5.8.6) are straightforward algebraic relations derived from the definitions of the k's and the above properties of the brackets. Thus the first one states that

$$\langle 11\rangle = (\langle 1\rangle)^2 - N^{-1}(\langle 2\rangle - \langle 11\rangle)$$

or

$$\begin{aligned} \frac{N-1}{N}\langle 11\rangle &= (\langle 1\rangle)^2 - N^{-1}\langle 2\rangle \\ &= N^{-2}S_1{}^2 - N^{-2}S_2 \end{aligned}$$

which is equivalent to Eq. (A.11.1).

A.12 Bernoulli Numbers and Sheppard's Corrections

The rth Bernoulli number is defined as the coefficient of $t^r/r!$ in the expansion of $t(e^t - 1)^{-1}$. Therefore,

(A.12.1) $$\sum_{r=0}^{\infty} B_r\frac{t^r}{r!} = t(e^t - 1)^{-1} = \frac{t}{2}\left(\coth\frac{t}{2} - 1\right)$$

where $\coth x = (e^x + e^{-x})/(e^x - e^{-x})$

The first few of these numbers are

$$B_0 = 1,\ B_1 = -\frac{1}{2},\ B_2 = \frac{1}{6},\ B_4 = -\frac{1}{30},\ B_6 = \frac{1}{42}$$

All the B's with odd subscript, except B_1, are zero.

In the grouping of a distribution into classes with class-interval c, any value X of the variate is recorded as the nearest class-mark X_i. The difference between X_i and X is not greater numerically than $c/2$. That is,

(A.12.2) $$X_i = X + \varepsilon$$

where ε may be supposed to have a uniform (rectangular) distribution on the interval $-c/2$ to $c/2$. This, of course, is not usually true and is an assumption for reasons of mathematical convenience, but if the grouping is reasonably fine (intervals short compared with the effective range) it is not likely to be very far out. If $K(h)$, $K_i(h)$ and $K_\varepsilon(h)$ are the cumulant generating functions for X, X_i and ε respectively,

(A.12.3) $$K_i(h) = K(h) + K_\varepsilon(h)$$

by the main property of such functions, § 2.12.

Now $K_i(h)$ is the c.g.f. for the grouped distribution and will give the uncorrected cumulants, while $K(h)$ is the c.g.f. for the true distribution and will give the true cumulants. Also, by § 2.10, Example 4,

(A.12.4) $$K_\varepsilon(h) = \log M_\varepsilon(h) = \log \sinh \frac{ch}{2} - \log \frac{ch}{2}$$
$$= \log \sinh(t/2) - \log(t/2)$$

where $t = ch$.
Differentiating with respect to t, we have

(A.12.5) $$\frac{dK_\varepsilon(h)}{dt} = \frac{1}{2} \coth\left(\frac{t}{2}\right) - \frac{1}{t}$$

But, by Eq. (1),

$$\frac{t}{2}\left(\coth \frac{t}{2} - 1\right) = 1 - \frac{1}{2} t + \sum_{r=2}^{\infty} B_r \frac{t^r}{r!}$$

so that, on dividing by t,

(A.12.6) $$\frac{1}{2} \coth\left(\frac{t}{2}\right) - \frac{1}{t} = \sum_{2}^{\infty} B_r \frac{t^{r-1}}{r!}$$

From Eqs. (5) and (6), therefore,

$$\frac{dK_\varepsilon(h)}{dt} = \sum_{2}^{\infty} B_r \frac{t^{r-1}}{r!}$$

from which, after integration, we obtain

$$K_\varepsilon(h) = C + \sum_{2}^{\infty} B_r \frac{t^r}{rr!}$$

Since $K_\varepsilon(h) \to 0$ as $t \to 0$ ($t = ch$), the constant C is zero, and therefore,

$$K(h) = K_i(h) - \sum_{2}^{\infty} B_r \frac{t^r}{rr!}$$

This is equivalent to

(A.12.7) $$(\kappa_r)_c = \kappa_r - c^r \frac{B_r}{r}, \qquad r = 2, 3 \ldots$$

In practice, the corrections are usually applied to the sample k-statistics, rather than to the cumulants, since the latter are seldom known except insofar as they are estimated by the former.

A.13 The Non-Central Chi-Square Distribution

The quantity x with probability density,

(A.13.1) $$f(x) = e^{-\lambda} x^{\frac{1}{2}k-1} e^{-x} \sum_{m=0}^{\infty} \frac{(\lambda x)^m}{m!\Gamma(m + k/2)}$$

is said to have the non-central chi-square distribution with k degrees of freedom and with parameter of non-centrality λ. When $\lambda = 0$ it reduces to the ordinary chi-square distribution, the first term of the series being interpreted as $1/\Gamma(k/2)$, and x taken as $\chi^2/2$.

If $X_1, X_2, \ldots X_k$ are independent normal variates, with unit variance and with expectations $\mu_1, \mu_2 \ldots \mu_k$, and if H_0 is the null hypothesis which specifies the values of the μ_i as $\mu_1{}^0, \mu_2{}^0 \ldots \mu_k{}^0$, then an unbiased test (see § 8.10) of the hypothesis H_0 is provided by the rule: reject H_0 if

(A.13.2) $$\sum_{i=1}^{k} (X_i - \mu_i{}^0)^2 > \chi_k{}^2(\alpha)$$

where $\chi_k{}^2(\alpha)$ is the tabular value of central χ^2 with k degrees of freedom, corresponding to the significance level α, i.e.,

(A.13.3) $$\frac{1}{\Gamma(k/2)} \int_{\frac{1}{2}\chi_k{}^2}^{\infty} u^{\frac{1}{2}k-1} e^{-u}\, du = \alpha$$

If H_0 is not true, but instead μ_i is *not* equal to $\mu_i{}^0$ for at least one value of i, (hypothesis H_1) the quantity on the left-hand side of (2) follows the non-central chi-square distribution with k degrees of freedom and parameter

(A.13.4) $$\lambda = \tfrac{1}{2} \sum_{i=1}^{k} (\mu_i - \mu_i{}^0)^2$$

The power function of this test is

(A.13.5) $$\begin{aligned} P(\lambda) &= Pr\{\textstyle\sum (X_i - \mu_i{}^0)^2 > \chi_k{}^2(\alpha) | H_1\} \\ &= e^{-\lambda} \sum_{m=0}^{\infty} \frac{\lambda^m}{m!\Gamma(m + k/2)} \int_{\frac{1}{2}\chi_k{}^2(\alpha)}^{\infty} x^{m+\frac{1}{2}k-1} e^{-x}\, dx \end{aligned}$$

In the tables prepared by Miss Evelyn Fix (reference [5] of Chapter 6) the quantity tabulated is the value of λ for certain assigned values of α and $P(\lambda)$, and for $k = 1(1)20(2)40(5)60(10)100$. (The λ of the tables is twice the λ defined above).

A.14 Some Theorems on Conditional Probability

Let X be a random variable which takes the value x_i with probability $p_i(Y)$, subject to the occurrence of the event Y. That is,

(A.14.1) $$\begin{aligned} p_i(Y) &= P(X = x_i | Y), \qquad i = 1, 2 \ldots n, \\ &= P\{(X = x_i) \cap Y\}/P(Y) \end{aligned}$$

Then the *conditional expectation* of X, given Y, is defined as

(A.14.2) $$E(X|Y) = \sum_{i=1}^{n} p_i(Y)x_i$$

This is the same definition as for the ordinary expectation, except that conditional probabilities are used.

If Y_j is the event that the random variable Y takes the value y_j,

(A.14.3) $$p_i(Y_j) = \frac{P(X = x_i \text{ and } Y = y_j)}{P(Y = y_j)}$$

THEOREM 1 *The expected value of X is equal to the expectation of the conditional expectation of X, given Y.* In symbols,

(A.14.4) $$\begin{aligned} E(X) &= E[E(X|Y)] \\ &= \sum_j p_j E(X|Y_j) \end{aligned}$$

where p_j is the probability that $Y = y_j$. *Proof*:

$$E(X|Y_j) = \sum_i x_i p_i(Y_j)$$

by the definition of conditional expectation. Also, by Eq. (3),

$$p_i(Y_j) = \frac{P(X = x_i, Y = y_j)}{p_j}$$

Therefore, the right-hand side of Eq. (4) is $\sum_i \sum_j x_i P(X = x_i, Y = y_j)$. But the sum over j covers all the possible values of Y, and some value must occur with every X; consequently,

$$\sum_j P(X = x_i, Y = y_j) = P(X = x_i)$$

and the right-hand side of Eq. (4) reduces to

$$\sum_i x_i P(X = x_i) \text{ which is } E(X)$$

The *conditional variance* of X given Y is defined by

(A.14.5) $$V(X|Y) = E\{[X - E(X|Y)]^2|Y\}$$

with a similar definition for conditional covariance.

THEOREM 2 *The variance of X can be regarded as made up of two parts, the expectation of the conditional variance and the variance of the conditional expectation.* Symbolically,

(A.14.6) $$V(X) = E[V(X|Y)] + V[E(X|Y)]$$

Proof: we may write $X - E(X)$ as $X - E(X|Y) + E(X|Y) - E(X)$, so that

(A.14.7) $$\begin{aligned} [X - E(X)]^2 = [X - E(X|Y)]^2 &+ 2[X - E(X|Y)][E(X|Y) - E(X)] \\ &+ [E(X|Y) - E(X)]^2 \end{aligned}$$

The variance of X is the expectation of this expression. By Eq. (4),

$$E[X - E(X|Y)]^2 = E\{E[X - E(X|Y)]^2|Y\}$$
$$= E[V(X|Y)], \qquad \text{by Eq.(5)}$$

Also, since $E(X) = E[E(X|Y)]$, the expectation of the last term of Eq. (7) is $V[E(X|Y)]$. It only remains to show that the expectation of the middle term of Eq. (7) is zero. This middle term is

(A.14.8) $$2\{XE(X|Y) - XE(X) + E(X)\cdot E(X|Y) - [E(X|Y)]^2\}$$

Now $E[XE(X|Y)] = E[E(X|Y)\cdot E(X|Y)]$ by Eq. (4), so that the expectations of the first and last terms of (8) cancel. Similarly,

$$E[XE(X)] = E[E(X|Y)\cdot E(X)]$$

so that the expectations of the two middle terms in (8) cancel. The variance of X is therefore the sum of expectations of the first and last terms of Eq. (7) and this gives Eq. (6).

A.15 **Extrema of a Function of Several Variables Connected by Given Relations**

For the sake of definiteness, let us think of a function of three variables, $f(x, y, z)$, for which we want a maximum or minimum subject to the given relation $\phi(x, y, z) = 0$.

This relation may be regarded as expressing z in terms of x and y. The conditions for an extremum of f are

(A.15.1) $$\begin{cases} \dfrac{\partial f}{\partial x} + \dfrac{\partial f}{\partial z}\cdot\dfrac{\partial z}{\partial x} = 0 \\[2ex] \dfrac{\partial f}{\partial y} + \dfrac{\partial f}{\partial z}\cdot\dfrac{\partial z}{\partial y} = 0 \end{cases}$$

and the partial derivatives $\dfrac{\partial z}{\partial x}, \dfrac{\partial z}{\partial y}$, are connected by the equations

(A.15.2) $$\begin{cases} \dfrac{\partial \phi}{\partial x} + \dfrac{\partial \phi}{\partial z}\cdot\dfrac{\partial z}{\partial x} = 0 \\[2ex] \dfrac{\partial \phi}{\partial y} + \dfrac{\partial \phi}{\partial z}\cdot\dfrac{\partial z}{\partial y} = 0 \end{cases}$$

Eliminating these partial derivatives from Eqs. (1) and (2), we get

(A.15.3) $$\begin{cases} \dfrac{\partial f}{\partial x}\cdot\dfrac{\partial \phi}{\partial z} - \dfrac{\partial f}{\partial z}\dfrac{\partial \phi}{\partial x} = 0 \\[2ex] \dfrac{\partial f}{\partial y}\dfrac{\partial \phi}{\partial z} - \dfrac{\partial f}{\partial z}\dfrac{\partial \phi}{\partial y} = 0 \end{cases}$$

which may be written as Jacobians (see § A.4)

$$(A.15.4) \qquad J\left(\frac{f, \phi}{x, z}\right) = 0, \qquad J\left(\frac{f, \phi}{y, z}\right) = 0$$

These, together with $\phi = 0$, determine the values of x, y and z corresponding to extrema. Equations (3) express the conditions that we can find a function λ to satisfy the three equations

$$(A.15.5) \qquad \begin{cases} \dfrac{\partial f}{\partial x} + \lambda \dfrac{\partial \phi}{\partial x} = 0, & \dfrac{\partial f}{\partial y} + \lambda \dfrac{\partial \phi}{\partial y} = 0, \\ \dfrac{\partial f}{\partial z} + \lambda \dfrac{\partial \phi}{\partial z} = 0 & \end{cases}$$

so that we may replace Eqs. (3) or (4) by the set (5), in which λ is an unknown auxiliary function. This set is given by equating to zero the partial derivatives of the function $f + \lambda\phi$ with respect to the variables x, y, z where λ is regarded for the purpose of this differentiation as a constant. The quantity is called an *undetermined multiplier*, and the method is due originally to Lagrange.

The method may be extended to n variables connected by h relations. The extrema of $f(x_1, x_2 \ldots x_n)$, subject to the conditions $\phi_1 = 0, \phi_2 = 0 \ldots \phi_h = 0$, are found by equating to zero the n partial derivatives of the function,

$$f + \lambda_1\phi_1 + \lambda_2\phi_2 + \ldots + \lambda_h\phi_h$$

the undetermined multipliers $\lambda_1 \ldots \lambda_h$ being regarded as constants. The actual values of these multipliers do not matter.

A.16 The Multinomial Theorem

The multinomial theorem gives the expansion of $(x_1 + x_2 + \ldots + x_k)^n$, where we suppose that n is a positive integer. Each term in the expansion is formed by picking an x_i from the set $x_1, x_2 \ldots x_k$, doing this n times and multiplying the results together. If a particular x_i is picked n_i times, each term in the product is of the form $x_1^{n_1}x_2^{n_2} \ldots x_k^{n_k}$, where $\sum_1^k n_i = n$. The number of terms like this which are identical in value is the number of ways of arranging n_1 objects of one kind (x_1), n_2 of another kind (x_2), and so on, and by Theorem 1.11 this number is $n!/(n_1!n_2! \ldots n_k!)$. We have, therefore,

$$(A.16.1) \qquad (x_1 + x_2 + \ldots + x_k)^n = \sum \frac{n!}{n_1!\,n_2! \ldots n_k!} x_1^{n_1} \ldots x_k^{n_k}$$

where the sum is taken over all sets of values of $n_1, n_2 \ldots n_k$ (all non-negative) such that $\sum n_i = n$.

As an illustration, if $n = 3$ and $k = 3$, the possible sets of values of (n_1, n_2, n_3) are (3, 0, 0), (0, 3, 0), (0, 0, 3), (2, 1, 0), (2, 0, 1), (1, 2, 0), (1, 0, 2), (0, 1, 2), (0, 2, 1) and (1, 1, 1), so that $(x_1 + x_2 + x_3)^3 = x_1^3 + x_2^3 + x_3^3 + 3(x_1^2x_2 + x_1^2x_3 + x_1x_2^2 + x_1x_3^2 + x_2x_3^2 + x_2^2x_3) + 6(x_1x_2x_3)$.

The terms in the sum on the right-hand side of Eq. (1) can be interpreted as the probabilities that a random sample of n objects, drawn from a population which is divided into k classes will have exactly n_1 in the first class, n_2 in the second class and so on. It is assumed that, *in the population*, the probability that an object falls in the ith class is x_i, where of course $x_i \geq 0$ and $\sum_{i=1}^{n} x_i = 1$.

Such a distribution of n objects among k classes is called a *multinomial distribution*. The binomial distribution is the special case $k = 2$.

A.17 The Multinomial Distribution and Chi-Square

The probability of the particular multinomial distribution with f_i objects in the ith class ($i = 1, 2 \ldots k$) is, from (A.16.1),

(A.17.1)
$$p = \frac{N!}{f_1!f_2! \ldots f_k!} \pi_1^{f_1} \pi_2^{f_2} \ldots \pi_k^{f_k}$$

where $\sum f_i = N$ and where π_i is the probability that any object in the population belongs to the ith class. Therefore

(A.17.2)
$$\log p = \log N! - \sum_i \log f_i! + \sum_i f_i \log \pi_i$$

Using Stirling's approximation (Appendix A.2), on the assumption that the f_i are all sufficiently large, we can replace $\log f_i!$ by $(f_i + \frac{1}{2})\log f_i - f_i + \frac{1}{2} \log (2\pi)$, and thus obtain

(A.17.3)
$$\begin{aligned} \log p &= (N + \tfrac{1}{2})\log N - N + \tfrac{1}{2} \log(2\pi) \\ &\quad - \sum (f_i + \tfrac{1}{2})\log f_i + \sum f_i - \frac{k}{2} \log(2\pi) + \sum f_i \log \pi_i \\ &= \tfrac{1}{2} \log N + \sum f_i \log\left(\frac{N\pi_i}{f_i}\right) - \frac{k-1}{2} \log(2\pi) - \tfrac{1}{2} \sum \log f_i \end{aligned}$$

since the term $N \log N$ may be written $\sum f_i \log N$. We therefore have

(A.17.4)
$$\log p = \log C + \sum_i (f_i + \tfrac{1}{2})\log\left(\frac{N\pi_i}{f_i}\right)$$

where

(A.17.5)
$$\log C = -\frac{k-1}{2} \log(2\pi N) - \tfrac{1}{2} \sum_i \log \pi_i$$

If we put $\phi_i = N\pi_i$ (which is the expected number in the ith class in a sample of size N) and let

(A.17.6)
$$z_i = (f_i - \phi_i)/\phi_i^{1/2}$$

we have

$$\text{(A.17.7)} \qquad \log p - \log C = \sum (f_i + \tfrac{1}{2})\log\left(\frac{\phi_i}{f_i}\right)$$
$$= -\sum (f_i + \tfrac{1}{2})\log(1 + \phi_i^{-1/2} z_i)$$
$$= -\sum (\phi_i + \phi_i^{1/2} z_i + \tfrac{1}{2})\log(1 + \phi_i^{-1/2} z_i).$$

Now if the ϕ_i are fairly large, the *differences* between the f_i and the ϕ_i will usually be small compared with the ϕ_i themselves, so that we would expect $\phi_i^{-1/2} z_i$ to be less than 1. If therefore we expand the logarithm in Eq. (7) in a series, namely,

$$\log(1 + \phi_i^{-1/2} z_i) = \phi_i^{-1/2} z_i - \tfrac{1}{2}\phi_i^{-1} z_i^2 + \ldots$$

and multiply by the preceding factor, we obtain

$$\text{(A.17.8)} \qquad \log p - \log C = -\sum_i \left[\tfrac{1}{2} z_i^2 + z_i \phi_i^{1/2} + O(\phi_i^{-1/2})\right].$$

But $\sum_i z_i \phi_i^{1/2} = \sum_i (f_i - \phi_i) = 0$, so that, to order $\phi_i^{-1/2}$,

$$\log p - \log C \approx -\tfrac{1}{2} \sum z_i^2$$

or, equivalently,

$$\text{(A.17.9)} \qquad p = Ce^{-\frac{1}{2}\Sigma z_i^2}, \qquad C = (2\pi N)^{-(k-1)/2} \prod_i \pi_i^{-1/2}$$

This means that the z_i are approximately normally distributed about zero with unit variance. They are not, however, independent, since they are subject to a linear constraint

$$\text{(A.17.10)} \qquad \sum z_i \phi_i^{1/2} = 0$$

We have seen in § 4.6 that the sum of squares of n independent standard normal variates is distributed as χ^2 with n degrees of freedom. If we make an orthogonal transformation (§ A.10) to new variables y_j, where $y_j = \sum_i C_{ji} z_i$, and let $C_{ki} = \phi_i^{1/2}$, the variable y_k will be zero by Eq. (10) and we shall have $k - 1$ *independent* variates, all normally distributed about zero with unit variance. Also $\sum z_i^2 = \sum_1^{k-1} y_j^2$, so that $\sum z_i^2$ is the sum of squares of $k - 1$ independent standard normal variates and is therefore distributed as χ^2 with $k - 1$ d.f.

It may be shown that if the variates are connected by l linear constraints, $\sum z_i^2$ is approximately χ^2 with $k - l$ d.f. This sum, by Eq. (6), may be written

$$\chi_s^2 = \sum_i \frac{(f_i - N\pi_i)^2}{N\pi_i}$$

and this is the ordinary definition of χ_s^2 for a variate grouped into classes.

A.18 **Matrix Algebra**

A set of *mn* elements arranged in a rectangular array of *m* rows and *n* columns is called a matrix of order *m* by *n*. When $m = n$ the matrix is said to be *square*.

The elements may be real or complex numbers, but in statistical applications they are usually real. The whole array is often denoted by a single letter, or by a typical element enclosed in parentheses. Thus

$$A = \begin{bmatrix} a_{11} & a_{12} & \cdots & a_{1n} \\ a_{21} & a_{22} & \cdots & a_{2n} \\ \cdot & \cdot & & \cdot \\ \cdot & \cdot & & \cdot \\ \cdot & \cdot & & \cdot \\ a_{m1} & a_{m2} & \cdots & a_{mn} \end{bmatrix} = (a_{ij})$$

The matrix is thought of as a single mathematical entity, which is subject to algebraic operations. These operations must, of course, be defined.

A matrix with a single row or a single column is called a *vector*.

A matrix is *zero* if and only if all its elements are zero.

Two matrices are *conformable* if they each have the same number of rows and also the same number of columns.

Equality. Two conformable matrices are equal if and only if each element in one is equal to the corresponding element in the other.

Addition. The sum of two conformable matrices A and B is a conformable matrix C such that

(A.18.1) $$c_{ij} = a_{ij} + b_{ij}$$

That is, elements in corresponding positions in the two matrices are simply added. Subtraction is similarly defined. The matrix denoted by $-A$ has all its elements opposite in sign to those of A.

Multiplication of a matrix and a number. If λ is any number (real or complex) the product λA is defined by

(A.18.2) $$\lambda A = (\lambda a_{ij})$$

Each element in A is multiplied by λ.

Multiplication. If A is an m by p matrix and B a p by n matrix, the product AB is an m by n matrix defined by

(A.18.3) $$AB = \left(\sum_{k=1}^{p} a_{ik} b_{kj}\right)$$

For this product to exist it is necessary that the number of rows in B should be equal to the number of columns in A. Each row in A is multiplied, term by term, with each column in B.

EXAMPLE 1

$$\begin{bmatrix} 3 & 1 & 4 \\ 2 & -4 & 5 \end{bmatrix} \cdot \begin{bmatrix} 1 & 5 \\ 8 & 2 \\ 3 & -1 \end{bmatrix} = \begin{bmatrix} 23 & 13 \\ -15 & -3 \end{bmatrix}$$

Here 3(1) + 1(8) + 4(3) = 23, and so for the other elements of the product.

Matrix addition satisfies the ordinary commutative and associative laws of elementary algebra, but this is not true of multiplication. If the matrices in Example 1 above are multiplied in the reverse order we get an entirely different product:

$$\begin{bmatrix} 1 & 5 \\ 8 & 2 \\ 3 & -1 \end{bmatrix} \begin{bmatrix} 3 & 1 & 4 \\ 2 & -4 & 5 \end{bmatrix} = \begin{bmatrix} 13 & -16 & 29 \\ 28 & 0 & 42 \\ 7 & 7 & 7 \end{bmatrix}$$

It is not therefore true in general that matrix multiplication is commutative. It is necessary to distinguish between "pre-multiplication" and "post-multiplication," or to speak of multiplying "on the left" or "on the right." If we multiply B by A on the left, we get AB, and if on the right, BA.

Another law of elementary algebra that does not hold with matrices is the *product law*. This asserts that if $ab = 0$, then either a or b must be zero. But AB can be a zero matrix without either A or B being zero.

EXAMPLE 2

$$\begin{bmatrix} 2 & -1 \\ 10 & -5 \end{bmatrix} \cdot \begin{bmatrix} 1 & 3 \\ 2 & 6 \end{bmatrix} = \begin{bmatrix} 0 & 0 \\ 0 & 0 \end{bmatrix}$$

On the other hand, the associative and distributive laws of algebra hold for matrices, provided of course that the matrices are properly conformable for the operations suggested. With this understanding we can write

(A.18.4)
$$(AB)C = A(BC) = ABC$$
$$A(B + C) = AB + AC$$
$$(A + B)C = AC + BC$$

A.19 Transposition

If the successive columns of a matrix A are written as successive rows of a new matrix A', then A' is called the *transpose* of A. If A is an m by n matrix, A' is an n by m matrix, and $a'_{ij} = a_{ji}$.

EXAMPLE 3

$$A = \begin{bmatrix} 3 & 6 & 2 \\ 2 & 1 & 0 \\ 5 & 9 & 7 \\ 1 & 0 & 6 \end{bmatrix}, \quad A' = \begin{bmatrix} 3 & 2 & 5 & 1 \\ 6 & 1 & 9 & 0 \\ 2 & 0 & 7 & 6 \end{bmatrix}$$

The transpose of a row vector is a column vector, and conversely.

A square matrix is *symmetric* if $A' = A$ and *skew-symmetric* if $A' = -A$. Thus $\begin{bmatrix} 2 & 3 \\ 3 & 2 \end{bmatrix}$ is symmetric and $\begin{bmatrix} 0 & 1 \\ -1 & 0 \end{bmatrix}$ is skew-symmetric. In any skew-symmetric matrix all the elements along the principal diagonal must be zero.

The following theorem is sometimes referred to as the *reversal rule*:

(A.19.1) $$(AB)' = B'A'$$

Proof: If c'_{ij} is the element in the ith row and j^{th} column of $(AB)'$,

$$\begin{aligned} c'_{ij} = c_{ji} &= \sum_k a_{jk} b_{ki} \\ &= \sum_k a'_{kj} b'_{ik} = \sum_k b'_{ik} a'_{kj} \end{aligned}$$

which is the $(i, j)^{\text{th}}$ element of $B'A'$. Similarly, $(ABC)' = C'B'A'$, etc.

A square matrix in which all the elements except those in the principal diagonal are zero is called a *diagonal matrix*. A diagonal matrix is symmetric, and commutes with any other diagonal matrix having the same number of rows.

EXAMPLE 4

$$\begin{bmatrix} 3 & 0 & 0 \\ 0 & 1 & 0 \\ 0 & 0 & 2 \end{bmatrix} \cdot \begin{bmatrix} -1 & 0 & 0 \\ 0 & -4 & 0 \\ 0 & 0 & 3 \end{bmatrix} = \begin{bmatrix} -3 & 0 & 0 \\ 0 & -4 & 0 \\ 0 & 0 & 6 \end{bmatrix}$$

$$= \begin{bmatrix} -1 & 0 & 0 \\ 0 & -4 & 0 \\ 0 & 0 & 3 \end{bmatrix} \cdot \begin{bmatrix} 3 & 0 & 0 \\ 0 & 1 & 0 \\ 0 & 0 & 2 \end{bmatrix}$$

Multiplication of a matrix by a number λ is equivalent to multiplying (either on the left or on the right) by a diagonal matrix with each non-zero element equal to λ.

EXAMPLE 5

$$\begin{bmatrix} \lambda & 0 & 0 \\ 0 & \lambda & 0 \\ 0 & 0 & \lambda \end{bmatrix} \cdot \begin{bmatrix} 2 & 1 & 7 \\ 9 & 3 & -2 \\ 1 & 4 & -5 \end{bmatrix} = \begin{bmatrix} 2\lambda & \lambda & 7\lambda \\ 9\lambda & 3\lambda & -2\lambda \\ \lambda & 4\lambda & -5\lambda \end{bmatrix} = \lambda \begin{bmatrix} 2 & 1 & 7 \\ 9 & 3 & -2 \\ 1 & 4 & -5 \end{bmatrix}$$

A diagonal matrix with each diagonal element 1 is called a *unit matrix*. Multiplication by a unit matrix, on the right or on the left, leaves any other matrix unchanged.

(A.19.2) $$AI = IA = A$$

where I is a unit matrix with the proper number of rows and columns.

A.20 The Determinant of a Matrix

If A is a square $n \times n$ matrix, the determinant of A, denoted by $d(A)$, is a polynomial of the nth degree in the elements of A. The terms of the polynomial are obtained by multiplying together all possible sets of n elements taken one from each row and one from each column, and giving the products alternate plus and minus signs. It is assumed that the reader is familiar with the elementary properties of determinants as found in most text-books of college algebra, but we recall briefly a few of these properties.

The usual notation for a determinant is

(A.20.1)
$$d(A) = \begin{vmatrix} a_{11} & \cdots & a_{1n} \\ \cdot & & \cdot \\ \cdot & & \cdot \\ \cdot & & \cdot \\ a_{n1} & \cdots & a_{nn} \end{vmatrix} = |a_{ij}|$$

The determinant of $n - 1$ rows obtained by omitting the ith row and j^{th} column of $d(A)$ is called the *minor* of a_{ij} and will be denoted by $d(A_{ij})$. The signed minor

(A.20.2)
$$C_{ij} = (-1)^{i+j}\, d(A_{ij})$$

is called the *cofactor* of a_{ij}. The formulas for the development of $d(A)$, in terms of the ith row and in terms of the jth column, are

(A.20.3)
$$d(A) = \sum_j a_{ij}C_{ij} = \sum_i a_{ij}C_{ij}$$

If in these formulas we replace the cofactors of the ith row (or the jth column) by the cofactors of a *different* row or column (what Aitken has called *alien cofactors*) the expressions reduce to zero. That is,

(A.20.4)
$$\begin{cases} \sum_j a_{ij}C_{kj} = 0, & i \neq k \\ \sum_i a_{ij}C_{ik} = 0, & j \neq k \end{cases}$$

A convenient symbol for expressing such pairs of relations as Eqs. (3) and (4) is the *Kronecker delta*, δ_{jk}, defined as equal to 1 when $j = k$, and equal to 0 when $j \neq k$. Equations (3) and (4), with this notation, may be written:

(A.20.5)
$$\begin{cases} \sum_j a_{ij}C_{kj} = \delta_{ik}\, d(A) \\ \sum_i a_{ij}C_{ik} = \delta_{jk}\, d(A) \end{cases}$$

Even if a matrix is not square, we can form determinants by crossing out rows and/or columns to leave square arrays. The determinants of these arrays are all determinants of the matrix. The *rank* of a matrix is the order of the largest non-zero determinant that can be formed in this way. A square n by n matrix is called *singular* if the determinant of the whole matrix is zero. Its rank is then of course less than n.

If A and B are square matrices of the same size, and if $C = AB$, then

$$d(C) = d(A) \cdot d(B)$$

Although the matrices AB and BA are in general different they have the same determinant.

A.21 The Inverse of a Matrix

If an n by n matrix A is non-singular, there exists a unique n by n matrix denoted by A^{-1}, such that

(A.21.1) $$AA^{-1} = A^{-1}A = I$$

This matrix A^{-1} is called the *inverse* of A.

The transpose of the matrix of cofactors of the elements of A is called the *adjoint* of A, denoted by adj A. That is

(A.21.2) $$\text{adj } A = (C_{ij})' = (C_{ji})$$

It follows that

(A.21.3) $$A \cdot \text{adj } A = \left(\sum_k a_{ik} C_{jk} \right) = (\delta_{ij}\, d(A))$$

But since $d(A)$ is a number and δ_{ij} is the $(i, j)^{\text{th}}$ element of the unit matrix I, we can express the matrix on the right of Eq. (3) as $d(A)I$. We have, then,

(A.21.4) $$A \cdot \frac{\text{adj } A}{d(A)} = I$$

so that the inverse of A is the adjoint of A divided by its determinant (supposed non-zero). We can therefore define, for non-singular square matrices, the operations of "pre-division" and "post-division," symbolised by $A^{-1}B$ and BA^{-1}.

The *reversal rule* applies to inversion, namely,

(A.21.5) $$(AB)^{-1} = B^{-1}A^{-1}$$

To prove this, we note that

$$\begin{aligned}(AB)(AB)^{-1} = I = AA^{-1} &= A(BB^{-1})A^{-1} \\ &= (AB)(B^{-1}A^{-1})\end{aligned}$$

where we have used the associative law for matrix multiplication.

The operations of transposition and inversion are commutative. That is,

(A.21.6) $$(A^{-1})' = (A')^{-1}$$

Proof: $A'(A^{-1})' = (A^{-1}A)' = I' = I$, which shows that $(A^{-1})'$ is the inverse of A'.

The elements of A^{-1} are conveniently written as a^{ij}, using superscripts instead of subscripts. From Eqs. (2) and (4) it follows that

(A.21.7) $$a^{ij} = \frac{C_{ji}}{d(A)}$$

Given the set of normal equations in § 12.1, namely, $Ab = g$, we can solve for b by multiplying both sides on the left by A^{-1}. This gives $b = A^{-1}g$, or, using the notation of Eq. (7),

(A.21.8) $$b_i = \sum_j a^{ij} g_j = \left(\sum_j C_{ji} g_j\right) \Big/ d(A)$$

Since $\sum_j C_{ji} g_j$ is the expanded form of $d(A)$ in which the ith column has been replaced by a column of g's, this equation is a statement of *Cramer's rule* (§ 12.3).

A.22 **Orthogonal Matrices**

A non-singular square matrix A is orthogonal if its transpose is equal to its inverse, that is, if

(A.22.1) $$AA' = I$$

The matrix of the coefficients of an orthogonal transformation (§ A.10) is orthogonal. Thus the transformation expressed by Eqs. (A.10.1) and (A.10.2) can be written in matrix form

(A.22.2) $$Y = CX, \qquad CC' = I$$

where Y and X are column vectors. An example is the transformation

$$\begin{cases} Y_1 = X_1 \cos\theta - X_2 \sin\theta \\ Y_2 = X_1 \sin\theta + X_2 \cos\theta \end{cases}$$

which corresponds to a clockwise rotation of the axes about the origin through an angle θ. The matrix $\begin{bmatrix} \cos\theta & -\sin\theta \\ \sin\theta & \cos\theta \end{bmatrix}$ is orthogonal.

A.23 **Calculation of the Inverse of a Matrix**

The solution of a set of normal equations (§ 12.1) is immediately expressible in terms of the inverse matrix of the coefficients. Moreover, some of the elements of the inverse matrix are required for testing the significance of the partial regression coefficients b_i. It is therefore sometimes worth while to invert a matrix. In practice the matrix is usually symmetric.

With a square matrix of three rows and columns, it is a simple matter to compute the cofactors and the whole determinant, and thus obtain the inverted matrix directly from its definition (§ A.21). However, for larger matrices a more compact and systematic method is desirable. The following method is known as *Jordan's*.

If we can find a matrix J such that

(A.23.1) $$J(A, I) = (I, J)$$

then J is A^{-1}. Here I is a unit matrix of the same number of rows as A, placed alongside A. The method consists in multiplying the augmented matrix (A, I) by successive matrices of the form $I + J_i$, where J_i differs from a zero matrix only in the ith column. By suitably choosing the non-zero elements of J_i we can build up a unit matrix I on the left of the product matrix and the rest of the product matrix is then the required inverse, J. For a 4×4 matrix we should have

$$\begin{aligned}(I + J_1)(A, I) \quad &= (A_1, K_1)\\ (I + J_2)(A_1, K_1) &= (A_2, K_2)\\ (I + J_3)(A_2, K_2) &= (A_3, K_3)\\ (I + J_4)(A_3, K_3) &= (I, A^{-1})\end{aligned}$$

The first column of $I + J_1$ is chosen so that the first column of A_1 has elements (read downwards) 1, 0, 0, 0, and so need not be recorded. This first column of $I + J_1$ becomes the first column of K_1 and is recorded there. The second column of $I + J_2$ is chosen so that the second column of A_2 becomes 0, 1, 0, 0, and the second column of K_2 becomes this second column of $I + J_2$, and so on. In this way the unit matrix I is built up in four steps, but need not be recorded.

If the elements of the first column of A (read downwards) are a_1, a_2, a_3, a_4, those of the first column of $I + J_1$ are $1/a_1$, $-a_2/a_1$, $-a_3/a_1$, $-a_4/a_1$. If the elements of the *second* column of A_1 are a'_1, a'_2, a'_3, a'_4, then those of the second column of $I + J_2$ are $-a'_1/a'_2$, $1/a'_2$, $-a'_3/a'_2$, $-a'_4/a'_2$. The other products are formed similarly.

Example 6 The steps in the calculation of the inverse of a 4×4 matrix are given below. At each step the matrix $I + J_i$ has been shown in full for the sake of clearness, but it is not actually necessary to set it down. Unit columns in the augmented matrix have, however, been omitted. The pivotal elements in the key columns of A, A_1, A_2, and A_3 are marked. Thus $a_1 = 1.0$, $a'_2 = 0.84$, $a''_3 = 0.7381$, $a'''_4 = 0.5903$. These are the elements whose reciprocals are used in forming the corresponding columns of $I + J_i$.

$$\overbrace{\begin{bmatrix} 1.0 & 0 & 0 & 0 \\ -0.4 & 1 & 0 & 0 \\ -0.5 & 0 & 1 & 0 \\ -0.6 & 0 & 0 & 1 \end{bmatrix}}^{I+J_i} \left[\begin{array}{cccc|cccc} \underline{1.0} & 0.4 & 0.5 & 0.6 & . & . & . & . \\ 0.4 & 1.0 & 0.3 & 0.4 & . & . & . & . \\ 0.5 & 0.3 & 1.0 & 0.2 & . & . & . & . \\ 0.6 & 0.4 & 0.2 & 1.0 & . & . & . & . \end{array}\right]$$

(upper braces: A over the first four columns, I over the last four)

$$\begin{bmatrix} 1 & -0.4762 & 0 & 0 \\ 0 & 1.1905 & 0 & 0 \\ 0 & -0.1190 & 1 & 0 \\ 0 & -0.1905 & 0 & 1 \end{bmatrix} \left[\begin{array}{cccccccc} . & 0.40 & 0.50 & 0.60 & 1.00 & . & . & . \\ . & \underline{0.84} & 0.10 & 0.16 & -0.40 & . & . & . \\ . & 0.10 & 0.75 & -0.10 & -0.50 & . & . & . \\ . & 0.16 & -0.10 & 6.64 & -0.60 & . & . & . \end{array}\right]$$

$$\begin{bmatrix} 1 & 0 & -0.6129 & 0 \\ 0 & 1 & -0.1612 & 0 \\ 0 & 0 & 1.3548 & 0 \\ 0 & 0 & 0.1612 & 1 \end{bmatrix} \left[\begin{array}{cccccccc} . & . & 0.4524 & 0.5238 & 1.1905 & -0.4762 & . & . \\ . & . & 0.1190 & 0.1905 & -0.4762 & 1.1905 & . & . \\ . & . & \underline{0.7381} & -0.1190 & -0.4524 & -0.1190 & . & . \\ . & . & -0.1190 & 0.6095 & -0.5238 & -0.1905 & . & . \end{array}\right]$$

$$\begin{bmatrix} 1 & 0 & 0 & -1.0108 \\ 0 & 1 & 0 & -0.3552 \\ 0 & 0 & 1 & 0.2731 \\ 0 & 0 & 0 & 1.6940 \end{bmatrix} \left[\begin{array}{cccccccc} . & . & . & 0.5967 & 1.4678 & -0.4033 & -0.6129 & . \\ . & . & . & 0.2097 & -0.4033 & 1.2097 & -0.1612 & . \\ . & . & . & -0.1612 & -0.6129 & -0.1612 & 1.3548 & . \\ . & . & . & \underline{0.5903} & -0.5967 & -0.2097 & 0.1612 & . \end{array}\right]$$

$$\left[\begin{array}{cccc|cccc} . & . & . & . & 2.0709 & -0.1913 & -0.7758 & -1.0108 \\ . & . & . & . & -0.1913 & 1.2842 & -0.2185 & -0.3552 \\ . & . & . & . & -0.7758 & -0.2185 & 1.3988 & 0.2731 \\ . & . & . & . & -1.0108 & -0.3552 & 0.2731 & 1.6940 \end{array}\right]$$

(lower braces: I under the first four columns, A^{-1} under the last four)

Since the original matrix A was symmetric, the inverse A^{-1} is also symmetric. It is therefore unnecessary to compute separately the terms of A^{-1} below the main diagonal, except as checks on the arithmetic.

A.24 Solution of a Set of Normal Equations by the Square Root Method

The given equation, $Ab = g$, can be solved if we can find a triangular matrix S (that is, a square matrix with all the elements below, or above, the principal diagonal equal to zero) such that

$$S'S = A \tag{A.24.1}$$

If so, we have

$$S'(Sb) = g$$

which is equivalent to the two matrix equations

$$Sb = k, \qquad S'k = g \tag{A.24.2}$$

Since S and S' are triangular, these equations are comparatively easy to solve.

The first step is to find the elements of S by solving the equations, equivalent to Eq. (1),

$$\left\{\begin{aligned} S_{11}^2 &= a_{11} \\ S_{11}S_{12} &= a_{12} \\ &\cdots\cdots\cdots\cdots \\ S_{12}^2 + S_{22}^2 &= a_{22} \\ S_{12}S_{13} + S_{22}S_{23} &= a_{23} \\ &\cdots\cdots\cdots\cdots\cdots\cdots\cdots\cdots \\ S_{1p}^2 + S_{2p}^2 + \ldots + S_{pp}^2 &= a_{pp} \end{aligned}\right. \tag{A.24.3}$$

These equations are solved in order, beginning with the first.

The second step is to find the elements of k by solving $S'k = g$, which, when written out, is:

$$\text{(A.24.4)} \quad \begin{cases} S_{11}k_1 = g_1 \\ S_{12}k_1 + S_{22}k_2 = g_2 \\ \cdots\cdots\cdots\cdots \\ S_{1p}k_1 + S_{2p}k_2 + \ldots + S_{pp}k_p = g_p \end{cases}$$

The final step is to solve the set equivalent to $Sb = k$, namely,

$$\text{(A.24.5)} \quad \begin{cases} S_{11}b_1 + S_{12}b_2 + \ldots + S_{1p}b_p = k_1 \\ S_{22}b_2 + \ldots + S_{2p}b_p = k_2 \\ \cdots\cdots\cdots\cdots \\ S_{p-1,p-1}b_{p-1} + S_{p-1,p}b_p = k_{p-1} \\ S_{pp}b_p = k_p \end{cases}$$

These are solved backwards, beginning with the last equation, obtaining first b_p, then b_{p-1}, and so on. The process may be illustrated by the following example:

$$58b_1 + 23b_2 + 43b_3 = 160$$

$$23b_1 + 78b_2 + 168b_3 = 1910$$

$$43b_1 + 168b_2 + 1096b_3 = 240$$

The various steps, including a check column, may be set out in a table:

TABLE A.24

				g	$\bar{g}$
A	58	23	43	160	284
	23	78	168	1910	2179
	43	168	1096	240	1547
				k	$\bar{k}$
S	7.616	3.020	5.646	21.008	37.290
		8.299	18.189	222.503	248.991
			27.079	−144.972	−117.893
b'	−8.557	38.545	−5.354		
$\bar{b}'$	−7.557	39.545	−4.354		

Step 1 $S_{11} = (58)^{1/2} = 7.616$

$$S_{12} = \frac{23}{7.616} = 3.020$$

$$S_{13} = \frac{43}{7.616} = 5.646$$

$$S_{22} = [78 - (3.020)^2]^{1/2} = (68.88)^{1/2} = 8.299$$

$$S_{23} = \frac{168 - (3.020)(5.646)}{8.299} = 18.189$$

$$S_{33} = [1096 - (5.646)^2 - (18.189)^2]^{1/2} = 27.079$$

Step 2 $k_1 = \dfrac{160}{7.616} = 21.008$

$$k_2 = \frac{1910 - (3.020)(21.008)}{8.299} = 222.503$$

$$k_3 = \frac{240 - (5.646)(21.008) - (18.189)(222.503)}{27.079} = -144.972$$

Step 3 $b_3 = \dfrac{-144.972}{27.079} = -5.354$

$$b_2 = \frac{222.503 - 18.189(-5.354)}{8.299} = 38.545$$

$$b_1 = \frac{21.008 - (3.020)(38.545) - (5.646)(-5.354)}{7.616} = -8.557$$

The check consists in computing a vector $\bar{g}$ whose elements are the sums of the corresponding rows in A and g. Thus

$$\bar{g}_1 = 58 + 23 + 43 + 160 = 284$$

Steps 2 and 3 are repeated with $\bar{g}$ instead of g, giving new vectors $\bar{k}$ and $\bar{b}$. Apart from rounding-off errors in the last decimal place, we should find

$$\bar{k}_j = S_{jj} + S_{j,j+1} + \ldots + S_{jp} + k_j$$

and

$$\bar{b}_j = b_j + 1.$$

In practice, each row of S in Table A.24 is completed, and the check applied, before the next row is started.

Appendix B — TABLES

TABLE B.1 RANDOM SAMPLING NUMBERS*

First Thousand

	1–4	5–8	9–12	13–16	17–20	21–24	25–28	29–32	33–36	37–40
1.	23 15	75 48	59 01	83 72	59 93	76 24	97 08	86 95	23 03	67 44
2	05 54	55 50	43 10	53 74	35 08	90 61	18 37	44 10	96 22	13 43
3	14 87	16 03	50 32	40 43	62 23	50 05	10 03	22 11	54 38	08 34
4	38 97	67 49	51 94	05 17	58 53	78 80	59 01	94 32	42 87	16 95
5	97 31	26 17	18 99	75 53	08 70	94 25	12 58	41 54	88 21	05 13
6	11 74	26 93	81 44	33 93	08 72	32 79	73 31	18 22	64 70	68 50
7	43 36	12 88	59 11	01 64	56 23	93 00	90 04	99 43	64 07	40 36
8	93 80	62 04	78 38	26 80	44 91	55 75	11 89	32 58	47 55	25 71
9	49 54	01 31	81 08	42 98	41 87	69 53	82 96	61 77	73 80	95 27
10	36 76	87 26	33 37	94 82	15 69	41 95	96 86	70 45	27 48	38 80
11	07 09	25 23	92 24	62 71	26 07	06 55	84 53	44 67	33 84	53 20
12	43 31	00 10	81 44	86 38	03 07	52 55	51 61	48 89	74 29	46 47
13	61 57	00 63	60 06	17 36	37 75	63 14	89 51	23 35	01 74	69 93
14	31 35	28 37	99 10	77 91	89 41	31 57	97 64	48 62	58 48	69 19
15	57 04	88 65	26 27	79 59	36 82	90 52	95 65	46 35	06 53	22 54
16	09 24	34 42	00 68	72 10	71 37	30 72	97 57	56 09	29 82	76 50
17	97 95	53 50	18 40	89 48	83 29	52 23	08 25	21 22	53 26	15 87
18	93 73	25 95	70 43	78 19	88 85	56 67	16 68	26 95	99 64	45 69
19	72 62	11 12	25 00	92 26	82 64	35 66	65 94	34 71	68 75	18 67
20	61 02	07 44	18 45	37 12	07 94	95 91	73 78	66 99	53 61	93 78
21	97 83	98 54	74 33	05 59	17 18	45 47	35 41	44 22	03 42	30 00
22	89 16	09 71	92 22	23 29	06 37	35 05	54 54	89 88	43 81	63 61
23	25 96	68 82	20 62	87 17	92 65	02 82	35 28	62 84	91 95	48 83
24	81 44	33 17	19 05	04 95	48 06	74 69	00 75	67 65	01 71	65 45
25	11 32	25 49	31 42	36 23	43 86	08 62	49 76	67 42	24 52	32 45

Second Thousand

	1–4	5–8	9–12	13–16	17–20	21–24	25–28	29–32	33–36	37–40
1	64 75	58 38	85 84	12 22	59 20	17 69	61 56	55 95	04 59	59 47
2	10 30	25 22	89 77	43 63	44 30	38 11	24 90	67 07	34 82	33 28
3	71 01	79 84	95 51	30 85	03 74	66 59	10 28	87 53	76 56	91 49
4	60 01	25 56	05 88	41 03	48 79	79 65	59 01	69 78	80 00	36 66
5	37 33	09 46	56 49	16 14	28 02	48 27	45 47	55 44	55 36	50 90
6	47 86	98 70	01 31	59 11	22 73	60 62	61 28	22 34	69 16	12 12
7	38 04	04 27	37 64	16 78	95 78	39 32	34 93	24 88	43 43	87 06
8	73 50	83 09	08 83	05 48	00 78	36 66	93 02	95 56	46 04	53 36
9	32 62	34 64	74 84	06 10	43 24	20 62	83 73	19 32	35 64	39 69
10	97 59	19 95	49 36	63 03	51 06	62 06	99 29	75 95	32 05	77 34
11	74 01	23 19	55 59	79 09	69 82	66 22	42 40	15 96	74 90	75 89
12	56 75	42 64	57 13	35 10	50 14	90 96	63 36	74 69	09 63	34 88
13	49 80	04 99	08 54	83 12	19 98	08 52	82 63	72 92	92 36	50 26
14	43 58	48 96	47 24	87 85	66 70	00 22	15 01	93 99	59 16	23 77
15	16 65	37 96	64 60	32 57	13 01	35 74	28 36	36 73	05 88	72 29
16	48 50	26 90	55 65	32 25	87 48	31 44	68 02	37 31	25 29	63 67
17	96 76	55 46	92 36	31 68	62 30	48 29	63 83	52 23	81 66	40 94
18	38 92	36 15	50 80	35 78	17 84	23 44	41 24	63 33	99 22	81 28
19	77 95	88 16	94 25	22 50	55 87	51 07	30 10	70 60	21 86	19 61
20	17 92	82 80	65 25	58 60	87 71	02 64	18 50	64 65	79 64	81 70
21	94 03	68 59	78 02	31 80	44 99	41 05	41 05	31 87	43 12	15 96
22	47 46	06 04	79 56	23 04	84 17	14 37	28 51	67 27	55 80	03 68
23	47 85	65 60	88 51	99 28	24 39	40 64	41 71	70 13	46 31	82 88
24	57 61	63 46	53 92	29 86	20 18	10 37	57 65	15 62	98 69	07 56
25	08 30	09 27	04 66	75 26	66 10	57 18	87 91	07 54	22 22	20 13

*Reproduced with the permission of Professor E. S. Pearson from M. G. Kendall and B. Babington Smith, *Tables of Random Sampling Numbers* (Tracts for Computers, No. 24), Cambridge Univ. Press.

TABLE B.1 *(cont.)*

RANDOM SAMPLING NUMBERS

Third Thousand

	1–4	5–8	9–12	13–16	17–20	21–24	25–28	29–32	33–36	37–40
1	89 22	10 23	62 65	78 77	47 33	51 27	23 02	13 92	44 13	96 51
2	04 00	59 98	18 63	91 82	90 32	94 01	24 23	63 01	26 11	06 50
3	98 54	63 80	66 50	85 67	50 45	40 64	52 28	41 53	25 44	41 25
4	41 71	98 44	01 59	22 60	13 14	54 58	14 03	98 49	98 86	55 79
5	28 73	37 24	89 00	78 52	58 43	24 61	34 97	97 85	56 78	44 71
6	65 21	38 39	27 77	76 20	30 86	80 74	22 43	95 68	47 68	37 92
7	65 55	31 26	78 90	90 69	04 66	43 67	02 62	17 69	90 03	12 05
8	05 66	86 90	80 73	02 98	57 46	58 33	27 82	31 45	98 69	29 98
9	39 30	29 97	18 49	75 77	95 19	27 38	77 63	73 47	26 29	16 12
10	64 59	23 22	54 45	87 92	94 31	38 32	00 59	81 18	06 78	71 37
11	07 51	34 87	92 47	31 48	36 60	68 90	70 53	36 82	57 99	15 82
12	86 59	36 85	01 56	63 89	98 00	82 83	93 51	48 56	54 10	72 32
13	83 73	52 25	99 97	97 78	12 48	36 83	89 95	60 32	41 06	76 14
14	08 59	52 18	26 54	65 50	82 04	87 99	01 70	33 56	25 80	53 84
15	41 27	32 71	49 44	29 36	94 58	16 82	86 39	62 15	86 43	54 31
16	00 47	37 59	08 56	23 81	22 42	72 63	17 63	14 47	25 20	63 47
17	86 13	15 37	89 81	38 30	78 68	89 13	29 61	82 07	00 98	64 32
18	33 84	97 83	59 04	40 20	35 86	03 17	68 86	63 08	01 82	25 46
19	61 87	04 16	57 07	46 80	86 12	98 08	39 73	49 20	77 54	50 91
20	43 89	86 59	23 25	07 88	61 29	78 49	19 76	53 91	50 08	07 86
21	29 93	93 91	23 04	54 84	59 85	60 95	20 66	41 28	72 64	64 73
22	38 50	58 55	55 14	38 85	50 77	18 65	79 48	87 67	83 17	08 19
23	31 82	43 84	31 67	12 52	55 11	72 04	41 15	62 53	27 98	22 68
24	91 43	00 37	67 13	56 11	55 97	06 75	09 25	52 02	39 13	87 53
25	38 63	56 89	76 25	49 89	75 26	96 45	80 38	05 04	11 66	35 14

Fourth Thousand

	1–4	5–8	9–12	13–16	17–20	21–24	25–28	29–32	33–36	37–40
1	02 49	05 41	22 27	94 43	93 64	04 23	07 20	74 11	67 95	40 82
2	11 96	73 64	69 60	62 78	37 01	09 25	33 02	08 01	38 53	74 82
3	48 25	68 34	65 49	69 92	40 79	05 40	33 51	54 39	61 30	31 36
4	27 24	67 30	80 21	48 12	35 36	04 88	18 99	77 49	48 49	30 71
5	32 53	27 72	65 72	43 07	07 22	86 52	91 84	57 92	65 71	00 11
6	66 75	79 89	55 92	37 59	34 31	43 20	45 58	25 45	44 36	92 65
7	11 26	63 45	45 76	50 59	77 46	34 66	82 69	99 26	74 29	75 16
8	17 87	23 91	42 45	56 18	01 46	93 13	74 89	24 64	25 75	92 84
9	62 56	13 03	65 03	40 81	47 54	51 79	80 81	33 61	01 09	77 30
10	62 79	63 07	79 35	49 77	05 01	30 10	50 81	33 00	99 79	19 70
11	75 51	02 17	71 04	33 93	36 60	42 75	76 22	23 87	56 54	84 68
12	87 43	90 16	91 63	51 72	65 90	44 43	70 72	17 98	70 63	90 32
13	97 74	20 26	21 10	74 87	88 03	38 33	76 52	26 92	14 95	90 51
14	98 81	10 60	01 21	57 10	28 75	21 82	88 39	12 85	18 86	16 24
15	51 26	40 18	52 64	60 79	25 53	29 00	42 66	95 78	58 36	29 98
16	40 23	99 33	76 10	41 96	86 10	49 12	00 29	41 80	03 59	93 17
17	26 93	65 91	86 51	66 72	76 45	46 32	94 46	81 94	19 06	66 47
18	88 50	21 17	16 98	29 94	09 74	42 39	46 22	00 69	09 48	16 46
19	63 49	93 80	93 25	59 36	19 95	79 86	78 05	69 01	02 33	83 74
20	36 37	98 12	06 03	31 77	87 10	73 82	83 10	83 60	50 94	40 91
21	93 80	12 23	22 47	47 95	70 17	59 33	43 06	47 43	06 12	66 60
22	29 85	68 71	20 56	31 15	00 53	25 36	58 12	65 22	41 40	24 31
23	97 72	08 79	31 88	26 51	30 50	71 01	71 51	77 06	95 79	29 19
24	85 23	70 91	05 74	60 14	63 77	59 93	81 56	47 34	17 79	27 53
25	75 74	67 52	68 31	72 79	57 73	72 36	48 73	24 36	87 90	68 02

TABLE B.1 *(cont.)*

RANDOM SAMPLING NUMBERS

Fifth Thousand

	1–4	*5–8*	*9–12*	*13–16*	*17–20*	*21–24*	*25–28*	*29–32*	*33–36*	*37–40*
1	29 93	50 69	71 63	17 55	25 79	10 47	88 93	79 61	42 82	13 63
2	15 11	40 71	26 51	89 07	77 87	75 51	01 31	03 42	94 24	81 11
3	03 87	04 32	25 10	58 98	76 29	22 03	99 41	24 38	12 76	50 22
4	79 39	03 91	88 40	75 64	52 69	65 95	92 06	40 14	28 42	29 60
5	30 03	50 69	15 79	19 65	44 28	64 81	95 23	14 48	72 18	15 94
6	29 03	99 98	61 28	75 97	98 02	68 53	13 91	98 38	13 72	43 73
7	78 19	60 81	08 24	10 74	97 77	09 59	94 35	69 84	82 09	49 56
8	15 84	78 54	93 91	44 29	13 51	80 13	07 37	52 21	53 91	09 86
9	36 61	46 22	48 49	19 49	72 09	92 58	79 20	53 41	02 18	00 64
10	40 54	95 48	84 91	46 54	38 62	35 54	14 44	66 88	89 47	41 80
11	40 87	80 89	97 14	28 60	99 82	90 30	87 80	07 51	58 71	66 58
12	10 22	94 92	82 41	17 33	14 68	59 45	51 87	56 08	90 80	66 60
13	15 91	87 67	87 30	62 42	59 28	44 12	42 50	88 31	13 77	16 14
14	13 40	31 87	96 49	90 99	44 04	64 97	94 14	62 18	15 59	83 35
15	66 52	39 45	96 74	90 89	02 71	10 00	99 86	48 17	64 06	89 09
16	91 66	53 64	69 68	34 31	78 70	25 97	50 46	62 21	27 25	06 20
17	67 41	58 75	15 08	20 77	37 29	73 20	15 75	93 96	91 76	96 99
18	76 52	79 69	96 23	72 43	34 48	63 39	23 23	94 60	88 79	06 17
19	19 81	54 77	89 74	34 81	71 47	10 95	43 43	55 81	19 45	44 07
20	25 59	25 35	87 76	38 47	25 75	84 34	76 89	18 05	73 95	72 22
21	55 90	24 55	39 63	64 63	16 09	95 99	98 28	87 40	66 66	66 92
22	02 47	05 83	76 79	79 42	24 82	42 42	39 61	62 47	49 11	72 64
23	18 63	05 32	63 13	31 99	76 19	35 85	91 23	50 14	63 28	86 59
24	89 67	33 82	30 16	06 39	20 07	59 50	33 84	02 76	45 03	33 33
25	62 98	66 73	64 06	59 51	74 27	84 62	31 45	65 82	86 05	73 00

Sixth Thousand

	1–4	*5–8*	*9–12*	*13–16*	*17–20*	*21–24*	*25–28*	*29–32*	*33–36*	*37–40*
1	27 50	13 05	46 34	63 85	87 60	35 55	05 67	88 15	47 00	50 92
2	02 31	57 57	62 98	41 09	66 01	69 88	92 83	35 70	76 59	02 58
3	37 43	12 83	66 39	77 33	63 26	53 99	48 65	23 06	94 29	53 04
4	83 56	65 54	19 33	35 42	92 12	37 14	70 75	18 58	98 57	12 52
5	06 81	56 27	49 32	12 42	92 42	05 96	82 94	70 25	45 49	18 16
6	39 15	03 60	15 56	73 16	48 74	50 27	43 42	58 36	73 16	39 90
7	84 45	71 93	10 27	15 83	84 20	57 42	41 28	42 06	15 90	70 47
8	82 47	05 77	06 89	47 13	92 85	60 12	32 89	25 22	42 38	87 37
9	98 04	06 70	24 21	69 02	65 42	55 33	11 95	72 35	73 23	57 26
10	18 33	49 04	14 33	48 50	15 64	58 26	14 91	46 02	72 13	48 62
11	33 92	19 93	38 27	43 40	27 72	79 74	86 57	41 83	58 71	56 99
12	48 66	74 30	44 81	06 80	29 09	50 31	69 61	24 64	28 89	97 79
13	85 85	07 54	21 50	31 80	10 19	56 65	82 52	26 58	55 12	26 34
14	08 27	08 08	35 87	96 57	33 12	01 77	52 76	09 89	71 12	17 69
15	59 61	22 14	26 09	96 75	17 94	51 08	41 91	45 94	80 48	59 92
16	17 45	77 79	31 66	36 54	92 85	65 60	53 98	63 50	11 20	96 63
17	11 26	37 08	07 71	95 95	39 75	92 48	99 78	23 33	19 56	06 67
18	48 08	13 98	16 52	41 15	73 96	32 55	03 12	38 30	88 77	17 03
19	76 27	72 22	99 61	72 15	00 25	21 54	47 79	18 41	58 50	57 66
20	98 89	22 25	72 92	53 55	07 98	66 71	53 29	61 71	56 96	41 78
21	88 69	61 63	01 67	61 88	58 79	35 65	08 45	63 38	69 86	79 47
22	12 58	13 75	80 98	01 35	91 16	18 36	90 54	99 17	68 36	85 06
23	08 86	96 36	14 09	43 85	51 20	65 18	06 40	52 17	48 10	68 97
24	33 81	05 51	32 48	60 12	32 44	08 12	89 00	98 82	79 17	97 22
25	05 15	99 28	87 15	07 08	66 92	53 81	69 42	02 27	65 33	57 69

TABLE B.1 *(cont.)*

RANDOM SAMPLING NUMBERS

Seventh Thousand

	1–4	5–8	9–12	13–16	17–20	21–24	25–28	29–32	33–36	37–40
1	80 30	23 64	67 96	21 33	36 90	03 91	69 33	90 13	34 48	02 19
2	61 29	89 61	32 08	12 62	26 08	42 00	31 73	31 30	30 61	34 11
3	23 33	61 01	02 21	11 81	51 32	36 10	23 74	50 31	90 11	73 52
4	94 21	32 92	93 50	72 67	23 20	74 59	30 30	48 66	75 32	27 97
5	87 61	92 69	01 60	28 79	74 76	86 06	39 29	73 85	03 27	50 57
6	37 56	19 18	03 42	86 03	85 74	44 81	86 45	71 16	13 52	35 56
7	64 86	66 31	55 04	88 40	10 30	84 38	06 13	58 83	62 04	63 52
8	22 69	58 45	49 23	09 81	98 84	05 04	75 99	27 70	72 79	32 19
9	23 22	14 22	64 90	10 26	74 23	53 91	27 73	78 19	92 43	68 10
10	42 38	59 64	72 96	46 57	89 67	22 81	94 56	69 84	18 31	06 39
11	17 18	01 34	10 98	37 48	93 86	88 59	69 53	78 86	37 26	85 48
12	39 45	69 53	94 89	58 97	29 33	29 19	50 94	80 57	31 99	38 91
13	43 18	11 42	56 19	48 44	45 02	84 29	01 78	65 77	76 84	88 85
14	59 44	06 45	68 55	16 65	66 13	38 00	95 76	50 67	67 65	18 83
15	01 50	34 32	38 00	37 57	47 82	66 59	19 50	87 14	35 59	79 47
16	79 14	60 35	47 95	90 71	31 03	85 37	38 70	34 16	64 55	66 49
17	01 56	63 68	80 26	14 97	23 88	59 22	82 39	70 83	48 34	46 48
18	25 76	18 71	29 25	15 51	92 96	01 01	28 18	03 35	11 10	27 84
19	23 52	10 83	45 06	49 85	35 45	84 08	81 13	52 57	21 23	67 02
20	91 64	08 64	25 74	16 10	97 31	10 27	24 48	89 06	42 81	29 10
21	80 86	07 27	26 70	08 65	85 20	31 23	28 99	39 63	32 03	71 91
22	31 71	37 60	95 60	94 95	54 45	27 97	03 67	30 54	86 04	12 41
23	05 83	50 36	09 04	39 15	66 55	80 36	39 71	24 10	62 22	21 53
24	98 70	02 90	30 63	62 59	26 04	97 20	00 91	28 80	40 23	09 91
25	82 79	35 45	64 53	93 24	86 55	48 72	18 57	05 79	20 09	31 46

Eighth Thousand

	1–4	5–8	9–12	13–16	17–20	21–24	25–28	29–32	33–36	37–40
1	37 52	49 55	40 65	27 61	08 59	91 23	26 18	95 04	98 20	99 52
2	48 16	69 65	69 02	08 83	08 83	68 37	00 96	13 59	12 16	17 93
3	50 43	06 59	56 53	30 61	40 21	29 06	49 60	90 38	21 43	19 25
4	89 31	62 79	45 73	71 72	77 11	28 80	72 35	75 77	24 72	98 43
5	63 29	90 61	86 39	07 38	38 85	77 06	10 23	30 84	07 95	30 76
6	71 68	93 94	08 72	36 27	85 89	40 59	83 37	93 85	73 97	84 05
7	05 06	96 63	58 24	05 95	56 64	77 53	85 64	15 95	93 91	59 03
8	03 35	58 95	46 44	25 70	31 66	01 05	44 44	62 91	36 31	45 04
9	13 04	57 67	74 77	53 35	93 51	82 83	27 38	63 16	04 48	75 23
10	49 96	43 94	56 04	02 79	55 78	01 44	75 26	85 54	01 81	32 82
11	24 36	24 08	44 77	57 07	54 41	04 56	09 44	30 58	25 45	37 56
12	55 19	97 20	01 11	47 45	79 79	06 72	12 81	86 97	54 09	06 53
13	02 28	54 60	28 35	32 94	36 74	51 63	96 90	04 13	30 43	10 14
14	90 50	13 78	22 20	37 56	97 95	49 95	91 15	52 73	12 93	78 94
15	33 71	32 43	29 58	47 38	39 96	67 51	64 47	49 91	64 58	93 07
16	70 58	28 49	54 32	97 70	27 81	64 69	71 52	02 56	61 37	04 58
17	09 68	96 10	57 78	85 00	89 81	98 30	19 40	76 28	62 99	99 83
18	19 36	60 85	35 04	12 87	83 88	66 54	32 00	30 20	05 30	42 63
19	04 75	44 49	64 26	51 46	80 50	53 91	00 55	67 36	68 66	08 29
20	79 83	32 39	46 77	56 83	42 21	60 03	14 47	07 01	66 85	49 22
21	80 99	42 43	08 58	54 41	98 05	54 39	34 42	97 47	38 35	59 40
22	48 83	64 99	86 94	48 78	79 20	62 23	56 45	92 65	56 36	83 02
23	28 45	35 85	22 20	13 01	73 96	70 05	84 50	68 59	96 58	16 63
24	52 07	63 15	82 30	66 23	14 26	66 61	17 80	41 97	40 27	24 80
25	39 14	52 18	35 87	48 55	48 81	03 11	26 99	03 80	08 86	50 42

TABLE B.2

ORDINATES AND AREAS OF THE NORMAL CURVE, $\phi(z) = \frac{1}{\sqrt{2\pi}} e^{-z^2/2}$

z	$\phi(z)$	$\int_0^z \phi(z)dz$	z	$\phi(z)$	$\int_0^z \phi(z)dz$	z	$\phi(z)$	$\int_0^z \phi(z)dz$
.00	.39894	.00000	.45	.36053	.17364	.90	.26609	.31594
.01	.39892	.00399	.46	.35889	.17724	.91	.26369	.31859
.02	.39886	.00798	.47	.35723	.18082	.92	.26129	.32121
.03	.39876	.01197	.48	.35553	.18439	.93	.25888	.32381
.04	.39862	.01595	.49	.35381	.18793	.94	.25647	.32639
.05	.39844	.01994	.50	.35207	.19146	.95	.25406	.32894
.06	.39822	.02392	.51	.35029	.19497	.96	.25164	.33147
.07	.39797	.02790	.52	.34849	.19847	.97	.24923	.33398
.08	.39767	.03188	.53	.34667	.20194	.98	.24681	.33646
.09	.39733	.03586	.54	.34482	.20540	.99	.24439	.33891
.10	.39695	.03983	.55	.34294	.20884	1.00	.24197	.34134
.11	.39654	.04380	.56	.34105	.21226	1.01	.23955	.34375
.12	.39608	.04776	.57	.33912	.21566	1.02	.23713	.34614
.13	.39559	.05172	.58	.33718	.21904	1.03	.23471	.34850
.14	.39505	.05567	.59	.33521	.22240	1.04	.23230	.35083
.15	.39448	.05962	.60	.33322	.22575	1.05	.22988	.35314
.16	.39387	.06356	.61	.33121	.22907	1.06	.22747	.35543
.17	.39322	.06749	.62	32918	.23237	1.07	.22506	.35769
.18	.39253	.07142	.63	.32713	.23565	1.08	.22265	.35993
.19	.39181	.07535	.64	.32506	.23891	1.09	.22025	.36214
.20	.39104	.07926	.65	.32297	.24215	1.10	.21785	.36433
.21	.39024	.08317	.66	.32086	.24537	1.11	.21546	.36650
.22	.38940	.08706	.67	.31874	.24857	1.12	.21307	.36864
.23	.38853	.09095	.68	.31659	.25175	1.13	.21069	.37076
.24	.38762	.09483	.69	.31443	.25490	1.14	.20831	.37286
.25	.38667	.09871	.70	.31225	.25804	1.15	.20594	.37493
.26	.38568	.10257	.71	.31006	.26115	1.16	.20357	.37698
.27	.38466	.10642	.72	.30785	.26424	1.17	.20121	.37900
.28	.38361	.11026	.73	.30563	.26730	1.18	.19886	.38100
.29	.38251	.11409	.74	.30339	.27035	1.19	.19652	.38298
.30	.38139	.11791	.75	.30114	.27337	1.20	.19419	.38493
.31	.38023	.12172	.76	.29887	.27637	1.21	.19186	.38686
.32	.37903	.12552	.77	.29659	.27935	1.22	.18954	.38877
.33	.37780	.12930	.78	.29431	.28230	1.23	.18724	.39065
.34	.37654	.13307	.79	.29200	.28524	1.24	.18494	.39251
.35	.37524	.13683	.80	.28969	.28814	1.25	.18265	.39435
.36	.37391	.14058	.81	.28737	.29103	1.26	.18037	.39617
.37	.37255	.14431	.82	.28504	.29389	1.27	.17810	.39796
.38	.37115	.14803	.83	.28269	.29673	1.28	.17585	.39973
.39	.36973	.15173	.84	.28034	.29955	1.29	.17360	.40147
.40	.36827	.15542	.85	.27798	.30234	1.30	.17137	.40320
.41	.36678	.15910	.86	.27562	.30511	1.31	.16915	.40490
.42	.36526	.16276	.87	.27324	.30785	1.32	.16694	.40658
.43	.36371	.16640	.88	.27086	.31057	1.33	.16474	.40824
.44	.36213	.17003	.89	.26848	.31327	1.34	.16256	.40988

TABLE B.2 *(cont.)*

ORDINATES AND AREAS OF THE NORMAL CURVE, $\phi(z) = \frac{1}{\sqrt{2\pi}} e^{-z^2/2}$

z	$\phi(z)$	$\int_0^z \phi(z)dz$	z	$\phi(z)$	$\int_0^z \phi(z)dz$	z	$\phi(z)$	$\int_0^z \phi(z)dz$
1.35	.16038	.41149	1.80	.07895	.46407	2.25	.03174	.48778
1.36	.15822	.41309	1.81	.07754	.46485	2.26	.03103	.48809
1.37	.15608	.41466	1.82	.07614	.46562	2.27	.03034	.48840
1.38	.15395	.41621	1.83	.07477	.46638	2.28	.02965	.48870
1.39	.15183	.41774	1.84	.07341	.46712	2.29	.02898	.48899
1.40	.14973	.41924	1.85	.07206	.46784	2.30	.02833	.48928
1.41	.14764	.42073	1.86	.07074	.46856	2.31	.02768	.48956
1.42	.14556	.42220	1.87	.06943	.46926	2.32	.02705	.48983
1.43	.14350	.42364	1.88	.06814	.46995	2.33	.02643	.49010
1.44	.14146	.42507	1.89	.06687	.47062	2.34	.02582	.49036
1.45	.13943	.42647	1.90	.06562	.47128	2.35	.02522	.49061
1.46	.13742	.42786	1.91	.06439	.47193	2.36	.02463	.49086
1.47	.13542	.42922	1.92	.06316	.47257	2.37	.02406	.49111
1.48	.13344	.43056	1.93	.06195	.47320	2.38	.02349	.49134
1.49	.13147	.43189	1.94	.06077	.47381	2.39	.02294	.49158
1.50	.12952	.43319	1.95	.05959	.47441	2.40	.02239	.49180
1.51	.12758	.43448	1.96	.05844	.47500	2.41	.02186	.49202
1.52	.12566	.43574	1.97	.05730	.47558	2.42	.02134	.49224
1.53	.12376	.43699	1.98	.05618	.47615	2.43	.02083	.49245
1.54	.12188	.43822	1.99	.05508	.47670	2.44	.02033	.49266
1.55	.12001	.43943	2.00	.05399	.47725	2.45	.01984	.49286
1.56	.11816	.44062	2.01	.05292	.47778	2.46	.01936	.49305
1.57	.11632	.44179	2.02	.05186	.47831	2.47	.01889	.49324
1.58	.11450	.44295	2.03	.05082	.47882	2.48	.01842	.49343
1.59	.11270	.44408	2.04	.04980	.47932	2.49	.01797	.49361
1.60	.11092	.44520	2.05	.04879	.47982	2.50	.01753	.49379
1.61	.10915	.44630	2.06	.04780	.48030	2.51	.01709	.49396
1.62	.10741	.44738	2.07	.04682	.48077	2.52	.01667	.49413
1.63	.10567	.44845	2.08	.04586	.48124	2.53	.01625	.49430
1.64	.10396	.44950	2.09	.04491	.48169	2.54	.01585	.49446
1.65	.10226	.45053	2.10	.04398	.48214	2.55	.01545	.49461
1.66	.10059	.45154	2.11	.04307	.48257	2.56	.01506	.49477
1.67	.09893	.45254	2.12	.04217	.48300	2.57	.01468	.49492
1.68	.09728	.45352	2.13	.04128	.48341	2.58	.01431	.49506
1.69	.09566	.45449	2.14	.04041	.48382	2.59	.01394	.49520
1.70	.09405	.45543	2.15	.03955	.48422	2.60	.01358	.49534
1.71	.09246	.45637	2.16	.03871	.48461	2.61	.01323	.49547
1.72	.09089	.45728	2.17	.03788	.48500	2.62	.01289	.49560
1.73	.08933	.45818	2.18	.03706	.48537	2.63	.01256	.49573
1.74	.08780	.45907	2.19	.03626	.48574	2.64	.01223	.49585
1.75	.08628	.45994	2.20	.03547	.48610	2.65	.01191	.49598
1.76	.08478	.46080	2.21	.03470	.48645	2.66	.01160	.49609
1.77	.08329	.46164	2.22	.03394	.48679	2.67	.01130	.49621
1.78	.08183	.46246	2.23	.03319	.48713	2.68	.01100	.49632
1.79	.08038	.46327	2.24	.03246	.48745	2.69	.01071	.49643

TABLE B.2 *(cont.)*

ORDINATES AND AREAS OF THE NORMAL CURVE, $\phi(z) = \frac{1}{\sqrt{2\pi}} e^{-z^2/2}$

z	$\phi(z)$	$\int_0^z \phi(z)dz$	z	$\phi(z)$	$\int_0^z \phi(z)dz$	z	$\phi(z)$	$\int_0^z \phi(z)dz$
2.70	.01042	.49653	3.15	.00279	.49918	3.60	.00061	.49984
2.71	.01014	.49664	3.16	.00271	.49921	3.61	.00059	.49985
2.72	.00987	.49674	3.17	.00262	.49924	3.62	.00057	.49985
2.73	.00961	.49683	3.18	.00254	.49926	3.63	.00055	.49986
2.74	.00935	.49693	3.19	.00246	.49929	3.64	.00053	.49986
2.75	.00909	.49702	3.20	.00238	.49931	3.65	.00051	.49987
2.76	.00885	.49711	3.21	.00231	.49934	3.66	.00049	.49987
2.77	.00861	.49720	3.22	.00224	.49936	3.67	.00047	.49988
2.78	.00837	.49728	3.23	.00216	.49938	3.68	.00046	.49988
2.79	.00814	.49736	3.24	.00210	.49940	3.69	.00044	.49989
2.80	.00792	.49744	3.25	.00203	.49942	3.70	.00042	.49989
2.81	.00770	.49752	3.26	.00196	.49944	3.71	.00041	.49990
2.82	.00748	.49760	3.27	.00190	.49946	3.72	.00039	.49990
2.83	.00727	.49767	3.28	.00184	.49948	3.73	.00038	.49990
2.84	.00707	.49774	3.29	.00178	.49950	3.74	.00037	.49991
2.85	.00687	.49781	3.30	.00172	.49952	3.75	.00035	.49991
2.86	.00668	.49788	3.31	.00167	.49953	3.76	.00034	.49992
2.87	.00649	.49795	3.32	.00161	.49955	3.77	.00033	.49992
2.88	.00631	.49801	3.33	.00156	.49957	3.78	.00031	.49992
2.89	.00613	.49807	3.34	.00151	.49958	3.79	.00030	.49992
2.90	.00595	.49813	3.35	.00146	.49960	3.80	.00029	.49993
2.91	.00578	.49819	3.36	.00141	.49961	3.81	.00028	.49993
2.92	.00562	.49825	3.37	.00136	.49962	3.82	.00027	.49993
2.93	.00545	.49831	3.38	.00132	.49964	3.83	.00026	.49994
2.94	.00530	.49836	3.39	.00127	.49965	3.84	.00025	.49994
2.95	.00514	.49841	3.40	.00123	.49966	3.85	.00024	.49994
2.96	.00499	.49846	3.41	.00119	.49968	3.86	.00023	.49994
2.97	.00485	.49851	3.42	.00115	.49969	3.87	.00022	.49995
2.98	.00471	.49856	3.43	.00111	.49970	3.88	.00021	.49995
2.99	.00457	.49861	3.44	.00107	.49971	3.89	.00021	.49995
3.00	.00443	.49865	3.45	.00104	.49972	3.90	.00020	.49995
3.01	.00430	.49869	3.46	.00100	.49973	3.91	.00019	.49995
3.02	.00417	.49874	3.47	.00097	.49974	3.92	.00018	.49996
3.03	.00405	.49878	3.48	.00094	.49975	3.93	.00018	.49996
3.04	.00393	.49882	3.49	.00090	.49976	3.94	.00017	.49996
3.05	.00381	.49886	3.50	.00087	.49977	3.95	.00016	.49996
3.06	.00370	.49889	3.51	.00084	.49978	3.96	.00016	.49996
3.07	.00358	.49893	3.52	.00081	.49978	3.97	.00015	.49996
3.08	.00348	.49897	3.53	.00079	.49979	3.98	.00014	.49997
3.09	.00337	.49900	3.54	.00076	.49980	3.99	.00014	.49997
3.10	.00327	.49903	3.55	.00073	.49981			
3.11	.00317	.49906	3.56	.00071	.49981			
3.12	.00307	.49910	3.57	.00068	.49982			
3.13	.00298	.49913	3.58	.00066	.49983			
3.14	.00288	.49916	3.59	.00063	.49983			

TABLE B.3

VALUES OF χ^2 CORRESPONDING TO GIVEN PROBABILITIES*

Degrees of freedom n	Probability of a deviation greater than χ^2						
	.01	.02	.05	.10	.20	.30	.50
1	6.635	5.412	3.841	2.706	1.642	1.074	.455
2	9.210	7.824	5.991	4.605	3.219	2.408	1.386
3	11.341	9.837	7.815	6.251	4.642	3.665	2.366
4	13.277	11.668	9.488	7.779	5.989	4.878	3.357
5	15.086	13.388	11.070	9.236	7.289	6.064	4.351
6	16.812	15.033	12.592	10.645	8.558	7.231	5.348
7	18.475	16.622	14.067	12.017	9.803	8.383	6.346
8	20.090	18.168	15.507	13.362	11.030	9.524	7.344
9	21.666	19.679	16.919	14.684	12.242	10.656	8.343
10	23.209	21.161	18.307	15.987	13.442	11.781	9.342
11	24.725	22.618	19.675	17.275	14.631	12.899	10.341
12	26.217	24.054	21.026	18.549	15.812	14.011	11.340
13	27.688	25.472	22.362	19.812	16.985	15.119	12.340
14	29.141	26.873	23.685	21.064	18.151	16.222	13.339
15	30.578	28.259	24.996	22.307	19.311	17.322	14.339
16	32.000	29.633	26.296	23.542	20.465	18.418	15.338
17	33.409	30.995	27.587	24.769	21.615	19.511	16.338
18	34.805	32.346	28.869	25.989	22.760	20.601	17.338
19	36.191	33.687	30.144	27.204	23.900	21.689	18.338
20	37.566	35.020	31.410	28.412	25.038	22.775	19.337
21	38.932	36.343	32.671	29.615	26.171	23.858	20.337
22	40.289	37.659	33.924	30.813	27.301	24.939	21.337
23	41.638	38.968	35.172	32.007	28.429	26.018	22.337
24	42.980	40.270	36.415	33.196	29.553	27.096	23.337
25	44.314	41.566	37.652	34.382	30.675	28.172	24.337
26	45.642	42.856	38.885	35.563	31.795	29.246	25.336
27	46.963	44.140	40.113	36.741	32.912	30.319	26.336
28	48.278	45.419	41.337	37.916	34.027	31.391	27.336
29	49.588	46.693	42.557	39.087	35.139	32.461	28.336
30	50.892	47.962	43.773	40.256	36.250	33.530	29.336

For larger values of n, the quantity $(2\chi^2)^{1/2} - (2n - 1)^{1/2}$ may be used as a normal deviate with unit standard deviation.

*This table is reproduced from "Statistical Methods for Research Workers," with the generous permission of the author, Sir Ronald A. Fisher, and the publishers, Messrs. Oliver and Boyd.

TABLE B.3 *(cont.)*

VALUES OF χ^2 CORRESPONDING TO GIVEN PROBABILITIES*

Degrees of freedom n	Probability of a deviation greater than χ^2					
	.70	.80	.90	.95	.98	.99
1	.148	.0642	.0158	.00393	.000628	.000157
2	.713	.446	.211	.103	.0404	.0201
3	1.424	1.005	.584	.352	.185	.115
4	2.195	1.649	1.064	.711	.429	.297
5	3.000	2.343	1.610	1.145	.752	.554
6	3.828	3.070	2.204	1.635	1.134	.872
7	4.671	3.822	2.833	2.167	1.564	1.239
8	5.527	4.594	3.490	2.733	2.032	1.646
9	6.393	5.380	4.168	3.325	2.532	2.088
10	7.267	6.179	4.865	3.940	3.059	2.558
11	8.148	6.989	5.578	4.575	3.609	3.053
12	9.034	7.807	6.304	5.226	4.178	3.571
13	9.926	8.634	7.042	5.892	4.765	4.107
14	10.821	9.467	7.790	6.571	5.368	4.660
15	11.721	10.307	8.547	7.261	5.985	5.229
16	12.624	11.152	9.312	7.962	6.614	5.812
17	13.531	12.002	10.085	8.672	7.255	6.408
18	14.440	12.857	10.865	9.390	7.906	7.015
19	15.352	13.716	11.651	10.117	8.567	7.633
20	16.266	14.578	12.443	10.851	9.237	8.260
21	17.182	15.445	13.240	11.591	9.915	8.897
22	18.101	16.314	14.041	12.338	10.600	9.542
23	19.021	17.187	14.848	13.091	11.293	10.196
24	19.943	18.062	15.659	13.848	11.992	10.856
25	20.867	18.940	16.473	14.611	12.697	11.524
26	21.792	19.820	17.292	15.379	13.409	12.198
27	22.719	20.703	18.114	16.151	14.125	12.879
28	23.647	21.588	18.939	16.928	14.847	13.565
29	24.577	22.475	19.768	17.708	15.574	14.256
30	25.508	23.364	20.599	18.493	16.306	14.953

*For larger values of n, the quantity $(2\chi^2)^{1/2} - (2n-1)^{1/2}$ may be used as a normal deviate with unit standard deviation.

TABLE B.4

VALUES OF t CORRESPONDING TO GIVEN PROBABILITIES*

Degrees of freedom n	Probability of a deviation greater than t					
	.005	.01	.025	.05	.1	.15
1	63.657	31.821	12.706	6.314	3.078	1.963
2	9.925	6.965	4.303	2.920	1.886	1.386
3	5.841	4.541	3.182	2.353	1.638	1.250
4	4.604	3.747	2.776	2.132	1.533	1.190
5	4.032	3.365	2.571	2.015	1.476	1.156
6	3.707	3.143	2.447	1.943	1.440	1.134
7	3.499	2.998	2.365	1.895	1.415	1.119
8	3.355	2.896	2.306	1.860	1.397	1.108
9	3.250	2.821	2.262	1.833	1.383	1.100
10	3.169	2.764	2.228	1.812	1.372	1.093
11	3.106	2.718	2.201	1.796	1.363	1.088
12	3.055	2.681	2.179	1.782	1.356	1.083
13	3.012	2.650	2.160	1.771	1.350	1.079
14	2.977	2.624	2.145	1.761	1.345	1.076
15	2.947	2.602	2.131	1.753	1.341	1.074
16	2.921	2.583	2.120	1.746	1.337	1.071
17	2.898	2.567	2.110	1.740	1.333	1.069
18	2.878	2.552	2.101	1.734	1.330	1.067
19	2.861	2.539	2.093	1.729	1.328	1.066
20	2.845	2.528	2.086	1.725	1.325	1.064
21	2.831	2.518	2.080	1.721	1.323	1.063
22	2.819	2.508	2.074	1.717	1.321	1.061
23	2.807	2.500	2.069	1.714	1.319	1.060
24	2.797	2.492	2.064	1.711	1.318	1.059
25	2.787	2.485	2.060	1.708	1.316	1.058
26	2.779	2.479	2.056	1.706	1.315	1.058
27	2.771	2.473	2.052	1.703	1.314	1.057
28	2.763	2.467	2.048	1.701	1.313	1.056
29	2.756	2.462	2.045	1.699	1.311	1.055
30	2.750	2.457	2.042	1.697	1.310	1.055
∞	2.576	2.326	1.960	1.645	1.282	1.036

The probability of a deviation numerically greater than t is twice the probability given at the head of the table.

*This table is reproduced from "Statistical Methods for Research Workers," with the generous permission of the author, Sir Ronald A. Fisher, and the publishers, Messrs. Oliver and Boyd.

TABLE B.4 *(cont.)*

VALUES OF t CORRESPONDING TO GIVEN PROBABILITIES*

Degrees of freedom n	Probability of a deviation greater than t					
	.2	.25	.3	.35	.4	.45
1	1.376	1.000	.727	.510	.325	.158
2	1.061	.816	.617	.445	.289	.142
3	.978	.765	.584	.424	.277	.137
4	.941	.741	.569	.414	.271	.134
5	.920	.727	.559	.408	.267	.132
6	.906	.718	.553	.404	.265	.131
7	.896	.711	.549	.402	.263	.130
8	.889	.706	.546	.399	.262	.130
9	.883	.703	.543	.398	.261	.129
10	.879	.700	.542	.397	.260	.129
11	.876	.697	.540	.396	.260	.129
12	.873	.695	.539	.395	.259	.128
13	.870	.694	.538	.394	.259	.128
14	.868	.692	.537	.393	.258	.128
15	.866	.691	.536	.393	.258	.128
16	.865	.690	.535	.392	.258	.128
17	.863	.689	.534	.392	.257	.128
18	.862	.688	.534	.392	.257	.127
19	.861	.688	.533	.391	.257	.127
20	.860	.687	.533	.391	.257	.127
21	.859	.686	.532	.391	.257	.127
22	.858	.686	.532	.390	.256	.127
23	.858	.685	.532	.390	.256	.127
24	.857	.685	.531	.390	.256	.127
25	.856	.684	.531	.390	.256	.127
26	.856	.684	.531	.390	.256	.127
27	.855	.684	.531	.389	.256	.127
28	.855	.683	.530	.389	.256	.127
29	.854	.683	.530	.389	.256	.127
30	.854	.683	.530	.389	.256	.127
∞	.842	.674	.524	.385	.253	.126

*The probability of a deviation numerically greater than t is twice the probability given at the head of the table.

TABLE B.5

5% (ROMAN TYPE) AND 1% (BOLDFACE TYPE) POINTS IN THE DISTRIBUTION OF F*

n_2	n_1 degrees of freedom (for greater mean square)																									n_2
	1	2	3	4	5	6	7	8	9	10	11	12	14	16	20	24	30	40	50	75	100	200	500	∞	
1	161 **4,052**	200 **4,999**	216 **5,403**	225 **5,625**	230 **5,764**	234 **5,859**	237 **5,928**	239 **5,981**	241 **6,022**	242 **6,056**	243 **6,082**	244 **6,106**	245 **6,142**	246 **6,169**	248 **6,208**	249 **6,234**	250 **6,258**	251 **6,286**	252 **6,302**	253 **6,323**	253 **6,334**	254 **6,352**	254 **6,361**	254 **6,366**	1
2	18.51 **98.49**	19.00 **99.01**	19.16 **99.17**	19.25 **99.25**	19.30 **99.30**	19.33 **99.33**	19.36 **99.34**	19.37 **99.36**	19.38 **99.38**	19.39 **99.40**	19.40 **99.41**	19.41 **99.42**	19.42 **99.43**	19.43 **99.44**	19.44 **99.45**	19.45 **99.46**	19.46 **99.47**	19.47 **99.48**	19.47 **99.48**	19.48 **99.49**	19.49 **99.49**	19.49 **99.49**	19.50 **99.50**	19.50 **99.50**	2
3	10.13 **34.12**	9.55 **30.81**	9.28 **29.46**	9.12 **28.71**	9.01 **28.24**	8.94 **27.91**	8.88 **27.67**	8.84 **27.49**	8.81 **27.34**	8.78 **27.23**	8.76 **27.13**	8.74 **27.05**	8.71 **26.92**	8.69 **26.83**	8.66 **26.69**	8.64 **26.60**	8.62 **26.50**	8.60 **26.41**	8.58 **26.35**	8.57 **26.27**	8.56 **26.23**	8.54 **26.18**	8.54 **26.14**	8.53 **26.12**	3
4	7.71 **21.20**	6.94 **18.00**	6.59 **16.69**	6.39 **15.98**	6.26 **15.52**	6.16 **15.21**	6.09 **14.98**	6.04 **14.80**	6.00 **14.66**	5.96 **14.54**	5.93 **14.45**	5.91 **14.37**	5.87 **14.24**	5.84 **14.15**	5.80 **14.02**	5.77 **13.93**	5.74 **13.83**	5.71 **13.74**	5.70 **13.69**	5.68 **13.61**	5.66 **13.57**	5.65 **13.52**	5.64 **13.48**	5.63 **13.46**	4
5	6.61 **16.26**	5.79 **13.27**	5.41 **12.06**	5.19 **11.39**	5.05 **10.97**	4.95 **10.67**	4.88 **10.45**	4.82 **10.27**	4.78 **10.15**	4.74 **10.05**	4.70 **9.96**	4.68 **9.89**	4.64 **9.77**	4.60 **9.68**	4.56 **9.55**	4.53 **9.47**	4.50 **9.38**	4.46 **9.29**	4.44 **9.24**	4.42 **9.17**	4.40 **9.13**	4.38 **9.07**	4.37 **9.04**	4.36 **9.02**	5
6	5.99 **13.74**	5.14 **10.92**	4.76 **9.78**	4.53 **9.15**	4.39 **8.75**	4.28 **8.47**	4.21 **8.26**	4.15 **8.10**	4.10 **7.98**	4.06 **7.87**	4.03 **7.79**	4.00 **7.72**	3.96 **7.60**	3.92 **7.52**	3.87 **7.39**	3.84 **7.31**	3.81 **7.23**	3.77 **7.14**	3.75 **7.09**	3.72 **7.02**	3.71 **6.99**	3.69 **6.94**	3.68 **6.90**	3.67 **6.88**	6
7	5.59 **12.25**	4.74 **9.55**	4.35 **8.45**	4.12 **7.85**	3.97 **7.46**	3.87 **7.19**	3.79 **7.00**	3.73 **6.84**	3.68 **6.71**	3.63 **6.62**	3.60 **6.54**	3.57 **6.47**	3.52 **6.35**	3.49 **6.27**	3.44 **6.15**	3.41 **6.07**	3.38 **5.98**	3.34 **5.90**	3.32 **5.85**	3.29 **5.78**	3.28 **5.75**	3.25 **5.70**	3.24 **5.67**	3.23 **5.65**	7
8	5.32 **11.26**	4.46 **8.65**	4.07 **7.59**	3.84 **7.01**	3.69 **6.63**	3.58 **6.37**	3.50 **6.19**	3.44 **6.03**	3.39 **5.91**	3.34 **5.82**	3.31 **5.74**	3.28 **5.67**	3.23 **5.56**	3.20 **5.48**	3.15 **5.36**	3.12 **5.28**	3.08 **5.20**	3.05 **5.11**	3.03 **5.06**	3.00 **5.00**	2.98 **4.96**	2.96 **4.91**	2.94 **4.88**	2.93 **4.86**	8
9	5.12 **10.56**	4.26 **8.02**	3.86 **6.99**	3.63 **6.42**	3.48 **6.06**	3.37 **5.80**	3.29 **5.62**	3.23 **5.47**	3.18 **5.35**	3.13 **5.26**	3.10 **5.18**	3.07 **5.11**	3.02 **5.00**	2.98 **4.92**	2.93 **4.80**	2.90 **4.73**	2.86 **4.64**	2.82 **4.56**	2.80 **4.51**	2.77 **4.45**	2.76 **4.41**	2.73 **4.36**	2.72 **4.33**	2.71 **4.31**	9
10	4.96 **10.04**	4.10 **7.56**	3.71 **6.55**	3.48 **5.99**	3.33 **5.64**	3.22 **5.39**	3.14 **5.21**	3.07 **5.06**	3.02 **4.95**	2.97 **4.85**	2.94 **4.78**	2.91 **4.71**	2.86 **4.60**	2.82 **4.52**	2.77 **4.41**	2.74 **4.33**	2.70 **4.25**	2.67 **4.17**	2.64 **4.12**	2.61 **4.05**	2.59 **4.01**	2.56 **3.96**	2.55 **3.93**	2.54 **3.91**	10
11	4.84 **9.65**	3.98 **7.20**	3.59 **6.22**	3.36 **5.67**	3.20 **5.32**	3.09 **5.07**	3.01 **4.88**	2.95 **4.74**	2.90 **4.63**	2.86 **4.54**	2.82 **4.46**	2.79 **4.40**	2.74 **4.29**	2.70 **4.21**	2.65 **4.10**	2.61 **4.02**	2.57 **3.94**	2.53 **3.86**	2.50 **3.80**	2.47 **3.74**	2.45 **3.70**	2.42 **3.66**	2.41 **3.62**	2.40 **3.60**	11
12	4.75 **9.33**	3.88 **6.93**	3.49 **5.95**	3.26 **5.41**	3.11 **5.06**	3.00 **4.82**	2.92 **4.65**	2.85 **4.50**	2.80 **4.39**	2.76 **4.30**	2.72 **4.22**	2.69 **4.16**	2.64 **4.05**	2.60 **3.98**	2.54 **3.86**	2.50 **3.78**	2.46 **3.70**	2.42 **3.61**	2.40 **3.56**	2.36 **3.49**	2.35 **3.46**	2.32 **3.41**	2.31 **3.38**	2.30 **3.36**	12
13	4.67 **9.07**	3.80 **6.70**	3.41 **5.74**	3.18 **5.20**	3.02 **4.86**	2.92 **4.62**	2.84 **4.44**	2.77 **4.30**	2.72 **4.19**	2.67 **4.10**	2.63 **4.02**	2.60 **3.96**	2.55 **3.85**	2.51 **3.78**	2.46 **3.67**	2.42 **3.59**	2.38 **3.51**	2.34 **3.42**	2.32 **3.37**	2.28 **3.30**	2.26 **3.27**	2.24 **3.21**	2.22 **3.18**	2.21 **3.16**	13

*Reproduced from *Statistical Methods* (5th ed., 1956), by kind permission of Dr. G. W. Snedecor and the Iowa State Univ. Press.

TABLE B.5 *(cont.)*

5% (ROMAN TYPE) AND 1% (BOLDFACE TYPE) POINTS IN THE DISTRIBUTION OF F

n_2	n_1 degrees of freedom (for greater mean square)																								n_2
	1	2	3	4	5	6	7	8	9	10	11	12	14	16	20	24	30	40	50	75	100	200	500	∞	
14	4.60 **8.86**	3.74 **6.51**	3.34 **5.56**	3.11 **5.03**	2.96 **4.69**	2.85 **4.46**	2.77 **4.28**	2.70 **4.14**	2.65 **4.03**	2.60 **3.94**	2.56 **3.86**	2.53 **3.80**	2.48 **3.70**	2.44 **3.62**	2.39 **3.51**	2.35 **3.43**	2.31 **3.34**	2.27 **3.26**	2.24 **3.21**	2.21 **3.14**	2.19 **3.11**	2.16 **3.06**	2.14 **3.02**	2.13 **3.00**	14
15	4.54 **8.68**	3.68 **6.36**	3.29 **5.42**	3.06 **4.89**	2.90 **4.56**	2.79 **4.32**	2.70 **4.14**	2.64 **4.00**	2.59 **3.89**	2.55 **3.80**	2.51 **3.73**	2.48 **3.67**	2.43 **3.56**	2.39 **3.48**	2.33 **3.36**	2.29 **3.29**	2.25 **3.20**	2.21 **3.12**	2.18 **3.07**	2.15 **3.00**	2.12 **2.97**	2.10 **2.92**	2.08 **2.89**	2.07 **2.87**	15
16	4.49 **8.53**	3.63 **6.23**	3.24 **5.29**	3.01 **4.77**	2.85 **4.44**	2.74 **4.20**	2.66 **4.03**	2.59 **3.89**	2.54 **3.78**	2.49 **3.69**	2.45 **3.61**	2.42 **3.55**	2.37 **3.45**	2.33 **3.37**	2.28 **3.25**	2.24 **3.18**	2.20 **3.10**	2.16 **3.01**	2.13 **2.96**	2.09 **2.89**	2.07 **2.86**	2.04 **2.80**	2.02 **2.77**	2.01 **2.75**	16
17	4.45 **8.40**	3.59 **6.11**	3.20 **5.18**	2.96 **4.67**	2.81 **4.34**	2.70 **4.10**	2.62 **3.93**	2.55 **3.79**	2.50 **3.68**	2.45 **3.59**	2.41 **3.52**	2.38 **3.45**	2.33 **3.35**	2.29 **3.27**	2.23 **3.16**	2.19 **3.08**	2.15 **3.00**	2.11 **2.92**	2.08 **2.86**	2.04 **2.79**	2.02 **2.76**	1.99 **2.70**	1.97 **2.67**	1.96 **2.65**	17
18	4.41 **8.28**	3.55 **6.01**	3.16 **5.09**	2.93 **4.58**	2.77 **4.25**	2.66 **4.01**	2.58 **3.85**	2.51 **3.71**	2.46 **3.60**	2.41 **3.51**	2.37 **3.44**	2.34 **3.37**	2.29 **3.27**	2.25 **3.19**	2.19 **3.07**	2.15 **3.00**	2.11 **2.91**	2.07 **2.83**	2.04 **2.78**	2.00 **2.71**	1.98 **2.68**	1.95 **2.62**	1.93 **2.59**	1.92 **2.57**	18
19	4.38 **8.18**	3.52 **5.93**	3.13 **5.01**	2.90 **4.50**	2.74 **4.17**	2.63 **3.94**	2.55 **3.77**	2.48 **3.63**	2.43 **3.52**	2.38 **3.43**	2.34 **3.36**	2.31 **3.30**	2.26 **3.19**	2.21 **3.12**	2.15 **3.00**	2.11 **2.92**	2.07 **2.84**	2.02 **2.76**	2.00 **2.70**	1.96 **2.63**	1.94 **2.60**	1.91 **2.54**	1.90 **2.51**	1.88 **2.49**	19
20	4.35 **8.10**	3.49 **5.85**	3.10 **4.94**	2.87 **4.43**	2.71 **4.10**	2.60 **3.87**	2.52 **3.71**	2.45 **3.56**	2.40 **3.45**	2.35 **3.37**	2.31 **3.30**	2.28 **3.23**	2.23 **3.13**	2.18 **3.05**	2.12 **2.94**	2.08 **2.86**	2.04 **2.77**	1.99 **2.69**	1.96 **2.63**	1.92 **2.56**	1.90 **2.53**	1.87 **2.47**	1.85 **2.44**	1.84 **2.42**	20
21	4.32 **8.02**	3.47 **5.78**	3.07 **4.87**	2.84 **4.37**	2.68 **4.04**	2.57 **3.81**	2.49 **3.65**	2.42 **3.51**	2.37 **3.40**	2.32 **3.31**	2.28 **3.24**	2.25 **3.17**	2.20 **3.07**	2.15 **2.99**	2.09 **2.88**	2.05 **2.80**	2.00 **2.72**	1.96 **2.63**	1.93 **2.58**	1.89 **2.51**	1.87 **2.47**	1.84 **2.42**	1.82 **2.38**	1.81 **2.36**	21
22	4.30 **7.94**	3.44 **5.72**	3.05 **4.82**	2.82 **4.31**	2.66 **3.99**	2.55 **3.76**	2.47 **3.59**	2.40 **3.45**	2.35 **3.35**	2.30 **3.26**	2.26 **3.18**	2.23 **3.12**	2.18 **3.02**	2.13 **2.94**	2.07 **2.83**	2.03 **2.75**	1.98 **2.67**	1.93 **2.58**	1.91 **2.53**	1.87 **2.46**	1.84 **2.42**	1.81 **2.37**	1.80 **2.33**	1.78 **2.31**	22
23	4.28 **7.88**	3.42 **5.66**	3.03 **4.76**	2.80 **4.26**	2.64 **3.94**	2.53 **3.71**	2.45 **3.54**	2.38 **3.41**	2.32 **3.30**	2.28 **3.21**	2.24 **3.14**	2.20 **3.07**	2.14 **2.97**	2.10 **2.89**	2.04 **2.78**	2.00 **2.70**	1.96 **2.62**	1.91 **2.53**	1.88 **2.48**	1.84 **2.41**	1.82 **2.37**	1.79 **2.32**	1.77 **2.28**	1.76 **2.26**	23
24	4.26 **7.82**	3.40 **5.61**	3.01 **4.72**	2.78 **4.22**	2.62 **3.90**	2.51 **3.67**	2.43 **3.50**	2.36 **3.36**	2.30 **3.25**	2.26 **3.17**	2.22 **3.09**	2.18 **3.03**	2.13 **2.93**	2.09 **2.85**	2.02 **2.74**	1.98 **2.66**	1.94 **2.58**	1.89 **2.49**	1.86 **2.44**	1.82 **2.36**	1.80 **2.33**	1.76 **2.27**	1.74 **2.23**	1.73 **2.21**	24
25	4.24 **7.77**	3.38 **5.57**	2.99 **4.68**	2.76 **4.18**	2.60 **3.86**	2.49 **3.63**	2.41 **3.46**	2.34 **3.32**	2.28 **3.21**	2.24 **3.13**	2.20 **3.05**	2.16 **2.99**	2.11 **2.89**	2.06 **2.81**	2.00 **2.70**	1.96 **2.62**	1.92 **2.54**	1.87 **2.45**	1.84 **2.40**	1.80 **2.32**	1.77 **2.29**	1.74 **2.23**	1.72 **2.19**	1.71 **2.17**	25
26	4.22 **7.72**	3.37 **5.53**	2.98 **4.64**	2.74 **4.14**	2.59 **3.82**	2.47 **3.59**	2.39 **3.42**	2.32 **3.29**	2.27 **3.17**	2.22 **3.09**	2.18 **3.02**	2.15 **2.96**	2.10 **2.86**	2.05 **2.77**	1.99 **2.66**	1.95 **2.58**	1.90 **2.50**	1.85 **2.41**	1.82 **2.36**	1.78 **2.28**	1.76 **2.25**	1.72 **2.19**	1.70 **2.15**	1.69 **2.13**	26

TABLE B.5 *(cont.)*

5% (ROMAN TYPE) AND 1% (BOLDFACE TYPE) POINTS IN THE DISTRIBUTION OF F

n_1 degrees of freedom (for greater mean square)

n_2	1	2	3	4	5	6	7	8	9	10	11	12	14	16	20	24	30	40	50	75	100	200	500	∞
27	4.21 **7.68**	3.35 **5.49**	2.96 **4.60**	2.73 **4.11**	2.57 **3.79**	2.46 **3.56**	2.37 **3.39**	2.30 **3.26**	2.25 **3.14**	2.20 **3.06**	2.16 **2.98**	2.13 **2.93**	2.08 **2.83**	2.03 **2.74**	1.97 **2.63**	1.93 **2.55**	1.88 **2.47**	1.84 **2.38**	1.80 **2.33**	1.76 **2.25**	1.74 **2.21**	1.71 **2.16**	1.68 **2.12**	1.67 **2.10**
28	4.20 **7.64**	3.34 **5.45**	2.95 **4.57**	2.71 **4.07**	2.56 **3.76**	2.44 **3.53**	2.36 **3.36**	2.29 **3.23**	2.24 **3.11**	2.19 **3.03**	2.15 **2.95**	2.12 **2.90**	2.06 **2.80**	2.02 **2.71**	1.96 **2.60**	1.91 **2.52**	1.87 **2.44**	1.81 **2.35**	1.78 **2.30**	1.75 **2.22**	1.72 **2.18**	1.69 **2.13**	1.67 **2.09**	1.65 **2.06**
29	4.18 **7.60**	3.33 **5.42**	2.93 **4.54**	2.70 **4.04**	2.54 **3.73**	2.43 **3.50**	2.35 **3.33**	2.28 **3.20**	2.22 **3.08**	2.18 **3.00**	2.14 **2.92**	2.10 **2.87**	2.05 **2.77**	2.00 **2.68**	1.94 **2.57**	1.90 **2.49**	1.85 **2.41**	1.80 **2.32**	1.77 **2.27**	1.73 **2.19**	1.71 **2.15**	1.68 **2.10**	1.65 **2.06**	1.64 **2.03**
30	4.17 **7.56**	3.32 **5.39**	2.92 **4.51**	2.69 **4.02**	2.53 **3.70**	2.42 **3.47**	2.34 **3.30**	2.27 **3.17**	2.21 **3.06**	2.16 **2.98**	2.12 **2.90**	2.09 **2.84**	2.04 **2.74**	1.99 **2.66**	1.93 **2.55**	1.89 **2.47**	1.84 **2.38**	1.79 **2.29**	1.76 **2.24**	1.72 **2.16**	1.69 **2.13**	1.66 **2.07**	1.64 **2.03**	1.62 **2.01**
32	4.15 **7.50**	3.30 **5.34**	2.90 **4.46**	2.67 **3.97**	2.51 **3.66**	2.40 **3.42**	2.32 **3.25**	2.25 **3.12**	2.19 **3.01**	2.14 **2.94**	2.10 **2.86**	2.07 **2.80**	2.02 **2.70**	1.97 **2.62**	1.91 **2.51**	1.86 **2.42**	1.82 **2.34**	1.76 **2.25**	1.74 **2.20**	1.69 **2.12**	1.67 **2.08**	1.64 **2.02**	1.61 **1.98**	1.59 **1.96**
34	4.13 **7.44**	3.28 **5.29**	2.88 **4.42**	2.65 **3.93**	2.49 **3.61**	2.38 **3.38**	2.30 **3.21**	2.23 **3.08**	2.17 **2.97**	2.12 **2.89**	2.08 **2.82**	2.05 **2.76**	2.00 **2.66**	1.95 **2.58**	1.89 **2.47**	1.84 **2.38**	1.80 **2.30**	1.74 **2.21**	1.71 **2.15**	1.67 **2.08**	1.64 **2.04**	1.61 **1.98**	1.59 **1.94**	1.57 **1.91**
36	4.11 **7.39**	3.26 **5.25**	2.86 **4.38**	2.63 **3.89**	2.48 **3.58**	2.36 **3.35**	2.28 **3.18**	2.21 **3.04**	2.15 **2.94**	2.10 **2.86**	2.06 **2.78**	2.03 **2.72**	1.98 **2.62**	1.93 **2.54**	1.87 **2.43**	1.82 **2.35**	1.78 **2.26**	1.72 **2.17**	1.69 **2.12**	1.65 **2.04**	1.62 **2.00**	1.59 **1.94**	1.56 **1.90**	1.55 **1.87**
38	4.10 **7.35**	3.25 **5.21**	2.85 **4.34**	2.62 **3.86**	2.46 **3.54**	2.35 **3.32**	2.26 **3.15**	2.19 **3.02**	2.14 **2.91**	2.09 **2.82**	2.05 **2.75**	2.02 **2.69**	1.96 **2.59**	1.92 **2.51**	1.85 **2.40**	1.80 **2.32**	1.76 **2.22**	1.71 **2.14**	1.67 **2.08**	1.63 **2.00**	1.60 **1.97**	1.57 **1.90**	1.54 **1.86**	1.53 **1.84**
40	4.08 **7.31**	3.23 **5.18**	2.84 **4.31**	2.61 **3.83**	2.45 **3.51**	2.34 **3.29**	2.25 **3.12**	2.18 **2.99**	2.12 **2.88**	2.07 **2.80**	2.04 **2.73**	2.00 **2.66**	1.95 **2.56**	1.90 **2.49**	1.84 **2.37**	1.79 **2.29**	1.74 **2.20**	1.69 **2.11**	1.66 **2.05**	1.61 **1.97**	1.59 **1.94**	1.55 **1.88**	1.53 **1.84**	1.51 **1.81**
42	4.07 **7.27**	3.22 **5.15**	2.83 **4.29**	2.59 **3.80**	2.44 **3.49**	2.32 **3.26**	2.24 **3.10**	2.17 **2.96**	2.11 **2.86**	2.06 **2.77**	2.02 **2.70**	1.99 **2.64**	1.94 **2.54**	1.89 **2.46**	1.82 **2.35**	1.78 **2.26**	1.73 **2.17**	1.68 **2.08**	1.64 **2.02**	1.60 **1.94**	1.57 **1.91**	1.54 **1.85**	1.51 **1.80**	1.49 **1.78**
44	4.06 **7.24**	3.21 **5.12**	2.82 **4.26**	2.58 **3.78**	2.43 **3.46**	2.31 **3.24**	2.23 **3.07**	2.16 **2.94**	2.10 **2.84**	2.05 **2.75**	2.01 **2.68**	1.98 **2.62**	1.92 **2.52**	1.88 **2.44**	1.81 **2.32**	1.76 **2.24**	1.72 **2.15**	1.66 **2.06**	1.63 **2.00**	1.58 **1.92**	1.56 **1.88**	1.52 **1.82**	1.50 **1.78**	1.48 **1.75**
46	4.05 **7.21**	3.20 **5.10**	2.81 **4.24**	2.57 **3.76**	2.42 **3.44**	2.30 **3.22**	2.22 **3.05**	2.14 **2.92**	2.09 **2.82**	2.04 **2.73**	2.00 **2.66**	1.97 **2.60**	1.91 **2.50**	1.87 **2.42**	1.80 **2.30**	1.75 **2.22**	1.71 **2.13**	1.65 **2.04**	1.62 **1.98**	1.57 **1.90**	1.54 **1.86**	1.51 **1.80**	1.48 **1.76**	1.46 **1.72**
48	4.04 **7.19**	3.19 **5.08**	2.80 **4.22**	2.56 **3.74**	2.41 **3.42**	2.30 **3.20**	2.21 **3.04**	2.14 **2.90**	2.08 **2.80**	2.03 **2.71**	1.99 **2.64**	1.96 **2.58**	1.90 **2.48**	1.86 **2.40**	1.79 **2.28**	1.74 **2.20**	1.70 **2.11**	1.64 **2.02**	1.61 **1.96**	1.56 **1.88**	1.53 **1.84**	1.50 **1.78**	1.47 **1.73**	1.45 **1.70**

TABLE B.5 *(cont.)*

5% (ROMAN TYPE) AND 1% (BOLDFACE TYPE) POINTS IN THE DISTRIBUTION OF F

n_2	n_1 degrees of freedom (for greater mean square)																								n_2
	1	2	3	4	5	6	7	8	9	10	11	12	14	16	20	24	30	40	50	75	100	200	500	∞	
50	4.03	3.18	2.79	2.56	2.40	2.29	2.20	2.13	2.07	2.02	1.98	1.95	1.90	1.85	1.78	1.74	1.69	1.63	1.60	1.55	1.52	1.48	1.46	1.44	50
	7.17	**5.06**	**4.20**	**3.72**	**3.41**	**3.18**	**3.02**	**2.88**	**2.78**	**2.70**	**2.62**	**2.56**	**2.46**	**2.39**	**2.26**	**2.18**	**2.10**	**2.00**	**1.94**	**1.86**	**1.82**	**1.76**	**1.71**	**1.68**	
55	4.02	3.17	2.78	2.54	2.38	2.27	2.18	2.11	2.05	2.00	1.97	1.93	1.88	1.83	1.76	1.72	1.67	1.61	1.58	1.52	1.50	1.46	1.43	1.41	55
	7.12	**5.01**	**4.16**	**3.68**	**3.37**	**3.15**	**2.98**	**2.85**	**2.75**	**2.66**	**2.59**	**2.53**	**2.43**	**2.35**	**2.23**	**2.15**	**2.06**	**1.96**	**1.90**	**1.82**	**1.78**	**1.71**	**1.66**	**1.64**	
60	4.00	3.15	2.76	2.52	2.37	2.25	2.17	2.10	2.04	1.99	1.95	1.92	1.86	1.81	1.75	1.70	1.65	1.59	1.56	1.50	1.48	1.44	1.41	1.39	60
	7.08	**4.98**	**4.13**	**3.65**	**3.34**	**3.12**	**2.95**	**2.82**	**2.72**	**2.63**	**2.56**	**2.50**	**2.40**	**2.32**	**2.20**	**2.12**	**2.03**	**1.93**	**1.87**	**1.79**	**1.74**	**1.68**	**1.63**	**1.60**	
65	3.99	3.14	2.75	2.51	2.36	2.24	2.15	2.08	2.02	1.98	1.94	1.90	1.85	1.80	1.73	1.68	1.63	1.57	1.54	1.49	1.46	1.42	1.39	1.37	65
	7.04	**4.95**	**4.10**	**3.62**	**3.31**	**3.09**	**2.93**	**2.79**	**2.70**	**2.61**	**2.54**	**2.47**	**2.37**	**2.30**	**2.18**	**2.09**	**2.00**	**1.90**	**1.84**	**1.76**	**1.71**	**1.64**	**1.60**	**1.56**	
70	3.98	3.13	2.74	2.50	2.35	2.23	2.14	2.07	2.01	1.97	1.93	1.89	1.84	1.79	1.72	1.67	1.62	1.56	1.53	1.47	1.45	1.40	1.37	1.35	70
	7.01	**4.92**	**4.08**	**3.60**	**3.29**	**3.07**	**2.91**	**2.77**	**2.67**	**2.59**	**2.51**	**2.45**	**2.35**	**2.28**	**2.15**	**2.07**	**1.98**	**1.88**	**1.82**	**1.74**	**1.69**	**1.62**	**1.56**	**1.53**	
80	3.96	3.11	2.72	2.48	2.33	2.21	2.12	2.05	1.99	1.95	1.91	1.88	1.82	1.77	1.70	1.65	1.60	1.54	1.51	1.45	1.42	1.38	1.35	1.32	80
	6.96	**4.88**	**4.04**	**3.56**	**3.25**	**3.04**	**2.87**	**2.74**	**2.64**	**2.55**	**2.48**	**2.41**	**2.32**	**2.24**	**2.11**	**2.03**	**1.94**	**1.84**	**1.78**	**1.70**	**1.65**	**1.57**	**1.52**	**1.49**	
100	3.94	3.09	2.70	2.46	2.30	2.19	2.10	2.03	1.97	1.92	1.88	1.85	1.79	1.75	1.68	1.63	1.57	1.51	1.48	1.42	1.39	1.34	1.30	1.28	100
	6.90	**4.82**	**3.98**	**3.51**	**3.20**	**2.99**	**2.82**	**2.69**	**2.59**	**2.51**	**2.43**	**2.36**	**2.26**	**2.19**	**2.06**	**1.98**	**1.89**	**1.79**	**1.73**	**1.64**	**1.59**	**1.51**	**1.46**	**1.43**	
125	3.92	3.07	2.68	2.44	2.29	2.17	2.08	2.01	1.95	1.90	1.86	1.83	1.77	1.72	1.65	1.60	1.55	1.49	1.45	1.39	1.36	1.31	1.27	1.25	125
	6.84	**4.78**	**3.94**	**3.47**	**3.17**	**2.95**	**2.79**	**2.65**	**2.56**	**2.47**	**2.40**	**2.33**	**2.23**	**2.15**	**2.03**	**1.94**	**1.85**	**1.75**	**1.68**	**1.59**	**1.54**	**1.46**	**1.40**	**1.37**	
150	3.91	3.06	2.67	2.43	2.27	2.16	2.07	2.00	1.94	1.89	1.85	1.82	1.76	1.71	1.64	1.59	1.54	1.47	1.44	1.37	1.34	1.29	1.25	1.22	150
	6.81	**4.75**	**3.91**	**3.44**	**3.14**	**2.92**	**2.76**	**2.62**	**2.53**	**2.44**	**2.37**	**2.30**	**2.20**	**2.12**	**2.00**	**1.91**	**1.83**	**1.72**	**1.66**	**1.56**	**1.51**	**1.43**	**1.37**	**1.33**	
200	3.89	3.04	2.65	2.41	2.26	2.14	2.05	1.98	1.92	1.87	1.83	1.80	1.74	1.69	1.62	1.57	1.52	1.45	1.42	1.35	1.32	1.26	1.22	1.19	200
	6.76	**4.71**	**3.88**	**3.41**	**3.11**	**2.90**	**2.73**	**2.60**	**2.50**	**2.41**	**2.34**	**2.28**	**2.17**	**2.09**	**1.97**	**1.88**	**1.79**	**1.69**	**1.62**	**1.53**	**1.48**	**1.39**	**1.33**	**1.28**	
400	3.86	3.02	2.62	2.39	2.23	2.12	2.03	1.96	1.90	1.85	1.81	1.78	1.72	1.67	1.60	1.54	1.49	1.42	1.38	1.32	1.28	1.22	1.16	1.13	400
	6.70	**4.66**	**3.83**	**3.36**	**3.06**	**2.85**	**2.69**	**2.55**	**2.46**	**2.37**	**2.29**	**2.23**	**2.12**	**2.04**	**1.92**	**1.84**	**1.74**	**1.64**	**1.57**	**1.47**	**1.42**	**1.32**	**1.24**	**1.19**	
1000	3.85	3.00	2.61	2.38	2.22	2.10	2.02	1.95	1.89	1.84	1.80	1.76	1.70	1.65	1.58	1.53	1.47	1.41	1.36	1.30	1.26	1.19	1.13	1.08	1000
	6.66	**4.62**	**3.80**	**3.34**	**3.04**	**2.82**	**2.66**	**2.53**	**2.43**	**2.34**	**2.26**	**2.20**	**2.09**	**2.01**	**1.89**	**1.81**	**1.71**	**1.61**	**1.54**	**1.44**	**1.38**	**1.28**	**1.19**	**1.11**	
∞	3.84	2.99	2.60	2.37	2.21	2.09	2.01	1.94	1.88	1.83	1.79	1.75	1.69	1.64	1.57	1.52	1.46	1.40	1.35	1.28	1.24	1.17	1.11	1.00	∞
	6.64	**4.60**	**3.78**	**3.32**	**3.02**	**2.80**	**2.64**	**2.51**	**2.41**	**2.32**	**2.24**	**2.18**	**2.07**	**1.99**	**1.87**	**1.79**	**1.69**	**1.59**	**1.52**	**1.41**	**1.36**	**1.25**	**1.15**	**1.00**	

TABLE B.6

CRITICAL VALUES FOR THE KOLMOGOROV TEST

Values of D_N such that $P[\max|S_N(x) - F(x)| > D_N] = \alpha$

N	α			
	0.20	0.10	0.05	0.01
5	0.446	0.510	0.565	0.669
6	0.410	0.470	0.521	0.618
7	0.381	0.438	0.486	0.577
8	0.358	0.411	0.457	0.543
9	0.339	0.388	0.432	0.514
10	0.322	0.368	0.410	0.490
11	0.307	0.352	0.391	0.468
12	0.295	0.338	0.375	0.450
13	0.284	0.325	0.361	0.433
14	0.274	0.314	0.349	0.418
15	0.266	0.304	0.338	0.404
16	0.258	0.295	0.328	0.392
17	0.250	0.286	0.318	0.381
18	0.244	0.278	0.309	0.371
19	0.237	0.272	0.301	0.363
20	0.231	0.264	0.294	0.356
25	0.21	0.24	0.27	0.32
30	0.19	0.22	0.24	0.29
35	0.18	0.21	0.23	0.27
>35	$1.07N^{-1/2}$	$1.22N^{-1/2}$	$1.36N^{-1/2}$	$1.63N^{-1/2}$

Adapted from Massey, F. J., Jr., "The Kolmogorov-Smirnov Test for Goodness of Fit," *J. Amer. Stat. Assoc.*, **46**, 1951, p. 70, with the kind permission of the author and the publisher.

TABLE B.7

CUMULATIVE BINOMIAL PROBABILITIES

Probability of r or fewer successes in N independent trials with $\theta = 0.5$

N \ r	0	1	2	3	4	5	6	7	8	9	10	11	12
5	0.0312	0.1875	0.5000										
6	0.0156	0.1094	0.3438	0.6562									
7	0.0078	0.0625	0.2266	0.5000									
8	0.0039	0.0352	0.1445	0.3633	0.6367								
9	0.0020	0.0195	0.0898	0.2539	0.5000								
10	0.0010	0.0107	0.0547	0.1719	0.3770	0.6230							
11	0.0005	0.0056	0.0327	0.1133	0.2744	0.5000							
12	0.0002	0.0032	0.0193	0.0730	0.1938	0.3872	0.6128						
13	0.0001	0.0017	0.0112	0.0461	0.1334	0.2905	0.5000						
14	0.0001	0.0009	0.0065	0.0287	0.0898	0.2120	0.3953	0.6047					
15		0.0005	0.0037	0.0176	0.0592	0.1509	0.3036	0.5000					
16		0.0003	0.0021	0.0106	0.0384	0.1051	0.2272	0.4018	0.5982				
17		0.0001	0.0012	0.0064	0.0245	0.0717	0.1662	0.3145	0.5000				
18		0.0001	0.0007	0.0038	0.0154	0.0481	0.1189	0.2403	0.4073	0.5927			
19			0.0004	0.0022	0.0096	0.0318	0.0835	0.1796	0.3238	0.5000			
20			0.0002	0.0013	0.0059	0.0207	0.0577	0.1316	0.2517	0.4119	0.5881		
21			0.0001	0.0007	0.0036	0.0133	0.0392	0.0946	0.1917	0.3318	0.5000		
22			0.0001	0.0004	0.0022	0.0085	0.0262	0.0669	0.1431	0.2617	0.4159	0.5841	
23				0.0002	0.0013	0.0053	0.0173	0.0466	0.1050	0.2024	0.3388	0.5000	
24				0.0001	0.0008	0.0033	0.0113	0.0320	0.0758	0.1537	0.2706	0.4194	
25				0.0001	0.0005	0.0020	0.0073	0.0216	0.0539	0.1148	0.2122	0.3450	0.5000

For $r > N/2$, subtract the probability for $N - r - 1$ from 1.

TABLE B.8

SIGNIFICANCE TEST FOR THE MEDIAN (THE WALSH TEST) AT STATED SIGNIFICANCE LEVELS*

N	α	Either	Or
5	0.125	$\frac{1}{2}(d_4 + d_5) < 0$	$\frac{1}{2}(d_1 + d_2) > 0$
	0.062	$d_5 < 0$	$d_1 > 0$
6	0.094	$\max[d_5, \frac{1}{2}(d_4 + d_6)] < 0$	$\min[d_2, \frac{1}{2}(d_1 + d_3)] > 0$
	0.062	$\frac{1}{2}(d_5 + d_6) < 0$	$\frac{1}{2}(d_1 + d_2) > 0$
	0.031	$d_6 < 0$	$d_1 > 0$
7	0.109	$\max[d_5, \frac{1}{2}(d_4 + d_7)] < 0$	$\min[d_3, \frac{1}{2}(d_1 + d_4)] > 0$
	0.047	$\max[d_6, \frac{1}{2}(d_5 + d_7)] < 0$	$\min[d_2, \frac{1}{2}(d_1 + d_3)] > 0$
	0.031	$\frac{1}{2}(d_6 + d_7) < 0$	$\frac{1}{2}(d_1 + d_2) > 0$
	0.016	$d_7 < 0$	$d_1 > 0$
8	0.086	$\max[d_6, \frac{1}{2}(d_4 + d_8)] < 0$	$\min[d_3, \frac{1}{2}(d_1 + d_5)] > 0$
	0.055	$\max[d_6, \frac{1}{2}(d_5 + d_8)] < 0$	$\min[d_3, \frac{1}{2}(d_1 + d_4)] > 0$
	0.023	$\max[d_7, \frac{1}{2}(d_6 + d_8)] < 0$	$\min[d_2, \frac{1}{2}(d_1 + d_3)] > 0$
	0.016	$\frac{1}{2}(d_7 + d_8) < 0$	$\frac{1}{2}(d_1 + d_2) > 0$
	0.008	$d_8 < 0$	$d_1 > 0$
9	0.102	$\max[d_6, \frac{1}{2}(d_4 + d_9)] < 0$	$\min[d_4, \frac{1}{2}(d_1 + d_6)] > 0$
	0.043	$\max[d_7, \frac{1}{2}(d_5 + d_9)] < 0$	$\min[d_3, \frac{1}{2}(d_1 + d_5)] > 0$
	0.020	$\max[d_8, \frac{1}{2}(d_5 + d_9)] < 0$	$\min[d_2, \frac{1}{2}(d_1 + d_5)] > 0$
	0.012	$\max[d_8, \frac{1}{2}(d_7 + d_9)] < 0$	$\min[d_2, \frac{1}{2}(d_1 + d_3)] > 0$
	0.008	$\frac{1}{2}(d_8 + d_9) < 0$	$\frac{1}{2}(d_1 + d_2) > 0$

For continuation of Table B. 8 see next page

*Adapted from Walsh, John E., "Applications of Some Significance Tests for the Median Which Are Valid Under Very General Conditions," *J. Amer. Stat. Assoc.*, **44**, 1949, p. 343, with the kind permission of the author and the publisher.

TABLE B.8 (*continued*)

N	α	Either	Or
10	0.111	$\max[d_6, \frac{1}{2}(d_4 + d_{10})] < 0$	$\min[d_5, \frac{1}{2}(d_1 + d_7)] > 0$
	0.051	$\max[d_7, \frac{1}{2}(d_5 + d_{10})] < 0$	$\min[d_4, \frac{1}{2}(d_1 + d_6)] > 0$
	0.021	$\max[d_8, \frac{1}{2}(d_6 + d_{10})] < 0$	$\min[d_3, \frac{1}{2}(d_1 + d_5)] > 0$
	0.010	$\max[d_9, \frac{1}{2}(d_6 + d_{10})] < 0$	$\min[d_2, \frac{1}{2}(d_1 + d_5)] > 0$
11	0.097	$\max[d_7, \frac{1}{2}(d_4 + d_{11})] < 0$	$\min[d_5, \frac{1}{2}(d_1 + d_8)] > 0$
	0.056	$\max[d_7, \frac{1}{2}(d_5 + d_{11})] < 0$	$\min[d_5, \frac{1}{2}(d_1 + d_7)] > 0$
	0.021	$\max[\frac{1}{2}(d_6 + d_{11}), \frac{1}{2}(d_8 + d_9)] < 0$	$\min[\frac{1}{2}(d_1 + d_6), \frac{1}{2}(d_3 + d_4)] > 0$
	0.011	$\max[d_9, \frac{1}{2}(d_7 + d_{11}) < 0$	$\min[d_3, \frac{1}{2}(d_1 + d_5)] > 0$
12	0.094	$\max[\frac{1}{2}(d_4 + d_{12}), \frac{1}{2}(d_5 + d_{11})] < 0$	$\min[\frac{1}{2}(d_1 + d_9), \frac{1}{2}(d_2 + d_8)] > 0$
	0.048	$\max[d_8, \frac{1}{2}(d_5 + d_{12})] < 0$	$\min[d_5, \frac{1}{2}(d_1 + d_8)] > 0$
	0.020	$\max[d_9, \frac{1}{2}(d_6 + d_{12})] < 0$	$\min[d_4, \frac{1}{2}(d_1 + d_7)] > 0$
	0.011	$\max[\frac{1}{2}(d_7 + d_{12}), \frac{1}{2}(d_9 + d_{10})] < 0$	$\min[\frac{1}{2}(d_1 + d_6), \frac{1}{2}(d_3 + d_4)] > 0$
13	0.094	$\max[\frac{1}{2}(d_4 + d_{13}), \frac{1}{2}(d_5 + d_{12})] < 0$	$\min[\frac{1}{2}(d_1 + d_{10}), \frac{1}{2}(d_2 + d_9)] > 0$
	0.047	$\max[\frac{1}{2}(d_5 + d_{13}), \frac{1}{2}(d_6 + d_{12})] < 0$	$\min[\frac{1}{2}(d_1 + d_9), \frac{1}{2}(d_2 + d_8)] > 0$
	0.020	$\max[\frac{1}{2}(d_6 + d_{13}), \frac{1}{2}(d_9 + d_{10})] < 0$	$\min[\frac{1}{2}(d_1 + d_8), \frac{1}{2}(d_4 + d_5)] > 0$
	0.010	$\max[d_{10}, \frac{1}{2}(d_7 + d_{13})] < 0$	$\min[d_4, \frac{1}{2}(d_1 + d_7)] > 0$
14	0.094	$\max[\frac{1}{2}(d_4 + d_{14}), \frac{1}{2}(d_5 + d_{13})] < 0$	$\min[\frac{1}{2}(d_1 + d_{11}), \frac{1}{2}(d_2 + d_{10})] > 0$
	0.047	$\max[\frac{1}{2}(d_5 + d_{14}), \frac{1}{2}(d_6 + d_{13})] < 0$	$\min[\frac{1}{2}(d_1 + d_{10}), \frac{1}{2}(d_2 + d_9)] > 0$
	0.020	$\max[d_{10}, \frac{1}{2}(d_6 + d_{14})] < 0$	$\min[d_5, \frac{1}{2}(d_1 + d_9)] > 0$
	0.010	$\max[\frac{1}{2}(d_7 + d_{14}), \frac{1}{2}(d_{10} + d_{11})] < 0$	$\min[\frac{1}{2}(d_1 + d_8), \frac{1}{2}(d_4 + d_5)] > 0$
15	0.094	$\max[\frac{1}{2}(d_4 + d_{15}), \frac{1}{2}(d_5 + d_{14})] < 0$	$\min[\frac{1}{2}(d_1 + d_{12}), \frac{1}{2}(d_2 + d_{11})] > 0$
	0.047	$\max[\frac{1}{2}(d_5 + d_{15}), \frac{1}{2}(d_6 + d_{14})] < 0$	$\min[\frac{1}{2}(d_1 + d_{11}), \frac{1}{2}(d_2 + d_{10})] > 0$
	0.020	$\max[\frac{1}{2}(d_6 + d_{15}), \frac{1}{2}(d_{10} + d_{11})] < 0$	$\min[\frac{1}{2}(d_1 + d_{10}), \frac{1}{2}(d_5 + d_6)] > 0$
	0.010	$\max[d_{11}, \frac{1}{2}(d_7 + d_{15})] < 0$	$\min[d_5, \frac{1}{2}(d_1 + d_9)] > 0$

Reject $H_0[\mu = 0]$ against two-tailed alternative $H_1[\mu \neq 0]$ at significance level α if either of the corresponding alternatives is true. For one-tailed test against $H_1[\mu < 0]$ use column 3 with significance level $\alpha/2$. For one-tailed test against $H_1[\mu > 0]$ use column 4 with significance level $\alpha/2$.

Table B.9

Critical Values of U for the Mann-Whitney Test

(a) $\alpha = 0.01$ (one-tailed test)

N_1 \ N_2	9	10	11	12	13	14	15	16	17	18	19	20
3	1	1	1	2	2	2	3	3	4	4	4	5
4	3	3	4	5	5	6	7	7	8	9	9	10
5	5	6	7	8	9	10	11	12	13	14	15	16
6	7	8	9	11	12	13	15	16	18	19	20	22
7	9	11	12	14	16	17	19	21	23	24	26	28
8	11	13	15	17	20	22	24	26	28	30	32	34
9	14	16	18	21	23	26	28	31	33	36	38	40
10	16	19	22	24	27	30	33	36	38	41	44	47
11	18	22	25	28	31	34	37	41	44	47	50	53
12	21	24	28	31	35	38	42	46	49	53	56	60
13	23	27	31	35	39	43	47	51	55	59	63	67
14	26	30	34	38	43	47	51	56	60	65	69	73
15	28	33	37	42	47	51	56	61	66	70	75	80

(b) $\alpha = 0.05$ (one-tailed test)

N_1 \ N_2	9	10	11	12	13	14	15	16	17	18	19	20
3	3	4	5	5	6	7	7	8	9	9	10	11
4	6	7	8	9	10	11	12	14	15	16	17	18
5	9	11	12	13	15	16	18	19	20	22	23	25
6	12	14	16	17	19	21	23	25	26	28	30	32
7	15	17	19	21	24	26	28	30	33	35	37	39
8	18	20	23	26	28	31	33	36	39	41	44	47
9	21	24	27	30	33	36	39	42	45	48	51	54
10	24	27	31	34	37	41	44	48	51	55	58	62
11	27	31	34	38	42	46	50	54	57	61	65	69
12	30	34	38	42	47	51	55	60	64	68	72	77
13	33	37	42	47	51	56	61	65	70	75	80	84
14	36	41	46	51	56	61	66	71	77	82	87	92
15	39	44	50	55	61	66	72	77	83	88	94	100

Abridged from Auble, D., "Extended Tables for the Mann-Whitney Statistic," *Bulletin of the Institute of Educational Research*, Indiana University, **1**, no. 2, 1953, with the kind permission of the author and the publisher.

TABLE B.10. JONCKHEERE'S k-SAMPLE TEST

Prob. ($S \geq S_0$) for k samples, each of size r

S_0	$k = 3$		S_0	$k = 3$	
	$r = 2$	$r = 4$		$r = 3$	$r = 5$
4	0.2889	0.4156	9	0.1940	0.3396
6	0.1667	0.3609	11	0.1387	0.3025
8	0.0889	0.3090	13	0.0946	0.2672
10	0.0333	0.2602	15	0.0613	0.2340
12	0.0111	0.2157	17	0.0369	0.2032
14		0.1756	19	0.0208	0.1748
16		0.1404	21	0.0107	0.1489
18		0.1099	23	0.0048	0.1256
20		0.0844	25	0.0018	0.1049
22		0.0632	27	0.0006	0.0867
24		0.0463	29		0.0708
26		0.0330	31		0.0572
28		0.0229	33		0.0456
30		0.0153	35		0.0359
32		0.0099	39		0.0214
40		0.0011	43		0.0120
			47		0.0063

S_0	$k = 4$			$k = 5$		$k = 6$
	$r = 2$	$r = 3$	$r = 4$	$r = 2$	$r = 3$	$r = 2$
10	0.1302	0.2659	0.3400	0.2110	0.3273	0.2699
12	0.0829	0.2220	0.3069	0.1625	0.2921	0.2265
14	0.0484	0.1823	0.2754	0.1213	0.2588	0.1871
16	0.0262	0.1472	0.2454	0.0878	0.2274	0.1521
18	0.0123	0.1166	0.2172	0.0613	0.1982	0.1215
20	0.0052	0.0907	0.1910	0.0412	0.1713	0.0953
22	0.0016	0.0691	0.1666	0.0265	0.1468	0.0734
24	0.0004	0.0515	0.1443	0.0162	0.1247	0.0553
26		0.0374	0.1241	0.0094	0.1049	0.0408
28		0.0266	0.1058	0.0051	0.0874	0.0294
30		0.0183	0.0895	0.0026	0.0721	0.0207
32		0.0123	0.0751	0.0012	0.0588	0.0142
34		0.0080	0.0624	0.0005	0.0475	0.0094
36		0.0050	0.0514		0.0379	0.0061
38		0.0030	0.0420		0.0299	0.0038
40		0.0017	0.0339		0.0234	0.0023
42			0.0272		0.0180	
44			0.0215		0.0137	
46			0.0168		0.0102	
48			0.0130		0.0075	
50			0.0100		0.0055	

Abridged from the tables in Jonckheere, A. R., "A Distribution-Free k-Sample Test Against Ordered Alternatives," *Biometrika*, **41,** 1954, 133–145, by kind permission of the author and the publishers.

TABLE B.11

VALUES OF Tanh $z' = r$ (FISHER'S TRANSFORMATION)

z'	0	1	2	3	4	5	6	7	8	9
.00	.0000	0010	0020	0030	0040	0050	0060	0070	0080	0090
.01	.0100	0110	0120	0130	0140	0150	0160	0170	0180	0190
.02	.0200	0210	0220	0230	0240	0250	0260	0270	0280	0290
.03	.0300	0310	0320	0330	0340	0350	0360	0370	0380	0390
.04	.0400	0410	0420	0430	0440	0450	0460	0470	0480	0490
.05	.0500	0510	0520	0530	0539	0549	0559	0569	0579	0589
.06	.0599	0609	0619	0629	0639	0649	0659	0669	0679	0689
.07	.0699	0709	0719	0729	0739	0749	0759	0768	0778	0788
.08	.0798	0808	0818	0828	0838	0848	0858	0868	0878	0888
.09	.0898	0907	0917	0927	0937	0947	0957	0967	0977	0987
.10	.0997	1007	1016	1026	1036	1046	1056	1066	1076	1086
.11	.1096	1105	1115	1125	1135	1145	1155	1165	1175	1184
.12	.1194	1204	1214	1224	1234	1244	1253	1263	1273	1283
.13	.1293	1303	1312	1322	1332	1342	1352	1361	1371	1381
.14	.1391	1401	1411	1420	1430	1440	1450	1460	1469	1479
.15	.1489	1499	1508	1518	1528	1538	1547	1557	1567	1577
.16	.1586	1596	1606	1616	1625	1635	1645	1655	1664	1674
.17	.1684	1694	1703	1713	1723	1732	1742	1752	1761	1771
.18	.1781	1790	1800	1810	1820	1829	1839	1849	1858	1868
.19	.1877	1887	1897	1906	1916	1926	1935	1945	1955	1964
.20	.1974	1983	1993	2003	2012	2022	2031	2041	2051	2060
.21	.2070	2079	2089	2098	2108	2117	2127	2137	2146	2156
.22	.2165	2175	2184	2194	2203	2213	2222	2232	2241	2251
.23	.2260	2270	2279	2289	2298	2308	2317	2327	2336	2346
.24	.2355	2364	2374	2383	2393	2402	2412	2421	2430	2440
.25	.2449	2459	2468	2477	2487	2496	2506	2515	2524	2534
.26	.2543	2552	2562	2571	2580	2590	2599	2608	2618	2627
.27	.2636	2646	2655	2664	2673	2683	2692	2701	2711	2720
.28	.2729	2738	2748	2757	2766	2775	2784	2794	2803	2812
.29	.2821	2831	2840	2849	2858	2867	2876	2886	2895	2904
.30	.2913	2922	2931	2941	2950	2959	2968	2977	2986	2995
.31	.3004	3013	3023	3032	3041	3050	3059	3068	3077	3086
.32	.3095	3104	3113	3122	3131	3140	3149	3158	3167	3176
.33	.3185	3194	3203	3212	3221	3230	3239	3248	3257	3266
.34	.3275	3284	3293	3302	3310	3319	3328	3337	3346	3355
.35	.3364	3373	3381	3390	3399	3408	3417	3426	3435	3443
.36	.3452	3461	3470	3479	3487	3496	3505	3514	3522	3531
.37	.3540	3549	3557	3566	3575	3584	3592	3601	3610	3618
.38	.3627	3636	3644	3653	3662	3670	3679	3688	3696	3705
.39	.3714	3722	3731	3739	3748	3757	3765	3774	3782	3791
.40	.3799	3808	3817	3825	3834	3842	3851	3859	3868	3876
.41	.3885	3893	3902	3910	3919	3927	3936	3944	3952	3961
.42	.3969	3978	3986	3995	4003	4011	4020	4028	4036	4045
.43	.4053	4062	4070	4078	4087	4095	4103	4112	4120	4128
.44	.4136	4145	4153	4161	4170	4178	4186	4194	4203	4211
.45	.4219	4227	4235	4244	4252	4260	4268	4276	4285	4293
.46	.4301	4309	4317	4325	4333	4342	4350	4358	4366	4374
.47	.4382	4390	4398	4406	4414	4422	4430	4438	4446	4454
.48	.4462	4470	4478	4486	4494	4502	4510	4518	4526	4534
.49	.4542	4550	4558	4566	4574	4582	4590	4598	4605	4613

*Reproduced from *Numerical Tables*, by J. W. Campbell, University of Alberta, by kind permission of Mrs. Campbell.

TABLE B.11 *(cont.)*

VALUES OF Tanh $z' = r$ (FISHER'S TRANSFORMATION)

z'	0	1	2	3	4	5	6	7	8	9
.50	.4621	4629	4637	4645	4653	4660	4668	4676	4684	4692
.51	.4699	4707	4715	4723	4731	4738	4746	4754	4762	4769
.52	.4777	4785	4792	4800	4808	4815	4823	4831	4839	4846
.53	.4854	4861	4869	4877	4884	4892	4900	4907	4915	4922
.54	.4930	4937	4945	4953	4960	4968	4975	4983	4990	4998
.55	.5005	5013	5020	5028	5035	5043	5050	5057	5065	5072
.56	.5080	5087	5095	5102	5109	5117	5124	5132	5139	5146
.57	.5154	5161	5168	5176	5183	5190	5198	5205	5212	5219
.58	.5227	5234	5241	5248	5256	5263	5270	5277	5285	5292
.59	.5299	5306	5313	5320	5328	5335	5342	5349	5356	5363
.60	.5370	5378	5385	5392	5399	5406	5413	5420	5427	5434
.61	.5441	5448	5455	5462	5469	5476	5483	5490	5497	5504
.62	.5511	5518	5525	5532	5539	5546	5553	5560	5567	5574
.63	.5581	5587	5594	5601	5608	5615	5622	5629	5635	5642
.64	.5649	5656	5663	5669	5676	5683	5690	5696	5703	5710
.65	.5717	5723	5730	5737	5744	5750	5757	5764	5770	5777
.66	.5784	5790	5797	5804	5810	5817	5823	5830	5837	5843
.67	.5850	5856	5863	5869	5876	5883	5889	5896	5902	5909
.68	.5915	5922	5928	5935	5941	5948	5954	5961	5967	5973
.69	.5980	5986	5993	5999	6005	6012	6018	6025	6031	6037
.70	.6044	6050	6056	6063	6069	6075	6082	6088	6094	6100
.71	.6107	6113	6119	6126	6132	6138	6144	6150	6157	6163
.72	.6169	6175	6181	6188	6194	6200	6206	6212	6218	6225
.73	.6231	6237	6243	6249	6255	6261	6267	6273	6279	6285
.74	.6291	6297	6304	6310	6316	6322	6328	6334	6340	6346
.75	.6351	6357	6363	6369	6375	6381	6387	6393	6399	6405
.76	.6411	6417	6423	6428	6434	6440	6446	6452	6458	6463
.77	.6469	6475	6481	6487	6492	6498	6504	6510	6516	6521
.78	.6527	6533	6539	6544	6550	6556	6561	6567	6573	6578
.79	.6584	6590	6595	6601	6607	6612	6618	6624	6629	6635
.80	.6640	6646	6652	6657	6663	6668	6674	6679	6685	6690
.81	.6696	6701	6707	6712	6718	6723	6729	6734	6740	6745
.82	.6751	6756	6762	6767	6772	6778	6783	6789	6794	6799
.83	.6805	6810	6815	6821	6826	6832	6837	6842	6847	6853
.84	6858	6863	6869	6874	6879	6884	6890	6895	6900	6905
.85	.6911	6916	6921	6926	6932	6937	6942	6947	6952	6957
.86	.6963	6968	6973	6978	6983	6988	6993	6998	7004	7009
.87	.7014	7019	7024	7029	7034	7039	7044	7049	7054	.7059
.88	.7064	7069	7074	7079	7084	7089	7094	7099	7104	7109
.89	.7114	7119	7124	7129	7134	7139	7143	7148	7153	7158
.90	.7163	7168	7173	7178	7182	7187	7192	7197	7202	7207
.91	.7211	7216	7221	7226	7230	7235	7240	7245	7249	7254
.92	.7259	7264	7268	7273	7278	7283	7287	7292	7297	7301
.93	.7306	7311	7315	7320	7325	7329	7334	7338	7343	7348
.94	.7352	7357	7361	7366	7371	7375	7380	7384	7389	7393
.95	.7398	7402	7407	7411	7416	7420	7425	7429	7434	7438
.96	.7443	7447	7452	7456	7461	7465	7469	7474	7478	7483
.97	.7487	7491	7496	7500	7505	7509	7513	7518	7522	7526
.98	.7531	7535	7539	7544	7548	7552	7557	7561	7565	7569
.99	.7574	7578	7582	7586	7591	7595	7599	7603	7608	7612

TABLE B.11 *(cont.)*

VALUES OF Tanh $z' = r$ (FISHER'S TRANSFORMATION)

z'	0	1	2	3	4	5	6	7	8	9
1.00	.7616	7620	7624	7629	7633	7637	7641	7645	7649	7653
1.01	.7658	7662	7666	7670	7674	7678	7682	7686	7691	7695
1.02	.7699	7703	7707	7711	7715	7719	7723	7727	7731	7735
1.03	.7739	7743	7747	7751	7755	7759	7763	7767	7771	7775
1.04	.7779	7783	7787	7791	7795	7799	7802	7806	7810	7814
1.05	.7818	7822	7826	7830	7834	7837	7841	7845	7849	7853
1.06	.7857	7860	7864	7868	7872	7876	7879	7883	7887	7891
1.07	.7895	7898	7902	7906	7910	7913	7917	7921	7925	7928
1.08	.7932	7936	7939	7943	7947	7950	7954	7958	7961	7965
1.09	.7969	7972	7976	7980	7983	7987	7991	7994	7998	8001
1.10	.8005	8009	8012	8016	8019	8023	8026	8030	8034	8037
1.11	.8041	8044	8048	8051	8055	8058	8062	8065	8069	8072
1.12	.8076	8079	8083	8086	8090	8093	8096	8100	8103	8107
1.13	.8110	8114	8117	8120	8124	8127	8131	8134	8137	8141
1.14	.8144	8148	8151	8154	8158	8161	8164	8168	8171	8174
1.15	.8178	8181	8184	8187	8191	8194	8197	8201	8204	8207
1.16	.8210	8214	8217	8220	8223	8227	8230	8233	8236	8240
1.17	.8243	8246	8249	8252	8256	8259	8262	8265	8268	8271
1.18	.8275	8278	8281	8284	8287	8290	8293	8296	8300	8303
1.19	.8306	8309	8312	8315	8318	8321	8324	8327	8330	8333
1.20	.8337	8340	8343	8346	8349	8352	8355	8358	8361	8364
1.21	.8367	8370	8373	8376	8379	8382	8385	8388	8391	8394
1.22	.8397	8399	8402	8405	8408	8411	8414	8417	8420	8423
1.23	.8426	8429	8432	8434	8437	8440	8443	8446	8449	8452
1.24	.8455	8457	8460	8463	8466	8469	8472	8474	8477	8480
1.25	.8483	8486	8488	8491	8494	8497	8500	8502	8505	8508
1.26	.8511	8513	8516	8519	8522	8524	8527	8530	8533	8535
1.27	.8538	8541	8543	8546	8549	8551	8554	8557	8560	8562
1.28	.8565	8568	8570	8573	8575	8578	8581	8583	8586	8589
1.29	.8591	8594	8596	8599	8602	8604	8607	8609	8612	8615
1.30	.8617	8620	8622	8625	8627	8630	8633	8635	8638	8640
1.31	.8643	8645	8648	8650	8653	8655	8658	8660	8663	8665
1.32	.8668	8670	8673	8675	8678	8680	8683	8685	8688	8690
1.33	.8692	8695	8697	8700	8702	8705	8707	8709	8712	8714
1.34	.8717	8719	8722	8724	8726	8729	8731	8733	8736	8738
1.35	.8741	8743	8745	8748	8750	8752	8755	8757	8759	8762
1.36	.8764	8766	8769	8771	8773	8775	8778	8780	8782	8785
1.37	.8787	8789	8791	8794	8796	8798	8801	8803	8805	8807
1.38	.8810	8812	8814	8816	8818	8821	8823	8825	8827	8830
1.39	.8832	8834	8836	8838	8840	8843	8845	8847	8849	8851
1.40	.8854	8856	8858	8860	8862	8864	8866	8869	8871	8873
1.41	.8875	8877	8879	8881	8883	8886	8888	8890	8892	8894
1.42	.8896	8898	8900	8902	8904	8906	8908	8911	8913	8915
1.43	.8917	8919	8921	8923	8925	8927	8929	8931	8933	8935
1.44	.8937	8939	8941	8943	8945	8947	8949	8951	8953	8955
1.45	.8957	8959	8961	8963	8965	8967	8969	8971	8973	8975
1.46	.8977	8978	8980	8982	8984	8986	8988	8990	8992	8994
1.47	.8996	8998	9000	9001	9003	9005	9007	9009	9011	9013
1.48	.9015	9017	9018	9020	9022	9024	9026	9028	9030	9031
1.49	.9033	9035	9037	9039	9041	9042	9044	9046	9048	9050

TABLE B.11 *(cont.)*

VALUES OF Tanh $z' = r$ (FISHER'S TRANSFORMATION)

z'	0	1	2	3	4	5	6	7	8	9
1.50	.9051	9053	9055	9057	9059	9060	9062	9064	9066	9068
1.51	.9069	9071	9073	9075	9076	9078	9080	9082	9083	9085
1.52	.9087	9089	9090	9092	9094	9096	9097	9099	9101	9103
1.53	.9104	9106	9108	9109	9111	9113	9114	9116	9118	9120
1.54	.9121	9123	9125	9126	9128	9130	9131	9133	9135	9136
1.55	.9138	9140	9141	9143	9144	9146	9148	9149	9151	9153
1.56	.9154	9156	9157	9159	9161	9162	9164	9165	9167	9169
1.57	.9170	9172	9173	9175	9177	9178	9180	9181	9183	9184
1.58	.9186	9188	9189	9191	9192	9194	9195	9197	9198	9200
1.59	.9201	9203	9205	9206	9208	9209	9211	9212	9214	9215
1.60	.9217	9218	9220	9221	9223	9224	9226	9227	9229	9230
1.61	.9232	9233	9235	9236	9237	9239	9240	9242	9243	9245
1.62	.9246	9248	9249	9251	9252	9253	9255	9256	9258	9259
1.63	.9261	9262	9263	9265	9266	9268	9269	9271	9272	9273
1.64	.9275	9276	9278	9279	9280	9282	9283	9284	9286	9287
1.65	.9289	9290	9291	9293	9294	9295	9297	9298	9299	9301
1.66	.9302	9304	9305	9306	9308	9309	9310	9312	9313	9314
1.67	.9316	9317	9318	9319	9321	9322	9323	9325	9326	9327
1.68	.9329	9330	9331	9332	9334	9335	9336	9338	9339	9340
1.69	.9341	9343	9344	9345	9347	9348	9349	9350	9352	9353
1.70	.9354	9355	9357	9358	9359	9360	9362	9363	9364	9365
1.71	.9366	9368	9369	9370	9371	9373	9374	9375	9376	9377
1.72	.9379	9380	9381	9382	9383	9385	9386	9387	9388	9389
1.73	.9391	9392	9393	9394	9395	9396	9398	9399	9400	9401
1.74	.9402	9403	9405	9406	9407	9408	9409	9410	9411	9413
1.75	.9414	9415	9416	9417	9418	9419	9421	9422	9423	9424
1.76	.9425	9426	9427	9428	9429	9431	9432	9433	9434	9435
1.77	.9436	9437	9438	9439	9440	9442	9443	9444	9445	9446
1.78	.9447	9448	9449	9450	9451	9452	9453	9454	9455	9457
1.79	.9458	9459	9460	9461	9462	9463	9464	9465	9466	9467
1.8	.9468	9478	9488	9498	9508	9517	9527	9536	9545	9554
1.9	.9562	9571	9579	9587	9595	9603	9611	9618	9626	9633
2.0	.9640	9647	9654	9661	9667	9674	9680	9687	9693	9699
2.1	.9705	9710	9716	9721	9727	9732	9737	9743	9748	9753
2.2	.9757	9762	9767	9771	9776	9780	9785	9789	9793	9797
2.3	.9801	9805	9809	9812	9816	9820	9823	9827	9830	9833
2.4	.9837	9840	9843	9846	9849	9852	9855	9858	9861	9863
2.5	.9866	9869	9871	9874	9876	9879	9881	9884	9886	9888
2.6	.9890	9892	9895	9897	9899	9901	9903	9905	9906	9908
2.7	.9910	9912	9914	9915	9917	9919	9920	9922	9923	9925
2.8	.9926	9928	9929	9931	9932	9933	9935	9936	9937	9938
2.9	.9940	9941	9942	9943	9944	9945	9946	9947	9949	9950
3.	.9951	9959	9967	9973	9978	9982	9985	9988	9990	9992
4.	.9993	9995	9996	9996	9997	9998	9998	9998	9999	9999

ANSWERS TO PROBLEMS

Chapter 1

A. **(1)** 80, $\frac{3}{7}$; **(3)** $\frac{1}{6}$, $\frac{2}{9}$, $\frac{1}{2}$; **(4)** $\frac{5}{6}$; **(7)** $\frac{5}{36}$, $\frac{1}{6}$, $\frac{1}{36}$, $\frac{5}{18}$, $\frac{1}{6}$, $\frac{1}{5}$.

B. **(1)** 3125, 1024; **(2)** 120; **(3)** 360; **(4)** 480; **(8)** 20; **(9)** $(4n^2 - 6n + 4)(2n - 2)!$

C. **(1)** 0.1055; **(2)** $\frac{4}{11}$; **(3)** $\frac{1}{17}$; **(4)** $\binom{950}{5}\Big/\binom{1000}{5} = 0.773$, $\binom{50}{2}\binom{950}{3}\Big/\binom{1000}{5} = 0.0423$; **(5)** 0.638; **(6)** 0.665; **(7)** $\frac{29}{30}$; **(10)** $1 - \sum_{k=2}^{n}(-1)^k/k!$; **(11)** $\sum_{k=0}^{10}(-1)^k\binom{10}{k}(50 - k)!/52! = 0.548$; **(12)** $e^{-1} = 0.368$, 0; **(13)** 0.096, 0.497, 0.407; **(14)** $\frac{15}{46}$; **(15)** $\frac{2}{14175}$.

D. **(1)** 37.5c; **(2)** \$1.44; **(3)** $(1/2)^{x+1}$, 1; **(4)** $(1 - p)/p$; **(5)** 13; **(6)** $7n/2$; **(7)** \$2.67; **(8)** \$6; **(9)** (a) 2 (b) 43 (c) 223 (d) 103 (e) 365 (f) 264.

E. **(1)** $\frac{1}{4}$; **(2)** 0.326; **(3)** $\frac{1}{2}$; **(5)** $\frac{1}{4}$; **(6)** $2V/3$.

Chapter 2

A. **(3)** 0.683; **(4)** $M = 7.26\%$, $Q_1 = 6.39\%$, $Q_3 = 8.22\%$, P.R. = 79.6; **(5)** 19.5d., 7.8d., 74.3; **(6)** 32, 5.2; **(8)** -0.104, 1.861, 0.111, $\bar{x} = 7.35$, $k_2 = 1.86$, $m_3/m_2^{3/2} = 0.27$.

B. **(1)** $\frac{3}{160}$, 2.16, 0.88, -0.14; **(2)** $\frac{5}{12}$, 2.184; **(3)** $16/\pi$, $\frac{3}{8}$, $\frac{3}{64}$; **(4)** $\frac{1}{2}$, 3, 3, $-3\log(1 - h)$; **(5)** $(e^h - 1)^2/h^2$; **(6)** $N\log[1 + p(e^h - 1)]$, Np, Npq, $Npq(q - p)$, $Npq(1 - 6pq)$; **(10)** $\alpha\beta/(\alpha - 1)$, $\alpha\beta^2/[(\alpha - 1)^2(\alpha - 2)]$; **(11)** $(b)^{-1/a}$, $[(a - 1)/\{b(a + 1)\}]^{1/a}$; **(12)** 2; **(13)** $\hat{\theta} = (1 + \bar{x})^{-1}$.

C. **(1)** $\frac{1}{3}$, $\frac{1}{3}$; **(2)** $\frac{1}{8}$; **(3)** (a) 0, (b) $e^{-3} = 0.05$, Chebyshev inequalities (a) $P \leq \frac{1}{48}$, (b) $P \leq \frac{1}{4}$.

Chapter 3

A. **(1)** 0.774; **(2)** (a) $\frac{121}{128}$, (b) $\frac{11}{64}$; **(4)** 0.063; **(5)** 0.468; **(6)** $n \geq 69$; **(8)** $C(n_1, n_2) = -n\theta(1 - \theta)$, $V(n_1/n - n_2/n) = 4\theta(1 - \theta)/n$.

B. **(1)** $\frac{4}{45}$; **(2)** $N \geq 6250$; **(5)** $E(X) = \frac{6}{5}$, $V(X) = \frac{9}{25}$.

C. **(1)** $2e^{-2} = 0.271$; **(2)** 2.303; **(3)** 0.423; **(4)** $P(11,5) = 0.0137$; **(5)** 0.0915, 0.216; **(6)** (a) 0.143, (b) 0.053, (c) 6 or more.

D. **(1)** 0.0863, 0.3251, 0.9808, 0.0515; **(2)** (a) 1.1505, (b) -0.1764; **(3)** 0.586; **(4)** (a) 75.8, (b) 1037; **(5)** 20; **(6)** 793, 4207; **(7)** 74.3, 3.23; **(8)** 125; **(10)** 0.219, 0.200, 0.214; **(11)** 0.217, 0.625.

E. **(3)** $V(X) = 14.2$, 14.3, 44.7, 47.5, 102.0, $V(A) = 9.73$, 11.38, 24.57, 22.02, 38.82.

Chapter 4

A. (**1**) $\frac{1}{2}$, $\frac{1}{12}$, 0, $-\frac{1}{120}$; (**2**) $Y = (X - 1)^2/4$; (**3**) e^{-u}, $0 < u < \infty$; (**4**) $f(v)/(2av)$, $v = [(u - b)/a]^{1/2}$; (**5**) $g(u) = 2u^{-1/2}/9\ (0 < u \leq 1)$, $g(u) = (1 + u^{-1/2})/9\ (1 \leq u < 4)$; (**6**) (a) $F(\log x)$, $0 < x < \infty$, (b) $\sum_{n=-\infty}^{\infty} [F(2n\pi + \sin^{-1}x) - F((2n - 1)\pi - \sin^{-1}x)]$, $-1 \leq x \leq 1$, (c) x, $0 \leq x \leq 1$.

B. (**3**) (a) $(\frac{2}{3})\,\Gamma(\frac{2}{3})$, (b) $e^6\,\Gamma(8)/(3 \cdot 6^7)$; (**4**) $[B(\frac{3}{2}, (N - 3)/2)]^{-1}$; (**9**) $(m - 1)/n$.

C. (**5**) $E(X) = e^{1/2}$, $V(X) = e(e - 1)$; (**6**) 0.2404.

Chapter 5

A. (**1**) $t_1 = 0.051\theta$, $t_2 = 1{,}553\theta$, 90% limits 0.026 to 0.784; (**2**) $t_1 = 0.736$, $t_2 = 1.264$, 95% limits 0.936 to 1.464; (**3**) $\bar{X} \pm 2.33\sigma N^{-1/2}$; (**8**) 5 to 25.3, compared with 5 to 39.2 from problem (7); (**9**) (b) $\binom{N-1}{m-1}\theta^m(1 - \theta)^{N-m}$

B. (**2**) 0.725, 0.753, 0.613, −0.046; (**3**) 0.061, 0.074, $\hat{C}(k_1, k_2) = 0.0031$; (**4**) 0.046, 0.091.

C. (**1**) No, $P(|\bar{X} - 20| \geq 1.8) = 0.027$; (**2**) 12.28 to 12.38 sec, 17; (**3**) Yes, $P(|\bar{x}_2 - \bar{x}_1| \geq 5) = 0.009$; (**4**) 0.08 to 0.17, 0.087 to 0.175; (**5**) No, $P = 0.17$; (**6**) 0.38 to 0.78; (**7**) 0.06 to 0.27; (**8**) hypothesis not rejected, $P = 0.11$; (**9**) 62 or more.

Chapter 6

A. (**8**) $V(s)/V(d\sqrt{\pi/2}) = 0.876$; (**11**) $\hat{\theta} = 2(m - 1)/m$; (**12**) $N\hat{\rho}(1 - \hat{\rho}^2) - \hat{\rho}(\sum x^2 + \sum y^2) + (1 + \hat{\rho}^2)\sum xy = 0$.

B. (**1**) 191; (**2**) reject H_0 if $m > 75 + 2.79N^{-1/2}$, 0.14, 0.57, 0.92, 23; (**3**) reject H_1 if $m > k$, where k is given by $P(10k, 10) \leq 0.05$, $k = 1.6$, power $= P(16, 20) = 0.84$; (**4**) $x > c$, where $\alpha \approx B(c, n, \theta_0)$, $1 - \beta \approx B(c, n, \theta_1)$, $n = 50$; (**5**) $|t| > c$, $t = N^{1/2}(m - \mu_0)/s$, s = sample standard deviation; (**6**) $\hat{\beta} = \sum (x_i - \bar{x})(y_i - \bar{y})/\sum (x_i - \bar{x})^2$, $\hat{\alpha} = \bar{y} - \hat{\beta}\bar{x}$, $N\hat{\sigma}^2 = \sum \{(y_i - \bar{y}) - \hat{\beta}(x_i - \bar{x})\}^2$; (**7**) 6561.

C. (**2**) $\sum_5^{10} \theta^3 p_\theta / \sum_1^{10} \theta^3 p_\theta$, $\frac{117}{121}$; (**3**) 0.898; (**5**) $\alpha(\xi) = 1 - \Phi(2c)$, $\beta(\xi) = \Phi(2c - 4)$, 0.46; (**7**) accept if $\int_{\theta_0}^1 \theta^r(1 - \theta)^{N-r}\xi(\theta)L_1(\theta)\,d\theta < \int_0^{\theta_0}\theta^r(1 - \theta)^{N-r}\xi(\theta)L_2(\theta)\,d\theta$.

Chapter 7

A. (**1**) 375, 625; (**2**) (a) 20, 23, 19, 17, 8, 6, 7, (b) 10, 18, 17, 19, 12, 9, 15, (a) 180%, (b) 216%; (**4**) $2 - 2x_0 = (2 - x_0)e^{-x_0}$, $x_0 = 0.644$.

B. (**1**) 1.58, 1.54; (**4**) $V(T/Nl) = 1.00$.

C. (**3**) $V(m_i) = 11.62$, variance within samples = 161.6, $V(m_r) = 30.66$; (**4**) 0.406.

D. (**3**) $V(\hat{T}) = 23.45 \times 10^6$; $V_r(\hat{T}) = 23.7 \times 10^6$; (**4**) $\sigma_1^2 h^2/C_1 = \sigma_2^2 k^2/C_2 = M\sigma^2 - M_1\sigma_1^2 - M_2\sigma_2^2)/(MC_0)$, $N = [M^2\sigma^2 + MM_1\sigma_1^2(h - 1) + MM_2\sigma_2^2(k - 1)]/(M\sigma^2 + \varepsilon^2)$; (**5**) $h = 1.23$, $k = 2.36$, $N = 487$, expected cost = 279, for random sample $N' = 302$, $C' = 319$.

E. (**1**) 55, 11; (**3**) 29 for H_0, 24 for H_1; (**7**) $E_0(n) = [\alpha \log A + (1 - \alpha)\log B]/[\mu_0 - \mu_1 + \mu_0 \log(\mu_1/\mu_0)]$, $E_1(n) = [(1 - \beta)\log A + \beta \log B]/[\mu_0 - \mu_1 + \mu_0 \log(\mu_1/\mu_0)]$.

Chapter 8

A. **(2)** 690 or more; **(3)** 0.308; **(4)** 62; **(6)** $K(h) = -(n/2)\log(1 - 2\sigma^2 h/n)$, $E(k_2) = \sigma^2$, $V(k_2) = 2\sigma^4/n$; **(7)** $n/(n + 2)$.

B. **(1)** 55.2 to 59.3; **(2)** 179.1 to 182.9; **(3)** Yes, at 5% level, $P = 0.04$; **(4)** Yes, $P = 0.10$; **(5)** -5.3 to 18.3 mm; **(6)** -3.2 to 7.4 lb; **(7)** No, $t = 1.56$; **(8)** Yes, $P = 0.02$; **(9)** (a) increase barely significant, $t = 2.11$, (b) highly significant, $t = 5.66$; **(12)** 0.69; **(13)** 8.4; **(15)** accept lot if $m + 3.36s \leq 3.0$, $\delta = 6.20$, $k = 3.36$.

C. **(1)** 12.6 to 168; **(2)** highly significant, $F = 23$; **(3)** No, $F = 2.0$; **(4)** Yes, $F = 2.10$; **(7)** 6.94; **(8)** 10.1; **(9)** $F = 1.36$, $\lambda = 1.59$.

D. **(1)** 276, 281; **(2)** Yes, $t < 0.005$; **(3)** Yes, $t < 0.005$; **(4)** (a) significant at 5% level but not at 1%, (b) not at 5%; **(5)** $\bar{R} = 12.5$; $\hat{\sigma} = 4.06$, **(7)** $g(R) = (N - 1)e^{-R}(1 - e^{-R})^{N-2}$; **(8)** 0.35 to 4.78; **(9)** $b(N - 1)/(N + 1)$, $2b^2(N - 1)/[(N + 1)^2(N + 2)]$; **(10)** 8.

Chapter 9

A. **(1)** homogeneity accepted, $M/c = 4.13$; **(4)** highly significant, $F = 8.3$; **(5)** No, $F = 3.33$.

B. **(1)** variety effect almost sig. at 1% level, $F = 3.11$, block effect sig. at 5%, $F = 2.76$; **(2)** F for fertilizers $= 71.4$; **(3)** F(makes) $= 34.1$, F(cities) $= 7.43$, F(interaction) $= 23.2$; **(4)** (a) -8.5, 0.5, 0.5, 7.5, (b) -1.07, -0.80, 1.90, -0.04; **(5)** (a) 0.002, -0.178, 0.140, 0.120, -0.058, 0.008, 0.100, -0.006, -0.128, (b) -0.086, 0.090, -0.053, -0.020, 0.069; **(6)** (a) 0.17, -1.05, 1.07, -1.62, 1.41, (b) 0.37, -0.54, 0.16 (c) $\hat{\gamma}_{1j} = -0.56$, 1.62, -1.05, $\hat{\gamma}_{2j} = -0.84$, 1.24, -0.40, $\hat{\gamma}_{3j} = 0.54$, -0.75, 0.22, $\hat{\gamma}_{4j} = -1.24$, 1.31, -0.06, $\hat{\gamma}_{5j} = 2.10$, -3.40, 1.31; **(8)** $\hat{\sigma}^2 = 25.9$, $\hat{\sigma}_\alpha{}^2 = 37.8$, correlation coefficient $= 0.59$, power ≈ 0.28; **(9)** 6.8, 9.3, 2.6, 4.5.

C. **(1)** 0.55, 0.195, 3.37, 0.455, F(makes) $= 1.47$; **(3)** city effect highly significant, $F = 1.56$, box effect non-sig., $F = 1.2$, $\hat{\sigma}_\alpha{}^2 = 0.197$, $\hat{\sigma}_\beta{}^2 = 0.0005$, $\hat{\sigma}^2 = 0.019$; **(5)** row and column effects non-sig., treatment effect highly sig., $F = 11.1$.

D. **(1)** F(detergents, elim. blocks) $= 27$, F(blocks, elim. detergents) $= 3.95$, both sig. at 1%; **(2)** 0.307, B and C only; **(3)** BC, BD and BG; **(4)** 5($r = 4$ almost sufficient).

Chapter 10

A. **(1)** $P = 0.40$; **(2)** No, $P = 0.66$; **(3)** not at 5% level, $P = 0.02$; **(4)** Yes, $P < 0.01$; **(5)** No, $P = 0.007$; **(6)** fit satisfactory, $P = 0.14$; **(7)** $P = 0.42$; **(8)** $P = 0.85$; **(9)** $P = 0.17$; **(10)** (a) $\chi^2 = 17.9$, $P = 0.003$, (b) $\chi^2 = 10.3$, $P = 0.035$, (c) $\chi^2 = 7.96$, $P = 0.046$.

B. **(1)** $D_N = 0.549$; **(2)** not quite sig. at 5%; **(3)** Yes, $N^{1/2}D_N = 0.86$, **(4)** Yes, max $|S_N(x) - F(x)| = 0.092$.

C. **(1)** Yes, $P = 0.51$; **(2)** No, $T = 18.5$; **(3)** No, $z \approx 0.50$; **(4)** null hypothesis rejected at about 2% level; **(5)** No, P(one-tailed) $= 0.26$.

D. **(1)** No; **(2)** randomness not rejected, $P > 0.20$; **(3)** number of runs = expected number (33), distribution of runs not random by χ^2 test; **(4)** $z \approx 3.52$, highly sig.; **(5)** Yes, $S = 51$, $P < 0.006$.

Chapter 11

A. **(1)** (a) $g(x) = 2(a - x)/a^2$, $0 \leq x < a$, $h(y) = 2y/a^2$, $0 \leq y < a$, $\eta_x = (a + x)/2$, $\xi_y = y/2$, $\mu_X = a/3$, $\mu_Y = 2a/3$, $\sigma_X{}^2 = a^2/18$, $\sigma_Y{}^2 = a^2/18$, $\sigma_{XY} = a^2/36$, $\rho = 1/2$; **(2)** necessary but not sufficient; **(9)** 1/2.

B. **(1)** $y_c = 0.886x - 0.57$, $x_c = 0.825y + 8.55$; **(2)** 0.855, 0.62 to 1.15, 0.58 to 1.07; **(3)** 125, 80, 15.1, 9.05, 75.0, 0.55; **(4)** $y_c = 0.070x + 58.2$, 0.052 to 0.087, 0.487; **(5)** No, $t = 1.91$; **(6)** 20; **(7)** No, relative accuracy = 1.19; **(8)** 64, 4.0; **(9)** 16, 13, 17, 0.60.

C. **(1)** $y_c = 16.58 - 1.27x$, 1923; **(2)** 33.9 in., 24.5 to 43.9 (approximation 26.7 to 41.1); **(3)** $\hat{\beta} = 3.214$, $\hat{\alpha} = 6.726$, $\hat{\sigma}^2 = 0.00062$; **(4)** (a) $y_c = 3.224x + 6.668$, (b) $y_c = 3.214x + 6.723$, (a) 3.199 to 3.263, (b) 3.182 to 3.246, $\hat{\sigma}_u{}^2 = 0.00071$.

D. **(1)** Yes, $t = -3.43$, No, $P = 0.07$; **(2)** 0.319 to 0.774, No ($P = 0.11$); **(3)** No, $P = 0.37$; **(4)** 0.705; **(5)** $\hat{\rho} \approx r(1 - 1/(2n))$.

E. **(1)** $r_S = 0.796$, $r_K = 0.554$; **(2)** $r_P = 0.636$, $r_S = 0.733$; **(3)** hypothesis of independent random rankings not rejected, $P = 0.14$; $r_S = 0.624$, $r_K = 0.422$, agreement not sig. at 10% level.

F. **(1)** $\chi_s{}^2 = 76$, highly sig., $C = 0.32$; **(2)** Yes, $P = 0.01$; **(3)** No, (a) $P = 0.24$, (b) $P = 0.23$; **(4)** No, $P \approx 0.10$; **(5)** Yes, $P < 0.01$.

Chapter 12

A. **(3)** $y_c = 3.37x + 0.00364z + 9.30$; **(4)** $y_c = 17.19 - 0.081x - 0.342z$; **(5)** $y_c = -1.0051 - 0.000065x_1 + 0.003894x_2 + 0.32505x_3$.

B. **(1)** rows of A^{-1} are $(0.3958, -0.00377, -0.03651)$, $(-0.00377, 0.000286, -0.000167)$, $(-0.03651, -0.000167, 0.00482)$, $\hat{\sigma}^2 = 2.34$; **(2)** $15.56 < \beta_0 < 18.82$, $-0.125 < \beta_1 < -0.037$, $-0.522 < \beta_2 < -0.163$; **(3)** $a^{11} = 1.134$, $a^{22} = -0.0000436$, $a^{33} = 0.0000633$, $a^{44} = 0.0355$, standard errors of b_1, b_2, b_3 = 0.0032, 0.0039, 0.092; **(4)** rows of A are (20, 98.2, 11,880), (98.2, 506.4, 57,284), (11,880, 57,284, 7,201,220), $\hat{\sigma}^2 = 8.235$, $V(y) = \sigma^2[8.62 - 1.112x - 0.0163z + 0.000874xz + 0.0603x^2 + 0.0000101z^2]$.

C. (1) $y_c = 0.778 + 0.557x + 0.1857x^2$; **(2)** not at 5% level, $F = 4.4$ with 1 and 4 d.f.; $r_c = 0.9974$, $r = 0.9854$; **(4)** $y_c = 196.54 + 2.918x - 0.0698x^2 + 0.000299x^3$; **(5)** cubic regression not sig., standard errors 1.56, 0.15, 0.0435, 0.0435.

D. **(1)** $r^2 = 0.481$, $E^2 = 0.584$, $F = 2.13$, linearity acceptable; **(2)** $E_{yx}{}^2 = 0.287$, $E_{xy}{}^2 = 0.271$, No, $F = 1.51$ and 1.32.

E. **(1)** (a) $5y' + x - 15 = 0$, (b) $y = 1000e^{-0.461x}$; **(2)** $100y = e^{2.3x} = 10^x$; **(3)** (a) $y_c = 0.643e^{1.043x}$, $y_c = 0.589e^{1.068x}$; **(5)** $a = 0.509$, $b = -2.036$; **(6)** $a = 200$, $b = 4.12$, $q = 0.733$, $y_c = a[1 + bq^u]^{-1}$, $u = (x - 1870)/10$; **(8)** $a = 0.768$, $b = 3.86$.

Chapter 13

A. **(1)** 0.67; **(2)** $r_{01,2} = 0.759$, $r_{02,1} = 0.097$, $r_{12,0} = -0.436$, $r_{0,12} = 0.802$, Yes, $F = 15.3$; **(3)** $r_{0,12} = 1.29$ (impossible); **(4)** $r_{02,1} = 0.707$, $r_{12,0} = -0.715$; **(5)** $r_{01} = 0.187$, $r_{02} = 0.165$, $r_{03} = 0.468$, $r_{12} = 0.732$, $r_{13} = 0.201$, $r_{23} = 0.083$, $r_{01,23} = 0.0035$, $r_{02,13} = 0.095$, $r_{03,21} = 0.454$, $r_{0,123} = 0.485$; **(6)** weighted mean $= -0.8975$.

B. **(1)** $f(x, y, z) = (2\pi)^{-3/2}[1 - \rho_{xy}^2 - \rho_{xz}^2 - \rho_{yz}^2 + 2\rho_{xy}\rho_{yz}\rho_{xz}]^{-1/2}$ $\exp(-Q/2)$, $Q = [x^2(1 - \rho_{yz}^2) + y^2(1 - \rho_{xz}^2) + 2xy(\rho_{xz}\rho_{yz} - \rho_{xy}) + 2xz(\rho_{xy}\rho_{yz} - \rho_{xz}) + 2yz(\rho_{xy}\rho_{xz} - \rho_{yz})]/[1 - \rho_{xy}^2 - \rho_{xz}^2 - \rho_{yz}^2 + 2\rho_{xy}\rho_{xz}\rho_{yz}]$; **(4)** $f(x_2|x_1) = (2\pi)^{-1/2}\,[C_{11}/(C_{11}C_{22} - C_{12}^2)]^{1/2}\exp(-Q/2)$, $Q = (C_{12}x_1 - C_{11}x_2)^2/[C_{11}(C_{11}C_{22} - C_{12}^2)]$; **(5)** $[(C_{12}C_{33} - C_{13}C_{23})x_2 + (C_{22}C_{13} - C_{12}C_{23})x_3]/[C_{22}C_{33} - C_{23}^2]$; $T^2 = 20.5$, hypothesis rejected at 1% level.

C. **(1)** $L = 0.890x_1 + 3.95x_2$, or $L = 9x_1 + 40x_2$ approx.; **(2)** $L = -x_1 + 40x_2 - 22x_3$, $L_1 = 1130$, $L_2 = 2123$, $L = 1487$; **(3)** $L = -0.0312x_1 - 0.1839x_2 + 0.2221x_3 + 0.3147x_4$, $L_1 = 0.669$, $L_2 = -0.384$, criterion $L < 0.142$ for (1); **(4)** $T_2^2 = 26.34$, highly sig.

D. **(1)** $\begin{pmatrix} 0.429, & 0{,}571 \\ 0.428, & 0.572 \end{pmatrix}$, [0.4286 0.5714]; **(2)** $(\frac{1}{3}, \frac{1}{2}, \frac{1}{12}, \frac{1}{12})$, $(\frac{1}{2}, \frac{1}{3}, \frac{1}{12}, \frac{1}{12})$, $(\frac{1}{3}, \frac{1}{2}, \frac{1}{12}, \frac{1}{12})$, matrix not regular; **(3)** (a) (0.4, 0.6), (b) $(\frac{2}{7}, \frac{3}{7}, \frac{2}{7})$; **(4)** $P(j \to j) = 2j(n - j)/n^2$, $P(j \to j - 1) = j^2/n^2$, $P(j \to j + 1) = (n - j)^2/n^2$, $0 < j < n$, expected number of black balls in (1) $= \frac{7}{4}$; **(6)** 0.55.

INDEX

A CATALOG OF SELECTED

DOVER BOOKS

IN SCIENCE AND MATHEMATICS

A CATALOG OF SELECTED

DOVER BOOKS

IN SCIENCE AND MATHEMATICS

QUALITATIVE THEORY OF DIFFERENTIAL EQUATIONS, V.V. Nemytskii and V.V. Stepanov. Classic graduate-level text by two prominent Soviet mathematicians covers classical differential equations as well as topological dynamics and ergodic theory. Bibliographies. 523pp. 5⅜ × 8½. 65954-2 Pa. $10.95

MATRICES AND LINEAR ALGEBRA, Hans Schneider and George Phillip Barker. Basic textbook covers theory of matrices and its applications to systems of linear equations and related topics such as determinants, eigenvalues and differential equations. Numerous exercises. 432pp. 5⅜ × 8½. 66014-1 Pa. $10.95

QUANTUM THEORY, David Bohm. This advanced undergraduate-level text presents the quantum theory in terms of qualitative and imaginative concepts, followed by specific applications worked out in mathematical detail. Preface. Index. 655pp. 5⅜ × 8½. 65969-0 Pa. $13.95

ATOMIC PHYSICS (8th edition), Max Born. Nobel laureate's lucid treatment of kinetic theory of gases, elementary particles, nuclear atom, wave-corpuscles, atomic structure and spectral lines, much more. Over 40 appendices, bibliography. 495pp. 5⅜ × 8½. 65984-4 Pa. $12.95

ELECTRONIC STRUCTURE AND THE PROPERTIES OF SOLIDS: The Physics of the Chemical Bond, Walter A. Harrison. Innovative text offers basic understanding of the electronic structure of covalent and ionic solids, simple metals, transition metals and their compounds. Problems. 1980 edition. 582pp. 6⅛ × 9¼. 66021-4 Pa. $15.95

BOUNDARY VALUE PROBLEMS OF HEAT CONDUCTION, M. Necati Özisik. Systematic, comprehensive treatment of modern mathematical methods of solving problems in heat conduction and diffusion. Numerous examples and problems. Selected references. Appendices. 505pp. 5⅜ × 8½. 65990-9 Pa. $12.95

A SHORT HISTORY OF CHEMISTRY (3rd edition), J.R. Partington. Classic exposition explores origins of chemistry, alchemy, early medical chemistry, nature of atmosphere, theory of valency, laws and structure of atomic theory, much more. 428pp. 5⅜ × 8½. (Available in U.S. only) 65977-1 Pa. $10.95

A HISTORY OF ASTRONOMY, A. Pannekoek. Well-balanced, carefully reasoned study covers such topics as Ptolemaic theory, work of Copernicus, Kepler, Newton, Eddington's work on stars, much more. Illustrated. References. 521pp. 5⅜ × 8½. 65994-1 Pa. $12.95

PRINCIPLES OF METEOROLOGICAL ANALYSIS, Walter J. Saucier. Highly respected, abundantly illustrated classic reviews atmospheric variables, hydrostatics, static stability, various analyses (scalar, cross-section, isobaric, isentropic, more). For intermediate meteorology students. 454pp. 6⅛ × 9¼. 65979-8 Pa. $14.95

CATALOG OF DOVER BOOKS

THE ELECTROMAGNETIC FIELD, Albert Shadowitz. Comprehensive undergraduate text covers basics of electric and magnetic fields, builds up to electromagnetic theory. Also related topics, including relativity. Over 900 problems. 768pp. 5⅜ × 8¼. 65660-8 Pa. $18.95

FOURIER SERIES, Georgi P. Tolstov. Translated by Richard A. Silverman. A valuable addition to the literature on the subject, moving clearly from subject to subject and theorem to theorem. 107 problems, answers. 336pp. 5⅜ × 8½. 63317-9 Pa. $8.95

THEORY OF ELECTROMAGNETIC WAVE PROPAGATION, Charles Herach Papas. Graduate-level study discusses the Maxwell field equations, radiation from wire antennas, the Doppler effect and more. xiii + 244pp. 5⅜ × 8½. 65678-0 Pa. $6.95

DISTRIBUTION THEORY AND TRANSFORM ANALYSIS: An Introduction to Generalized Functions, with Applications, A.H. Zemanian. Provides basics of distribution theory, describes generalized Fourier and Laplace transformations. Numerous problems. 384pp. 5⅜ × 8½. 65479-6 Pa. $9.95

THE PHYSICS OF WAVES, William C. Elmore and Mark A. Heald. Unique overview of classical wave theory. Acoustics, optics, electromagnetic radiation, more. Ideal as classroom text or for self-study. Problems. 477pp. 5⅜ × 8½. 64926-1 Pa. $12.95

CALCULUS OF VARIATIONS WITH APPLICATIONS, George M. Ewing. Applications-oriented introduction to variational theory develops insight and promotes understanding of specialized books, research papers. Suitable for advanced undergraduate/graduate students as primary, supplementary text. 352pp. 5⅜ × 8½. 64856-7 Pa. $8.95

A TREATISE ON ELECTRICITY AND MAGNETISM, James Clerk Maxwell. Important foundation work of modern physics. Brings to final form Maxwell's theory of electromagnetism and rigorously derives his general equations of field theory. 1,084pp. 5⅜ × 8½. 60636-8, 60637-6 Pa., Two-vol. set $21.90

AN INTRODUCTION TO THE CALCULUS OF VARIATIONS, Charles Fox. Graduate-level text covers variations of an integral, isoperimetrical problems, least action, special relativity, approximations, more. References. 279pp. 5⅜ × 8½. 65499-0 Pa. $7.95

HYDRODYNAMIC AND HYDROMAGNETIC STABILITY, S. Chandrasekhar. Lucid examination of the Rayleigh-Benard problem; clear coverage of the theory of instabilities causing convection. 704pp. 5⅜ × 8¼. 64071-X Pa. $14.95

CALCULUS OF VARIATIONS, Robert Weinstock. Basic introduction covering isoperimetric problems, theory of elasticity, quantum mechanics, electrostatics, etc. Exercises throughout. 326pp. 5⅜ × 8½. 63069-2 Pa. $8.95

DYNAMICS OF FLUIDS IN POROUS MEDIA, Jacob Bear. For advanced students of ground water hydrology, soil mechanics and physics, drainage and irrigation engineering and more. 335 illustrations. Exercises, with answers. 784pp. 6⅛ × 9¼. 65675-6 Pa. $19.95

HANDBOOK OF MATHEMATICAL FUNCTIONS WITH FORMULAS, GRAPHS, AND MATHEMATICAL TABLES, edited by Milton Abramowitz and Irene A. Stegun. Vast compendium: 29 sets of tables, some to as high as 20 places. 1,046pp. 8 × 10½. 61272-4 Pa. $24.95

MATHEMATICAL METHODS IN PHYSICS AND ENGINEERING, John W. Dettman. Algebraically based approach to vectors, mapping, diffraction, other topics in applied math. Also generalized functions, analytic function theory, more. Exercises. 448pp. 5⅜ × 8¼. 65649-7 Pa. $9.95

A SURVEY OF NUMERICAL MATHEMATICS, David M. Young and Robert Todd Gregory. Broad self-contained coverage of computer-oriented numerical algorithms for solving various types of mathematical problems in linear algebra, ordinary and partial, differential equations, much more. Exercises. Total of 1,248pp. 5⅜ × 8½. Two volumes. Vol. I 65691-8 Pa. $14.95
Vol. II 65692-6 Pa. $14.95

TENSOR ANALYSIS FOR PHYSICISTS, J.A. Schouten. Concise exposition of the mathematical basis of tensor analysis, integrated with well-chosen physical examples of the theory. Exercises. Index. Bibliography. 289pp. 5⅜ × 8½.
65582-2 Pa. $8.95

INTRODUCTION TO NUMERICAL ANALYSIS (2nd Edition), F.B. Hildebrand. Classic, fundamental treatment covers computation, approximation, interpolation, numerical differentiation and integration, other topics. 150 new problems. 669pp. 5⅜ × 8½. 65363-3 Pa. $15.95

INVESTIGATIONS ON THE THEORY OF THE BROWNIAN MOVEMENT, Albert Einstein. Five papers (1905–8) investigating dynamics of Brownian motion and evolving elementary theory. Notes by R. Fürth. 122pp. 5⅜ × 8½.
60304-0 Pa. $4.95

CATASTROPHE THEORY FOR SCIENTISTS AND ENGINEERS, Robert Gilmore. Advanced-level treatment describes mathematics of theory grounded in the work of Poincaré, R. Thom, other mathematicians. Also important applications to problems in mathematics, physics, chemistry and engineering. 1981 edition. References. 28 tables. 397 black-and-white illustrations. xvii + 666pp. 6⅛ × 9¼.
67539-4 Pa. $16.95

AN INTRODUCTION TO STATISTICAL THERMODYNAMICS, Terrell L. Hill. Excellent basic text offers wide-ranging coverage of quantum statistical mechanics, systems of interacting molecules, quantum statistics, more. 523pp. 5⅜ × 8½. 65242-4 Pa. $12.95

ELEMENTARY DIFFERENTIAL EQUATIONS, William Ted Martin and Eric Reissner. Exceptionally clear, comprehensive introduction at undergraduate level. Nature and origin of differential equations, differential equations of first, second and higher orders. Picard's Theorem, much more. Problems with solutions. 331pp. 5⅜ × 8½. 65024-3 Pa. $8.95

STATISTICAL PHYSICS, Gregory H. Wannier. Classic text combines thermodynamics, statistical mechanics and kinetic theory in one unified presentation of thermal physics. Problems with solutions. Bibliography. 532pp. 5⅜ × 8½.
65401-X Pa. $12.95